Defects in
Solids
Modern Techniques

NATO ASI Series

Advanced Science Institutes Series

*A series presenting the results of activities sponsored by the NATO Science Committee,
which aims at the dissemination of advanced scientific and technological knowledge,
with a view to strengthening links between scientific communities.*

The series is published by an international board of publishers in conjunction with the
NATO Scientific Affairs Division

A	**Life Sciences**	Plenum Publishing Corporation
B	**Physics**	New York and London
C	**Mathematical and Physical Sciences**	D. Reidel Publishing Company Dordrecht, Boston, and Lancaster
D	**Behavioral and Social Sciences**	Martinus Nijhoff Publishers
E	**Engineering and Materials Sciences**	The Hague, Boston, Dordrecht, and Lancaster
F	**Computer and Systems Sciences**	Springer-Verlag
G	**Ecological Sciences**	Berlin, Heidelberg, New York, London,
H	**Cell Biology**	Paris, and Tokyo

Recent Volumes in this Series

Series B: Physics

Defects in Solids

Modern Techniques

Edited by

A. V. Chadwick

University of Kent
Canterbury, England

and

M. Terenzi

University of Calabria
Renole, Italy

Springer Science+Business Media, LLC

Proceedings of a NATO Advanced Study Institute,
held September 16–27, 1985,
in Cetraro, Calabria, Italy

Library of Congress Cataloging in Publication Data

NATO Advanced Study Institute (1985: Calabria, Italy)
  Defects in solids.

  (NATO ASI series. Series B, Physics; v. 147)
  Proceedings of a NATO Advanced Study Institute, held September 16–27, 1985,
in Cetraro, Calabria, Italy"—T.p. verso.
  "Published in cooperation with NATO Scientific Affairs Division."
  Includes bibliographies and index.
  1. Solids—Congresses. 2. Crystals—Defects—Congresses. I. Chadwick, A. V.
(Alan V.) II. Terenzi, M. (Mario) III. North Atlantic Treaty Organization. Scientific
Affairs Division. IV. Series.
QC176.A1N3 1985   530.4′1   86-25417
ISBN 978-1-4757-0763-2   ISBN 978-1-4757-0761-8 (eBook)
DOI 10.1007/978-1-4757-0761-8

© 1986 Springer Science+Business Media New York
Originally published by Plenum Press, New York in 1986

PREFACE

        Defects in solids are of fundamental interest to scientists in
many disciplines and are important to the development of numerous
industrial technologies. Knowledge of the nature and behaviour of
defects is vital for the detailed understanding of solid
state phenomena in chemistry, physics, metallurgy, geology and in the
production of catalysts, ceramics, semiconductor components and solid
electrolytes. The relevance of defects to many areas of science has
long been recognized and a great variety of techniques has evolved for
their investigation. However, the last decade has witnessed the
introduction of new techniques, both experimental and theoretical,
which together with major developments in existing methods have greatly
increased the understanding of the defective solid state. These
advances, which were often developed for specialized problems, are
rapidly growing in their usage and are being applied to a greater
diversity of materials. As a consequence, defect research is making
considerable progress in both the understanding of long-standing
problems and in the complexity of problems that can be investigated.

        This volume is based on the lectures presented at the NATO
Advanced Study Institute held at Cetraro, Italy in 1985. The aim is to
survey the recent developments in the techniques used in defect
investigations. These can broadly be divided into three classes.
Firstly, there are the laboratory-based experiments where developments
in electronics, lasers and spectrometers have led to new or improved
methods which allow the research scientist to undertake sophisticated
studies in a university or company laboratory. Typical examples are
the positron annihilation technique (PAT), various forms of magnetic
resonance spectroscopy and complex impedance spectroscopy. The second
class of techniques rely on the special experimental probes provided by
large, centralized facilities such as neutron and synchrotron radiation
sources. Examples here include the range of neutron scattering
techniques, extended X-ray absorption fine structure (EXAFS), X-ray
topography and a variety of surface experiments. The number of these
facilities has increased (and is still growing) and many scientists now
have access to the instruments. Thus the horizons for the scientist
interested in defects has been widened. Finally, there have been rapid
developments in the applications of computer simulation to defect
studies. These have been possible partly due to the improvements in
computers (in speed and memory capabilities) and to the methods that
have been evolved to represent interatomic potentials.

        In order to present a structured coverage of the topics the
chapters have been arranged so that the contents follow the general
sequence:

(i)   introductory background to defects;

  (ii)   techniques used to investigate the structure of defects;

 (iii)   computer simulation techniques;

  (iv)   techniques used to investigate defects on surfaces;

   (v)   techniques used to investigate the transport of atoms via
          defects.

     The boundaries between these five sections should not be taken as
rigorous and there is naturally some overlap which is in fact
desirable.

     In Chapter 1 Corish surveys the variety of types of defects that
are found in solids and lays the foundation for the rest of the volume.
This is followed by two short introductory chapters. In Chapter 2
Chadwick outlines some of the objectives of point defect
investigations in ionic solid. Cormack considers the objectives of
theoretical studies and the links to experiments in Chapter 3.

     The ability of transmission electron microscopy to directly
observe the structure of individual defects makes it an extremely
powerful technique and it is described by Hobbs in Chapter 4. Stewart
in Chapter 5 reviews neutron scattering by defects in solids and shows
how the various experiments can be used to probe a wide range of defect
types. In Chapter 6 Beech concentrates on the specific uses of
neutrons to probe static defects in complex materials and in studies of
superionic solids. The theme of defect structures is maintained in
Chapter 7 where Eldrup covers the applications of the positron
annihilation technique. Sauvage-Simkin describes the use of
synchrotron radiation in the study of defects in Chapter 8. This
includes X-ray topography and related methods and points out the novel
applications in the study of surfaces. In Chapter 9 Spaeth reviews the
exciting developments in the spectroscopic studies of defects. This
includes the applications to complex, but important materials used in
microelectronics. Nuclear magnetic resonance spectroscopy has been
extensively used to study point defects in ionic crystals and the basic
principles are outlined by Strange in Chapter 10. The specific
applications of this technique in diffusion studies are treated in
Chapter 11 by Terenzi.

     Ten years ago the computer simulation of defects was in its
infancy, but it has matured very rapidly. In Chapter 12 Catlow reviews
the range of simulation methods that can now be regarded as
well-established. This is followed by a short contribution by Harding
in Chapter 13 which covers the recent developments in the calculation
of defect entropies. Chapter 14 by Henrich moves to the area of
surface defects and their investigation by electron spectroscopy. In
contrast to bulk defects relatively little is known about surface
defects. However, this chapter shows where progress is being made and
the promise offered by new techniques for an increased understanding of
this complex topic. One of the new techniques is computer simulation
of surface defects and this is specifically reviewed in Chapter 15 by
Colbourn.

     The remaining chapters are concerned with the migration of
defects. Philibert in Chapter 16 reviews atomic diffusion in metals.
This is probably the most thoroughly investigated area of defect

physics and has helped to develop many of the general ideas concerning atomic transport in materials. An area that has been growing over recent years is the study of diffusion in oxides and this is surveyed by Monty in Chapter 17. A short contribution by Chadwick and Corish in Chapter 18 considers the specific case of ionic conductance measurements in solids, both single crystals and ceramic pellets. The final chapter by Capelletti reviews the thermally stimulated depolarization technique. This essentially simple but very powerful technique for studying impurity-defect clusters in ionic solids is now being used to investigate a wide range of materials.

The appendix contains short contributions which were presented at the Advanced Study Institute. They have been included to exemplify the types and range of defect studies that are presently in progress.

It has not been possible to include a detailed treatment of every technique and there are some important omissions of which we are fully aware. Nevertheless, this book should give the reader a clear view of the defect techniques that are currently available and serve as a very useful reference source. A feature that we hope will prove attractive to the reader is that in addition to outlining the basic principles of techniques the contributors have described the applications and the scope for extending them to various types of material and problems. We thank the authors for taking this very helpful approach. Finally, we hope that the book conveys the enthusiasm that is being created by the new techniques.

A.V. Chadwick
M. Terenzi

ACKNOWLEDGEMENTS

We would like to thank NATO for their support of the Advanced
Study Institute which formed the basis of this book.

In addition, we would like to thank the following organisations
for their support: the British Council, British Petroleum plc, Cassa
di Risparmadi Calabria e Lucania, Consiglio Nazionale delle Ricerche,
ENEA, United States Army European Research Office, University of
Calabria and the University of Kent.

We are indebted to the organising committee and the lecturers at
the ASI. They put in a considerable effort in making the meeting
successful and also worked extremely hard on the chapters presented in
this book.

Finally we would like to thank Mrs. Heather Harrow for general
secretarial assistance and we are especially grateful to
Mrs. Eileen Stoydin for her efficient work in the preparation of the
manuscripts.

Alan Chadwick
Mario Terenzi

CONTENTS

INTRODUCTION TO DEFECTS IN SOLIDS

J. Corish

Chemistry Department
Trinity College
Dublin 2
Ireland

INTRODUCTION

It is a truism that the more we learn about anything the more complex that subject becomes. This is especially true of our understanding of the exact arrangement of atoms or ions which make up solid materials. The rapid development and improvement of the techniques by which defects in solids are detected and characterized, and which form the subject matter of this book, have necessitated constant revision of our concepts of perfect and defective solids. It is, for example, no longer adequate to confine our attention to isolated point defects without also considering their aggregation to form more complex defect centres and eventually even separate phases, their relationship with non-stoichiometry and their interactions with each other and with the linear and planar defects with which they co-exist. Similarly, the precise structural information now available for many materials reveals that non-stoichiometric phases may exist as modulated structures in which the composition may vary in a continuous fashion to yield commensurate, semi-commensurate and incommensurate structures. The presence of shear planes also changes the composition of non-stoichiometric phases.

These more complex defect structures will be considered in detail in the following chapters but this introduction will be concerned principally with a survey of the occurrence and nature of the fundamental defects, atomistic and extended, which are present in a range of materials. Thus it will provide necessary background information but, where appropriate, the treatment will include the interaction and aggregation of defects, particularly where these serve as suitable models for what follows. The

survey first describes point defects the presence of which is essential
to explain observed matter transport behaviour in crystalline solids as
well as oxide film growth and other high-temperature corrosion phenomena
in metals and alloys.   Defects at which electronic carriers are trapped,
and of which the best known is the F-centre which colours alkali halide
crystals, will also be discussed.   Many advanced battery systems as well
as solid state sensors rely for their operation on an important class of
compounds which have the ability to provide very rapid migration of ions.
The principal characteristics of these fast-ion conductors, or FIC as they
are now known, many of which depend on the presence of a surfeit of suitable sites or of channels or planar pathways through their structures,
rather than on defects, will also be described.   Dislocations or linear
defects which play vital roles in the determination of the strengths of
materials, in providing non-lattice diffusion pathways and in crystal
growth kinetics will then be considered.   The survey will conclude with
a discussion of planar imperfections although the special nature of the
defects present on solid surfaces will be treated in detail in the Chapter
by Henrich.

POINT DEFECTS

   Point defects are atomistic in nature and the principal types are
vacant sites, interstitial atoms or ions and substitutional impurity atoms
or ions.   They occur thermally in many materials, including metals and
ionic and molecular crystals, but can also arise because of the presence
of impurities or as a result of radiation damage, ion implantation treatment or the application of a stress.   Because their presence in ionic
solids is often associated with a charge they have been most extensively
studied in, and are consequently perhaps best characterized in, simple
ionic crystals.   These systems will be used here to illustrate the nature
of point defects and the thermodynamic considerations which govern their
formation and movement.

Point Defects in Ionic Crystals

   Point defects in ionic crystals must comply with the structural constraints of the lattice and it is also necessary to maintain charge neutrality within the bulk of the crystal.   Intrinsic defects are equilibrium
imperfections and, as we shall see are thermally produced: the principal
types are illustrated in Figure 1.   A Schottky defect occurs when corresponding cations and anions are missing from their lattice sites: in
Figure 1(a) which represents a rocksalt lattice this results in one cation
and one anion vacancy.   A Frenkel defect is formed when an ion moves from

2

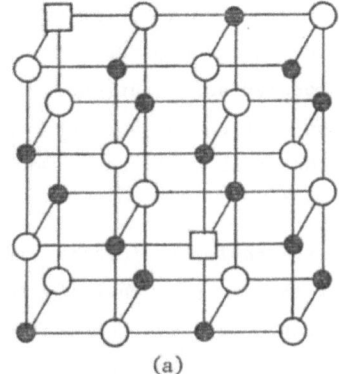 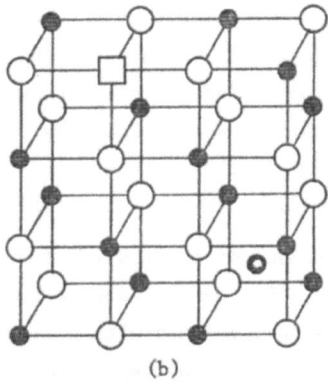

(a)   (b)

Figure 1   (a) Schottky defect in NaCl;   (b) Frenkel Defect in AgCl;
● and ○ represent lattice cations and anions, respectively,
☐ represents a vacancy and ◉ a cation interstitial.

its normal lattice position, thereby creating a vacancy, and occupies an
interstitial position: it is illustrated on the cation sublattice in
Figure 1(b).

Because it is the most adaptable and has gained widest official acceptance the Kroger-Vink (1956) notation will be used here where practicable.
In this notation defect is represented by a major symbol which signifies
the species and a subscript which signifies the site which that species
occupies. A vacancy is given the symbol V and an interstitial site the
subscript i. Charges may be shown either as real charges in the usual way
by superscripts n+ or n- or as virtual (or effective) charges which are
defined as the charges on the defect with respect to the ideal unperturbed
crystal. Positive and negative virtual charges are indicated by dot and
prime symbols, respectively, as superscripts: defects with no effective
charge relative to the unperturbed lattice may be indicated by a super-
script asterisk. The constituents of a Schottky defect in NaCl are
therefore V_{Na}' and $V_{Cl}^{\bullet}$ whereas a cation Frenkel defect in AgCl would comprise V_{Ag}' and $Ag_i^{\bullet}$ species. Although both types of intrinsic defects are
known to occur in simple ionic crystals it is usual for the concentration
of one type to greatly exceed that of the other. Thus Schottky defects
predominate in the rocksalt alkali halides, cation Frenkel defects in AgCl
and AgBr and anion Frenkel defects, $V_F^{\bullet}$ and F_i' in calcium fluoride and the
fluorites generally.

In addition to these intrinsic imperfections point defects also ensue
when foreign ions enter the crystal. The presence of such impurities may

be adventitious or they may be added purposely as dopants during growth. Aliovalent ions which enter the lattice must be charge compensated. For strontium in sodium chloride, which is accommodated substitutionally as $Sr_{Na}^{\cdot}$, this is accomplished by the introduction of a cation vacancy, $V_{Na}^{'}$, elsewhere in the lattice while trivalent lanthanum in CaF_2, $La_{Ca}^{\cdot}$, would be compensated by an interstitial fluoride ion $F_i^{'}$.

Because of their virtual charges attractive coulombic forces exist between point defects and at sufficiently low temperatures they may form associated complexes with characteristic binding energies. Thus the constituents of a Schottky defect in KCl may come together to occupy adjacent lattice sites and form a vacancy pair, $(V_K^{'} V_{Cl}^{\cdot})$. In the divalently-doped alkali halides the most basic of such species is a vacancy-impurity pair, $(V_K^{'} Sr_K^{\cdot})$ in $KCl:Sr^{2+}$, in which the constituents can be either in nearest-neighbour (nn) or next-nearest-neighbour (nnn) positions depending, in the main, on the degree of mismatch between the dopant ion and the host cation. In $CaF_2:Er^{3+}$ the corresponding pair is the impurity ion with nn or nnn interstitials, $(Er_{Ca}^{\cdot} F_i^{'})$, whereas in $MgO:Cr^{3+}$ the fully-compensated complex

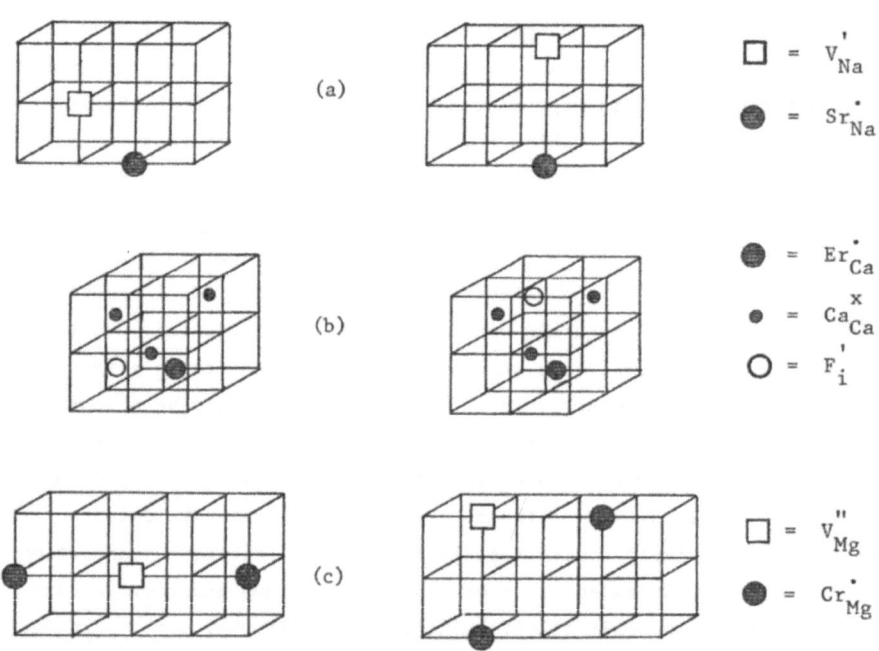

Figure 2   (a) nn and nnn vacancy-impurity complex in $KCl:Sr^{2+}$

             (b) nn and nnn impurity-interstitial complex in $CaF_2:Er^{3+}$

             (c) linear and bent $(Cr_{Mg}^{\cdot} V_{Mg}^{''} Cr_{Mg}^{\cdot})$ dimer in $MgO:Cr^{3+}$

4

contains two dopant ions with one magnesium vacancy ($Cr_{Mg}^{\bullet} V_{Mg}^{''} Cr_{Mg}^{\bullet}$) and can take a linear or bent shape. These defects are illustrated in Figure 2: larger aggregates containing larger numbers of impurity ion will be considered in a later section.

Before concluding this survey it should be noted that in a number of systems substitutional ions do not occupy the lattice site as might be expected but move to off-centre positions. Such defects, in addition to electric dipoles such as OH^- and CN^-, produce low frequency resonances and their reorientation may be studied. Off-centre behaviour has been observed for both cations and anions in the alkali halides and is explained as being the result of two effects. First a small substitutional ion can move closer to its immediate neighbours before their repulsive interaction becomes effective. Secondly there is an attractive interaction between the charged defect and dipoles induced in its neighbours. This enhanced attraction and reduced repulsion can lead to asymmetric sites being favoured. Thus off-centre behaviour is observed with small substitutional ions in which the repulsion is reduced ($KCl:Li^+$) and with heavy ions with large polarizabilities ($RbCl:Ag^+$). There is good agreement between theoretical predictions and the experimental observations (Catlow et al 1978).

Concentrations of Point Defects

The concentrations of intrinsic defects may be estimated using statistical thermodynamic arguments to minimize the Gibbs free energy of a defect-containing crystal which is subject to the necessary constraints of structure and electroneutrality (Allnatt and Jacobs 1961; Howard and Lidiard 1964). The free energy of a 1:1 monovalent crystal MX containing Schottky defects may be written as

$$G(n,T,P) = G^o(T,P) + n_{V_M'} \, g_{V_M'} + n_{V_X^{\bullet}} \, g_{V_X^{\bullet}} - Ts_{V_M'} - Ts_{V_X^{\bullet}} \qquad (1)$$

where $G^o(T,P)$ is the free energy of the defect free crystal, n is the number, g the free energy of formation, and s the configurational free entropy, respectively, of the defects specified. Each of the entropy terms has the form

$$s = k \, \ln\omega = k \, \ln \left[\frac{(N + n)!}{N! \; n!} \right] \qquad (2)$$

where N is the number of ion pairs in the crystal. The constraint of electrical neutrality implies that

$$n_{V_M'} - n_{V_X^{\bullet}} = 0 \qquad (3)$$

and a constrained minimization with respect to $n_{V_M'}$ and $n_{V_X^{\bullet}}$ with the use of Stirling's theorem leads to

$$\frac{n_{V_M'}\, n_{V_X^\bullet}}{N^2} = \exp(-g_S/kT) \tag{4}$$

Here $g_S = g_{V_M'} + g_{V_X^\bullet}$ is the Gibbs free energy change for the formation of a Schottky pair with $g_{V_X^\bullet}$ not necessarily being equal to $g_{V_M'}$. A more convenient variable than the number of species, r, is the site fraction which is defined as

$$x_r = \frac{n_r}{\sum\limits_{r} n_r} \tag{5}$$

and the summation is over all the sites on the appropriate sublattice. Thus using the mass-action formalism of chemical thermodynamics the Schottky defect equilibrium may be written

$$K_S = x_{V_M'}\, x_{V_X^\bullet} = \exp\left[\frac{-g_S}{kT}\right] \tag{6}$$

where K_S is the equilibrium constant. Using similar arguments an analogous expression may be derived for the Frenkel equilibrium constant:

$$K_F = x_{V_M'}\, x_{M_i^\bullet} = \xi \exp\left[\frac{-g_F}{kT}\right] \tag{7}$$

Here ξ is a configurational number, of order unity, which may arise when the numbers of lattice and interstitial sites available are not equal: in the rocksalt lattice $\xi = 2$.

This formation may be extended to short-range defect interactions and the equilibrium constant K_{vp} for the formation of a vacancy pair for which the reaction is

$$V_M' + V_X^\bullet \; \underset{\longleftarrow}{\longrightarrow} \; (V_M'\; V_X^\bullet) \tag{8}$$

is written as

$$K_{vp} = \eta \exp\left(\frac{-g_{vp}}{kT}\right) = \frac{x_{(V_M'\, V_X^\bullet)}}{x_{V_M'}\, x_{V_X^\bullet}} \tag{9}$$

where η is again configurational and g_{vp} is the Gibbs free energy change on forming the pair. Similarly for the association of an Sr^{2+} ion with its charge compensating vacancy in KCl we have

$$K_a = 12 \exp\left(\frac{-g_a}{kT}\right) = \frac{x_{(Sr_K^\bullet\, V_K')}}{x_{Sr_K^\bullet}\, x_{V_K'}} \tag{10}$$

Long range Coulombic interactions between point defects are also important because they alter the effective concentrations of the defect species. Such interactions are usually treated collectively in the Debye-

Hückel-Lidiard approximation (Lidiard; 1954, 1957). This is an extension
of the Deby-Hückel theory for solutions to the solid state and each charged
species is regarded as being surrounded by a diffuse cloud of defects with
a net charge which cancels the charge on the species. Such non-ideal behaviour is included in the analysis of defect equilibria by replacing the
defect concentrations by their activities, a_r, where

$$a_r = \gamma_r x_r \tag{11}$$

and the activity coefficient γ_r is given by

$$\gamma_r = \exp \left\{ \frac{-q_r \kappa}{2\epsilon kT(1 + \kappa R_o)} \right\} \tag{12}$$

q_r is the virtual charge on the species r, κ^{-1} the Debye length, ϵ the
permittivity of the crystal and R_o the distance of closest approach. The
Debye length is calculated from

$$\kappa^2 = \frac{4\pi}{V\epsilon kT} \sum_r c_r q_r^2 \tag{13}$$

where V is the volume of the unit cell and the summation is taken over the
site fractions of the interacting species. More sophisticated defect interaction models based on statistical mechanical theories have been developed,
especially by Allnatt and co-workers (Allnatt and Loftus; 1973 a,b; Allnatt and Yuen; 1975 a, b). However the DHL theory has proved to be re-
markably good considering the assumption on which it depends and, because
it is also convenient, it is still widely used and is particularly important in estimating defect concentrations at higher temperatures in the
classical ionic conductors.

Thus to evaluate the defect concentrations in a typical crystal containing impurity ions it is necessary to treat all the relevant defect
equilibria as simultaneous. A solution to the equation which results
will yield accurate defect concentrations provided that the level of impurity in the crystal be known as well as the values of the thermodynamic
parameters which govern defect formation and association, and that long-
range interactions are included in the treatment. Such thermodynamic
defect parameters have been determined experimentally for many years and,
more recently, the importance of reliable theoretical evaluations using
atomistic simulation techniques has greatly increased. Before leaving this
topic it is advisable to set down the relationship between such experimental and theoretical parameters, since they are often compared with each
other, and, in particular, to consider the temperature dependence of
these quantities.

The experimentally determined parameters have, in the main, been extracted from ionic conductance and diffusion measurements, as will be discussed in the Chapter by Chadwick and Corish and those by Monty and Philibert. Experiments are carried out under constant pressure and the results are interpreted using the Arrhenius formalism which assumes that the enthalpy, h_p, and the entropy, s_p, where the subscript p refers to constant pressure conditions, are independent of temperature. This has been the case despite the fact that it has long been realized that such parameters ought to be temperature dependent (Mott and Gurney; 1948). Because the high temperature Arrhenius curves are complex, since they often contain contributions from a variety of migration mechanisms as well as showing the effects of defect interactions, it is usually not possible, even in the most favourable cases (Corish and Jacobs; 1975) to quantify the temperature variation of the defect parameters from experimental data alone.

The theoretical methods used to calculate defect energies will be described later by Catlow. The calculation of vibrational entropies using a large crystallite method has also recently become feasible on a routine basis (Harding; 1985). These defect calculations refer to constant volume processes and yield values for u_v and f_v, but taken together they allow the free energy at constant volume, f_v to be calculated. They can also be used to determine the temperature dependence of these quantities by using the quasi-harmonic approximation in which temperature effects are described entirely in terms of the dependence on temperature of the lattice parameter.

The equations which relate thermodynamic properties at constant pressure with those at constant volume have been discussed extensively in the literature (Chadwick and Glyde; 1977) and the most useful results for our purpose have been collected by Catlow et al (1981). To first order in the defect volume, v_p, which is given by

$$v_p = -\kappa_T V (\partial f_v / \partial V)_T \tag{14}$$

where κ_T is the isothermal compressibility Gillan (1981) has shown that

$$g_p = f_v \tag{15}$$

Also the constant pressure and constant volume energy and entropy terms are related by the equations

$$h_p = u_v + \left(\frac{T\beta_p}{\kappa_T} \right) v_p \tag{16}$$

and $$s_p = s_v + \left(\frac{\beta_p}{\kappa_T} \right) v_p \tag{17}$$

8

where β_p is the volume thermal expansion coefficient. These equations show how the results of theoretical calculations may be used in conjunction with experimental transport data to determine a comprehensive set of temperature-dependent defect parameters. However, this approach, which requires extensive and accurate experimental data as well as reliable theoretical calculations has to date been used only for AgCl (Corish and Mulcahy; 1980), AgBr (Devlin; 1982) and KCl (Hooten and Jacobs; 1983).

Point Defects in Molecular Solids

Molecular crystals typically adopt close-packed structures with the volume available to accommodate interstitial species being very small. As a consequence the energy required to introduce interstitial defects is several times greater than that needed to form vacancies and the latter predominate in rare gas solids and in both non-plastic and plastic molecular crystals. The thermodynamics of the introduction of these intrinsic defects is analogous to that presented above for ionic crystals, though with the simplifications which come in treating a material containing only one species: they have been discussed in detail in the literature (Chadwick and Glyde; 1977). The vacancies can also associate to form divacancies with a binding energy and again this process may be treated in the same way as was the formation of vacancy pairs in Schottky defect containing ionic crystals.

The thermodynamic parameters for the formation of vacancies and for the Arrhenius energies which accompany diffusion in molecular crystals have been evaluated theoretically and using a variety of experimental techniques. In particular, theoretical calculations on the rare-gas solids have provided insights into the nature and magnitude of many-body forces since two-body potentials predict the hexagonal-close-packed structure rather than the face-centred cubic structure which is observed. Some defect entropies have also been calculated for these substances (Card and Jacobs; 1977). Atom-atom potentials (Kitaigorodski; 1973) have been used to calculate a variety of properties of organic solids. These potentials assume that the interaction energies between molecules are equal to the sum of those which operate between the atoms which make up the molecules, that central forces are adequate, and that the potential can be treated as a van der Waals potential, which dominates and which can be transferred between materials, and a small electrostatic contribution.

Experimental defect data have been reviewed recently by Chadwick (1983). Vacancy formation energies are derived from measurements of the extra contribution made by the defect creation to the specific heat of the perfect crystal, c_p^o. The heat capacity is then given by

$$c_p = c_p^o + \frac{nh^2}{kT^2} \qquad (18)$$

where n is the number of vacancies and h the enthalpy of vacancy formation. They can also be evaluated by comparing the macroscopic linear expansion with an X-ray determination of changes in the unit cell (Simmons; 1962) and from density data. The Positron Anihilation Technique (PAT), to be described in the Chapter by Eldrup, has also been used. Defect mobilities are determined using tracer diffusion techniques, measurements of creep, which requires dislocation climb and thus the diffusion of vacancies, and where suitable nuclei are available, by the nmr techniques which will be described by Strange and Terenzi.

In addition to intrinsic defects much interest has arisen in recent years in a number of polymeric materials which have been found to be able to incorporate impurities which greatly alter their electrical properties. For example, polyacetylene can be doped by diffusion or electrochemically with both anionic and cationic species (Heeger A.M. and MacDiarmid; 1979) so that its electrical conductivity can be varied over twelve orders of magnitude. Such materials can be used as electrodes in ionic battery systems and have many other applications. The exact nature of the doping and the location of the impurity ions within the structure of the polymers have yet to be determined.

Point Defects in Metals

The predominant intrinsic point defect found in metals is the vacancy and diffusion then occurs through the movement of thermally created vacancies. Such isolated defects may associate to form di- and tri-vacancies and eventually larger aggregates. The energies of formation of the isolated vacancies, which are typically of the order of 1 eV per defect, have been determined from measurements of the unit cell and macroscopic linear expansions, as for molecular solids, and also from investigations of the electrical resistivity. The total resistivity, $\rho(T)$, may be represented as

$$\rho(T) = \rho_o + \rho_i(T) + A \exp(-g_S/kT) \qquad (19)$$

where ρ_o is a residual temperature independent resistivity due to static lattice defects, $\rho_i(T)$ is the intrinsic restivity due to the thermal vibrations of the lattice and the final term is a contribution from the thermally produced defects the number of which depend on the free energy of defect formation g_S. The most accurate assessments of vacancy formation energies using this technique are made when the resistivity is meas-

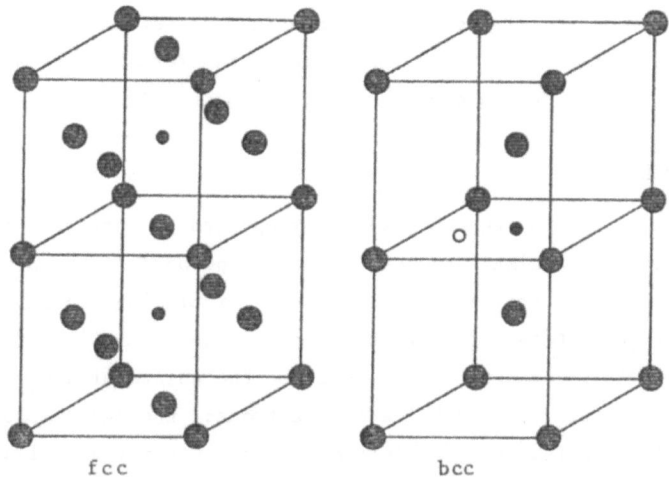

Figure 3   octahedral ● and tetrahedral ○ interstitial sites
in fcc and bcc lattices.

ured on samples in which the defect concentrations appropriate to higher
temperatures are quenched in by rapid cooling. In addition, vacancy
migration energies may be determined from measurements of the rate at
which the quenched resistivity is annealed out of the specimens. Care
must be taken with the maximum temperatures attained and the rate of
quenching to avoid complications due to cluster formation. Also the
analysis of data from the annealing may have to include higher order processes if divacancies are involved (Kimura; 1977). Positron anihilation
has in recent years proved to be a most useful technique in the study of
vacancy formation and migration in metals.

Radiation damage can produce self Frenkel pairs in metals and a
number of impurity atoms such as C, N, O and H can also take up interstitial positions. Available interstitial sites of interest are shown
for the fcc and bcc lattices in Figure 3. Of these metal-impurity
systems carbon in iron is the most important. Carbon atoms occupy octahedral interstitial sites in fcc high-temperature austenitic iron to a
concentration of 1.8 wt. % even though they are somewhat larger than the
interstice which has a radius of only 0.05 nm and their presence necessi-
tates an increase in the size of the unit cell. In the low temperature
bcc α-iron lattice, in which only 0.1 wt. % of carbon dissolves, the
carbon atoms also take up the octahedral interstice. The tetrahedral
site is, in fact, larger and the octahedral carbons are accommodated by
the anistropy of the bcc lattice in the z direction. Detailed information
on both carbon and nitrogen (which behave very similarly) interstitials

11

with respect to defect clustering and occupancy of tetrahedral and octahedral interstitial sites in the martensitic transformation of steels
has been provided by Mössbauer spectroscopy (Fujita; 1977).

COLOUR CENTRES

The name colour centre was originally applied to intrinsic electronic
defect centres in the alkali halides. For the purpose of the present discussion we will apply the term to all defects, including impurity ions,
that give rise to optical transitions which colour insulating crystals.
It is our principal concern here to describe the nature of those electronic
defects which contain trapped holes or electrons: the aggregation of
isolated defects of this type to form more complex centres will also be
described. The most important techniques used to study colour centres
are optical absorption and luminescence and, in suitable cases, electron
spin resonance (ESR) and electron nuclear double resonance (ENDOR) and the
special features introduced into these spectra as a consequence of the
defect being embedded in an ionic crystal will be discussed briefly. The
experimental techniques, and a number of more sophisticated investigative
methods which are used to elicit more detailed information on the structure
of the defect will be the subject of the Chapter by Spaeth.

The best known and most widely studied colour centre is the F-centre
in crystals with the rocksalt structure. These centres are produced when
alkali halides are irradiated with X-rays, γ-rays, or ultraviolet light
or exposed to energetic particles, but can also be created in a more
controlled manner by heating the crystal in a vapour of its metal. As
already discussed, vacancies in ionic crystals carry virtual charges and
these may trap both positive holes and electrons which are produced by
the incident photons. Photons of sufficient energy remove electrons
from the halide ions and the process involves promotion of an electron,
typically through a band gap of $\sim$ 10 eV, from the valence to the conduction band to yield a positive hole. Photons of smaller energies can promote the electrons to exciton states which lie just below the conduction
band and consist of an excited state electron bound to a hole.

The principal types of colour centres are illustrated in Figure 4
in a two-dimensional representation of the halide lattice. Although it
is possible to use the Kroger-Vink notation it is more usual to retain
the generic names introduced by the original workers: these names will
be used here with cognisance also being taken of a consistent nomen-
clature scheme suggested by Sonder and Sibley (1972). The F-centre
comprises an electron trapped at an anion vacancy and is, therefore,

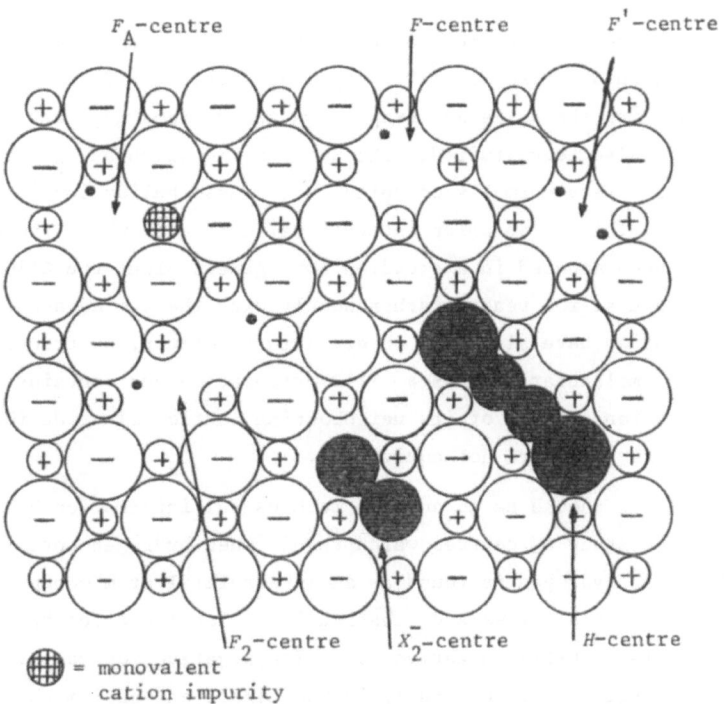

F_A-centre F-centre F'-centre

F_2-centre X_2^--centre H-centre

⊞ = monovalent
cation impurity

Figure 4. Models of common colour centres in the alkali halides
(after Henderson; 1972)

overall neutral in charge. F-centres may trap a further electron to give
F'-centres: they may also aggregate to form F_2-centres, which contain
two F-centres in nearest-neighbour sites along a <110> direction, or F_3^-
centres in which three F-centres occupy adjacent sites in the lattice
and therefore lie on a (111) plane. If the anion vacancy trapping the
electron has one of its six neighbouring cations substituted by an impurity ion then the centre is called an F_A-centre. In the alkaline
earth oxides, which are isostructural with the alkali halides and in
which colour centres have also been thoroughly investigated, (Henderson
and Wertz; 1977) an anion vacancy requires two electrons to give a
neutral defect. The original papers referred to this as the F'-centre
(by analogy with the alkali halides) but many workers now prefer the
term F-centre and then refer to an anion vacancy with one trapped electron in these systems as the F^+-centre.

Trapped hole centres are not as well understood as their antimorphs
containing trapped electrons. Positive holes may be trapped at cation
vacancies: they may also become self-trapped holes which are known as
V-centres. The V_K-centre in the alkali halides is shown in Figure 4 and
is comprised of an X_2^- molecular ion replacing two nearest-neighbour
anions in a <110> direction. This X_2^- molecular ion is stable (Stoneham;

13

1975).   Trapped hole centres in which the hole is trapped on a single anion have been observed in II-VI compounds such as oxides: two hole centres in which the holes are trapped on two O^- ions adjacent to the vacancy have also been studied.   The H-centre in which an X_2^- molecular ion occupies a site normally occupied by a single halide ion is also shown in Figure 4.   This is an anion interstitial centre in which the extra halogen is accommodated in a crowdion arrangement along the <110> direction. The X_2^- molecular ion reacts rather weakly with the two adjacent halide ions.   H-centres have also been observed in the fluorite structure but here the X_2^- molecular ion has a <111> orientation and contains one interstitial ion and one of its neighbouring lattice fluoride ions: the two anions are therefore not equivalent.

Finally it should be noted that much useful information has been derived from the study of centres which result when hydrogen enters ionic crystals.   The hydrogen is found as an interstitial atom occupying body-centred interstitial sites and substitutional lattice sites but may also occur as an interstitial negative ion.   The negative ion can also enter the anion lattice substitutionally in the alkali and cesium halides and its optical absorption has been observed in the alkaline earth fluorides.

Optical and Magnetic Properties

The characteristic features of optical absorption and emission and of ESR and ENDOR spectra from colour centres and other point defects and their analysis to reveal the structure of the defect causing them will be reviewed later by Spaeth: only the very basic principles will be outlined here. Impurity atoms and intrinsic colour centres with energy levels in the band gap give rise to selective absorption and the crystals are said to be 'coloured'.   The two factors which are peculiar to the crystal environment are the effect of the static crystalline field and the interaction of the electrons with the lattice phonons.   For transition ion impurities, e.g. the $MgO:Cr^{3+}$ system, the 3d energy levels are split by the crystal field.   The symmetry and magnitude of this splitting and hence the wavelength and intensity of the colouration depend on the host lattice.   The simplest model of the F-centre in the alkali halides is to regard it as a hydrogen-like atom with a single electron moving in a three-dimensional potential well created by the Coulombic field of the surrounding cations.   The potential energy is rather crudely assumed to be zero inside the vacancy and infinite everywhere else and following the usual 'particle in the box' formalism the Schrödinger equation for the wave function, ψ, is

$$\nabla^2\psi \;+\; \frac{2m_e E_n}{\hbar^2} \;=\; 0 \qquad\qquad (20)$$

14

The dimensions of the well are $x = y = z = a$, where a is the lattice constant, and the energy of the first transition from 1s to 2s-like states is

$$E_F = \frac{3\pi^2 \hbar^2}{2m_e a^2} \tag{21}$$

This model may be modified to alter the boundary conditions in line with ENDOR measurements which show the probability of finding the electron is not always zero outside the vacancy. Despite this and other problems the model is useful and F-band data are found to fit reasonably well to the type of relationship implied by eqn. (21) with the lattice spacing of the host crystal (Dawson and Pooley; 1969).

The transitions observed in the spectra of solids do not produce sharp absorption and emission lines typical of atomic spectroscopy but are broadened because of coupling to the phonon modes of the vibrating lattice. The processes of absorption and emission are understood in terms of the configuration-coordinate diagram illustrated in Figure 5: this shows the total potential energy for two different electronic states and also the vibrational states. Following the Franck-Condon principle the electronic transitions are represented by vertical lines since the time taken is very short compared with the frequency of the atomic vibrations. A typical absorption process goes from P, since at low temperatures the system is most likely to be in its lowest state, to N which is the excited state with the same configuration. The system then relaxes to R with the non-radiative release of the Huang-Rhys number of phonons. The emission

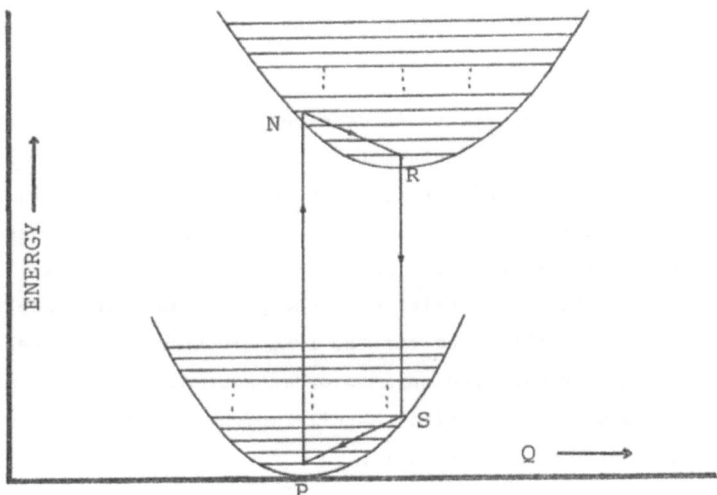

Figure 5. Configurational Coordinate diagram

15

spectrum results from the transition from R to S and this is followed by a further emission of phonons to return to the initial state. The difference between the absorption and emission energies, which is the sum of the energies of the emitted phonons, is called the Stokes shift. The extent to which the absorption and emission lines are broadened differs widely for different centres and depends on the degree of interaction between the vibrating crystal and the electronic energy levels in the optically active centre. In the absence of coupling a single sharp line, the zero-phonon line (P to R in Figure 5), occurs: the width and intensity of the sidebands reflect on the density of phonon modes in the crystal which are coupled to the electronic transition. Rare-earth ions tend to be very weakly coupled whereas F-centres interact strongly to give broad band spectra without sharp features.

ESR is used to study defects with unpaired electrons by observing the magnetic fields at which they resonantly absorb energy from applied microwave radiation of definite frequency. An unpaired electron possesses spin angular momentum and interacts with a magnetic field $\underline{B}$ so that its energy is

$$E = m_s g \mu_B \underline{B} \tag{22}$$

where $m_s = \pm\frac{1}{2}$ are the spin states, g is the spectroscopic splitting factor and μ_B is the Bohr magneton. The separation between the two spin states for the free electron is

$$\Delta E = 2.0023 \mu_B \underline{B} \tag{23}$$

and for electromagnetic radiation of frequency ν absorption will occur under suitable magnetic field conditions when

$$h\nu = \Delta E \tag{24}$$

A steady state resonant absorption is detected because electrons undergoing transitions from the slightly more populated ($m_s = -\frac{1}{2}$) lower level can relax from the ($m_s = \frac{1}{2}$) higher state by spin-lattice relaxation. Low temperatures are used because they increase the population difference between the states and increase the spin-lattice relaxation time. Experimental g values are characteristic of the particular paramagnetic species and differ from the value for the free electron. Internal interactions couple the electron spin to other sources of angular momenta. In a crystalline environment orbital angular momentum is mixed into the pure spin states by spin-orbit coupling.

The electron spin will also interact with the nuclear magnetic moments of central or neighbouring atoms. Nuclei with a spin quantum

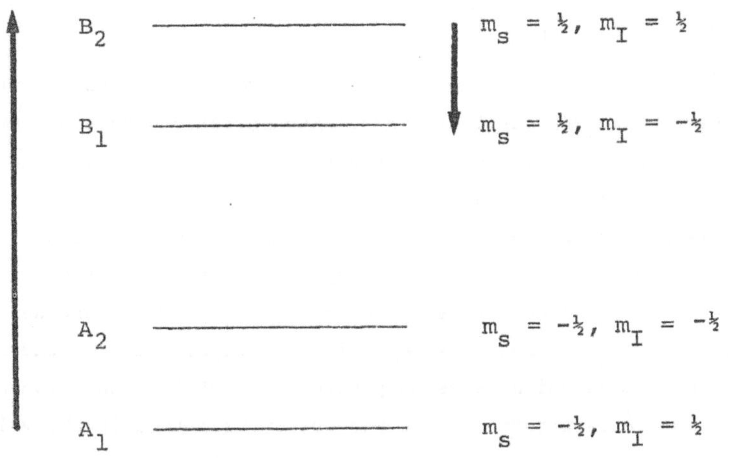

B_2 ——————————— $m_s = \frac{1}{2}, \; m_I = \frac{1}{2}$

B_1 ——————————— $m_s = \frac{1}{2}, \; m_I = -\frac{1}{2}$

A_2 ——————————— $m_s = -\frac{1}{2}, \; m_I = -\frac{1}{2}$

A_1 ——————————— $m_s = -\frac{1}{2}, \; m_I = \frac{1}{2}$

Figure 6.   Energy level scheme for ENDOR experiment

number I may split electron energy levels into (2I + 1) levels and the
interactions give rise to hyperfine and super hyperfine structure in the
spectrum, respectively.   This structure is particularly important in
deducing detailed information on the nature and environment of the defect.
Super hyperfine spectra are often very complex because the electron spins
may interact with nuclei several atom spaces away from the defect centre.
The ENDOR technique, which utilizes the fact that the selection rule for
a nuclear transition of a singular nucleus is $\Delta m_I = \pm 1$, can be used to
analyze overlapping components of the ESR spectrum.   The energy level
scheme is shown schematically in Figure 6.   First sufficient microwave
power is supplied at the electron-spin resonant condition to saturate the
major transition $A_1 \rightarrow B_2$ and the absorption then effectively ceases.   If
now the transition $B_2 \rightarrow B_1$ is stimulated using a strong radio frequency
field then the microwave absorption will recommence at resonance.   This
technique also allows analysis of nuclear quadrupole effects.

17

HIGHLY DEFECTIVE SOLIDS

The concentrations of point defects introduced thermally into ionic solids were shown earlier to increase exponentially with the temperature but the free energies of defect formation in the classical ionic conductors ensure that these concentrations are small, typically much less than 1%, even at temperatures approaching the melting points. In many instances it is also impossible to introduce large concentrations of dopant ions. However, there are other types of solids which are much more heavily disordered. The most common cause of this is a departure from the stoichiometric composition which is usually brought about in transition metal, actinide and certain rare earth compounds by variation in the valency of the cations involved. The presence in the lattice of cations with different valencies mean that certain sites carry effect changes and must be compensated, just as in the case of aliovalent impurity ions, by suitable defects elsewhere in the lattice.

The oxides, which predominate these types of solids, have been classified by Sørensen (1983) both on the basis of the nature of their deviation from stoichiometry and on the type of defects which they contain. The defects which form will depend on the possible valence states which the cation can adopt. For example in $Fe_{1-x}O$ which has the rocksalt structure and is metal deficient, iron vacancies, V_{Fe}'', are compensated for by oxidized cations, $Fe_{Fe}^{\bullet}$. In CeO_{2-x}, which has the fluorite structure and is oxygen deficient, the compensating defect is the reduced cation M_M'. Defect interactions may be written and the formation of $V_O^{\bullet\bullet}$ is represented by

$$2M_M + O_O \longrightarrow V_O^{\bullet\bullet} + 2M_M' + \tfrac{1}{2}O_2 \qquad (25)$$

Thus the stoichiometry and hence the defect concentration is seen to depend on the oxygen partial pressure, p_{O_2}. In general it is found that

$$[V_O^{\bullet\bullet}] \quad \alpha \quad p_{O_2}^{-1/n} \qquad (26)$$

where n is a factor which depends on the compound and on the type of defect being formed. In this way the dependence of properties such as ionic conductivity on the p_{O_2} can be useful in identifying the defect species which is effecting the transport.

Modes of Stabilization

Perhaps the most interesting feature of these systems is the manner in which their large defect concentrations are stabilized. The first of these stabilization modes involves the aggregation of the defect species

18

to form large complex clusters. This occurs in oxides such as UO_{2+x} and $Fe_{1-x}O$ and the contribution made to the understanding of these complexes by atomistic modelling techniques will be discussed by Catlow. The second method of stabilization is the formation of extended planar faults called crystallographic shear planes during which process defects are eliminated. In some systems, e.g. the trivalently-doped alkaline earth fluorides and $ZrO_2:Ca^{2+}$ it has been suggested that stabilization may occur through the formation of microdomains which are considered to be regions of the crystal in which the structure is very close to that of a known ordered phase and which extend over several lattice spacings. Catlow (1983) has given a classification of the more common systems on the basis of the stabilization mode which they exhibit.

The aggregation processes are analogous to those which occur when numbers of the associated pairs shown in Figure 2 join together to form more complex defect species. In the divalently-doped alkali halides system, Figure 2(a), larger clusters containing two, three and four

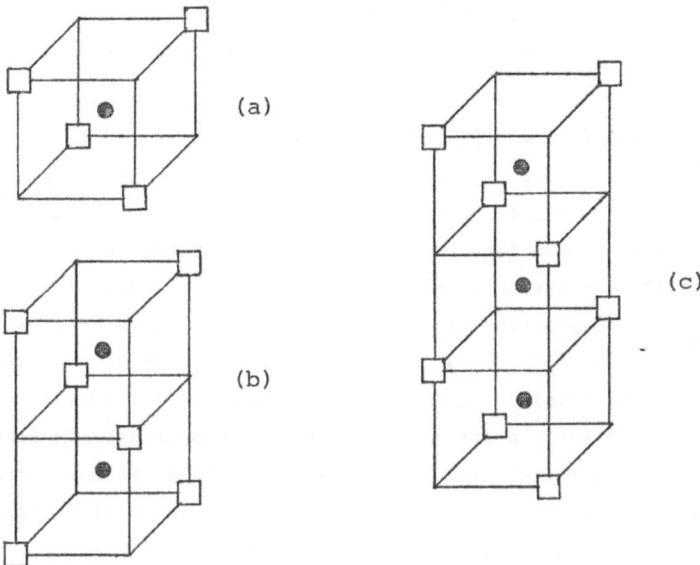

Figure 7. The 4-1 (a), 6-2 (b) and 8-3 (c) clusters in $Fe_{1-x}O$; □ is a cation vacancy and ● an Fe^{3+} interstitial.

impurity ions have been shown to be stable (Corish et al; 1981) and the process of aggregation can continue to give a superlattice called the Suzuki phase (Suzuki; 1955, 1958, 1961), in which compensating divalent dopant ions and cation vacancies are incorporated systematically on the cation sublattice and which has been identified by X-ray diffraction. In the trivalently-doped fluorite systems, Figure 2(b), which will be described by Chadwick, neutron diffraction studies (Catlow et al; 1983) as well as EXAFS measurements (Catlow et al; 1984) and theoretical calculations (Corish et al; 1982, Bendall et al; 1984) have shown that impurity-interstitial pairs can come together to form $2|0|2|2_1$ and $4|1|2|4_1$ clusters. These clusters are analogous to the clusters found in anion excess UO_{2+x} where the oxidation of the cation to give $U_u^{\cdot}$ is compensated for by $O_i^{''}$. The stability of these clusters is aided by large relaxations particularly of the anion interstitials and by some of the neighbouring lattice anions. Some of the clusters in $Fe_{1-x}O$ are shown in Figure 7. Here the vacancy aggregation occurs with cation interstitials and a surrounding distribution of Fe^{3+} ions and the basic arrangement is a tetrahedral aggregation in which an interstitial iron is surrounded by four $V_{Fe}^{''}$. This is known as the 4:1 cluster and these units can then share edges to give 6:2 and 8:3 clusters. Other larger aggregates such as the 16:5, which is corner sharing and an element of the inverse spinel structure and the 12:4 cluster which has been reported on the basis of microscopy [Lebreton and Hobbs; 1983] are also shown to result in a stabilization of the system by computer simulation calculations Tomlinson et al; 1984].

Crystallographic shear planes are planar faults in a crystal that separate two parts of the crystal which are displaced with respect to each other. As planar defects they might more properly be considered in the section dealing with that topic but they will rather be discussed here to complete the treatment of highly defective solids. Each shear plane causes the composition of the crystal to be changed by a small increment and oxides such as TiO_{2-x}, VO_{2-x}, WO_{3-x} and MoO_{3-x} are stabilized. in this way by the elimination of oxygen vacancies. The formation of a crystallographic shear plane is illustrated schematically in Figure 8 for a $<100>$ section through the ReO_3 structure. Shear in the direction indicated will eliminate the vacancies from the oxygen sublattice but leads to a fault in the stacking sequence of the metal sublattice. Again extensive relaxations, in this instance of the cations neighbouring the shear plane, are essential if the structure is to be stabilized. Because the composition of the crystal is altered slightly by each shear plane

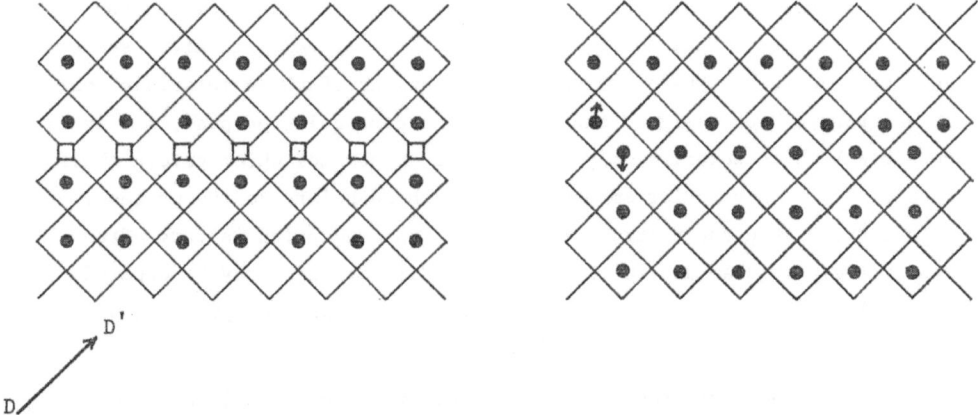

Figure 8.   Elimination of point defects by the formation of a
crystallographic shear plane. DD' is the direction
of shear and the relaxation of the cations are marked
by arrows (after Catlow; 1983).

the overall composition of a crystal which contains crystallographic shear
planes is dependent on the number and orientation of these planes. Disordered planes will produce a stoichiometric variation whereas planes
which are ordered in a parallel array give rise to a stoichiometric phase
with a complex formula. Such ordering of shear planes, sometimes with
repeat distances of 10 nm is common and has been shown to be thermodynamic
in nature and to be due to long range shear plane interaction energies
(Cormack et al; 1982). A change in the separation of such ordered planes
will produce a new phase with a new composition and the series of phases
produced in this way is called a homologous series. Examples are Ti_nO_{2n-1}
with n between 4 and 9 and $(Mo,W)_nO_{3n-1}$ with n between 8 and 14.

FAST ION CONDUCTORS

The contribution to the conductance of a solid, σ_r, by a species r is
given by

$$\sigma_r = |q_r|\mu_r x_r \tag{27}$$

where q_r and μ_r are the charge and mobility, respectively. The ability of
a mobile ionic species to conduct well relies therefore on it being present
in very large concentrations or being very highly mobile or on a combination of both of these factors. Fast-ion conductors is the name given to
a range of compounds, developed mainly during the last two decades, which
have the facility to offer very high ionic conductivities of specific ions.

They have aroused considerable interest because of their realized and potential usage in fuel cells, advanced battery systems, chemical sensors and other devices. There are now a great number of these materials, many of which have been designed and made with a particular purpose, and it is possible here to describe only the general principles which govern their operation with particular emphasis on how these relate to the presence of defects within their structures. Fast ion conduction is very often associated with disorder in the crystal. This disorder may be created thermally or it may result from non-stoichiometry and the presence of large concentrations of impurity ions.

Many of the early fast ion conductors of which α-AgI is a prototype exhibited the phenomenon only through a limited range of temperature below their melting points. AgI goes through a first-order structural phase transition at 419 K which results in the available cation sites, which are arranged in a bcc configuration, being only one-third occupied. Similarly α-Ag_2S has six mutually exclusive sites per unit cell for the four cations and therefore shows an effective occupancy of only two-thirds. Attempts to stabilize such high temperature phases at lower temperatures have been successful with the best known example being provided by the MAg_4I_5 range of compounds where M is a monovalent cation. It is important to emphasize that these surplus vacancies are a part of the crystal structure and are not thermally created point defects as are found in the classical ionic conductors. For this reason the Arrhenius energies for ionic motion between these equivalent and readily available sites are low and this gives the high mobilities typical of fast ion conduction.

Certain fluorites, e.g. PbF_2, also undergo phase transitions that are associated with the generation of disorder at high temperatures and gradually, as the temperature is increased, exhibit fast anion conduction (Chadwick; 1983). Although the mechanism by which this happens is not yet fully explained it is clear that the disorder is limited and that most of the anions remain close to their lattice sites. One possible explanation is the formation of large vacancy-interstitial clusters which would be analogous to those described earlier for the doped systems.

The β- and β''-alumina solid electrolytes, whose structures are comprised of spinel blocks separated by bridging oxygen atoms as shown schematically in Figure 9, are prepared as non-stoichiometric compounds which provide fast ion transport for a range of monovalent cations. Their general formula is $M_2O.nM_2'O_3$ where M is a monovalent cation and M' may be Al^{3+}, Fe^{3+} or Ga^{3+}: the β- and β''-forms differ, as shown in the figure,

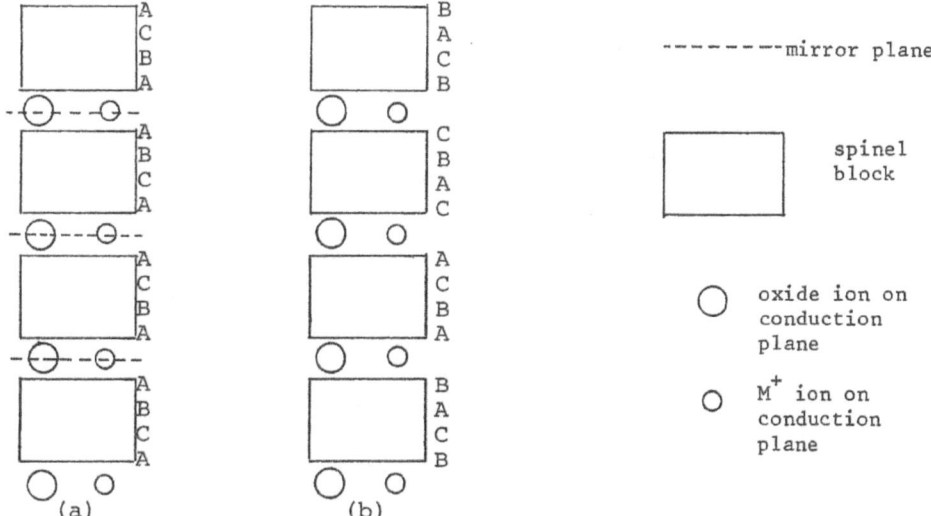

Figure 9. Schematic representation of the structures of (a)
beta- and (b) beta"- alumina.

in the stacking sequence of the layers. The most widely studied system is
sodium β-alumina (M=Na, M' = Al) since it has been used as the electrolyte
and separator in sodium sulphur cells. Commercial samples of sodium β-
alumina usually contain deliberate additions of Li_2O or MgO which allow
excess sodium to be incorporated: additions of MgO are also necessary to
stabilize the β"-form. These solid electrolytes are two-dimensional conductors with the conduction taking place in the planes between the spinel
blocks. In β-alumina these planes contain a variety of sites for the monovalent conducting ions. The sites are known as Beevers-Ross, anti Beevers-
Ross and mid-oxygen positions and the sodium ions can move readily between
them with an activation energy of ∿ 0.16 eV. The mechanistic details of
this process have been successfully modelled (Walker and Catlow; 1981).

The most widely used oxygen ion conductors rely on doping oxides such
as zirconia, hafnia and ceria with di- and trivalent cations to produce a
large concentration of charge-compensating oxygen vacancies. The resulting
fluorite structured electrolytes show quite complex behaviour, particularly
in respect of the effects of dopant-defect interactions. These systems
typically show an optimum dopant level which is characteristic of the
system. For $CeO:Y^{3+}$ the maximum conductivity is found at ∿ 8% Y^{3+} and
then decreases by some three orders of magnitude as additional dopant is

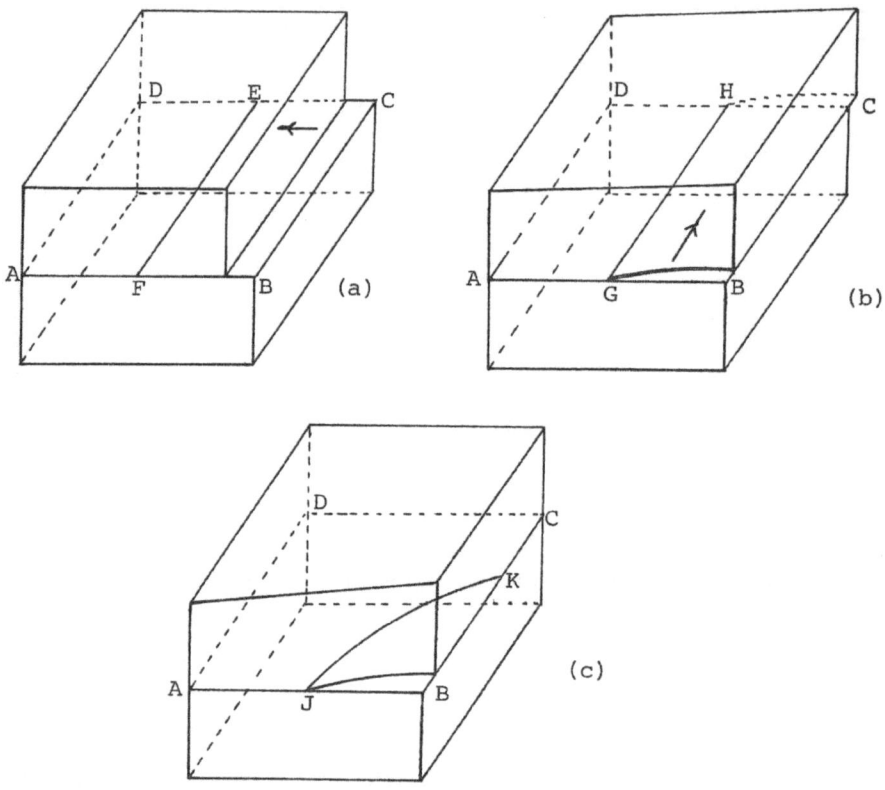

Figure 10.   (a) edge, (b) screw and (c) mixed dislocations.

added. These effects have been explained on the basis that the maximum occurs because when the dopant concentration is further increased the anion sites become associated with more than one dopant ion and the vacancies are then immobilized by trapping so that they can no longer contribute to the conduction process.

DISLOCATIONS

Dislocations are stoichiometric linear defects and are present in large numbers, typically $\sim 10^7$ cm^{-2}, even in well annealed metals. Their presence has a marked effect on the strength of materials and may also greatly influence the rates of crystal growth and dissolution. Edge and screw dislocations are the two ideal types from which an understanding of the nature of dislocations can begin. Figure 10(a) shows an edge dislocation. The displacement of part of the crystal has occurred in the direction shown by the arrow in the slip plane ABCD and the line FE, at which the disregister between the atoms in the planes immediately above

and below the slip plane is a/2 where a is the lattice spacing, is called
the dislocation line. It is evident that a similar configuration would
result if an extra half plane of atoms, terminating along the slip plane,
was inserted into the upper part of the crystal. By convention this is
called a positive edge dislocation: an extra half plane below the slip
plane would be termed negative. The dislocation line is perpendicular
to the direction of slip. Starting with the same piece of crystal a
screw dislocation is produced by displacing the material as shown by the
arrow in Figure 10(b). The dislocation line, GH, marks the boundary
between slipped and unslipped parts of the crystal and is now parallel to
the direction of slip. The atoms around the dislocation line are arranged
in a spiral ramp which advances, in this case, when rotated in a clock-
wise direction and the dislocation is therefore termed right-handed. This
spiral when it intersects the crystal surface will produce a step which
cannot be eliminated by the addition of atoms to form a complete atomic
layer. In general, real dislocations are neither pure edge or pure
screw in nature but have characteristics of both as shown in Figure 10(c).
At J the dislocation line is parallel to the slip direction so that it is
screw while at K it is perpendicular and therefore edge: in between it is
a mixed dislocation.

The feature of a dislocation which distinguishes it from other line
imperfections is that it cannot be eliminated by replacing defect-containing or 'bad' material with perfect of 'good' material. Burgers (1939) has
shown that it is possible to characterize a dislocation uniquely using a
vector. This vector, now called the Burgers vector b, is the closure failure of an atom-to-atom loop executed entirely in good material around the
dislocation: the same circuit must form a closed path in a perfect crystal.
The Burgers circuit around an edge dislocation is illustrated in Figure 11.

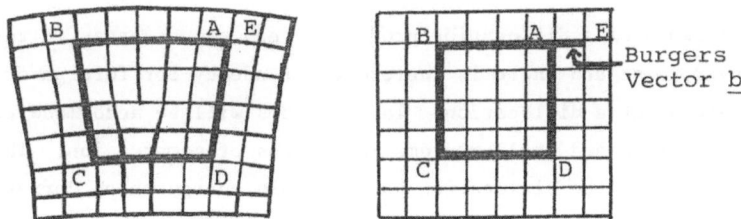

Figure 11. The Burgers Vector b of a dislocation represented
            as a closure failure in a dislocation free crystal.

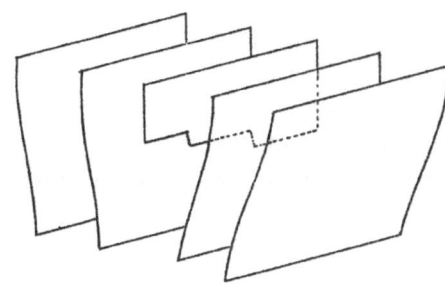

Figure 12. Jogs on an edge dislocation lie.

In this case $\underline{b}$ is perpendicular to the dislocation line, for a screw dislocation it is parallel to the dislocation line and in the more general
case of a mixed dislocation they enclose an arbitrary angle. Dislocation
lines may end at grain boundaries or at surfaces but never inside a
crystal. This implies that dislocations either branch into a number of
other dislocations at a node or form closed loops (Read; 1953).

The mechanism through which a dislocation greatly increases the ease
with which slip occurs in a crystal is most readily understood by considering the edge dislocation EF in Figure 10(a). The presence of the extra
half plane of atoms creates strain within the crystal and means that many
of the atoms in the plane immediately above the slip plane ABCD are displaced from their equilibrium positions with respect to the atoms in the
plane below them. Hence, on an atomic level, it is evident that movement
in the slip plane will be substantially less demanding of energy than in
a dislocation free crystal. Edge dislocations may also move out of their
slip planes but this process, which is called climb, requires the addition
or removal of atoms along the dislocation line and is therefore dependent
on the diffusion of interstitials or vacancies through the lattice. An
entire row of atoms is unlikely to be removed or extended in this way and
the dislocation lines are more likely to develop jogs as illustrated in
Figure 12.

The presence of a dislocation produces areas of unevenness in the
crystal environment and there is therefore a tendency for foreign atoms
to accumulate about a dislocation: larger atoms will be accommodated in
regions of dilation and smaller atoms in regions of compression. The
stress field around the dislocation has a strain energy which corresponds
to the work done in forming the dislocation. This energy may be divided
as:

$$E_{total} = E_{core} + E_{elastic\ strain} \qquad (28)$$

26

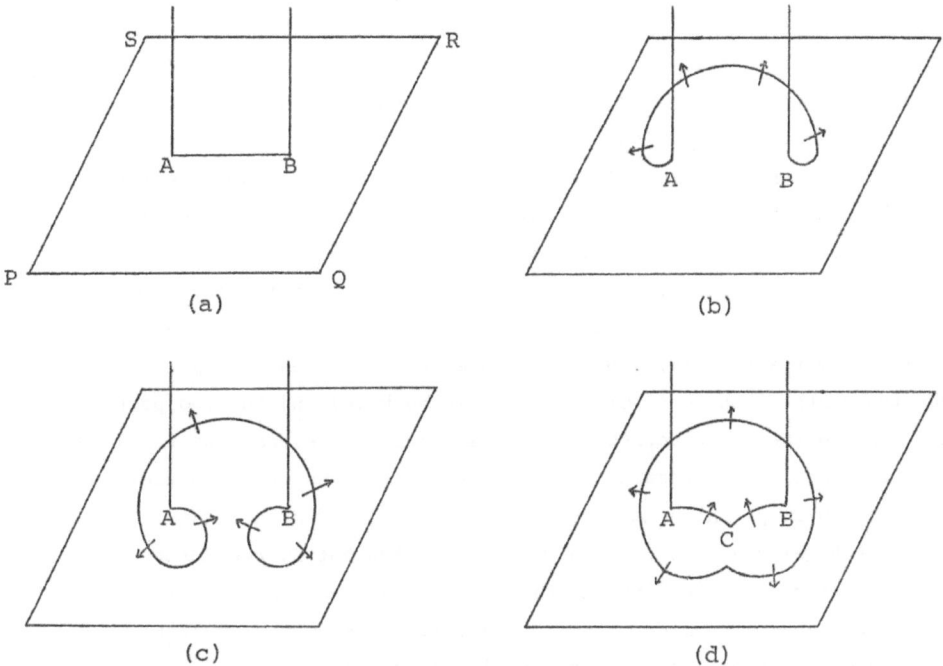

Figure 13. Frank Read source: the regenerative multiplication process
is described in the text.

where the core is the central region around the dislocation with a radius
r_o which is typically of 0.5 to 1.0 nm. The elastic strain may be simulated by distorting a cylindrical ring of material and for a screw dislocation the energy is calculated as (Hull; 1965)

$$E_{elastic\ strain} = \frac{\mu b^2}{4\pi}\ \ln\ (\frac{r_1}{r_o}) \tag{29}$$

where μ is the shear modules, b the magnitude of the Burgers vector and
r_1 is the radius over which the strain field acts. In principle the integration should be taken to the crystal boundaries but the result is
relatively insensitive to the magnitude of r_1 and the elastic strain
energy is typically 6 – 7 eV per atom length threaded by the dislocation
line. The corresponding energies calculated for edge and mixed dislocations are quite similar and estimates for the energies of the cores,
which are of necessity approximate, are of the order of 1 eV per atom
length. Formation energies of this magnitude ensure that the number of
dislocations which are in thermal equilibrium is negligible and that the
normal thermal fluctuations are not responsible for the very large numbers
of dislocations present in almost all materials. Dislocations are produced
during crystal growth through heterogeneous nucleation by internal stresses
at any source of stress concentration such as adherence of the growing

27

crystal to its containing vessel, thermal gradients, compositional variation or the presence of precipitates or other inclusions. They may also result from the formation and subsequent movement of dislocation loops. The concentration of dislocations in a crystal is known to increase progressively when the material is strained and two mechanisms are important for their regenerative multiplication. The first is the Frank-Read type source in Figure 13 in which a dislocation is held at two points A and B in the slip plane PQRS by obstacles (a) such as precipitates, composite jogs or intersections of nodes. The length AB will bow out under an applied shear stress (b) and the dislocation continues to expand until the two parts of the unstable loop double back (c) and the two parts of the slipped area come together. The sections of dislocations involved here are opposite in nature and will annihilate each other to yield a dislocation loop, which may continue to expand, and a section of dislocation, ACB (Figure 13(d), which will straighten out to regenerate the original pinned dislocation and can begin the process again. The second mechanism (Koehler; 1952, Orawan; 1954) is that of multiple cross glide and occurs when a dislocation line expanding under an applied stress glides across from one plane to another. Figure 14 shows double cross glide from the slip plane ABCD to EFGH. The composite jogs AE and BF are edge in character and so are essentially immobile while the sections of dislocations lying in the planes may continue to expand and are, in fact, Frank-Read sources. However, in metals with high stacking fault energies and in ionic solids cross slip occurs readily and is likely to happen again before a dislocation loop can be completed. Multiple cross glide ensues and results in a thick glide band with seg-

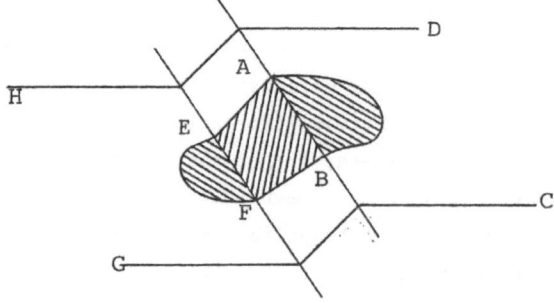

Figure 14. Double cross glide as described in the text.

ments lying on many parallel slip planes (Gilman and Johnson; (1962).

Applications

Attention has been drawn earlier to the fact that a screw dislocation which intersects a crystal surface produces a step which cannot be eliminated by the addition of atoms to form a perfect surface. This consequence was recognized by Frank (1959) as an explanation for the ready growth and dissolution of crystals at low super- and undersaturation, respectively. If the surface of a crystal is assumed to be perfect then growth or dissolution requires the formation of critical nuclei or voids, respectively, for which super- or undersaturations of up to 40% are calculated to be necessary. The spiral step on the surface as a consequence of the emerging dislocation removes the necessity for these nucleation processes and provides the explanation for crystal growth at low supersaturations. Circular growth spirals have been observed on many materials as have spiral etch pits during crystal dissolution and these provided one of the best experimental confirmations of the existence of dislocations.

The interactions between dislocations which, in general, reduce their mobilities has important consequences for the mechanical strength of solids. Thus the additional dislocations which are introduced into a crystal when strain is applied subsequently contribute to the strength of the material and this process is called work hardening. Many different strengthening mechanisms have been described (Hull; 1965) and the work hardening depends on the crystal structure, on the temperature, on the distribution and nature of the dislocations and also on the testing conditions. In general, the process is effected by the piling up of dislocations at boundaries and other obstacles.

Another important consequence of the interaction of dislocations with each other through their long-range elastic strain fields is the formation of low-angle or tilt boundaries. When edge dislocations attempt to minimize their energy and are arranged in an array at equally spaced intervals they give rise to a slight misorientation which is described by a small rotation, Θ, about the axis common to the material on either side of the boundary (Vogel et al; 1953). The energy per unit area of such a boundary is given by

$$E = \frac{\mu b}{4\pi(1-\nu)} \Theta \ (A - \ln\Theta) \tag{30}$$

where ν is Poisson's ratio and A a constant. This equation successfully accounts for experimental observations in a number of materials at small values of Θ. More complex tilt boundaries result when the boundary plane

29

is no longer a symmetry plane for the two parts of the crystal while a
twist boundary is produced from a cross grid of pure screw dislocations.
When a specimen which has been strained is annealed at moderate temperatures it will first reduce its energy by forming low angle boundaries: this
process is called recovery. Annealing at higher temperatures will give
rise to recrystallization in which larger grains with larger angle grain
boundaries form and subsequently grow. The overall energy is now further
reduced by the reduction in the area of the grain boundaries.

PLANAR DEFECTS

Many important properties of materials such as their catalytic
activity, their decomposition or corrosion and their growth rates relate
to their external surfaces which may formally be considered as planar
defects. However, the properties of these surfaces depend, in turn, on
the fact that they are not perfect but instead exhibit a variety of point
and linear imperfections which greatly affects their reactivity and use-
fulness. Materials also have internal surfaces such as stacking faults,
crystallographic shear planes, which were considered earlier, and grain
boundaries which were described briefly in the section on dislocations.
Henrich, in a later chapter, will discuss in detail the nature of defects
on external surfaces with emphasis on their electronic structures: the
principal physical features of these defects will be described here. The

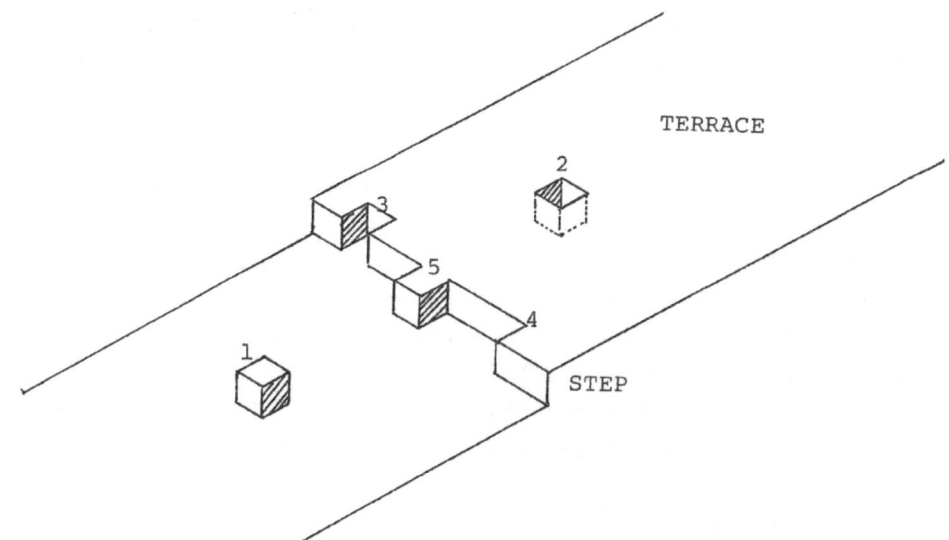

Figure 15. Some defects on surfaces: 1 an adatom on a terrace;
2 a vacancy on a terrace; 3 a vacancy in a step
4 a kink; and 5 an adatom on a step.

extent to which a surface can modify the distribution of intrinsic and extrinsic point defects in sub-surface regions will also be considered as will the phenomenon of defect segregation at grain boundaries.

The main point and extended defects which may occur on a surface are shown in Figure 15. As discussed earlier the intersections of dislocation lines with a surface also produce a characteristic defect type which is important in dissolution and growth of the material. The Madelung constants relevant to each of the defect sites shown in Figure 15 have been calculated and show, for the (100) rocksalt surface, a variation from 0.07 for the adatom to 1.68 for an atom in the terrace: the internal value in the bulk crystal is 1.747 (Magill et al; 1980). While these values serve as a useful guide as to the relative stabilities of the various sites it is also necessary to include a detailed treatment of relaxation and of lattice polarization into a complete calculation. The static lattice simulation methods to be described by Catlow have been adapted to enable them to be used in a variety of calculations relating to surfaces and to grain boundaries (Mackrodt and Stewart; 1977, Tasker; 1978). The modelling of defects on surfaces will be described later in the chapter by Colbourn. These types of calculations have been extremely valuable in that they can be made to relate to situations which are not always amenable to experimental investigation.

The atomistic modelling of surfaces has provided a detailed picture of the surface structure of ionic crystals. In the absence of defects the calculations show that when the surface layer contains both anions and cations a rumpling effect occurs in which the anions move slightly outwards and the cations slightly inwards (Tasker; 1979). This is in agreement with experimental observations. More recently the ionic displacements around a step in MgO have been modelled and indicate substantially modified step geometry but with the step height being reduced only slightly by the relaxations (Tasker; 1983). The explanation of these relaxations is that they result from polarization of the ions: this will occur in the opposite sense on the cations and anions. Impurities may segregate towards surfaces if their segregation energy, the difference between their substitutional energy in the bulk and at the surface, is favourable and may then greatly influence surface properties such as catalytic activity. Calculations for impurities in MgO (Mackrodt and Stewart; 1977, Colbourn et al; 1983) indicate that extensive segregation, dependent on ion size, should occur and this is again in line with experimental observations. Finally, the energies of vacancy formation near the (001) surface of MgO have been evaluated theoretically and it is found

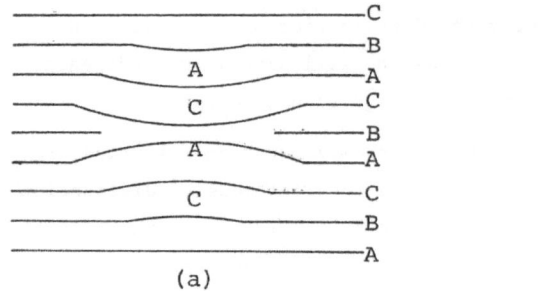 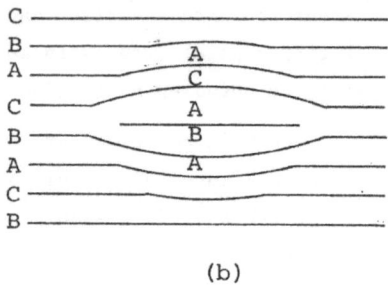

(a)   (b)

Figure 16. (a) intrinsic and (b) extrinsic stacking faults.

to have a maximum just below the surface (Duffy et al; 1984). This maximum results from competition between the variation with depth of surface
Madelung potential and the polarization energy and the variation should
clearly be taken into account when calculating the equilibrium defect
distribution.

Turning now to internal surfaces the first of these to be considered
is the stacking fault. This is a planar defect in a local region of the
crystal where the regular stacking sequence of the layers of atoms has
been interrupted: it has particular reference to fcc and hexagonal close-
packed metals. When a large supersaturation of vacancies exists, for
example after quenching or irradiation, these may aggregate to give a
disc as shown in Figure 16(a). The planes on either side will collapse
inwards and the result is termed an intrinsic stacking fault and shows a
single break in the stacking sequence. An extrinsic fault is formed by
the insertion locally of a partial layer of interstitial atoms as shown
in Figure 16(b). Here there are two breaks in the stacking sequence so
that the extrinsic stacking fault energy per unit area is greater than
that for the intrinsic fault. It is also possible for multiple layers
of vacancies to condense on top of each other and so produce double
faults: all of these structures have been observed in electron microscopy.
The faults are surrounded by partial dislocations. Indeed dislocations
may seek to reduce their energies during slip by dissociating into two
partial dislocations. These dislocations will then produce a sheet of
stacking fault between them the energy of which prevents the dislocations
from becoming too separated (Seeger; 1957). When stacking fault energies
are high, e.g. in hexagonal close-packed metals the defect structure may

32

stabilize by nucleating a partial dislocation which sweeps through the fault and changes the stacking sequence in the layers above it.

Grain boundaries occur between the crystallites which make up the crystal and thus separate regions of the crystal which have different orientations. The relationships between low-angle tilt and twist-boundaries and the presence of dislocations in the crystal have already been described. Recently, a series of simulation studies on grain boundaries in ionic solids has revealed details of the relaxation of the ions near to the boundaries, of the segregation of impurity ions to the vicinity of the boundaries and of the variation in the energies needed to form intrinsic point defects in moving from the bulk crystal to near to the grain boundaries. These are of particular interest here because they serve to set down the principles which govern the interactions of point defects with grain boundaries and also to explain how grain boundaries can act as a preferential pathway for matter transport through solids.

The calculations refer to the rocksalt structured oxides with most work relating to NiO (Duffy and Tasker; 1985). All seven $\langle 001\rangle$ and six $\langle 011\rangle$ tilt boundaries which were studied were found to be stable but the $\Sigma 5m$ [001] twist boundary was unstable with respect to its dissociation into two free surfaces unless Schottky defects were introduced at the interface. The boundary configurations were more open than the corresponding configurations in metals and they should have an enhanced diffusion capability as has been reported experimentally for NiO by Atkinson and Taylor (1981). Both neutral defects and solutes are known to segregate to grain boundaries both in metals and in ionic crystals and a number of experimental techniques such as SIMS and STEM have demonstrated this effect. Whereas the concentration of neutral defects may be simply related to their energies of segregation and to the strain relief available in the distorted lattice there is also a spread in the Madelung potentials available at the interface. This is analogous to the situation described earlier for external surfaces and it makes certain sites more favourable for aliovalent impurity ions. The formation energies of intrinsic defects near grain boundaries in NiO have been calculated and show that for each type of defect interface ionic sites exist with defect formation energies which are lower than the corresponding bulk values (Duffy and Tasker; 1984a). The concentration of singly charged vacancies was found to exceed that of compensating holes with the net negative charge in the boundary region being compensated by a positive space charge layer. Again the vacancy concentration near the interface was predicted to be greater than that in the bulk crystal and this modified distribution

33

would also be expected to increase diffusion. Finally, substitutional energies calculated for a range of impurity ions (Duffy and Tasker; 1984b) showed that defects with a high effective charge experience a larger decrease in energy in selected sites with suitably modified Madelung potentials at the boundary. Therefore the concentrations of both impurities and vacancies are strongly enhanced at these grain boundaries in NiO: the estimation of the distribution of aliovalent impurities is complicated by space charge effects.

SUMMARY

This chapter has aimed to provide a necessarily brief overview of the different types of defects which are found in solid materials. The discussion started with point defects, including colour centres, and progressed to more highly defective solids, to linear defects and finally to a short introduction to defects on surfaces. Where space has permitted the relationships between these different defect types have been described. In all cases the treatment has had to be selective and in many instances it has also been superficial. Hopefully it will have at least introduced each of the defect types whose detailed structures and properties are to be revealed by the experimental techniques which are the subject matter of the following chapters.

REFERENCES

Allnatt, A.R., and Jacobs, P.W.M., 1961, Proc. Roy. Soc. A260, 350.
Allnatt, A.R., and Loftus, E., 1973a, J. Chem. Phys., 59, 2541.
Allnatt, A.R., and Loftus E., 1973b, J. Chem. Phys., 59, 2550.
Allnatt, A.R., and Yuen, P.S., 1975a, J. Phys. C., 8, 2199.
Allnatt, A.R., and Yuen, P.S., 1975b, J. Phys. C., 8, 2213.
Atkinson, A., and Taylor, R.I., 1981, Phil. Mag. A43, 979.
Bendall, P.J., Catlow, C.R.A., Corish, J., and Jacobs, P.W.M., 1984,
        J. Solid State Chem., 51, 159.
Burgers, J.M., 1939, Proc. Koninkl. Ned. Akad. Wetenschap., 42, 293, ibid.
        42, 399.
Card, D.N., and Jacobs, P.W.M., 1977, Mol. Phys., 34, 1.
Catlow, C.R.A., Chadwick, A.V., Greaves, G.N., and Moroney, L.M., 1984,
        Nature, 312, 601.
Catlow, C.R.A., Chadwick, A.V., and Corish, J., 1983, J. Solid State Chem.,
        48, 65.
Catlow, C.R.A., 1983, 'Mass Transport in Solids', eds., F. Beniere and
        C.R.A. Catlow, (Plenum Press: New York), Ch. 16.
Catlow, C.R.A., Corish, J., Jacobs, P.W.M., and Lidiard, A.B., 1981,
        J. Phys. C., 14, L121.
Catlow, C.R.A., Diller, K.M., Norgett, M.J., Corish, J., Parker, B.M.C.,
        and Jacobs, P.W.M., 1978, Phys. Rev. B, 18, 2739.
Chadwick, A.V., and Glyde, H.R., 1977, 'Rare Gas Solids', Vol. 2, eds.,
        M.L. Klein and J.A. Venables (Academic: New York) Ch. 19.
Chadwick, A.V., 1983, 'Mass Transport in Solids', eds., F. Bénière and
        C.R.A. Catlow, (Plenum Press: New York), Ch. 11.
Chadwick, A.V., 1983, Radiation Effects, 74, 17.
Colbourne, E.A., Mackrodt, W.C., and Tasker, P.W., 1983, J. Mat. Sci.
        17, 1917.

Corish, J., Catlow, C.R.A., Jacobs, P.W.M., and Ong, S.H., 1982,
Phys. Rev. B, 6425.

Corish, J., Quigley, J.M., Jacobs, P.W.M., and Catlow, C.R.A., 1981,
Phil. Mag. A44, 13.

Corish, J., and Mulcahy, D.C.A., 1980, J. Phys. C., 13, 6459.

Corish, J., and Jacobs, P.W.M., 1975, Phys. Status Solidi (b), 67, 263.

Cormack, A.N., Tasker, P.W., Jones, R., and Catlow, C.R.A., 1982,
J. Solid State Chem., 44, 174.

Dawson, R.K., and Pooley, D., 1969, Phys. Status solidi, 35, 95.

Devlin, B.A., 1982, Ph.D. Thesis, Nat. Univ. of Ireland.

Duffy, D.M., and Tasker, P.W., 1985, 'Advances in Ceramics', eds. W.D.
Kingery, (Amer. Ceram. Soc.;Columbus), Vol. 10, p. 275.

Duffy, D.M., and Tasker, P.W., 1984a, Phil. Mag., A50, 143.

Duffy, D.M., and Tasker, P.W., 1984b, Phil. Mag. A, 50, 155.

Duffy, D.M., Hoare, J.P., and Tasker, P.W., 1984, J. Phys. C., 17, L195.

Frank, F.C., 1949, Proc. Phys. Soc., A62, 131.

Fujita, F.E., 1977, 'Progress in the Study of Point Defects', eds. M.
Doyama and S. Yoshida, (University of Tokyo Press : Tokyo) p.71.

Gillan, M.J., 1981, Phil. Mag. A43, 301.

Gilman, J.J., and Johnson, W.G., 1962, 'Solid State Physics', 13, 147,
(Academic Press: New York).

Harding, J.H., 1985, Phys. Rev. (in press) also A.E.R.E. Harwell Report
TP.1113.

Heeger, A.J., and MacDiarmid, A.G., 'The Physics and Chemistry of Low
Dimensional Solids', Ed. L. Alacer (Reidel:Dordrecht). p.353.

Henderson, B., 1972, 'Defects in Crystalline Solids', (Arnold:London).

Henderson, B., and Wertz, J.E., 1977, 'Defects in the Alkaline Earth
Oxides', (Taylor and Francis : London).

Hooten, I.E., and Jacobs, P.W.M., 1983, Radiation Effects 73, 233.

Howard, R.E., and Lidiard, A.B., 1964, Reports Prog. Phys., 27, 161.

Hull, D., 1965, 'Introduction to Dislocations', (Pergammon : Oxford).

Kimura, H., 1977, 'Progress in the Study of Point Defects', eds.,
M. Doyama and S. Yoshida, (University of Tokyo Press: Tokyo), p.119.

Kitaigorodski, A.I., 1973, 'Molecular Crystals and Molecules', (Academic
Press : New York).

Koehler, J.S., 1952, Phys. Rev. 86, 52.

Kroger, F.A., and Vink, H.J., 1956, Solid State Physics, 3, 307.

Lebreton, C., and Hobbs, L.W., 1983, Radiation Effects, 74, 227.

Lidiard, A.B., 1954, Phys. Rev., 94, 29.

Lidiard, A.B., 1957, 'Handbuch der Physik', Vol. 20, (Springer-Verlag:
Berlin), p.264.

Mackrodt, W.C., and Stewart, J., 1977, J. Phys. C., 10, 1431.

Magill, J., Long, K.A., Brooks, M.S., and Ohse, R.W., 1980, Proc. 4th Int.
Conf. Solid Surfaces 3rd European Conf. Surface Sci., eds., D.A.
Degras and M. Costa.

Mott., N.F., and Gurney, R.W., 1948, 'Electronic Processes in Ionic
Crystals', 2nd edn. (OUP:Oxford), Ch. II.

Orowan, E., 1954, 'Dislocations in Metals', (A.I.M.E.: New York), p.69.

Read, W.T., Jr., 1953, 'Dislocations in Crystals', (McGraw-Hill:New York).

Seeger, A., 1957, 'Dislocations and Mechanical Properties of Crystals',
(Wiley : New York), p.243.

Simmons, R.O., 1962, Proc. of Int. School of Physics 'Enrico Fermi', eds.,
D.S. Billington, XVIII Course, (Academic Press : New York), p.568.

Sonder, E., and Sibley, W.A., 1972, 'Defects in Crystalline Solids',
eds., J.H. Crawford, Jr., and L.M. Slifkin, (Plenum Press : New
York).

Sørensen, O.T., 1983, 'Mass Transport in Solids', eds. F. Bénière and
C.R.A. Catlow, (Plenum Press : New York), Ch. 15.

Stoneham, A.M., 1975, 'Theory of Defects in Solids', (Clarendon Press :
Oxford).

Suzuki, K., 1955, J. Phys. Soc. Japan, 10, 794; 1958, Ibid. 13, 179;
1961, Ibid. 16, 67.

Tasker, P.W., 1978, AERE Harwell Report, R-9130.

Tasker, P.W., 1979, Phil. Mag., A39, 119.

Tasker, P.W., 1983, 'Mass Transport in Solids', eds., F. Bénière and C.R.A. Catlow, (Plenum Press : New York).

Tomlinson, S.M., Catlow, C.R.A., and Harding, J.H., 1984, Harwell Report TP1095.

Vogel, F.L., Pfann, W.G., Corey, H.E., and Thomas, E.E., 1953, Phys. Rev. 90, 489.

Walker, J.R., and Catlow, C.R.A., 1981, J. Phys. C., 14, L979.

EXPERIMENTAL STUDIES OF POINT DEFECTS: THE EXAMPLE OF CRYSTALS WITH THE FLUORITE STRUCTURE

A.V. Chadwick

University Chemical Laboratory
University of Kent
Canterbury, Kent, CT2 7NH, U.K.

INTRODUCTION

In the preceeding chapter Corish presented a broad survey of the types of defect that are found in solids. Later chapters will describe in depth the techniques that are currently available for the investigation of these defects. The objective of this chapter is to take a general view of point defects in ionic solids; to consider some of the typical problems and to show how they can be resolved by modern techniques. This will include a discussion of the fundamental questions of the nature of the defects, their concentrations and energetics and also subsiduary questions such as the effects of impurities on the defects and the mechanisms of defect migration. In order to give some framework to the discussion the case of defects in crystals with the fluorite structure will be taken as the example. These materials exhibit a broad spectrum of defect behaviour, from simple to complex, and have been studied by a very wide range of techniques, both experimental and theoretical.

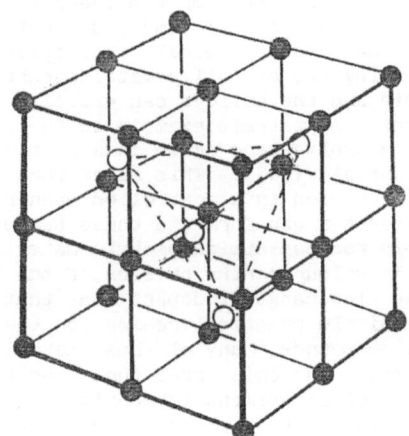

Figure 1. The fluorite lattice. The open circles represent the cations and the filled circles the anions.

The fluorite structure, shown in figure 1, is very simple, consisting of a simple cubic array of anions with alternate cubes occupied by divalent cations. The presence of the unoccupied anion cubes means that it is a very "open" structure and this leads to very interesting defect behaviour. It is a relatively common structure, being adopted by halides (e.g. CaF_2, SrF_2, BaF_2, CdF_2, β-PbF_2 and $SrCl_2$) and oxides (e.g. CeO_2, ThO_2, UO_2 and stabilized ZrO_2). In addition, a number of oxides (e.g. Li_2O) and sulphides (e.g. Na_2S) adopt the related anti-fluorite structure. Several fluorite-structured materials are technologically important and this has helped to generate interest in their defect properties. Examples include the use of UO_2 as a fuel in nuclear reactors, stabilized ZrO_2 as an electrolyte in fuel cells and oxygen sensors (Dell and Hooper, 1978) and the fluorides in solid state lasers (Sorokin and Stevenson, 1961) and as fluoride ion conductors (Reau and Portier, 1978). An extensive literature is available on these materials and they have been the subject of a book (Hayes, 1974).

There are a number of features of the fluorite-structured compounds that make them particularly attractive for defect investigations and the systems on which to focus this chapter. Firstly, the strongly ionic compounds at low to moderate temperatures show the normal defect behaviour of a simple ionic solid; in pure and lightly doped crystals the defect and transport properties can be understood (Lidiard, 1974) on the basis of the models developed for the alkali and silver halides (Lidiard, 1957). Thus in this case they are an example of classical behaviour. However, at high temperatures (usually above 0.8 times the melting temperature) the electrical conductivity is abnormally high, although still purely ionic in nature. This places these materials in the class of solids referred to as "fast-ion conductors", "solid electrolytes" or "superionic conductors" (Hagenmuller and van Gool, 1978; Salamon, 1979; Geller, 1977) which are characterized by ionic conductivities comparable to those found in molten salts. Fast-ion conductors have a number of potential, technological applications (Kleitz et al, 1983), notably as battery electrolytes. However, the phenomenon is usually associated with complex crystal structures (e.g. the β-aluminas (Kennedy, 1977)). The fluorite structure is the simplest crystal structure to show fast-ion conduction and therefore the alkaline-earth fluorides have been extensively employed as test-beds for the theoretical models. A third feature of many fluorite-structured compounds that leads to interesting defect behaviour is that their chemical composition can be considerably changed without altering the overall crystal structure. The oxides can be non-stoichiometric and the halides can dissolve high concentrations of aliovalent cations. An extreme example is the alkaline-earth fluorides which can dissolve rare-earth ions up to a level of 50 mole per cent (Ippolitov et al, 1967). This means that point defect concentrations can be varied in a controlled manner, various impurity-defect aggregates can form and these basically simple systems can be used as models for non-stoichiometric materials. It is also worth noting that the colour centre physics of the halides is also extensive due to the wide range of dopant ions that can be employed. Finally, there is a simple practical reason for the interest in the fluorite-structured compounds. Many of these materials can be prepared as large single crystals and this makes them amenable to study by the widest possible range of experimental techniques. Some of the techniques used in defect investigations include diffraction methods (X-ray, electron and neutron), electrical conductivity, radiotracer diffusion, neutron scattering, light scattering, nuclear magnetic resonance (NMR), electron paramagnetic resonance (EPR), dielectric relaxation spectroscopy, thermal depolarization spectroscopy and extended X-ray absorption fine structure (EXAFS).

The application of some of the above techniques will be included in the following discussion. This discussion will concentrate on the fluorite-structured halides and will consider three major defect problems; (i) the nature of the intrinsic defects and the mechanisms of matter transport, (ii) the defect structures in heavily doped crystals, and (iii) the high-temperature fast-ion conductivity.

INTRINSIC POINT DEFECTS

The basic concepts concerning point defects in ionic crystals were outlined in the chapter by Corish and a detailed discussion can be found in the review by Lidiard (1957). The early work on the thermodynamics and kinetics of point defects in fluoride-structured materials has been surveyed by Lidiard (1974).

Simple theoretical models predict that there are two major possibilities to be considered for the dominant point defect structures in pure CaF_2; anion-Frenkel pairs (anion vacancies and anions on the cube-centre interstitial sites) and Schottky triplets (cation and anion vacancies). The effective charges of these defects with respect to the perfect lattice are +1e, -1e, and -2e (where e is the charge of a proton) for the anion vacancy, anion interstitial and the cation vacancy, respectively. If the dominant defects are anion-Frenkel pairs then doping the cation sub-lattice with M^{3+} ions (effective charge +1e) will increase the anion interstitial concentration and doping with M^+ ions (effective charge -1e) will increase the anion vacancy concentration. Alternatively, if Schottky triplets dominate M^{3+} doping will increase the cation vacancy concentration and M^+ doping will again increase the anion vacancy concentration. Since the aliovalent dopant ions and the defects they create have opposite effective charges there are long-range Coulomb interactions to be considered. At high dopant concentrations or low temperatures these lead to the formation of impurity-defect complexes with the two species on neighbouring sites. Examples of simple complexes expected at low impurity concentrations (less than 1 mole per cent) are shown in figure 2.

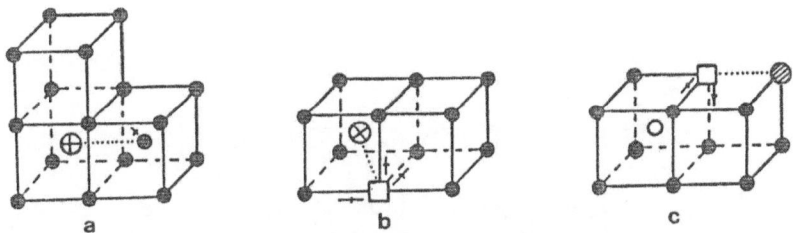

Figure 2. Simple impurity-defect complexes in CaF_2. The filled circles represent the anions, the open circles the cations and the open squares the anion vacancies. The arrows indicate the direction in which the defects move when the complex reorients. (a) M^{3+} - interstitial anion nearest neighbour complex ($\oplus$ = M^{3+} ion); (b) M^+ - vacancy nearest neighbour complex ($\otimes$ = M^+ ion); (c) O^{2-} - vacancy nearest neighbour complex ($\oslash$ = O^{2-} ion).

Given the possible defect structures outlined above a number of
mechanisms could be expected for the long-range migration of ions. The
most probable process for cation migration is via a vacancy mechanism.
There are four possible mechanisms for the migration of anions; the
vacancy, direct interstitial, non-collinear interstitialcy and the
exchange mechanisms. These are shown in figure 3. The first three
mechanisms involve defects as they give rise to a net movement of
charge will contribute to the ionic conductivity as well as diffusion.
The exchange mechanism does not involve defects and will only
contribute to the self-diffusion. It should also be noted that the
defects in an impurity-defect complex can also undergo a localized
motion around the impurity via the above defect-assisted mechanisms.

The considerable wealth of experimental information now available
on these materials has removed many of the uncertainties concerning the
nature of the defects and ionic transport. The identification of
anion-Frenkel pairs as the dominant defects is based on a number of
studies. Very early studies of the transport numbers in BaF_2 and PbF_2
(Tubandt, 1932) showed an anion transport number of unity. This is
indicative of the presence of very low levels of disorder on the cation
sub-lattice. More recent radiotracer studies of the cation diffusion
in CaF_2 (Matzke and Lindner, 1964) and $SrCl_2$ (Hood and Morrison, 1967)
have shown that the diffusion coefficients are negligible compared to
those for the anions. The classical method of identifying the
intrinsic disorder in solids is by comparison of the bulk density and
the density calculated from the X-ray lattice parameter (Simmons,
1962). Density studies of CaF_2-YF_3 were consistent with the dopant
creating F^- interstitials rather than Ca^{2+} vacancies (Short and Roy,
1963). Rare-earth doping allows a study of the defect structures with
a variety of sophisticated spectroscopic techniques, including optical,

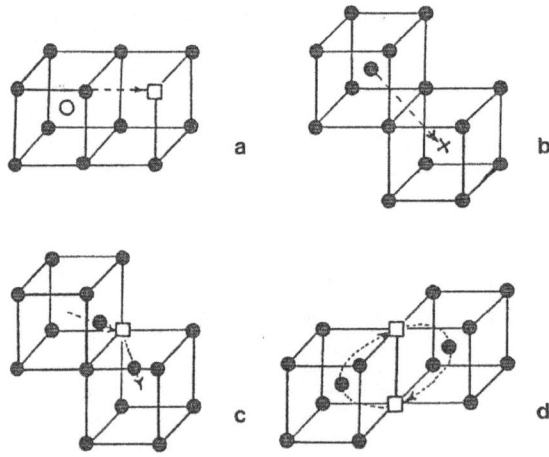

Figure 3. Diffusion mechanism for anions in fluorite
lattices. The filled circles represent the anions, the
open circles the cations and the open squares the anion
vacancies. (a) the vacancy mechanism; (b) the direct
interstitial mechanism; (c) the non-collinear
interstitialcy mechanism; (d) the exchange mechanism.

EPR and ENDOR measurements (Baker, 1974). These show that the F^- interstitials that are formed occupy the cube-centre interstitial site adjacent to the dopant, i.e. as shown in figure 2a. It is reasonable to extrapolate from these studies that the unbound interstitial also occupies the cube-centre site.

A powerful experimental method for testing defect models and determining defect energies in ionic crystals is a.c. electrical conductivity, σ, measurements. The technique is described in a later chapter by Chadwick and Corish and there are several reviews which cover this topic (see, for example Lidiard, 1957, 1974; Barr and Lidiard 1970; Corish and Jacobs 1973; Jacobs 1983). The conductivity plots for the fluorite-structured halides all exhibit the same general shape and the data for β-PbF_2 are shown as examples in figure 4. At this stage we will concentrate on the region of "normal" behaviour at moderate to low temperatures (i.e. 1000K/T < 1.6 for β-PbF_2) in lightly doped crystals and from which we can extract the defect parameters. This region divides into two roughly linear parts; the low temperature extrinsic part due to impurity created defects and the higher temperature intrinsic part due to thermally created defects. Over the past decade there have been high-precision conductivity measurements for all the fluorite systems and the data have been analysed by computer non-linear least-squares fitting techniques. This method of analysis has to be used with caution i.e. several crystals with different dopant types and levels should be studied, checks must be made to avoid false minima, etc., however, it does allow a means of simply incorporating a correction for the long-range Coulomb interactions between defects. The defect model that is used in the fitting <u>assumes</u> anion-Frenkel disorder, substitutional doping by impurities M^+ and M^{3+}, impurity - defect association at low temperatures and defect migration by vacancy and non-collinear

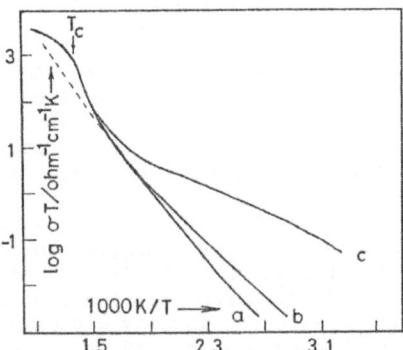

Figure 4. The a.c. conductivity plot for β-PbF_2 crystals; (a) nominally pure, (b) 100 p.p.m. La^{3+}, (c) 1000 p.p.m. Na^+. The dashed line is an extrapolation of the "normal", low temperature regime. (After Azimi et al, 1984.)

interstitialcy mechanisms (Azimi et al, 1984). This gives an expression with nine variables; h_F and s_F, the anion-Frenkel formation enthalpy and entropy; Δh_v and Δs_v, the vacancy migration enthalpy and entropy; Δh_i and Δs_i, the interstitial migration enthalpy and entropy; h_{ass} and s_{ass} the impurity-defect association enthalpy and entropy; c_T, the dopant concentration. Values of the formation and migration parameters for β-PbF$_2$ are shown in table 1. Similar data sets, with accuracies of ~ ± 0.1 eV in the enthalpies and ~ ± 1 k in the entropies, are available for CaF$_2$, SrF$_2$, BaF$_2$ and SrCl$_2$ and the enthalpies are in very good agreement with those derived from nuclear magnetic resonance (NMR) measurements and computer simulation calculations (Chadwick, 1983a, b).

TABLE 1

Defect parameters for β-PbF$_2$ from conductivity measurements (after Azimie et al, 1984)

Dopant	h_f/eV	s_F/k	Δh_v/eV	Δs_v/eV	Δh_i/eV	Δs_i/k
100 ppm La^{3+}	1.14	3.72	0.23	1.49	0.48	5.18
100 ppm Na$^+$	1.03	3.60	0.22	1.14	0.65	7.22
1000 ppm Na$^+$	1.04	5.14	0.23	0.75	0.47	3.53

A feature of the data in table 1, and common to the other fluorite--structured halides, is that $\Delta h_i > \Delta h_v$ and $\Delta s_i > \Delta s_v$. Thus in a pure crystal the vacancies will be the more mobile defects at low temperatures. However, at high temperatures the interstitials will have mobilities comparable to, or greater than, the vacancies.

A combination of measurements of the conductivity and the self-diffusion coefficient, D_T, of an isotopic tracer (by radiotracer or nmr methods) can be used to identify transport mechanisms in ionic crystals (LeClaire, 1970). The two parameters are related as σ depends on the net movement of charge by the defects and D_T on the movement of the tracer via the defects. The precise relationship must include the details of the mechanisms of the defect migration. This will enter as a correlation factor, f, expressing the statistical correlations between successive jumps of the tracer. These correlations are most easily visualized for a vacancy mechanism. The jumps of the vacancy will be in random directions. However, for a tracer which has just exchanged with a vacancy, the next most likely jump will be back to its original site as there is a vacancy there to accept it. Values of f have been calculated for the various mechanisms, crystal structures and tracer techniques. Diffusion mechanisms are identified by comparing these values with the "experimental" f, i.e. the value required to relate the measured σ and D_T. In the fluorides the most convenient means of measuring D_T for the anion is by ^{19}F NMR as there is no appropriate radiotracer. In K$^+$ doped BaF$_2$ the combination of NMR and σ measurements confirmed the expected vacancy mechanism of F$^-$ diffusion in the extrinsic region (Figueroa et al, 1978). However, the same study did not confirm the expected non-collinear interstitialcy mechanism for F$^-$ diffusion in La^{3+} doped BaF$_2$. A later study of La^{3+} doped SrF$_2$ did show the expected behaviour (Kirkwood, 1980). Radiotracer studies of ^{36}Cl diffusion in the extrinsic region of M$^+$ doped SrCl$_2$ were consistent with the expected vacancy mechanism (Saghafian, 1980). A similar study of pure SrCl$_2$ could not be interpreted in terms of vacancy or interstitialcy mechanisms and it was proposed that an additional mechanism, which only contributed to diffusion and not charge transport, was operative (Bénière et al, 1979).

There is extensive information on the impurity-defect complexes in the lightly doped fluorite-structured halides. In addition to the structural studies using spectroscopic techniques there have also been a large number of investigations of the reorientation of the complex. These have employed dielectric relaxation spectroscopy or ionic thermo-current methods, techniques which will be described in a later chapter by Capelletti.

In summary, there is now a good understanding of the major features of the point defects in lightly doped materials at moderate to low temperatures. The dominant defects have been identified, the thermodynamic parameters for defect formation and migration are accurately known and the sites occupied by many impurities have been established. There are still a number of points that require further testing by experiment. For example, it is usually assumed that alkali metal cations dissolve only substitutionally although there is evidence for a partitioning between substitutional and interstitial sites in $SrCl_2$ (Gervais et al, 1976). The most important question to be resolved is the mechanism of the anion interstitial migration.

HEAVILY RARE-EARTH DOPED CRYSTALS

The simple rare-earth cation-anion interstitial complex shown in figure 2a is found at dopant levels <0.1 mole per cent. However, it has already been mentioned that the rare-earth concentration can be increased to levels around 50 mole per cent. A consequence of increasing the dopant concentration is that the simple complexes will aggregate to form dimers, trimers, etc. Evidence for the formation of these clusters of two or more rare-earth ions and their charge-compensating F^- interstitials is available from laser spectroscopy (Moore and Wright, 1981; Tallant and Wright, 1975) and neutron diffraction (Cheetham et al, 1971). A great deal of interest has been focussed on the structure of these clusters as they provide amenable experimental systems on which to model problems found in non-stoichiometric materials. However, there has been considerable debate over the cluster structures and some of the confusion has only been removed in recent work.

The early neutron diffraction data on Y^{3+} doped CaF_2 of Cheetham et al (1971) gave evidence of clusters with F^- interstitial ions displaced from the cube-centre site. These authors proposed a small cluster was present in the 5 mole per cent doped sample with the "2:2:2" structure originally suggested by Willis (1964) in studies of non-stoichiometric UO_{2+x}. This is illustrated in figure 5 and contains two types of F^- interstitials; F^- interstitials that have relaxed from the cube-centre sites along the <110> direction and F^- interstitials formed by relaxation of normal cube-corner anions along <111> directions to leave behind vacancies. This cluster is designated $2|0|2|2|_1$ in the recently proposed systematic notation (Corish et al, 1982; Bendall et al, 1984); number of dopant ions|number of vacancies| number of relaxed lattice ions|number of neighbouring interstitials| and the subscript denotes the interstitial position (1 for nearest-neighbours). Support for the models of Cheetham et al was provided by the static-defect energy calculations using the HADES code performed by Catlow (1973, 1976). Other workers (for example, Gettman and

Greiss, 1978) on the basis of X-ray studies of more heavily doped CaF_2 have suggessted larger clusters containing six rare-earth cations packed around an interstitial site - the cubo-octahedral clusters based on the unit shown in Figure 5c. Support for this cluster is provided by the mineral tveitite, $Ca_{14}Y_5F_{43}$, which is an ordered system based on this arrangement of ions (Bevan et al, 1982).

A combination of neutron diffraction, computer simulation and EXAFS studies has shown that the nature of the clusters formed depends on the size of the dopant ion. Neutron diffraction studies on 5 mole per cent La^{3+} doped CaF_2 (Catlow et al, 1983a) supported the formation of the small $2|0|2|3|_1$ cluster (sometimes referred to as the "gettered" 2:2:2) which is shown in figure 5b. A similar study of Er^{3+} doped CaF_2 (Catlow et al, 1983b) showed that with this smaller dopant the cubo-octahedral cluster gave the best fit to the data. Defect energy calculations (Bendall et al, 1982; Catlow et al, 1984b) showed that small clusters were more stable for the larger rare-earth dopants and the stability of the hexameric clusters increased as the dopant size decreased; the cross-over point occurring at Gd^{3+} where hexamers are preferred. EXAFS can provide information on the local environments of dopants (Hayes and Boyce, 1982), in contrast to diffraction methods which give the contents of the average unit cell. This powerful technique has been elegantly employed to provide experimental proof of the theoretical predictions. EXAFS spectra were taken (Catlow et al, 1984a) for 10 rare-earths at 10 mole per cent dopant levels in CaF_2. The spectra showed a gradual, qualitative change of shape as the lanthanide series was traversed. Detailed analysis were carried out for the spectra of Nd^{3+} doped CaF_2 and Er^{3+} doped CaF_2. In the former case the data were fitted best by the $2|0|2|3_1$ cluster and the latter by the $6|0|8|6_1$ <111> cluster.

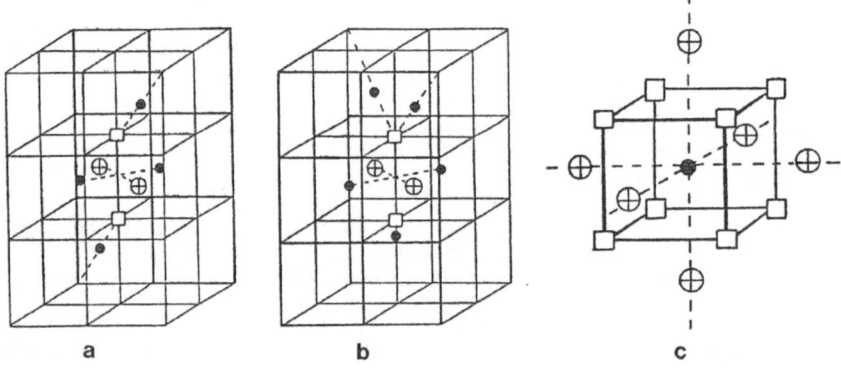

a   b   c

Figure 5. Defect cluster models for heavily rare-earth doped CaF_2. The filled circles represent anions, the crossed circles the lanthanide cation, and the open squares the vacancies (the cube-corner ions are omitted for clarity). (a) the $2|0|2|2|_1$ (or the "2:2:2") cluster; (b) the $2|0|2|3|_1$ (or the "gettered 2:2:2") clusters; (c) the arrangement of lanthanide cations in the cubo-octahedral cluster (the interstitial anions lie above each cube edge).

Finally, we should note in this section that other factors, in addition to dopant size, could play a role in the cluster structure. The experimental studies described in the previous paragraph used samples that were essentially rapidly cooled and measurements taken at relatively low temperatures. Annealing for long periods at high temperatures could give rise to other cluster structures. In addition, the possible ranges of dopant concentration and temperature have not been fully explored.

HIGH TEMPERATURE FAST-ION REGION

The conductivity plot shown in figure 4 for β-PbF$_2$ is typical for the fluorite-structured halides in that beyond the normal behaviour as the temperature increases there is a steeply rising portion followed by a region with an activation energy of only a few tenths of an eV (Derrington and O'Keeffe, 1973; Carr et al, 1978). The conductivities in this region are around 1 ohm^{-1} cm^{-1}, comparable to those in the melt and leading to the classification of these materials as fast-ion conductors. At about the same temperature as the deviation from normal behaviour becomes apparent in the conductivity there is the onset of a broad peak in the isoparic specific heat, Cp curve (Dworkin and Bredig, 1968). The Cp data for β-PbF$_2$ are shown in figure 6. The temperature of the maximum in the Cp peak, Tc, corresponds to the temperature at which the conductivity plots begin to flatten-off. Clearly, the thermal and conductivity behaviour originate from a common cause which is generally accepted as a disordering of the anion sub-lattice. However, the nature and degree of disorder have been the subject of considerable debate which is still not fully resolved. There have been a number of reviews and analyses of the behaviour of the fluorite-structured materials at high temperatures (Chadwick, 1983a, b; Catlow, 1980; Schoonman, 1980; Lidiard, 1980; Hayes, 1978; Hutchings et al, 1984) and a survey will be given here of the major features of this intriguing topic.

An early model that was used to explain the conductivity and Cp behaviour was based on the anion sub-lattice "melting". It was assumed that the effects were due to the generation of massive levels of anion-Frenkel defects, to the extent that the anions were randomly

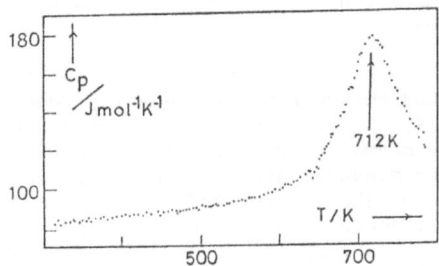

Figure 6. The isobaric specific heat of β-PbF$_2$ (after Schröter, 1979).

distributed between the normal and interstitial anion sites. This model provided a qualitative explanation of the effects. However, the current view is that the level of disorder is only a few per cent.

It is possible to estimate the defect concentrations in the fast-ion region from a variety of experiments provided the defects are assumed to be simple anion-Frenkel pairs with the same structure as those found at low temperatures, i.e. vacancies and cube-centre interstitial anions. On this basis, the analysis of the Cp data (Schröter and Nölting, 1980) taking the excess specific heat to be due to defect formation leads to anion-Frenkel pair concentrations in BaF_2, $SrCl_2$ and $\beta-PbF_2$ of 2 to 7 per cent at T_c increasing to 4 to 13% at the melting point, with no major differences between the three materials. The analysis has been refined (Oberschmidt, 1981) to include repulsive interactions between the defects by allowing the number of sites available to defects to vary as the number of defects increases. This leads to slightly higher defect concentrations at T_c. Catlow et al (1978) used lattice-energy calculations to estimate the energy required to form a fluorite lattice with all the interstitial sites occupied. These yielded values of ~ 1eV whereas the enthalpy change at T_c indicated from the Cp data is ~ 0.1eV. This implies that at this temperature this sytem is a $^1/10$ of the way to complete anion disorder, i.e. only 10% interstitials. Another approach is to assume the rise in the conductivity is solely due to defect formation (Chadwick, 1983a, b). The excess conductivity can be evaluated by extrapolating the low temperature plots and this leads to anion-Frenkel pair concentrations in the fast-ion region of, at most, 10% in $\beta-PbF_2$ and $SrCl_2$. Massive disorder would be expected to lead to a low correlation factor. However, the combination of conductivity and NMR measurements for $\beta-PbF_2$ (Carr et al, 1978) in the fast-ion region yield values typical for normal defect-assisted mechanisms in lowly defective solids. Further evidence for a level of disorder less than 10% is found from Raman scattering (Elliott et al, 1978), Brillouin scattering (Catlow et al, 1978) and acoustic phonon life-time measurements (Dickens et al, 1979).

The evidence from static defect energy calculations (Gillan and Richardson, 1979), molecular dynamics computer simulations (Dixon and Gillan, 1978), X-ray (Koto et al, 1980) and neutron diffraction (Shapiro and Reidinger, 1979) diffraction studies is that the cube-centre interstitial site is not appreciably occupied at high temperatures. In fact, the calculations suggest this site is not a stable position for the interstitial anion. The discussion of the nature of the defects and their concentrations in the fast-ion phase in the last few years has been dominated by two techniques - neutron scattering and molecular dynamics simulations - and the major points will be outlined below.

The extensive neutron studies of the fluorite-structured halides by Hayes, Hutchings and co-workers have recently been summarized (Hutchings et al, 1984). The diffraction data for $\beta-PbF_2$ show that a large fraction of the normal anion sites are unoccupied, ~ 25% at T_c. These data and similar measurements for $SrCl_2$ have been interpreted in terms of the presence of short-lived defect clusters being formed at high temperatures. Evidence for these clusters comes from the appearance and growth of a peak in the coherent diffuse quasi-elastic scattering in the region ot T_c. Thus the view was that, although at any instant of time a large fraction of anion sites are not occupied, this can be the result of the anions relaxing off the sites rather than moving into the cube-centre sites. In an attempt to visualize these clusters various static models were investigated and tested against the experimental data with the realization that the real system involved complex dynamic disorder. The type of model that gave good fits is shown in figure 7a, which is similar to the $2|0|2|2_1$ type cluster discussed for rare-earth doped crystals and shown in figure 5a. The difference is that charge compensation is now by the outer vacancies. Additional support for this model has been provided by defect energy calculations (Catlow and Hayes, 1982) which show this cluster is stable with respect to the isolated defects. Neutron incoherent quasi-elastic scattering measurements for $SrCl_2$ have been used to measure diffusion of Cl^- and indicate jumps between regular lattice sites as would be expected for vacancy diffusion.

. The formation of neutral clusters would not be expected to lead to the generation of mobile defects and therefore would not lead to enhanced conduction. If the clusters carried an extra charge (i.e. an extra vacancy or interstitial) then they would be charge compensated by isolated defects in the rest of the lattice. Thus the formation of charged clusters could provide a model for the onset of fast-ion conduction. In an attempt to test this possibility the stability of a wide range of clusters was calculated (Allnatt et al, 1985) and the cluster with an extra interstitial anion, shown in Figure 7c, was found to be particularly stable. Thus fast-ion conduction could be explained by a model involving the generation of these clusters plus mobile, isolated vacancies.

The molecular dynamics computer simulation technique will be described in a later chapter by Catlow. Early work by Jacucci and Rahman (1978) indicated the power of this technique for the investigation of the fluorite materials and it has been very successfully employed by Gillan and co-workers. Their thorough study of $SrCl_2$ (Gillan and Dixon, 1980; Dixon and Gillan, 1980) was particularly revealing as to the nature of the defects and their motion. Firstly, the simulation was able to qualitatively reproduce the shape of the conductivity plot; the difference in magnitude was assigned to the simplicity of the interatomic potentials. The anions were found to

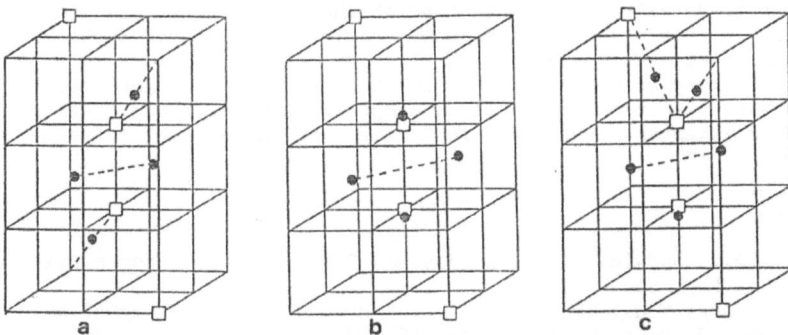

Figure 7. Defect cluster models in pure alkaline-earth fluorides used in the calculations of Allnatt et al. Filled circles represent interstitial anions and open squares the vacancies. (a) starting configuration for a $2|2|2_1^1$ cluster; (b) the minimum energy configuration for the cluster shown in (a); (c) the minimum energy configuration for a $2|2|3_1^1$ cluster.

move predominantly along the <100> cube-edge direction. The residence time of an anion on a regular site was an order of magnitude greater than the flight time, indicating a solid-like jump process rather than a motion in a liquid. Analysis of the hopping events showed that even at the highest temperatures studied there were only about ~3% of anion sites vacant. In addition, the simulation showed that the cube-centre sites had no appreciable occupancy. Essentially the same results were found in the simulation of PbF_2 (Walker et al, 1982) where the high temperature vacancy concentration was found to be 2%. The apparent conflict between the simulations and the neutron data has been discussed by Gillan and co-workers (Walker et al, 1982; Gillan, 1985). It was shown that many features of the neutron data could be reproduced by the simulations and the discrepancies between the defect concentrations, i.e. the number of anion sites vacant could result from the methods used to analyse the neutron data. Thus it was suggested that defect clusters did not need to be invoked and ions in motion could give rise to the apparent ion densities at positions assigned as ion sites in a cluster.

In summary, the evidence points to the fast-ion region as containing only a few per cent defects which are highly mobile and which move by a solid-like jump process. The cube-centre interstitial site is not appreciably occupied. The system is complex and it is difficult to describe the dynamic situation that exists in familiar solid state terms. The transient cluster model has the advantages of being simple to visualize and it can be simply built into models of the experimental properties, such as the conductivity. However, as pointed out by Hutchings et al (1984) it ultimately may be shown to be an over-simplification.

SUMMARY

This chapter has concentrated on the fluorite-structured halides because of the broad range of point defect properties that they exhibit. Some of the typical problems found in ionic crystals have been covered and a broad outline given of the techniques that can be used for their study. Later chapters will consider a number of these techniques in detail. An important point that it is hoped has been made clear in this chapter is that the resolution of defect problems very often involves the use of a combination of techniques, a particularly good example of this being the heavily rare-earth doped materials where neutron diffraction, computer simulation and EXAFS provide complementary structural information on the defect clusters.

References

Allnatt, A.R., Chadwick, A.V. and Jacobs, P.W.M., 1985, paper submitted to Proc. Roy. Soc.
Azimi, A., Carr, V.M., Chadwick, A.V., Kirkwood, F.G. and Saghafian, R., 1984, J. Phys. Chem. Solids, 45, 23.
Baker, J.M., 1974, in "Crystals with the Fluorite Structure", ed. W. Hayes, (Clarendon Press, Oxford), p. 341.
Barr, L.W. and Lidiard, A.B., 1970, in "Physical Chemistry - An Advanced Treatise", Vol. X, eds. H. Eyring, D. Henderson and W. Jost (Academic Press, New York), p. 151.
Bendall, P.J., Catlow, C.R.A., Corish, J. and Jacobs, P.W.M., 1984, J. Solid State Chem., 51, 159.
Bénière, M., Chemla, M. and Bénière, F., 1979, J. Phys. Chem. Solids, 40, 729.

Bevan, D.J.M., Strähle, J. and Greiss, O., 1982, J. Solid State Chem., 44, 75.

Carr, V.M., Chadwick, A.V. and Saghafian, R., 1978, J. Phys. C., 11, L637.

Catlow, C.R.A., 1973, J. Phys. C., 6, L24.

Catlow, C.R.A., 1976, J. Phys. C., 9, 1845.

Catlow, C.R.A., 1980, Comments Solid State Phys., 9, 157.

Catlow, C.R.A., Comins, J.D., Germano, F.A., Harley, R.T. and Hayes, W., 1978, J. Phys. C., 11, 3197.

Catlow, C.R.A. and Hayes, W., 1982, J. Phys. C., 15, L9.

Catlow, C.R.A., Chadwick, A.V. and Corish, J., 1983a, J. Solid State Chem., 48, 65.

Catlow, C.R.A., Chadwick, A.V. and Corish, J., 1983b, Radiation Effects, 75, 61.

Catlow, C.R.A., Chadwick, A.V., Greaves, G.N. and Moroney, L.M., 1984a, Nature, 312, 601.

Catlow, C.R.A., Corish, J. and Jacobs, P.W.M., 1984b, J. Solid State Chem., 51, 159.

Chadwick, A.V., 1983a, Solid State Ionics, 8, 209.

Chadwick, A.V., 1983b, Radiation Effects, 74, 17.

Cheetham, A.K., Fender, B.E.F. and Cooper, M., 1971, J. Phys. C., 4, 3107.

Corish, J. and Jacobs, P.W.M., 1973, in "Surface and Defect Properties of Solids", Vol. 2, eds. M.W. Roberts and J.M. Thomas (Chemical Society, London), p. 160.

Corish, J., Catlow, C.R.A., Jacobs, P.W.M. and Ong, S.H., 1982, Phys. Rev., B25, 6425.

Dell, R.M. and Hooper, A., 1978, in "Solid Electrolytes", eds. P. Hayenmuller and W. van Gool, (Academic Press, New York), ch. 18.

Derrington, C.E. and O'Keeffe, M.A., 1973, Nature, 246, 44.

Dickens, M.H., Hayes, W., Hutchings, M.T. and Kleppmann, W.G., 1979, J. Phys. C., 12, 17.

Dixon, M. and Gillan, M.J., 1978, J. Phys. C. 11, L165.

Dixon, M. and Gillan, M.J., 1980, J. Phys. C. 13, 1919.

Dworkin, A.S. and Bredig, M.A., 1968, J. Phys. Chem., 72, 1277.

Elliott, R.J., Hayes, W., Kleppmann, W.G., Rushworth, A.J. and Ryan, J.F., 1978, Proc. Roy. Soc., A360, 317.

Figueroa, D.R., Chadwick, A.V. and Strange, J.H., 1978, J. Phys. C., 11, 55.

Geller, S., 1977, ed., "Solid Electrolytes" (Springer-Verlag, Berlin).

Gervais, A., Jacquet, M. and Bathier, M., 1976, J. Phys. (Paris), C7, 281.

Gettman, P. and Greiss, O., 1978, J. Solid State Chem., 26, 255.

Gillan, M.J., 1985, Physica, 131B, 157.

Gillan, M.J. and Dixon, M., 1980, J. Phys. C., 13, 1901.

Gillan, M.J. and Richardson, D.D., 1979, J. Phys. C., 12, L61.

Hagenmuller, P. and Van Gool, W., 1978, eds. "Solid Electrolytes" (Academic Press, New York).

Hayes, J.B. and Boyce, T.M., 1982, Solid State Phys., 37, 173.

Hayes, W., 1974, ed. "Crystals with the Fluorite Structure" (Clarendon Press, Oxford).

Hayes, W., 1978, Contemp. Phys. 19, 469.

Hood, G.M. and Morrison, J.A., 1967, J. Appl. Phys., 38, 4796.

Hutchings, M.T., Clausen, K., Dickens, M.H., Hayes, W., Kjems, J.K., Schnabel, P.G. and Smith, C., 1984, J. Phys. C., 17, 3903.

Ippolitov, E.G., Garashina, L.S. and Maklasklov, A.G., 1967, Russ. Inorg. Mater., 3, 59.

Jacobs, P.W.M., 1983, in "Matter Transport in Solids", eds. F. Bénière and C.R.A. Catlow (Plenum Press, New York), p. 81.

Jacucci, G. and Rahman, A., 1978, J. Chem. Phys. 69, 4117.

Kirkwood, F.G., 1980, Ph.D. Thesis, (University of Kent).

Koto, K., Schulz, H. and Huggins, R.A., 1980, Solid State Ionics, 1, 355.

LeClaire, A.D., 1970, in "Physical Chemistry - An Advanced Treatise", Vol. X, eds. H. Eyring, D. Henderson and W. Jost (Academic Press, New York), p. 261.

Lidiard, A.B., 1957, Handbuch der Physik, Vol. XX, (Springer-Verlag, Berlin), p. 246.

Lidiard, A.B., 1974, in "Crystals with the Fluorite Structure", ed. W. Hayes (Clarendon Press, Oxford), p. 101.

Lidiard, A.B., 1980, AERE Harwell Report TP 841.

Kennedy, J.H., 1977, in "Solid Electrolytes", ed. S. Geller (Springer-Verlag, Berlin) ch. 5.

Kleitz, M., Sapoval, B. and Ravaine, D., 1983, eds., "Solid State Ionics - 83" (North Holland, Amsterdam).

Matzke, H.J. and Lindner, R., 1964, Z. Naturforsch., 19a, 1178.

Moore, S. and Wright, J.C., 1981, J. Chem. Phys., 74, 1626.

Oberschmidt, J., 1981, Phys. Rev., 23, 5038.

Reau, J.M. and Portier, J., 1978 in "Solid Electrolytes", eds. P. Hagenmuller and W. van Gool (Academic Press, New York), p. 313.

Saghafian, R., 1980, Ph.D. Thesis (University of Kent).

Salamon, M.B., 1979, ed., "Physics of Superionic Conductors" (Springer-Verlag, Berlin).

Schoonman, J., 1980, Solid State Ionics, 1, 123.

Schröter, W., 1979, Ph.D. Thesis (University of Göttingen).

Schröter, W. and Nolting, J., 1980, J. Phys., (Paris), 41, C6-20.

Shapiro, S.M. and Reidinger, F., 1979 in "Physics of Superionic Conductors", ed. M.B. Salamon (Springer-Verlag, Berlin), p. 45.

Short, R. and Roy, R., 1963, J. Phys. Chem. 67, 1860.

Simmons, R.O., 1962, Proc. Int. School of Physics, 'Enrico Fermi' XVIII Course, ed., D.S. Billington (Academic Press, New York), p. 568.

Sorokin, P.P. and Stevenson, M.J., 1961, IBM J. Res. Develop., 5, 56.

Tallant, D.R. and Wright, J.C., 1975, J. Chem. Phys., 63, 2074.

Tubandt, C., 1932, Handbuch der Experimental Physik, 12, 381.

Walker, A.B., Dixon, M. and Gillan, M.J., 1980, J. Phys. C., 15, 4061.

Willis, B.T.M., 1964, Proc. Brit. Ceram. Soc., 1, 9.

OBJECTIVE OF THEORETICAL STUDIES AND THEIR LINKS WITH EXPERIMENT

A. N. Cormack

NYS College of Ceramics
Alfred University
Alfred, NY 14802

INTRODUCTION

"It appears possible to correlate the quite unusual
transport properties exhibited ..... with a model
that does not violate physical intuition too
flagrantly" *

Although theory and experiment are often portrayed to be two different disciplines (in whatever field), it is nevertheless the case that progress in one would not be made without the other. Objectively, at least in the physical sciences, the causa principali is the advancement of our understanding and knowledge of the "real world".

From a theoretician's point of view, this is best done by providing a basis for quantitative predictions for events under a given set of conditions. The experimentalist, on the other hand, is primarily engaged in quantifying the behavior under these given conditions. At the same time, of course, theory is being tested. This takes two main forms. Firstly, in designing an experiment, one is using previously established guidelines to collect the data and analyse it, to give the information required from the experiment. This is generally done nowadays in terms of some model, which is one aspect of theory and to which we should return later. The second, and probably more conventional idea of theory, is concerned with calculating quantities from a theoretical basis and comparing the answer with experimentally obtained data.

We may take as a very simple example of these two notions of theory, the vacancy formation energies (or the Schottky energy) in MgO. From a knowledge of (or assumption about) the structure and interatomic forces, it is possible to calculate these quantities (1). These can also be measured using a number of different techniques, but which involve defect migration. Wuensch (2) has described, for example, how diffusion and conductivity data could be, and indeed was, interpreted in two different ways, one assuming that only the extrinsic region was being measured and the other assuming that a change of slope in the Arrhenius plot could be seen. Apart from saying something about the quality of the data, quite different numbers for the defect parameters, H_s, Schottky energy and H_m the migration energy were obtained from these two approaches. Wuensch also pointed out that appealing to

---

* From Sher et. al (ref. 9).

theory for guidance offered no benefit, as the range of calculated values fell in between the two extremes!

Happily, we are now at the stage where both theory and experiment have converged. The most recent theoretical work, based on HADES computer codes (3-5) and using reliable interatomic potentials, predicts large Schottky energies and hence the absence of any intrinsic behavior in the temperature ranges covered; and systematic conductivity studies by Sempolinski and Kingery (6), with a variety of dopants and dopant concentrations, were able to rationalize the previous data and produced cation migration energies, for example, that were in excellent agreement with those predicted by Mackrodt and Stewart (7).

The moral of this example is that assumptions about the theoretical framework or model in which the experimental data are interpreted may not be justified, no matter how carefully the experiment is done, and this will lead to wrong (or misleading) information being added to the database which is being used to develop a theoretical understanding.

INTERPRETATION OF EXPERIMENT: THEORETICAL MODELS

> "... requires that diffusion occurs because of one of these alternatives... (3) Frenkel [sic] defects are formed, with the Schottky defects being bound, which is unlikely since it adds another degree of complication to an already complicated system" *

From our discussion in the previous section, it is apparent that experimentalists rely on theory rather more heavily than perhaps they think. This is increasingly the case as more and more experiments become more and more sophisticated and require extensive data analysis. It is a tribute to experimentalists that this is happening because it means that they are making use of advances in theory (as well as, in many cases, breakthroughs in engineering and technology). However, a note of caution has to be introduced, in that analysis and interpretation of these experiments, such as A.C. impedance measurements for conductivity and EXAFS studies of disorder in solids, rely heavily on some theoretical model, be it an equivalent circuit or a preconceived idea of the nature of disorder.

Commonly nowadays, computer fitting of data is carried out to extract as much information as possible from the data. This, of course, requires a model to which to fit. If only one model is tried, some confusion may result. This may be because some earlier worker proposed a model (on rather dubious grounds, perhaps) and this has been tacitly accepted by later authors, since the data appears to fit reasonably adequately that model.

A case in point here would seem to be the nature of intrinsic disorder in LaF_3, a fast ionic conductor with important technological applications. LaF_3 has a hexagonal symmetry (although its space group is still under discussion, it is probably rhombohedral (8)) and its structure is quite complex. Early on, it was suggested (9) that Schottky disorder (i.e. vacancies) would dominate. Later authors seem to have merely accepted this to be so. Conductivity data (10) were fit using the assumption of vacancy disorder, and numbers extracted for the Schottky energy. (This is possible if, as was the case here, the temperature range covers the intrinsic regions as well as the lower temperature extrinsic region - see chapter (1).) However, at the same time, a number for the anion Frenkel energy was also quoted, giving the impression that this had also been determined. In fact all that had

---

* From Sher et. al (ref 9).

been determined from the experiment were the activation energy for conductivity (the energy barrier for the migrating species) and a contribution to the Arrhenius energy in the intrinsic regions from the formation energy of the migrating defect. The Schottky and Frenkel energies had been derived assuming $E_f = 1/2\ E_{Frenkel} = 1/4\ E_{Schottky}$, which is not necessarily correct, the Frankel and Schottky energies being unrelated.

In fact recent calculations (11) find that the anion Frenkel energy is about 0.5-0.6 eV per defect, but that Schottky energies are much higher. So here we have a situation where predictions from theory, in the form of computer simulations (see the chapter by Catlow in this volume), appear to conflict with experimental results, when in fact this is apparently only because of the model that was used to interpret the experimental data. Use of an alternative model for disorder in LaF_3 may produce a more complete understanding of defect behavior in this material.

As a postscript to this tale it turns out that Sher et al (9) based their determination of Schottky disorder on a combination of dilatometry and x-ray measurements of thermal expansion. However, they obtained an activation energy of Schottky formation of 0.07 eV! No attempt was made to justify this extraordinarily low value and they were led to a rather tortuous discussion in an attempt to explain it (hence the quotation at the beginning of the chapter).

An interesting example of how the analysis and interpretation of an experiment in terms of one model may show that model to be incomplete arises from a.c. impedance measurements of an insertion electrode material, $Li_{0.65}CoO_2$. A.C. impedance experiments are invariably analysed in terms of "equivalent circuits" (see the chapter by Chadwick in this volume) from which it is possible to separate out the electrical properties, especially resistivity of each component in the cell.

Now, in the case of a polycrystalline insertion electrode such as was studied by Thomas et al, (12) it can be shown theoretically that the behavior can be described in terms of a modified Randles circuit, as shown in Fig. (1). The output from this circuit is predicted to be a single, flattened semicircle.

The actual response, measured by Thomas et al (12) is shown in Fig (2). Two points arise here. Firstly, it is not inconceivable that an unsuspecting experimenter would try to fit this data to a single semicircle, since this is what was expected. This would obviously lead to errors in the information extracted and hence an erroneous interpretation of the experiment. Fortunately, however, it was recognized that there were two semicircles in the data and this then

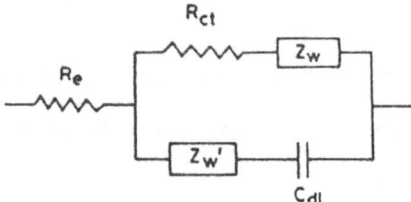

Figure 1. Modified Randles Circuit used to model a.c. response of polycrystalline insertion electrode.

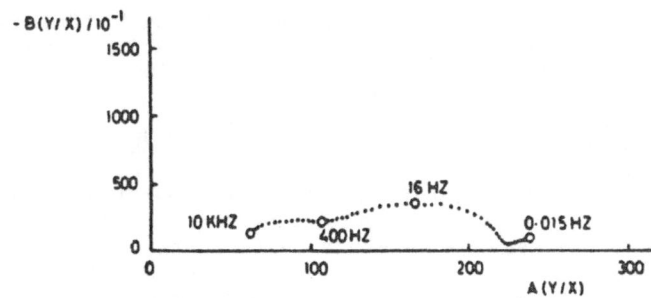

Figure 2.   Actual a.c. response, found by Thomas et al., for $Li_{0.5}CoO_2$.

raised the question of the nature of the additional process that was giving rise to the extra feature. Three alternatives were identified, two of which are variations on the same theme. These were:

(i) adsorption of some species (Li^+ or BF_4^- ions, or possibly propylene carbonate on to the electrode surface, without charge transfer.

(ii) essentially as (i), but assuming the adsorption to be diffusion controlled.

(iii) the formation of an ionically conducting, but electronically insulating layer at the electrode surface.

These surface processes can be modelled by the equivalent circuits shown in Fig. (3). Notice that these look somewhat different in nature, although, of course, they are predicted to give rise to a similar a.c. response. Just how similar was not envisaged until Thomas et al tried to fit the data to these different models. In fact, they were totally unable to distinguish between these alternative processes by fitting the data. Even the use of statistical significance tests did not help, and recourse to other experimental methods proved necessary. It turned out that the additional process was (iii), the formation of the ionically conducting but electronically insulating surface layer, which was identified through electron microscopy.

This example serves to illustrate how central theory can be to the design and interpretation of an experiment. On the one hand, theory in the form of equivalent circuits, is needed to interpret the data, but on the other hand, the data did not conform to that predicted from the theoretical considerations that went into the experiment. The interplay between theory and experiment is also clear in this example, since an additional process, not initially considered, was identified.

BASIC THEORETICAL CONSIDERATIONS: COMPUTATIONAL TECHNIQUES

In the examples discussed above, the role of theory has been to provide a model framework, within which to analyse and to fit the experimental data. This model was seen to be an integral part of the design considerations of the experiment, and the fit of the data to this model is intended to extract numerical values for the (physical) quantities of interest, in our particular case at this meeting, the various defect parameters.

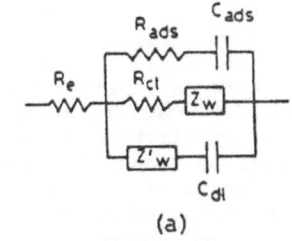

(a)

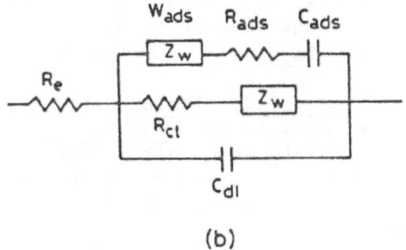

(b)

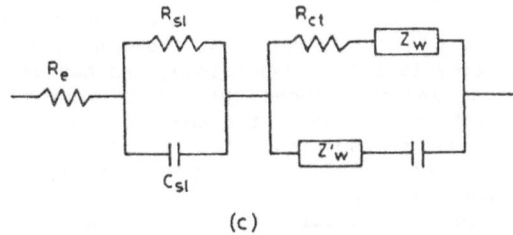

(c)

Figure 3. Equivalent circuits for different processes thought to be
occurring in $Li_{0.5}CoO_2$. (a) surface adsorption, without
charge transfer. (b) as (a), but assuming diffusion
controlled adsorption, (c) formation of an ionically
conducting, but electronically insulating layer at electrode
surface.

However, it is also possible to calculate values for these defect parameters directly from theory and these can then be compared with the experimental values. The computational techniques used in this application of theory, which, it must be emphasized, is different to that described earlier, have been discussed elsewhere in the volume by Catlow, and will not be dealt with in detail here.

Now, the primary aim of solid state physics and chemistry is to provide a fundamental framework in which to discuss the properties (thermodynamic, structural, defect, etc.) of materials at an atomistic level, and computational methods are a way of obtaining quantitative predictions from this framework. The ultimate goal is to find a parameter-free, ab-initio basis (i.e. description of the potential surface in the solid) from which all else stems. At present, theory has advanced to the stage where only parameterized models are generally available. Using these models, however, significant progress has been made in our understanding of atomistic processes (1)(13).

In this section, I want to focus attention on the case of ionic conductivity in heavily disordered systems, in particular, yttrium doped cerias. It was well established that there may be additional contributions to the Arrhenius energy in conductivity experiments depending on the temperature region of interest (see chapter 1), and that the conductivity plots, as a function of temperature, show several distinct regions, which we may summerize as follows:

Stage I is the <u>intrinsic</u> region in which defects are thermally generated; $\Delta H_\sigma = \Delta H_m + \Delta H_f$

Stage II is the <u>dissociated, extrinsic</u> region; $\Delta H_\sigma = \Delta H_m$

Stage III is the <u>associated, extrinsic</u> region in which the conducting species may interact with other ions; $\Delta H_\sigma = \Delta H_m + \Delta H_a$

where ΔH_σ is the Arrhenius enthalpy; ΔH_m is the free migration enthalpy (in the pure, perfect crystal) and ΔH_a is the association enthalpy between different defects.

In the intrinsic region, the conducting ions are created thermally, whereas in the extrinsic region these ions are present due to the incorporation of other defect species. In the case of Y/CeO_2, Y^{3+} dopants are compensated by anion vacancies, which are the conducting ions. In dilute systems, the additional term, ΔH_a, in the Arrhenius enthalpy is easily understood, and has been demonstrated (14) to be due to a binding between the oxygen vacancy and the trivalent dopant (it is not restricted to the case of Y_2O_3 doping). This is fine as long as one can see isolated defect pairs which do not interact with other defect pairs. But what happens when the level of doping increases to the stage when this is no longer true? On the one hand, the ionic conductivity should rise because of an increaesd number of mobile species, but what of the association term? A simple binding energy between two defect species is no longer appropriate because the defect pairs can not be considered in isolation. More complex defect clusters may be invoked and this has been done for some other systems (15), but even this description must become inadequate because the stoichiometry of the clusters is fixed and thus new clusters must be invoked as the dopant concentration changes to the extent that the density of clusters becomes inappropriate. In addition, of course, the vacancies are mobile and this may also alter the structure of the defect complex, possibly involving a dynamic relationship between different coexisting complexes.

The actual shape of the conductivity curve as a function of dopant concentration for Y/CeO_2 (16) (Fig. (4)) highlights these comments since it is apparent that after the maximum, the defect structure must be changing rapidly as the concentration increases. It becomes difficult to envisage what these clusters may be, a prior, and so the definition of ΔH_a becomes less and less clear as does that of ΔH_m since

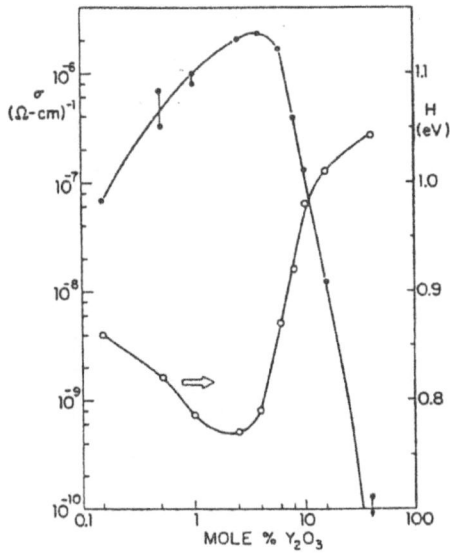

Figure 4. Conductivity as a function of dopant concentration for Y/CeO$_2$ (from ref 16).

regions of undoped crystal become more and more scarce as the level of dopants increases. At this point then, it is no longer fruitful to consider the individual terms ΔH_m and ΔH_a, so whilst it is feasible to calculate these terms for dilute solutions (17), or low concentrations of defects, a different theoretical strategy is required for the more highly doped cases.

An appropriate approach seems to be that offered by the Monte Carlo technique, often used in statistical physics. This is not a simulation method in the sense that molecular dynamics or static lattice calculations are, but is nevertheless a theoretical way of obtaining thermodynamic and related quantities (18). The basic technique is described in detail elsewhere (18,19). Here, we will only remark that it works by sampling over a large number of points in configuration space; averaging over these points, one obtains the petit canonical partition function. A move from one point to another is allowed if the relative energies of the two points are within a Boltzmans factor which is generated randomly in the interval [0, 1]. Monte Carlo calculations have been widely applied to transport problems in disordered systems (18), and recently Murch has suggested that improvement would result if the energies in configuration space could be accessed via computer simulation techniques (20).

In particular, if the migration energy of a vacancy in doped CeO$_2$ could be calculated as a function of the number of dopants immediately surrounding the jump path (see Fig. (5)), then Monte Carlo simulation could calculate the conductivity as a function of dopant concentration. Detailed analysis of the results and comparison with experiment should highlight the theoretical aspects of this unusual behavior. This work

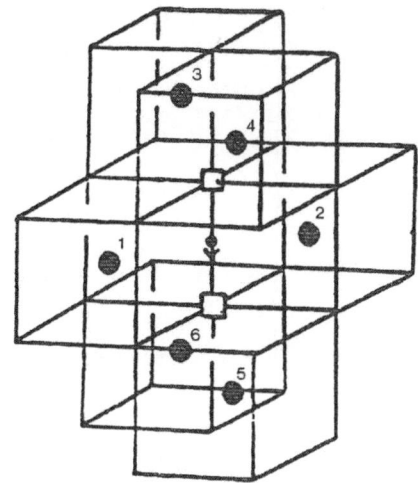

Figure 5. Structure of vacancy jump in CeO_2, showing possible
distribution of dopants in nearest neighbor cation
positions. There are 30 different configurations to be
considered.

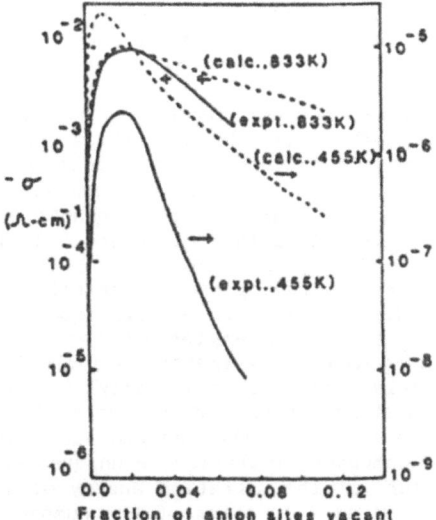

Figure 6. Comparison between conductivity curves calculated via Monte
Carlo/HADES techniques and the experimental curve in Fig. 4.

58

has been initiated (20) and Fig. (6) shows comparison of the theoretical and experimental results, from which it is seen that the drastic change in conductivity found experimentally is reproduced by the theory. The only input from experiment into the theory, it should be noted, is the parameterization of the interatomic potentials for Y/CeO_2.

From calculations done at different temperatures it is possible to derive Arrhenius energies and analysis of the configuration of dopants and vacancies (which has not yet been done) should provide considerable insight into the different kinds of defect cluster types and their interaction, as a function of composition. This would strengthen our understanding of the defect quantities ΔH_m and ΔH_a, and the role that structure plays in the migration of defect species in heavily disordered systems.

SUMMARY

In this chapter, we have attempted to describe the objective of theoretical studies and show, by discussing particular examples, how they are linked with experimental studies. We may identify two principle aspects. Firstly, there is the theory upon which experiments to extract defect parameters are designed. These give little insight into the actual mechanisms of defect behavior — although there are dynamic probes, such as NMR, these cannot be fully exploited since a complete theory is not yet available. Secondly, there are the theoretical techniques which may be used to calculate directly, with minimal input from experiment (through interatomic potential parameterization, and even this can be avoided), the various defect parameters. Since an initial atomistic mechanism from the basis for the calculation, alternatives can be examined systematically, giving a reasonably comprehensive predictive capability. Contact with experiment allows us to resolve, in many cases, situations which remain ambiguous experimentally. In essence, the simulation techniques may be considered as applied theory or indeed as computational experiments: the link between them is then immediately apparent.

In conclusion, theory and experiment really go hand in hand and the most progress may be expected if the symbiotic relationship is maintained.

REFERENCES

1) C.R.A. Catlow and W.C. Mackrodt, eds, "Computer Simulation of Solids", Lecture Notes in Physics, 116, Springer-Verlag, Berlin (1982)
2) B.J. Wuensch, in: "Mass Transport in Solids", F. Beniere and C.R.A. Catlow, eds, Plenum Press, New York (1983)
3) C.R.A. Catlow, R. James, W.C. Mackrodt and R.F. Stewart, Phys. Rev., B25, 1006 (1982)
4) M.J. Norgett, Reports AERE-R7015, UKAEA Harwell (1972) and AERE-R7650, UKAEA, Harwell (1974)
5) M. Leslie, SERC Daresbury Laboratory Technical Memorandum DL/SCI/TM31T (1982)
6) D.R. Sempolinski and W.D. Kingery, J. Amer. Ceram. Soc., 63, 664 (1980)
7) W.C. Mackrodt and R.F. Stewart, J. Phys. C: Solid State, 10, 1431, (1977) ibid, 12, 431 (1979)
8) B. Maximov and H. Schultz, Acta Crystallogr. B41, 88 (1985); A. Zalkin and D.H. Templeton, Acta Crystallogr. B41, 91 (1985)
9) A. Sher, R. Solomon, K. Lee and M.W. Muller, Phys. Rev., 144, 593 (1966)

10) A.V. Chadwick, D.S. Hope, G. Jaroskiewicz and J.H. Strange, in: Fast Ion Transport in Solids. eds. P. Vashishta, J.N. Mundy and G.K. Shenay, p. 683, North Holland, New York (1979)

11) W.M. Jordan and C.R.A. Catlow, unpublished work

12) M.G.S.R. Thomas, P.G. Bruce and J.B. Goodenough, in "Transport—Structure Relation, in Fast Ion and Mixed Conductors", eds. F.W. Poulsen, N. Hessel Anderson, K. Clausen, S. Skaarup and O. Toft Sorensen, Riso National Laboratory, Roskilde, Denmark (1985)

13) Computer Simulation of Condensed Matter, eds. C.R.A. Catlow and W.C. Mackrodt, Physica 131B (1985)

14) R. Gerhardt Anderson and A.S. Nowick, Solid State Ionics, 5, 547 (1981)

15) C.R.A. Catlow in "Mass Transport in Solids", eds. F. Beniere and C.R.A. Catlow, Plenum Press (1983)

16) A.S. Nowick, Da Yu Wang, D.S. Parkand J. Griffith in: Fast Ion Transport in Solids, eds. P. Vashishta, J.N. Mundy and G.K. Shenoy (North Holland, New York) 637 (1979)

17) V. Butler, C.R.A. Catlow, B.E.F. Fender and J.H. Harding, Solid State Ionics, 8, 109 (1983)

18) G.E. Murch, "Atomic Diffusion Theory in Highly Defective Solids" Trans. Tech. Aedemansdorf, Switzerland (1980)

19) N. Metropolis, A.W. Rosenbluth, M.N. Rosenbluth, A.H. Teller and E. Teller, J. Chem. Phys. 21, 1087 (1953)

20) A.D. Murray, G.E. Murch and C.R.A. Catlow, Proceedings of the Solid State Ionics '85 Conference, Lake Tahoe, August 1985

TRANSMISSION ELECTRON MICROSCOPY OF DEFECTS IN SOLIDS

Linn W. Hobbs

Massachusetts Institute of Technology
Cambridge, MA 02139

THE CHOICE OF ELECTRON MICROSCOPY

Microscopy is based on the ability to retrieve spatial information about an object, usually in magnified form, from diffraction of radiation by the object. It differs from indirect structural probes such as spectroscopy, scattering, magnetic resonance, electrical or thermal conductivity measurements in that information about internal structure or composition of the object is averaged over at most one dimension of the object, providing at least two-dimensional mapping of structural features of interest, such as atomic arrangements or structural defects. Microscopy is therefore a *direct* technique for observing structure and sometimes composition of individual defects.

The fundamental requirements for a useful microscopical radiation probe are resolution, sensitivity and magnification. The relevant parameters which relate to these requirements are given in Table 1 for light, X-rays and electrons.

Resolution is the ability to distinguish two spatially-separated points in the object and is related to the wave character of radiation, possessed by moving particles (as well as electromagnetic radiation) through the de Broglie relation which relates particle momentum p to a wavelength λ by

$$\underline{p} = h\underline{k}$$

$$|\underline{k}| = 1/\lambda \, , \tag{1}$$

with the wave vector $\underline{k}$ in the direction of wave (and particle) propagation. For an electron of mass m, velocity v and kinetic energy $T = p^2/2m$ accelerated through potential $V_a = T/e$

$$\lambda = h/\sqrt{\{2meV_a(1+eV_a/2mc^2)\}} \, . \tag{2}$$

An electron accelerated through $V_a = 100$ V travels about 0.1 × the speed of light c and has a wavelength about 0.1 nm. Relativistic effects become important for electrons accelerated beyond 100 kV ($\lambda = 3.7$ pm, $v \simeq c/2$).

61

Table 1. Properties of Microscopical Probes

Wave	λ	n	f_{min}	Scattered wave amplitude	Scattering length
light	500 nm	1.5	5 mm		1 nm $\rightarrow \infty$
X-rays	0.1 nm	$1-10^{-5}$	10^5 m	10^{-13} m	700 μm
electrons	1 pm	$1+10^{-4}$	10^4 m	10^{-9} m	200 nm
neutrons	0.1 nm	$1-10^{-6}$	10^6 m	10^{-14} m	30 mm

The fundamental limit to spatial information that can be extracted
from an object using wave phenomena is classically a consequence of
Huygens' principle and quantum-mechanically a consequence of the Heisenberg
uncertainty principle. This limit is easily illustrated for a narrow slit
of width δx examined using plane-wave radiation of wavelength λ and wave
vector $\underline{k} = k_x$ (Fig. 1). If the slit were infinitely narrow, Huygens'
principle holds that each point along the slit defines a point source
of new waves which propagate radially in all directions in the x-z plane,
so that an observer some distance away would not deduce anything about
the spatial width of the slit by collecting information over any range of
observation positions defined by angle θ. Quantum-mechanically, the slit
introduces a change $\delta p_x = h \delta k_x$ in the wave momentum, and therefore an
uncertainty δx in the width of the slit given by

$$\delta x \cdot \delta p = h$$

$$\delta x = \frac{h}{\delta p_x} = \frac{1}{\delta k_x} = \frac{1}{2k \, \sin\theta} = \frac{\lambda}{2 \, \sin\theta} \qquad . \qquad (3)$$

The spatial resolution definable in any system using wave-like probes is
therefore limited to $\sim\lambda/2$; this limits the resolution of light microscopes
to 0.2μm but yields the potential for subatomic resolution with electrons.

The interaction strength of the microscopical probe with the object
governs both the distance in the third dimension along which it is necessary to integrate the interaction and the probe intensity required for
acceptable collection statistics. One measure of the *elastic* interaction is the refractive index n of the object.

$$n = \frac{v_{vac}}{v_{obj}} = \frac{(\nu/k)_{vac}}{(\nu/k)_{obj}} \simeq k_{obj}/k_{vac} \qquad . \qquad (4)$$

For light, the interaction is largely with the least-tightly bound
valence or conduction electrons whose distribution can be collectively
polarized by the electric field of the incoming photon. The index of
refraction is then given by $n \simeq \sqrt{\epsilon}$ where ϵ is the high-frequency dielectric constant; a typical value for a transparent solid is n = 1.5.
Conduction electrons respond so strongly that the photon is wholly
absorbed in the first few atom layers, precluding light microscopy of
metals and many semiconductors, except in reflection. Higher-energy
X-ray photons interact with more tightly bound electrons as well. The
collective response is characterized by $n \simeq 1 - \nu_p^2/\nu^2 = 1 - 10^{-5}$,
where ν_p is the natural collective response (plasmon) frequency. Much
of this interaction is incoherent, and the coherent scattering of X-rays
required for microscopy is smaller still.

Fast electrons interact with all the atomic electrons and additionally
with the nucleus. The interaction is electrostatic and overall is best

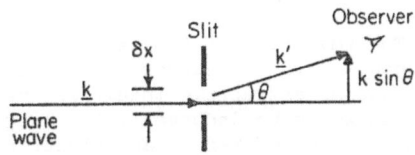

Figure 1. Resolution limit as a consequence of the Uncertainty Principle.

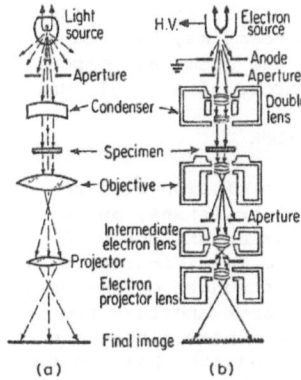

*Figure 2. Configuration of (b) conventional transmission electron micro-
scope (CTEM) and (a) its optical analogue.*

gauged from the atomic *potential* V(r), which is of the order of a hundred volts. The electron feels the additional electrostatic potential field of the atoms and responds to a combined potential $V = V_a + V(r)$. Since $k^2 = 2\,meV/h^2$ from (1),

$$n = \frac{k_{obj}}{k_{vac}} = \sqrt{\{(V(r) + V_a)/V_a\}}$$

$$\simeq 1 + \frac{1}{2}\frac{V(r)}{V_a} \simeq 1 + 10^{-4} \tag{5}$$

because $V(r) \ll V_a$. The smallness of the deviation of the refractive index from unity for both X-rays and electrons precludes the use of material lenses making use of refractive index differences; the focal distances for such lenses would be inconveniently large (Table 1); photoelectric effect absorption of X-rays and electron deceleration from inelastic interactions also preclude use of the thick lenses which would be required. X-rays can be crudely focussed by grazing-incidence Bragg diffraction from crystalline surfaces, but such lenses are not of sufficient quality to realize the potential resolving power of X-ray wavelengths, and X-ray topographic images (see Sauvage-Simkin, this volume) are generally obtained without lenses. Inhomogeneous electric or magnetic fields possess focusing properties for electrons, however, and magnetic lenses of sufficient quality can be made to realize the resolution potential inherent in fast electron wavelengths, at least down to atomic dimensions.

Other measures of the comparative interaction parameter are the elastic atomic scattering amplitude and elastic scattering lengths, both indicated in Table 1. The large comparative strength of the interaction of the electron with matter itself is due to the Coulombic nature of the interaction. The strength of the interaction is such that electron beams may be used as probes for local determination of structure and chemical composition down to very small volumes, even of the order of a single unit cell.

INSTRUMENT CONFIGURATIONS

A *conventional transmission electron microscope* (CTEM) is configured with two or more condenser lenses, an objective lens, and two or more projector lenses, in direct analogy with a transmission light microscope (Fig. 2). Overall image magnification from 10^2 to 10^6 is typical, controlled by varying the strength of the projector lenses. The condenser system and second condenser aperture provide an illuminated area down to about 1μm diameter with a beam divergence of about 1 mrad. The electron source can be a simple tungsten hairpin filament thermionic emitter or, increasingly, indirectly-heated pointed tips of ceramics, such as LaB_6, with low work function. Since the entire area of interest is illuminated at once, information about each point in the object is acquired *in parallel*.

A second configuration is shown in Fig. 3b, in which the condenser system is used to focus the electron beam to a fine spot as small as 0.5 nm diameter on the specimen object. This fine probe is scanned over the specimen, by means of scanning coils, in synchronization with the beam of a cathode ray tube whose intensity is modulated by the signal from a detector collecting electrons transmitted through the specimen. The magnification is simply the geometrical ratio of the two scanned areas; the ultimate resolution is clearly set by the probe diameter. This instrument is called a *scanning transmission electron microscope* (STEM) and the image is clearly acquired *serially*. An

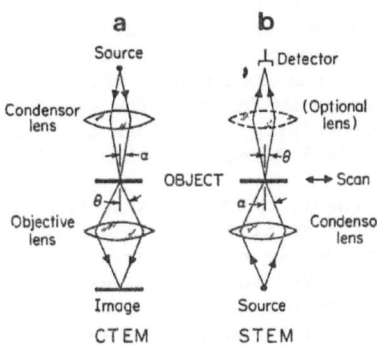

Figure 3. *Equivalence of (a) parallel (CTEM) and (b) serial (STEM) image formation configurations for the electron microscope upon electron reversal.*

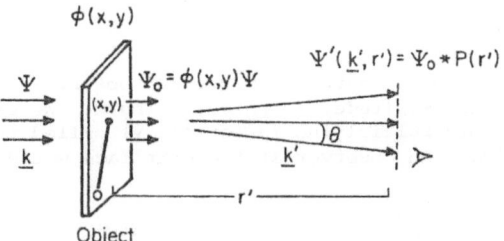

Figure 4. *Wave-optical formulation of diffraction from an object.*

important requirement for STEM is sufficient current density in the small probe in order to provide adequate pixel statistics while scanning; this requirement is most easily fulfilled with a field-emission source whose effective source size is much smaller than for thermionic emitters and therefore more efficiently demagnified to a probe of namometer dimensions.

CTEMs can be turned into STEMs by addition of scanning coils, an electron detector and a probe forming lens. The most recent generations of CTEMs are equipped with this capability, for the most part not for image formation but for the high-resolution microanalytical information about elemental chemistry which can be derived from the characteristic X-rays generated in the specimen by the electron probe or the complementary energy losses suffered by the electrons in traversing the specimen. Local microchemistry in regions as small as 2 nm across can be deduced with a minimal detectable mass $\sim 10^{-24}$ kg.

Because the theory of TEM imaging has been worked out for the CTEM, it is useful to have an algorithm to connect CTEM and STEM images. Fig. 3 shows that the two configurations can be made equivalent *if the electron paths are reversed*. Since neither instrument distinguishes the direction of electron propagation, the theories which govern image formation are identical, provided the configurations of the two instruments as portrayed in Fig. 3 are equivalent, *viz* the illumination half-angle α in CTEM is identical to the collection half-angle θ in STEM, and the illumination half-angle α in STEM is identical with the objective lens acceptance angle θ in CTEM.

WAVE OPTICS OF IMAGING

Suppose an object (a thin object for electrons, as we shall see) is illuminated by a plane, parallel monochromatic propagating wave (Fig. 4). We can represent the *spatial* variation of the wave by a wave function

$$\Psi = \psi \cos (2\pi \underline{k} \cdot \underline{r}) + i \sin (2\pi \underline{k} \cdot \underline{r}) = \psi \exp(2\pi i \underline{k} \cdot \underline{r}) \qquad (6)$$

whose observable *intensity* is

$$\Psi\Psi = \psi\psi^* \quad . \qquad (7)$$

Imposition of an object between source and observer clearly alters the observed wave--in amplitude, in phase, or both. We can represent the effect by a simple multiplying factor $\phi(x,y)$ called the *object transmission function*. An observer at the exit face would perceive a wave amplitude

$$\Psi_O = \phi(x,y)\Psi \quad . \qquad (8)$$

An *amplitude object* alters the amplitude ψ of Ψ by a factor $\phi(x,y) = \psi(x,y)/\psi$. For a phase object, $\phi(x,y) = \exp\{2\pi i \chi(x,y)\}$, introducing a phase shift $\chi(x,y)$. An observer some distance r' from the object will *not* perceive the exit-face wave function $\Psi_O = \phi\Psi$, however, but will instead see a more complicated distribution of amplitude and phase given by

$$\Psi'(k',r') = \Psi_O * P(r') \qquad (9)$$

where

$$P(r') = \frac{ik}{z} \exp(2\pi ikz) \exp 2\pi ik \frac{(x^2+y^2)}{2z} \qquad (10)$$

66

is called the *Fresnel propagator* and k' is post-object wave propagation vector. For a *very* distant observer (r' → ∞), the convolution in (9) reduces to a particularly simple result for an incident plane wave

$$\Psi(k', r' \to \infty) = i\lambda \frac{\exp(2\pi i k' r')}{r'} \int_{-\infty}^{\infty} \phi(\underline{r}) \exp\{-2\pi i(\underline{k}' - \underline{k}) \cdot \underline{r}\} \, d\underline{r} \quad (11)$$

The result (11) represents a spherical wave propagating from the distant object whose amplitude

$$\int_{-\infty}^{\infty} \phi(\underline{r}) \exp(-2\pi i \kappa \cdot r) d\underline{r} = \tilde{\phi}(\kappa) \quad (12)$$

is the *Fourier transform* of the transmission function and whose intensity is just the *Fraunhofer diffraction pattern* of the object. The vector $\underline{\kappa} = \underline{k}' - \underline{k} = 2k \sin(\theta/2) = (2/\lambda) \sin(\theta/2)$ is called the *diffraction vector* and is essentially a measure of the angle θ at which the distant observer observes the object. As an example, an object consisting of two pinholes spaced a distance a apart along the x-axis generates the familiar cosine fringes

$$I(\underline{\kappa}) = \cos^2(\pi \kappa_x a) \quad (13)$$

of the Young's double-slit experiment. The wave $\Psi(\underline{\kappa})$ is described in a space *reciprocal* to real space, and the result (13) shows that information about *spatial correlations* in real space are recoverable in the reciprocal-space diffraction pattern intensities.

A *lens system* as used in CTEM involves both finite propagation distances and propagation through the lenses (Fig. 5). A lens introduces a phase shift, and its transmission function is

$$\phi_\ell(x, y) = \exp\left\{-2\pi i k \frac{(x^2 + y^2)}{2f}\right\}$$

for a symmetrical lens of focal length f. If we place the object approximately in the front focal plane, $u \simeq f$, and follow the wave propagation from the object through the lens system illustrated in Fig. 5, we find at the back focal plane, z = f, the result

$$\Psi_f(x_f, y_f) = \frac{i\lambda}{f} \exp\{2\pi i k(u+f)\} \tilde{\Psi}_o(\kappa_x, \kappa_y) \quad (14)$$

where $\tilde{\Psi}_o = \tilde{\phi}$ for a plane incident wave Ψ. This is *exactly* the Fraunhofer diffracted amplitude found in (11) for z = ∞ without lenses. If we let the wave continue to propagate a further distance v-f, we find at z = v (called the *Gaussian image plane*)

$$\Psi_v(x_v, y_v) = -\frac{1}{M} \exp\left\{\frac{2\pi i k(x_v^2 + y_v^2)}{2fM}\right\} \exp\{2\pi i k(u+v)\}$$

$$\cdot \Psi_o\left(-\frac{x_v}{M}, -\frac{y_v}{M}\right) \quad (15)$$

where M = v/u and 1/u + 1/v = 1/f. Apart from a phase shift and an amplitude diminished by M (and thus intensity by M^2), we find the original object exit face wave $\Psi_o = \phi\Psi$ inverted and magnified by M, i.e. we generate a real magnified image of the object with all its internal phase relationships intact. The lens has, in fact, Fourier transformed the object information twice, once at the back focal plane z = f and again at the image plane z = v. Subsequent projector lenses can be made to focus on either the back focal plane of the first (objective) lens, to yield a magnified reciprocal-space diffraction pattern, or on the Gaussian image plane, to yield a still further magnified real-space image.

By analogy to the relationship between time (s) and frequency (s^{-1} or Hz), we can speak of correlations in real space (nm) appearing as *spatial frequencies* (nm^{-1}) in reciprocal space, the lens transferring spatial frequencies associated with object to spatial frequencies comprising the image. The fidelity with which this transfer occurs is a measure of the *spatial resolution* of the imaging system and can be characterized by a lens transfer function $T(\kappa_x, \kappa_y)$ operating in the back focal plane. What appears in the back focal plane is (apart from the propagation factors appearing in (14))

$$\Psi_f(x_f, y_f) = \tilde{\Psi}_o(\kappa_x, \kappa_y) \cdot T(\kappa_x, \kappa_y) \quad . \tag{16}$$

For a perfect lens, T = 1. At the Gaussian image plane

$$\Psi_v(x_v, y_v) = F[\tilde{\Psi}_o(\kappa_x, \kappa_y) \cdot T(\kappa_x, \kappa_y)]$$

$$= \Psi_o(-\frac{x_v}{M}, -\frac{y_v}{M}) * \tilde{T}(x, y) \tag{17}$$

because a Fourier transform F of a product is the convolution of the two transforms. $\tilde{T}(x,y)$ is called the *impulse response function* because the image of a point object (impulse) becomes a fuzzy point (Fig. 6). $\tilde{T}$ is thus a measure of resolution in real space, the ability to separate images of two point objects, and can be normalized to resolution in the object by dividing by M. Resolution in STEM is related to the width of the probe, which can only be as small as the impulse response function of the probe-forming lens.

The form of the lens transfer function T is dictated by the lens *aberrations*, the most important of which for electron lenses are defocus, spherical aberration, chromatic aberration, partial coherence and finite acceptance angle. The first two introduce phase shifts

$$\chi(\kappa) = 2\pi(\frac{1}{2} \Delta f \lambda \kappa^2 + \frac{1}{4} C_s \lambda^3 \kappa^4) \tag{18}$$

which can be made to nearly cancel over limited range of κ for a given spherical aberration coefficient C_s by deliberately selecting a small negative (underfocus) defocus value Δf typically <100 nm. The last three represent damping or truncation of the spatial frequency information available at the back focal plane and so may be subsumed into a *pupil* function $P(\kappa)$, so that the overall transfer function may be expressed as

$$T(\kappa) = P(\kappa) \exp\{i\chi(\kappa)\} \quad . \tag{19}$$

Examples of the form of $T(\kappa)$ for a modern objective lens at two values of defocus are shown in Fig. 7, from which it is clear that the image is under general circumstances *not* a straightforward representation of the object transmission function. For a given C_s, the appropriate compensating underfocus is called the Scherzer defocus and yields an approximately linear transfer response for the $P(\kappa) \sin\chi(\kappa)$ part of the lens transfer function to all spatial frequencies out to some critical value κ^*. The corresponding FWHM value for the impulse

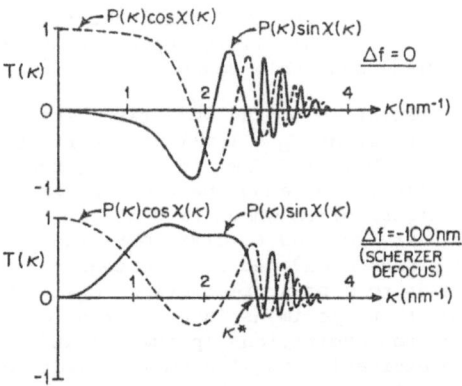

Figure 5. Wave optics of a single lens system.

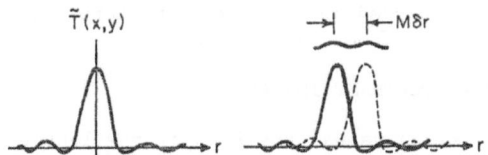

Figure 6. (a) An impulse response function for a lens and (b) its relation to resolving separate images of two points spaced δr apart in the object.

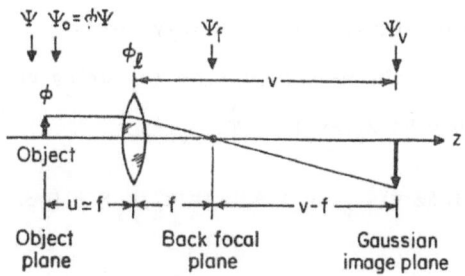

Figure 7. Form of typical lens transfer function at two values of defocus.

response function is $\sim 1/2\kappa*$, and the Rayleigh criterion gives a point resolution limit of $\sim 1/\kappa*$. This limit currently stands at just under 0.2 nm in the best CTEMs available today.

The chromatic aberration of the objective lens (characterized by a chromatic aberration constant $C_c \sim 1$ mm) is important because electrons lose energy (at a rate about 1 GeV/m) in passing through the specimen, and magnetic lenses for electrons cannot be made achromatic. It is the *width* of the energy loss distribution which is critical, because electrons of different energies U are focused to different image points, and each point in the image is therefore broadened as a consequence of the energy spread ΔU. This effect leads to a point resolution $\sim C_c \lambda \kappa* (\Delta U/U)$. The principal energy loss mode for thin specimens is excitation of long-wavelength collective valence electron excitations (plasmons) at ~ 10-30 eV, the mean free path for which excitation is ~ 100 nm with width $\Delta U \sim 10$ eV. This means that resolution for a 100 nm-thick specimen is limited to ~ 2 nm from chromatic aberration effects alone. Therefore, to achieve a useful resolution ~ 5 nm for microstructural studies, specimens must be no more than several hundred nm thick; for the ultimate point resolution given by spherical aberration alone, a specimen must be no more than ~ 10 nm thick, an exceedingly stringent constraint. Except for the case of deliberately deposited thin films, the latter condition is achieved only at the extreme edges of specimens thinned in some way (fracture, mechanical or chemical thinning, ion etching) from bulk samples.

PHASE AND AMPLITUDE CONTRAST IMAGES

In certain special cases, the form of the lens transfer function simplifies enough to permit unambiguous interpretation of the image in terms of the object transmission function down to the point resolution limit. One such case is when the specimen is sufficiently thin (thickness δz) to alter, to first approximation, only the *phase* of the incoming electron wave. In this case, using (5), the exit face wave can be written

$$\psi(x,y) = \exp\{2\pi i k \cdot r(x,y)\} \ \exp\{i\beta(x,y)\} \ ,$$

where the phase shift $\beta(x,y)$ is given by

$$\beta(x,y) = (\pi k/V_a) V(x,y) \ \delta z = \sigma \ V(x,y) \ \delta z \tag{20}$$

where $\sigma = (\pi k/V_a) = 2\pi m e/h^2$. For a sufficiently *weak* phase object

$$\phi(x,y) = \exp \ i\beta(x,y) \simeq 1 - i \sigma V(x,y) \ \delta z \ . \tag{21}$$

At the image plane, the wave amplitude (assuming unit magnification) is

$$\psi_v = 1 - i\sigma\delta z \ V(-x_v,-y_v) \ * \ \tilde{T}(-x_v,-y_v)$$

$$= 1 - \ i\sigma\delta z \ V(-x_v,-y_v) \ * \ F[P(\kappa_x,\kappa_y) \cdot \{\cos\chi(\kappa) + i \ \sin\chi(\kappa)\}] \ . \tag{22}$$

Because of κ^2 and κ^4 terms in $\chi(\kappa)$, both $\cos\chi(\kappa)$ and $\sin\chi(\kappa)$ are symmetric, so their transforms are real. To first order, neglecting terms in V^2, the intensity becomes

$$I_v = \psi_v\psi_v* = 1 + 2 \ \sigma \ \delta z \ V(-x_v,-y_v)$$

$$* \ F[P(\kappa_x,\kappa_y) \ \sin\chi(\kappa_x,\kappa_y)] \tag{23}$$

and only the $\sin\chi(\kappa)$ portion of the lens transfer function is important. The quantity $P(\kappa) \sin\chi(\kappa)$ is therefore called the *phase contrast transfer function* and can be made to have a reasonably flat response over the range $\kappa < \kappa^*$ near the Scherzer defocus, as in Fig. 7b. This result (23) is probably applicable for a single layer of atoms (Fig. 8).

For a thicker specimen, the phase shift must be integrated over the specimen thickness t

$$\beta(x,y) = \sigma \int_0^t V(x,y,z) \, dz \qquad (24)$$

and also the approximation (21) fails. Considering only the defocus part of the lens aberration alone, with Δf chosen to optimize the $\sin\chi(\kappa)$ transfer response, the approximate result analogous to (23) is

$$I_v \simeq 1 - \frac{\Delta f}{2\pi k} \nabla^2 \beta(-x_v, -y_v)$$

$$= 1 + \frac{\Delta f}{2V_a \varepsilon} \int_0^t \rho(-x_v, -y_v, z) \, dz \qquad (25)$$

using Poisson's relation $\nabla^2 V(r) = -\rho(r)/\varepsilon$, where $\rho(r)$ is the local *charge density*. For $\Delta f < 0$ (underfocus), (25) yields an image contrast which is proportional to the *projected charge density* of the object (Fig. 9). The approximation fails when the electrons scattered (through angles ~ 10 mrad) by atoms at the top of the object no longer sample the same part of the object potential at the bottom of the object. This failure limits specimen thickness to ~ 10 nm if images are to remain naively interpretable.

For still larger thickness, information about local *periodicities* in crystalline specimens can still be extracted, but without any necessarily one-to-one correspondence with the projected charge density. Provided the effective pupil function is still large enough to embrace strongly localized scattered information about spatial correlations, the correlations will be reproduced, but with in most cases indeterminate phase shifts in the back focal plane and spatial shifts in the image plane. A useful special case is that of periodic objects (e.g. crystals) whose periodicities lie within the image resolution deriving from the pupil function. A periodic object can be described by a (real) lattice of regularly-repeated *lattice points* (Fig. 10), each associated with the regularly recurring feature (*viz* a unit cell). This information appears in the back focal plane as the Fourier transform of the real lattice which, for a sufficiently large real lattice, is a sharply-peaked distribution known as the *reciprocal lattice*. This is a consequence of what is essentially a two-dimensional Young's double-slit experiment, the intersection of sets of interference fringes defining the reciprocal lattice points.

The electrons, of course, effectively sample along a two-dimensional projection of the real lattice; at the back focal plane, the corresponding sampling of reciprocal space is given by the *Laue condition*

$$\underline{\kappa} = \underline{g} \qquad (26)$$

where $\underline{\kappa} = \underline{k}' - \underline{k}$ as before and $\underline{g}$ is a *reciprocal lattice vector*. The sampling is best seen with the aid of the *Ewald sphere construction*

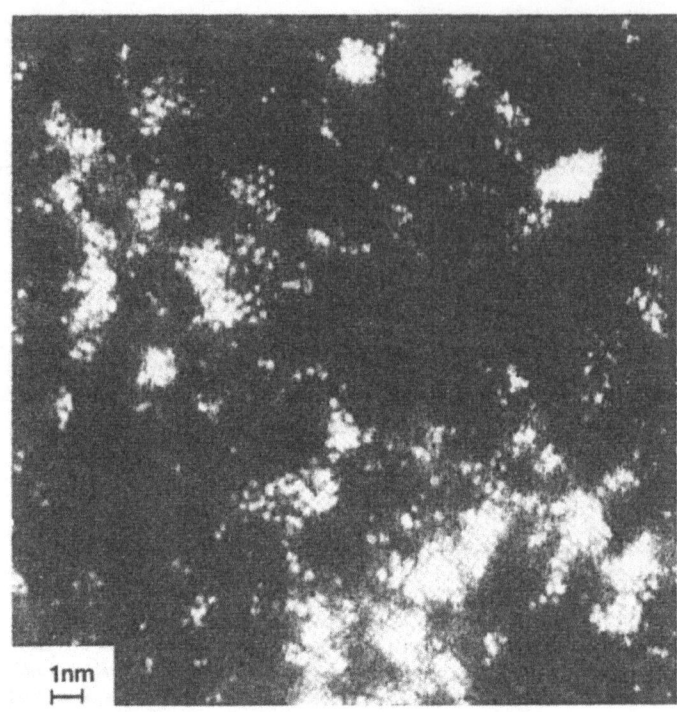

Figure 8. *Phase contrast image from single and clustered uranium atoms (derived from an evaporated solution of uranyl acetate) on a very thin carbon substrate (micrograph courtesy of A. V. Crewe; see [1]).*

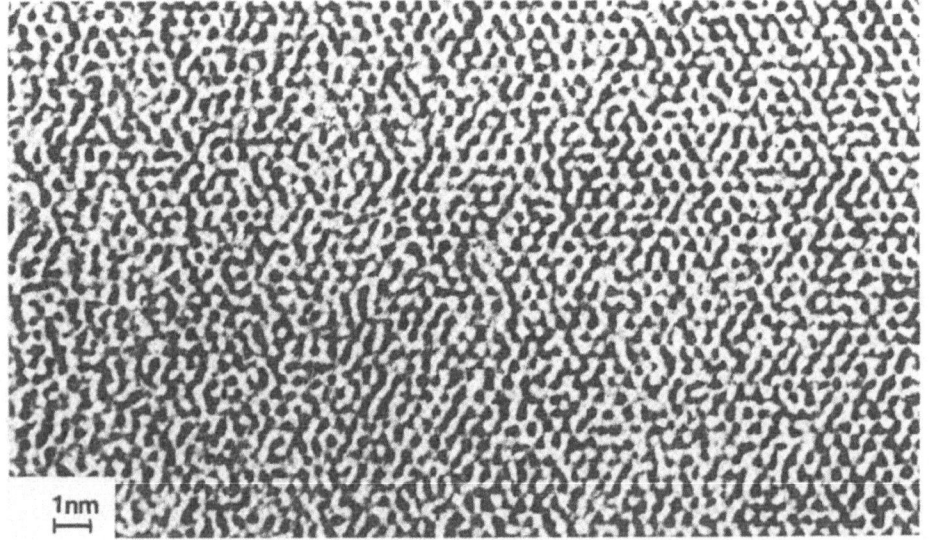

Figure 9. *Projected charge density image of quartz showing the hexagonal arrangement of low-density c-axis channels in the structure [2]. The specimen is undergoing radiolysis in the electron beam and transforming to an aperiodic glass in the lower left corner of the micrograph. The distance between channels is about 0.4 nm.*

72

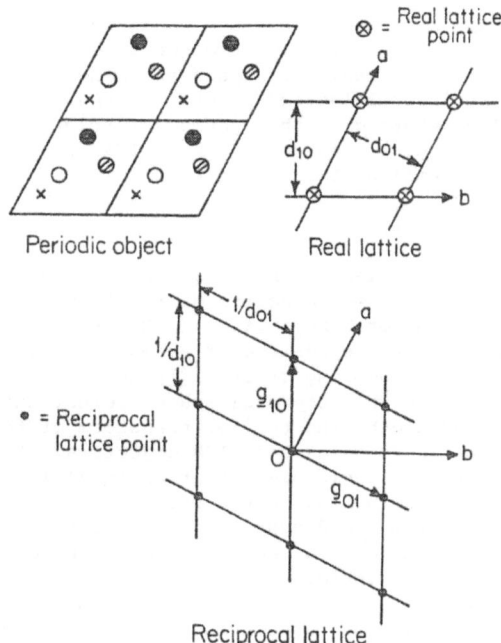

Figure 10. *A periodic object represented by an array of lattice points,*
and the corresponding reciprocal lattice which is the Fourier
transform of the real lattice. The spacing between reciprocal
lattice points is proportional to the reciprocal of the 'planar'
spacings d in the real lattice, and the reciprocal lattice
vector g is normal to the corresponding set of such planes.

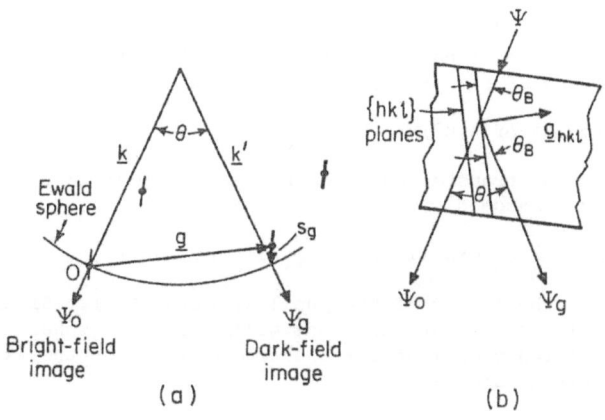

Figure 11. *(a) Ewald sphere reciprocal-space construction for depicting*
sampling of reciprocal space, and (b) corresponding Bragg
construction in real space for a periodic structure.

(Fig. 11) which represents the Laue condition (26). *Strongly* scattered intensity in other then the forward scattered direction is detected whenever a sphere of radius $1/\lambda$ passes through the origin of reciprocal space and a reciprocal lattice point. By the Fourier transform relationship $|\underline{g}| = 1/d$, where d is the spacing of "planes" of real space lattice points normal to g, so that using $|\underline{k}| = 1/\lambda$, the Laue condition (26) reduces to the familiar *Bragg condition*

$$\lambda = 2 \ d \ \sin(\theta/2) \qquad\qquad (27)$$

where $\theta/2 = \theta_B$ is called the Bragg angle. Because $\lambda \simeq 1\text{-}4$ pm, Bragg angles are very small for electron diffraction, only ~ 10 mrad, compared to ~ 1 radian for X-rays with $\lambda \sim 0.1$ nm. Note that the information mation is sampled on the surface of a two-dimensional sphere, which makes simultaneous sampling of strongly scattered information unlikely; however, the maxima in reciprocal space are not infinitely sharply peaked but broadened by the Fourier transform of the shape of the object. Since, as we have seen, TEM objects are thin in the direction of the electron beam incidence, the reciprocal lattice points are *extended* into spikes normal to the specimen surfaces and thus approximately normal to the Ewald sphere. This happy circumstance means that much of the information available at the reciprocal lattice points, which represents the various spatial frequencies of the periodic object, can be collected all at once. Thus we see that the projected charge density approximation derived earlier for very thin objects is essentially a flat Ewald sphere (high-energy) approximation, a consequence of the very thin specimen dimension along the electron beam trajectory. For very thin crystals and sufficient resolving power, it is in principle possible to obtain atomically-correct projections of crystal structure (*structure images*), but anomalies can easily arise for larger thickness or when operating beyond the point resolution limit [3].

For thicker specimens, it must be remembered that the specimen functions as an *amplitude* object as well as a phase object, so the phase object approximation fails. In the case of a periodic object, we can assign to each Bragg maximum (including the forward-scattered electron wave) an amplitude ψ_g and phase χ_g

$$\Psi_g = \psi_g \ \exp{(i \chi_g)} \ . \qquad\qquad (28)$$

If we suppose that the lens transfer function $T(\kappa)$ includes only the effects of defocus Δf as before, then from (18)

$$T(\kappa) = \exp{(i\pi \ \Delta f \ \lambda \ \kappa^2)} \qquad\qquad (29)$$

and the information transferred at each $\underline{\kappa} = \underline{g}$ maximum in the back-focal plane $z = f$ becomes

$$\Psi_f \ (x_f, y_f) = \sum_g \psi_g \ \exp{(i\chi_g)} \cdot \exp{(i\pi \ \Delta f \ \lambda \ g^2)} \qquad\qquad (30)$$

for all $\underline{g}$ out to the limits of the pupil function $P(\kappa)$. Since the reciprocal lattice point intersections with the Ewald sphere are nearly points (δ-functions), the transform of (30) for the axial case is particularly simple

$$\Psi_v(x_v, y_v) = \psi_o - \sum_{g \neq 0} \psi_g \ \exp{[2\pi i (g_x \ x + g_y \ y)]}$$

$$\cdot \exp{(i\pi \ \Delta f \ \lambda \ g^2)} \ . \qquad\qquad (31)$$

Consider the two-beam Bragg condition illustrated in Fig. 11 in which the forward-scattered beam Ψ_o and only one diffracted maximum Ψ_g are strong and included within an aperture (or virtual aperture) at the objective lens back focal plane. The image intensity obtained from (31) is

$$I_v(x_v) = \underbrace{\psi_o^2 + \psi_g^2}_{\text{background}} + \underbrace{2\,\psi_o\,\psi_g\ \sin(2\pi gx - \pi\Delta f\lambda g^2)}_{\text{fringes}}\ . \qquad (32)$$

The resulting sinusoidal interference fringes are called *lattice fringes* because their spacing 1/g corresponds to the real-space lattice plane periodicies d that gave rise to the diffraction maximum at g. These fringes can be resolved down to spacings well beyond the point resolution limit, since the amplitude information is truncated only by the pupil function $P(\kappa)$. There is the temptation to associate such fringes with actual "lattice planes", but the *position* of the fringes is not unique and depends on Δf (as well as on specimen thickness, specimen tilt and incident beam tilt), so it is wise to remember that what is being observed is an interference pattern which bears no necessarily direct correlation to structural features (like atomic planes) in the object. Nevertheless, local information about defect microstructure can often be deduced, such as the presence of dislocations (Fig. 12).

The non-axial *tilted* two-beam condition (Fig. 13a) is useful, because the *same* instrumental phase shift now applies to each beam, since $\chi(\kappa)$ is an even function, and spherical aberration is minimized. Also, to a first approximation, such images are independent of Δf, but it is also quite clear from the geometry that what is being observed is still an interference pattern and not a projection of "lattice planes". The *three-beam axial* condition (Fig. 13b), for which the image intensity is

$$I_v(x_v) = \psi_o + \psi_g + \psi_{-g} + 2\,\psi_g\,\psi_{-g}\underbrace{\cos(4\pi gx)}_{\text{half period}}$$

$$+\ 4\,\psi_o\,\psi_g\underbrace{\cos(2\pi gx)}_{\text{full period}}\ \sin(\pi\,\Delta f\,\lambda\,g^2) \qquad (33)$$

retains a true "lattice plane" projection, but interference between the two extreme beams produces a modulation with *half spacing*. While the half-period fringes are insensitive to defocus Δf, the full-period fringes nevertheless still do depend on defocus (and other instrumental phase shifts). Any naive interpretation of structural anomalies (such as local variation of lattice parameter with composition or atomic details of defects) through use of lattice-fringe images must therefore be approached with considerable caution and an awareness of instrumental influences [5].

SINGLE-BEAM IMAGES FROM AMPLITUDE OBJECTS

For specimens sufficiently thick (∿100 nm) that chromatic aberration precludes resolution of lattice fringes or atomic details, information can be derived only from inhomogeneities which scatter electrons differently from the surrounding matrix. Contrast is obtained by aperturing the forward-scattered beam alone (*bright-field* imaging) or some other portion of the scattered electron distribution, e.g. a Bragg-diffraction maximum (*dark-field* imaging), in the objective lens back

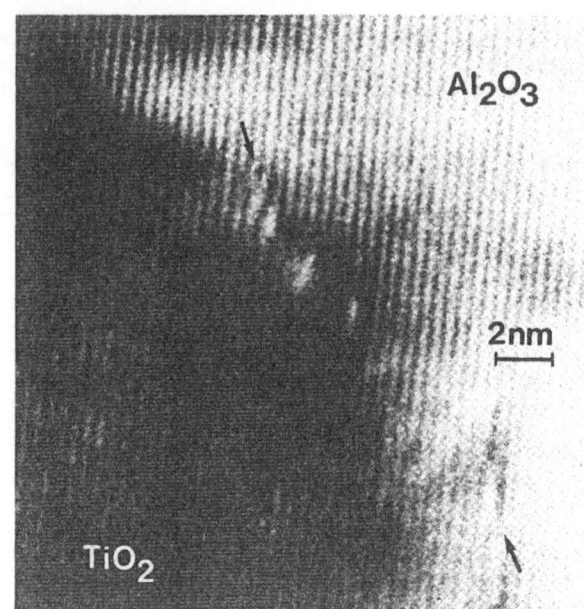

Figure 12. Two-beam lattice fringe image of a portion of the semicoherent interface between α-Al₂O₃ and TiO₂ (rutile) in star sapphire [4]. The lattice mismatch is accommodated by an array of interface dislocations. The latter are most easily seen by viewing the micrograph obliquely along the fringes.

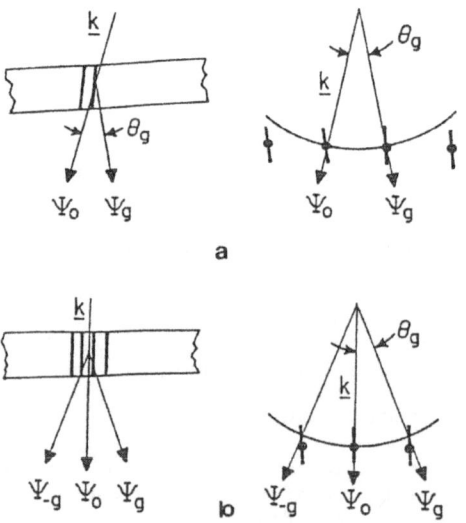

Figure 13. Imaging geometries for optimum generation of lattice fringes. (a) Tilted two-beam condition, (b) axial three-beam condition.

focal plane; electrons scattered outside the aperture do not contribute to the image, reducing the local image intensity.

The transmission function for an atom is its atomic potential $V_{atom}(r)$, so that the amplitude elastically scattered by a single atom, known as the *electron scattering amplitude*, is using (12)

$$f^{el}(\underline{\kappa}) = (2\pi me/h^2)\, \tilde{V}_{atom}(\underline{\kappa}) = \sigma\, \tilde{V}_{atom}(\underline{\kappa}) \quad . \tag{34}$$

f^{el} can also be related to the *X-ray scattering factor* f^X via

$$f^{el}(\kappa) = (me^2/2\pi h^2 \epsilon)\,[Z - f^X(\kappa)]/\kappa^2 \quad . \tag{35}$$

Since $f^X \sim Z$, (35) shows that electron scattering is much stronger and confined to much smaller angles than for X-rays (Table 1). Calculations [6], based on (35) and Hartree-Fock atomic wave functions, provide reasonable agreement with experimental determinations. Because of the Z-dependence of (35), elements of high atomic number scatter elastically more strongly to high angles (and, in fact, throughout the angular range) than do lower-Z elements. For this reason, biologists, working with specimens containing H, C, O, N and other elements of low atomic number, customarily selectively *stain* their films with salts of Os, U, Th, or other high-Z elements which scatter electrons more effectively outside the objective aperture and thus provide contrast where these elements locally segregate. *Inelastic scattering*, particularly from phonons, provides another source of high-angle scattering (typically to > 10 mrad). Mean-free paths for elastic scattering and inelastic phonon scattering in solids are of the order 100 nm and 1000 nm, respectively; both effects thus produce contrast from thicker (or denser) regions of the specimen.

For crystalline specimens, the amplitude scattered by the repeating unit, the unit cell, is given by

$$F(\underline{\kappa}) = \sum_j f_j^{el} \exp(-2\pi i \underline{\kappa} \cdot \underline{r}_j) \tag{36}$$

which serves an analogous role to the structure factor in X-ray diffraction. The partition of amplitude into sharply-peaked Bragg beams for a periodic arrangement of unit cells can then be represented by

$$\psi'(\underline{\kappa}) = F(\underline{\kappa}) \sum_{\underline{g}} \delta(\underline{\kappa} - \underline{g}) * \tilde{S}(\underline{\kappa}) \tag{37}$$

where $\delta(\underline{\kappa} - \underline{g})$ just represents the Laue condition (26) for those reciprocal lattice vectors $\underline{g}$ which intersect the Ewald sphere, and $S(\underline{r})$ represents the shape of the crystal (= 1 inside, = 0 outside). The *transform* of the crystal shape $\tilde{S}(\underline{\kappa})$ accounts for the extension of the reciprocal lattice points alluded to earlier. If we represent by s_g the deviation of the Ewald sphere from the reciprocal lattice point (Fig. 11), so that

$$\underline{k}' - \underline{k} = \underline{\kappa} = \underline{g} + \underline{s}_g \tag{38}$$

is the modified Laue condition, the diffracted intensity from a thin extensive crystal of thickness t, for unit incident intensity, can be derived from (37) and (38) as

$$I'(g) = \left(\frac{\pi t}{\xi_g}\right)^2 \frac{\sin^2(\pi t s_g)}{(\pi t s_g)^2} \tag{39}$$

where, for unit cell volume $\Omega \simeq 10^{-28}$ m^3,

$$\xi_g = \pi\Omega/\lambda F(g) \tag{40}$$

is a length of order 50 nm. $I'(g)$ clearly becomes unphysical for $t \sim \xi_g$ because it neglects the possibility of subsequent Bragg scattering of the Bragg-diffracted wave Ψ_g (Fig. 14). $I'(g)$ in (39) is therefore termed *kinematical* scattering; a treatment of multiple scattering, applying (39) differentially to incremental thicknesses, yields the two-beam *dynamical* scattering result

$$I_g = (\xi_g{}^{eff}/\xi_g) \sin^2 (\pi t/\xi_g{}^{eff})$$

$$I_o = 1 - I_g \tag{41}$$

where $\xi_g{}^{eff} = \xi_g/\sqrt{(1 + s_g^2\xi_g^2)}$. When $s_g = 0$ (Bragg Condition), I_g and I_o are seen to oscillate sinusoidally with depth (Fig. 15) with periodicity ξ_g; ξ_g is thus appropriately known as the *extinction distance*. Wedge specimens, typical of those produced by most thinning methods, consequently exhibit *thickness fringes* (Fig. 16) in bright-field or dark-field near-Bragg conditions, which become weaker and more closely spaced as deviation s from the Bragg condition increases. Fortunately, inelastic scattering effects damp out these oscillations in crystals 5-10 ξ_g thick. Similar intensity variations for constant thickness but with varying s, as occurs in slightly bent (by even $\sim$ 1 mrad) thin foils, are called *bend contours* (Fig. 17).

Owing to the flatness of the Ewald sphere, particularly at higher voltage, the two-beam approximation inherent in (41) provides useful predictions only for V_a < 200 kV. In general, the best that can be arranged is the Ewald sphere intersecting a single row of reciprocal lattice points -2g, -g, 0, g, 2g, etc., comprising a set of *systematic* reflections $\sum_n \Psi_{ng}$. The extra diffracted beams perturb the simple result (41) because of the many dynamical interactions amongst the whole set of Bragg beams.

Point and extended defects introduce further changes in the distribution of these diffracted intensities, but the continual oscillations in the amplitudes of Bragg-scattered plane waves along their propagation directions make it less than straightforward to assess altered scattering properties of defective regions and thus to characterize a defect from its image. A description of electron propagation in crystals heuristically more attractive than the plane-wave description (28) is in the form of a set of *Bloch waves*

$$\psi^{(j)} (\underline{r}) = C^{(j)} (\underline{r}) \exp (2\pi i \underline{k}^{(j)} \cdot \underline{r}) \tag{42}$$

whose amplitudes $C(\underline{r})$ are spatially periodic with the same periodicities as the crystal potential $V(\underline{r})$.

Because $V(\underline{r})$ and $C(\underline{r})$ are periodic, both may be expanded in Fourier series

$$V(\underline{r}) = \sum_g V_g \exp (2\pi i \underline{g} \cdot \underline{r})$$

$$\tag{43}$$

$$C(\underline{r}) = \sum_g C_g \exp (2\pi i \underline{g} \cdot \underline{r})$$

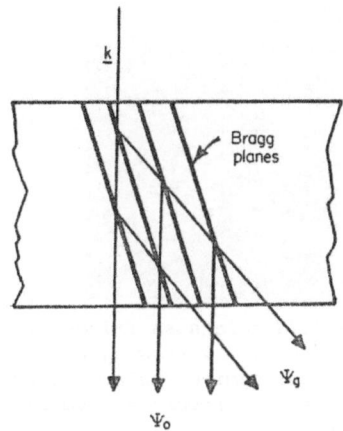

Figure 14. *Multiple Bragg scattering for crystals with* $t \geq \xi_g$.

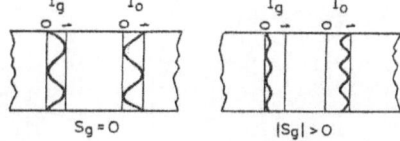

Figure 15. *Oscillation of forward-scattered and Bragg beam intensities with depth for two different values of deviation parameter* s_g.

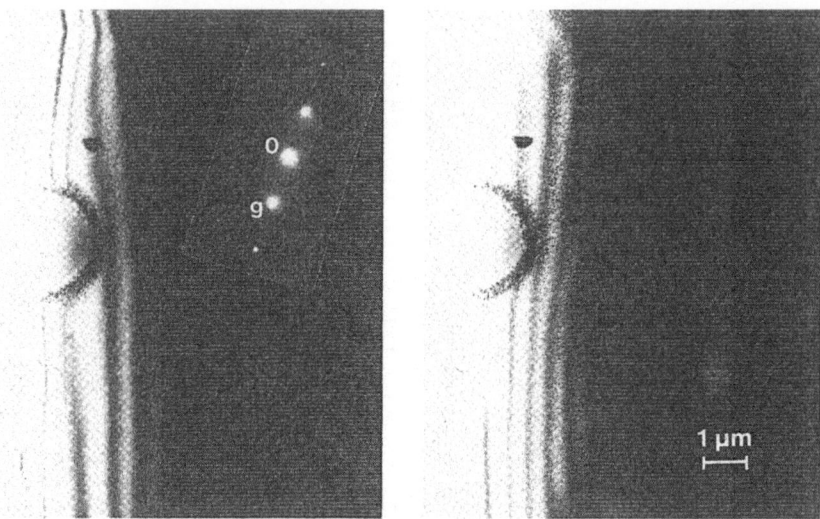

Figure 16. *Bright-field thickness fringes in a wedge crystal of quartz for a) s ≃ 0, b) s > 0 showing closer fringe spacing. Lack of contrast in the semi-circular non-crystalline region, produced by radiolysis in a focussed electron beam, confirms that dynamical intensity oscillations are the property of electron propagation in crystalline material. Approximate diffraction conditions inset.*

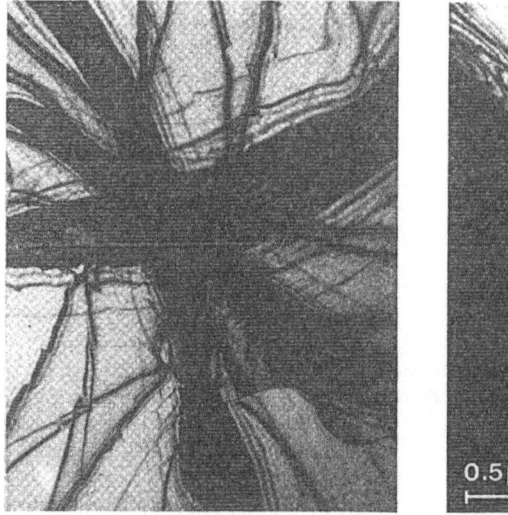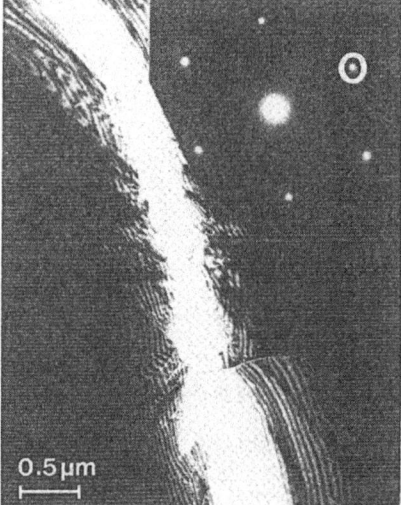

Figure 17. *Bend contours in a dynamically-thick buckled crystal across which s varies appreciably. a) Bright-field images shows contours from all Bragg diffraction processes, b) dark-field image using a single Bragg beam (circled) yields a single contour along the axis of Bragg plane rotation in the bent crystal. The electron beam is incident along an axis of high symmetry (zone axis), and the crystal is locally dome-shaped.*

evaluated over the reciprocal lattice vectors $\underline{g}$ which represent the periodicities. The Bloch waves can then be written in the form

$$\psi^{(j)}(\underline{r}) = \sum_g C_g^{(j)} \exp[2\pi i(\underline{k}^{(j)} + \underline{g}) \cdot \underline{r}] \quad . \tag{44}$$

Substituting (44) and $V(\underline{r})$ given by (43) into the time-independent Schrödinger equation

$$\nabla^2 \Psi + \frac{8\pi^2 me}{h^2}[V_a + V(\underline{r})]\Psi = \nabla^2 \Psi + 4\pi^2 k^2 \Psi + 4\pi^2 \sigma V(\underline{r}) \Psi = 0 \tag{45}$$

yields an eigenvalue equation

$$[k^2 - |\underline{k}^{(j)} + \underline{g}|^2]C_g^{(j)} + \sum_{h \neq g} U_{g-h} C_h^{(j)} = 0 \quad . \tag{46}$$

For the two-beam case $h = 0, \underline{g}$ considered before, (46) can be written in matrix form as

$$\begin{pmatrix} k^2 - k^{(j)2} & \sigma V_{-g} \\ \\ \sigma V_g & k^2 - |\underline{k}^{(j)} + \underline{g}|^2 \end{pmatrix} \begin{pmatrix} C_o^{(j)} \\ \\ C_g^{(j)} \end{pmatrix} = 0 \tag{47}$$

whose determinant reduces to the simple quadratic result

$$(k - k^{(j)})(k - |\underline{k}^{(j)} + \underline{g}|) = \sigma^2 V_g V_{-g}/4k^2 \tag{48}$$

with two roots $k^{(1)}$ and $k^{(2)}$. This result is depicted schematically in Fig. 18 in the Ewald construction. The origins of the Bloch wave vectors are constrained to lie on surfaces, called *dispersion surfaces*, which embody the relationship between energy and momentum of the wave. Wave matching at the crystal/vacuum interface requires that the transverse components of all wave vectors be equal, at a value set by the deviation parameter s_g. The longitudinal z-components clearly differ, and the fact that the Bloch wave vectors have different lengths implies that the associated electron kinetic energies are also different, so that the different Bloch waves must be localized in different parts of the periodic crystal potential (Fig. 19). The Bloch waves can therefore be viewed as *standing waves*, into which the overall electron wave is partitioned, whose *mean* amplitudes $C_o^{(j)}$ are set at the entrance face of the crystal, and which propagate *without scattering* to the exit face of the crystal.

Both Bloch wave and plane wave descriptions of electrons in the crystal must yield equivalent total electron wave functions, so that

$$\Psi(r) = \sum_g \Psi_g = \sum_g \psi_g \exp[2\pi i(\underline{k}+\underline{g}) \cdot \underline{r}]$$

$$= \sum_j \alpha^{(j)} \psi^{(j)} = \sum_j \alpha^{(j)} \sum_g C_g^{(j)} \exp[2\pi i(\underline{k}^{(j)}+\underline{g}) \cdot \underline{r}] \quad . \tag{49}$$

Initial boundary conditions require that the partitioning coefficients

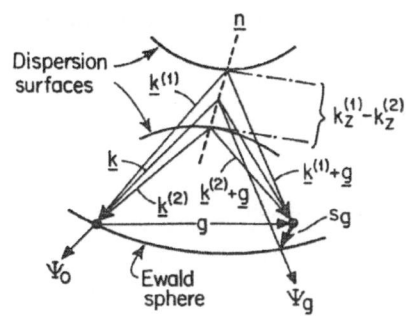

Figure 18.   *Bloch-wave representation in the Ewald sphere construction for the two-beam case (h = 0,g; j = 1,2). In the many-beam case further dispersion surfaces would lie below dispersion surfaces (1) and (2). The transverse components of all Bloch wave vectors are identical and set by the relative orientations of incident beam direction k and the crystal (and thus its reciprocal lattice). Beating between the z-components gives rise to interference fringe contrast with depth periodicity $1/|k_z^{(j)} - k_z^{(\ell)}|$.*

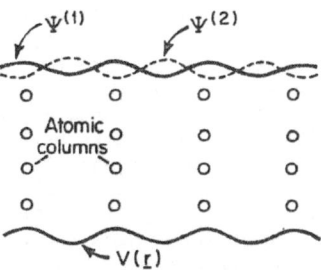

Figure 19.   *Bloch waves as propagating standing waves localized in different parts of the crystal potential V(r). Bloch wave (1) localization in the vicinity of the atomic cores leads to its strong absorption (electrons scattered outside aperture system) from inelastic processes.*

$\alpha^{(j)}$, known as *Bloch wave excitations*, are given by the $C_0^{(j)}$, whence, using (49), the Bragg beam amplitudes and intensities can be expressed in terms of Bloch wave parameters

$$\psi_g(z) = \sum_j C_o^{(j)} C_g^{(j)} \exp\left[2\pi i (k_z^{(j)} - k_z) z\right] \tag{50}$$

$$I_g(z) = \psi_g(z) \psi_g(z)*$$

$$= \sum_{j \neq \ell} \sum C_o^{(j)} C_g^{(j)} C_o^{(\ell)*} C_g^{(\ell)*} \exp\left[2\pi i \left| k_z^{(j)} - k_z^{(\ell)} \right| z\right] \quad .$$

The difference in the longitudinal components of the Bloch wave vectors is just the separation of the (j) and (ℓ) dispersion surfaces at the excitation points; at the exact Bragg condition ($s_g = 0$), the separation is

$$\left| k_z^{(j)} - k_z^{(\ell)} \right| = \frac{\sigma V_g}{k_z} = \frac{\lambda F(g)}{\pi \Omega} = 1/\xi_g \quad ! \tag{51}$$

The oscillatory effects in the Bragg beam intensities in dynamical conditions are thus seen to be due to beating between Bloch waves of very slightly different wave vector with a depth periodicity ξ_g related to Bloch wave parameters.

Inelastic scattering can be included by making $V(\underline{r})$ complex, incorporating an imaginary component $iV^i(\underline{r})$. The effect on the Bloch waves is to add imaginary components $iq^{(j)}$ to the Bloch wave vectors,

$$\psi^{(j)} = \sum_g C_g \exp\left[2\pi i (\underline{k}^{(j)} + \underline{g} + iq^{(j)}) \cdot \underline{r}\right]$$

$$= \exp(-2\pi q^{(j)} z) \sum_g C_g \exp\left[2\pi i (\underline{k}^{(j)} + \underline{g}) \cdot \underline{r}\right] \tag{52}$$

which leads to selective damping of certain Bloch waves. For example, for the two-beam case $h = 0, g$; $j = 1, 2$

$$q^{(1)} = (\sigma/2k_z)(V_o^i + V_g^i)$$

$$q^{(2)} = (\sigma/2k_z)(V_o^i - V_g^i) \quad , \tag{53}$$

showing that Bloch wave (1) is much more strongly damped than Bloch wave (2). This is why oscillatory contrast from e.g. thickness variations damps out in thick crystals, as one of the important Bloch waves is progressively removed from consideration. Bloch wave (1) is particularly susceptible because it peaks at the atomic positions where inelastic interactions are most probable. Other consequences are that image intensity is considerably enhanced over what it would be in an equivalent thickness of non-crystalline material if the specimen is oriented so as to favor strong excitation of the less heavily damped Bloch waves; this effect, known as *anomalous transmission*, extends the useful thickness for specimens comprising light or medium atomic number atoms to $\sim$0.5 μm for $V_a = 100$ kV.

IMAGE CONTRAST FROM DEFECTS

Probably the single most important contribution TEM has made to solid-state science has been in the characterization defects and extended defect microstructures which are not adequately differentiated by spatial-ly averaging analysis techniques. Defects provide contrast because they alter two important properties of the medium in which they are embedded; they <u>replace</u> the host material, exhibiting different electron scattering properties, altering f^{el}, $V(\underline{r})$, $F(\underline{g})$, ξ_g and often for larger defective regions even the reciprocal lattice itself; they may also <u>displace</u> atoms surrounding the defect to new equilibrium positions characterized by a displacement field $\underline{u}(\underline{r})$ or strain field $\nabla\underline{u}$. Contrast from these replacement or displacement fields occurs by several mechanisms, depending on the image mode selected.

In the high resolution mode, an altered projected charge density (PCD) can be detected. Although it is not likely that *single* point defects can be detected, atomic columns with >10% defect content prob-ably can be distinguished. Fig. 20, for example, shows a structure image of ordered vacancy ($f^{el} = 0!$) superlattices in Fe_9S_{10} (iron-deficient FeS) with about 12% defect content in the atomic columns. Fig. 21 illustrates PCD contrast from columns of a more complex vacancy-interstitial complex in $Fe_{0.89}O$ containing about eight *net* vacancies. The principle of con-verting a phase difference to an intensity change embodied in structure imaging can also be applied to more extensive defect regions, e.g. voids or precipitates. Fig. 22 illustrates phase contrast imaging of small inclusions in kinematical bright-field conditions using large underfocus ($\Delta f = -1\ \mu m$). Inclusions with lower charge density than the matrix (including voids) show up brighter; those with higher density show reversed contrast.

For larger inhomogeneities, amplitude contrast may be used as well. Crystalline precipitates or inclusions with different lattice parameters, orientation or structure from the matrix will generate a different set of diffraction maxima in the back focal plane. This information may be utilized to distinguish the precipitate from the matrix by forming a dark-field image of the precipitate using one (or more) of the precipitate Bragg maxima (Fig. 23). Aperiodic (glassy) inclusions usually exhibit a broad diffraction ring in the back focal plane due to the nearest-neighbor atomic correlations characteristic of any condensed phase, which can be used for identification and sometimes to provide distinguishable dark-field contrast. Distinguishing crystalline inclusions in a glassy matrix (Fig. 24) is an even easier exercise.

Imaging less voluminous one- or two-dimensional defects such as dislocations, faults, interfaces, or very small inclusions, requires a more sophisticated application of many-beam dynamical theory. The Bloch-wave approach is particularly powerful here, because Bloch waves propagate through perfect crystal without scattering, and it is only the defect and its surrounding distortions which scatter Bloch waves. A simple illustration is shown in Fig. 25 for an inclined stacking fault characterized by a rigid body displacement $\underline{u}$ across the fault plane. In two-beam h=0,g dynamical conditions, Bloch waves (1) and (2) will be strongly excited, as in Fig. 19. Upon crossing the fault, however, Bloch wave (1) will suddenly find itself in an environment appropriate to Bloch wave (2) and vice versa; in other words, the fault has scattered one Bloch wave into another. Even in thick crystal, where thickness fringes due to interference of Bloch waves (1) and (2) have been dampened out, the sudden reappearance of Bloch wave (1) at the bottom of the fault reintroduces interference fringes; by reciprocity, a similar argument can be made at the top of the fault, and inclined faults will exhibit fringes with depth periodicity ξ_g^{eff}.

A more rigorous treatment solves the Schrödinger equation as before

84

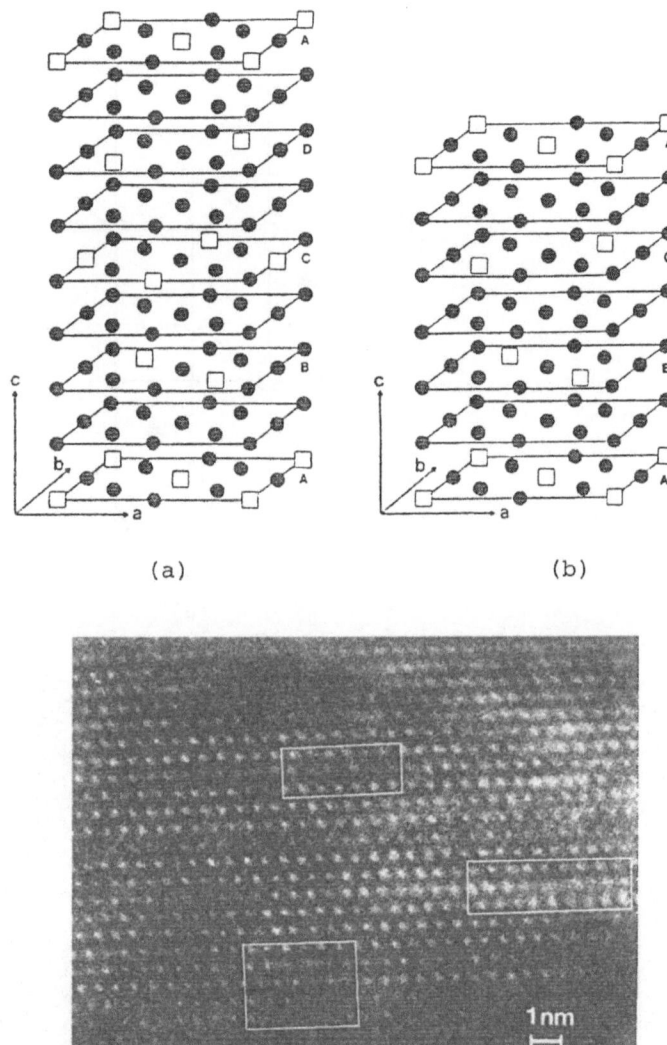

Figure 20. (a) 4C and (b) 3C variants of vacancy ordering in the Fe sub-
lattice of Fe_9S_{10}, and (c) high resolution structure image
showing more intensity from vacancy-containing atomic columns
[7].   Within boxed regions, there is a transition in the local
vacancy superlattice ordering, and columns with transitional
vacancy content are less distinct.   Separation between rows of
vacancies is 0.56 nm.

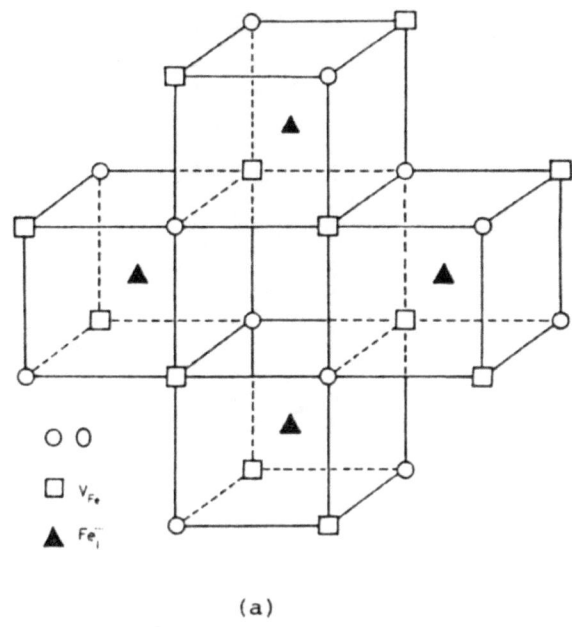

O O O

□ V_{Fe}

▲ $Fe_i^{\cdot\cdot}$

(a)

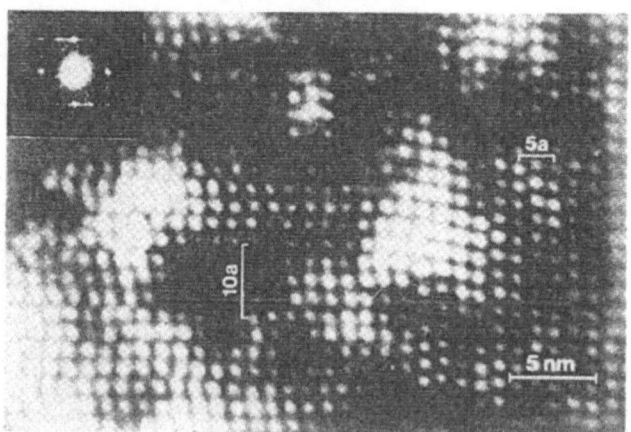

Figure 21. *(a) Probable vacancy-interstitial cluster in $Fe_{1-x}O$ containing eight net vacancies, and (b) structure image of $Fe_{0.89}O$ showing contrast from columns of cluster complexes arranged in a super-lattice [8]. Optical diffractogram from structure image (inset) reveals superperiod maxima around central maximum.*

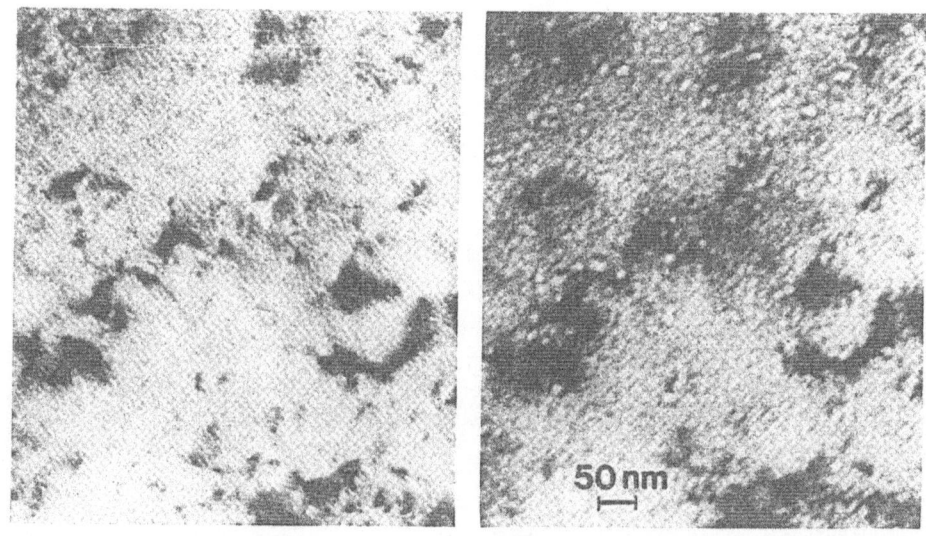

Figure 22. *Colloidal inclusions of Na metal in electron-irradiated NaCl,
imaged (a) at Gaussian focus and (b) kinematical underfocus
conditions. The black line contrast is from segments of dislocation produced by the radiolysis which are pinned by the
Na colloids.*

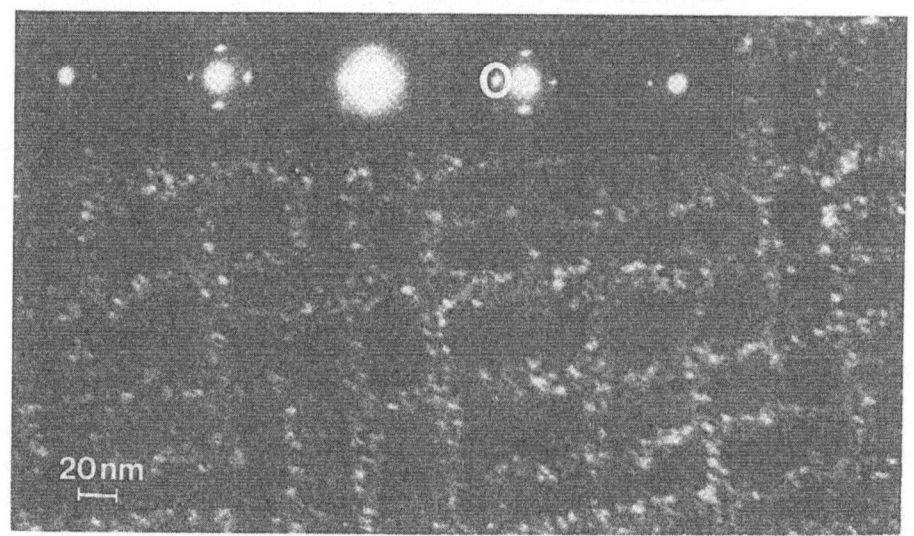

*ordered-defect complexes of Figure 21. A superlattice maximum
(indicated in inset electron diffraction pattern) has been used
to image the domains in dark-field mode. Spacing of rafts
averages 40 nm.*

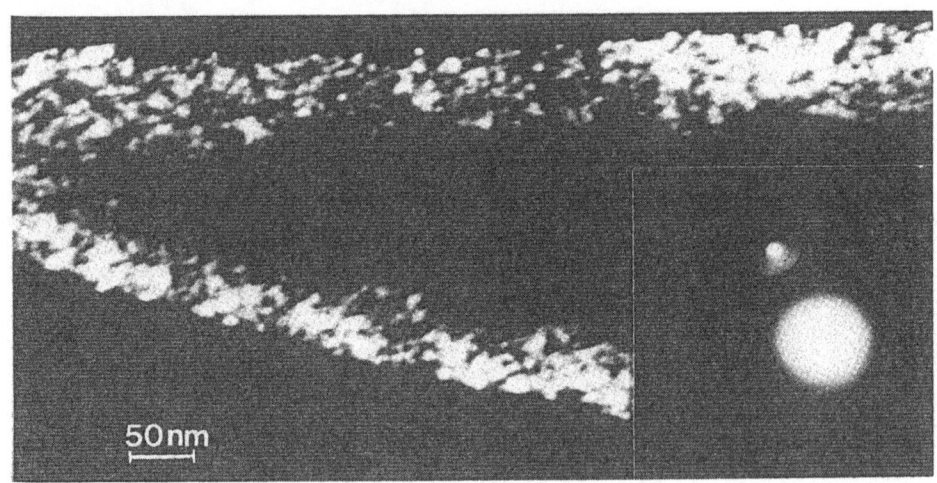

Figure 24. *Remanent crystalline islands in CaPuTi$_2$O$_7$ fracture sherd, most of which has been disordered to a glassy state from recoil of ^{238}Pu nuclei following α decay [10]. The crystallites have been imaged in dark field, using surviving Bragg beams (insert), and retain the approximate orientation of the originally crystalline matrix.*

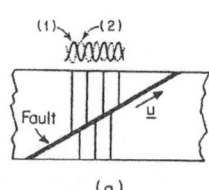

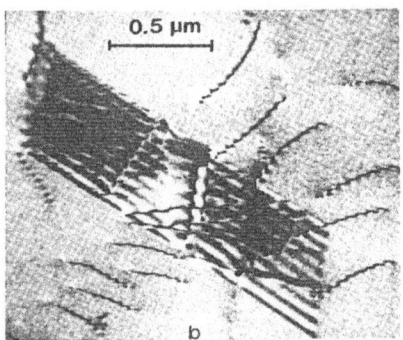

Figure 25. *(a) Bloch wave behavior at an inclined fault, and (b) corresponding stacking fault contrast in dynamical bright-field conditions. The oscillatory contrast along neighboring dislocation lines also arises from Bloch wave interferences with the same depth periodicity.*

but for a potential locally perturbed by the defect influence. For a small change ΔV in $V(\underline{r})$, the change in the Bloch wave excitation for a given Bloch wave j is to first order

$$\Delta\alpha^{(j)} = \pi i \sum_{\ell} \alpha^{(\ell)}(z) \int \psi^{(j)*} \sigma \Delta V \ \psi^{(\ell)} \ dz \quad . \tag{54}$$

For a displacement field $\underline{u}(\underline{r})$, the displaced potential is

$$V^D(\underline{r}) = \sum_g V_g \exp [2\pi i \ \underline{g} \cdot (\underline{r}-\underline{u})]$$

$$= \sum_g V_g \exp (-2\pi i \ \underline{g} \cdot \underline{u}) \exp (2\pi i \ \underline{g} \cdot \underline{r}) \tag{55}$$

so that

$$\Delta V = V^D(\underline{r}) - V(\underline{r}) = [\exp (-2\pi i \ \underline{g} \cdot \underline{u}) - 1] \exp (2\pi i \ \underline{g} \cdot \underline{r}) \tag{56}$$

and (54) becomes

$$\Delta\alpha^{(j)} = \pi i \sum_{\ell} \alpha^{(\ell)}(z) \sum_{g,h} C_h^{(j)*} C_{h-g}^{(\ell)} (\sigma V_g / 2k_z^{(j)})$$

$$\cdot \int_o^t [\exp (-2\pi i \ \underline{g} \cdot \underline{u}) - 1] \exp \{2\pi i \ |k_z^{(\ell)} - k_z^{(j)}| \ z\} dz \quad . \tag{57}$$

It should be noted that the Bloch wave excitations $\alpha^{(\ell)}$ are no longer constant and now depend upon the depth at which the defect influence is felt.

The change $\Delta\alpha^{(j)}$ in the excitation of a single Bloch wave has particular significance where it is possible to arrange that principally one Bloch wave, say (j), is well transmitted--for example, in thick foils under two-beam conditions $\underline{h}=0, \underline{g}$ with $s_g \sim 0$. It can then be shown that

$$I_o \simeq \alpha^{(j)3}(0) [\Delta\alpha^{(j)*} + \Delta\alpha^{(j)}] = 2 \alpha^{(j)3}(0) \ Re[\Delta\alpha^{(j)}] \tag{58}$$

and $Re[\Delta\alpha^{(j)}]$ is a direct measure of contrast. It is also apparent from (57) that *no* contrast is produced when $\underline{g} \cdot \underline{u} = 0$ for all g, a condition that can be achieved under systematic diffraction conditions $h=0, g, 2g....,ng$. This property of the contrast enables the direction of *uniaxial* displacement fields to be identified.

For less sudden displacements, it is often easier to envisage the strain field $d\underline{u}/dz$ than the displacement field $\underline{u}(z)$ because it represents the bending of lattice planes around the imperfection (Fig. 26). The corresponding form of (57) in this case is

$$\Delta\alpha^{(j)} = 2\pi i \sum_{\ell} \alpha^{(\ell)}(0) \sum_g C_g^{(j)*} C_g^{(\ell)} \int_o^t (\underline{g} \cdot d\underline{u}/dz)$$

$$\cdot \exp \{2\pi i \ |k_z^{(\ell)} - k_z^{(j)}| \ z \} dz \tag{59}$$

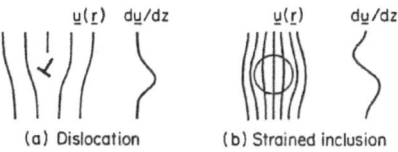

Figure 26. *Displacement fields* u(r) *and corresponding strain fields* du/dz *for two common extended defects (a) dislocation and (b) strained inclusion.*

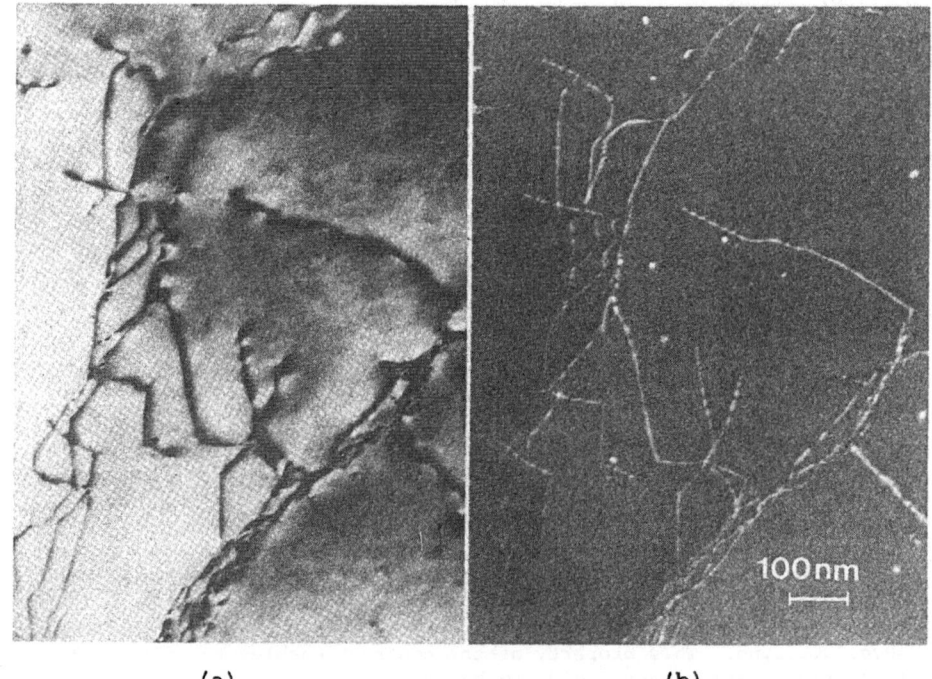

(a)   (b)

Figure 27. *Images of dislocations in stainless steel under (a) two-beam dynamical bright-field conditions and (b) weak-beam dark-field conditions. The decreased image width and high contrast in weak-beam images favor this imaging mode for revealing high-resolution detail of extended defects.*

90

Because $\Delta\alpha^{(j)}$ is seen to depend on the Fourier transform of the quantity $(\underline{g} \cdot d\underline{u}/dz)$ in (59), for transitions to occur $\underline{g} \cdot d\underline{u}/dz$ must have a Fourier component at the relevant scattering vector $|k_z^{(\ell)} - k_z^{(j)}|$. The contribution to $\alpha^{(j)}$ when $\ell = j$ is called *intrabranch* scattering because it represents transitions from one dispersion surface to another point on the same dispersion surface; in this case the Bloch waves retain their identity by following the curvature of the lattice plane bending. Intraband transitions give rise to weak contrast, because the change $|k_z^{(j')} - k_z^{(j)}|$ in the Bloch wave vector is small, and such transitions occur only in the weakest parts of the strain field. When the lattice plane bending becomes too sudden, the Bloch waves cannot follow the lattice planes and *interbranch* ($\ell \neq j$) scattering occurs, as was the case for the sudden fault. Under two-beam dynamical conditions ($s_g \sim 0$), $|k_z^{(\ell)} - k_z^{(j)}| \simeq 1/\xi_g$; thus the part of the strain field sampled must have a wavelength $\sim\xi_g$, and dynamical images from defects will have a similar spatial extent. Dynamical images of dislocations (Fig. 27a) typically have widths $\sim\xi_g/2$.

To detect more rapidly varying strain fields, for example, closer to dislocation cores or the details of very small defect aggregates, larger values of $|k_z^{(\ell)} - k_z^{(j)}|$ must be utilized. This means operating well away from dynamical conditions, in kinematical conditions where the separation of the dispersion surface branches is larger and ψ_g is necessarily weaker. To obtain adequate signal-to-noise ratio under these *weak-beam* conditions, a *dark-field* rather than the bright-field imaging mode is chosen. The change in Bragg-beam amplitude is, using (50),

$$|\Delta\psi_g| = \left| \sum_j \Delta\alpha^{(j)} c_g^{(j)} \exp\left[2\pi i \left(k_z^{(j)} - k_z\right) z\right] \right| . \qquad (60)$$

To keep the image peak as narrow as possible and to maximize contrast, one must ensure that only one Bloch wave transition principally occurs-- from that Bloch wave, say (ℓ), with the largest $\alpha^{(\ell)}(0)$ to the Bloch wave, say j, with the largest $c_g^{(j)}$. It turns out that this transition can be assured by making s_g for all g large and different. Maximum contrast occurs where $\underline{g} \cdot d\underline{u}/dz = -|k_z^{(j)} - k_z|$ at a turning point of $d\underline{u}/dz$ (Fig. 26). Kinematically, it is only at this point where the defect strain is able to rotate the lattice locally into the Bragg condition ($s_g \sim 0$). Considerably narrower images ($\sim\xi_g^{eff}/2$) result (Fig. 27b).

Replacement fields can be treated analogously starting with the resulting perturbation of the potential ΔV in (54), with the result that the quantity $(\underline{g} \cdot d\underline{u}/dz)$ appearing in (59) is replaced (or augmented) by a term comprising the appropriate ξ_g and s_g changes. Contrast from very small (<5 nm) inclusions therefore resembles contrast from small strain centers, such as small dislocation loops, and it is sometimes difficult to distinguish details of defect aggregates on this scale. For larger inclusions image intensity may be below or above background, depending on depth in the foil, foil thickness, s_g, etc. Nearly coherent inclusions or interfaces in which there are small lattice or rotational mismatches can additionally give rise to moiré fringe contrast (Fig. 28).

Since the range of defects and available contrast mechanisms in TEM is large, the reader is encouraged to consult the articles and books cited in the Appendix for further information.

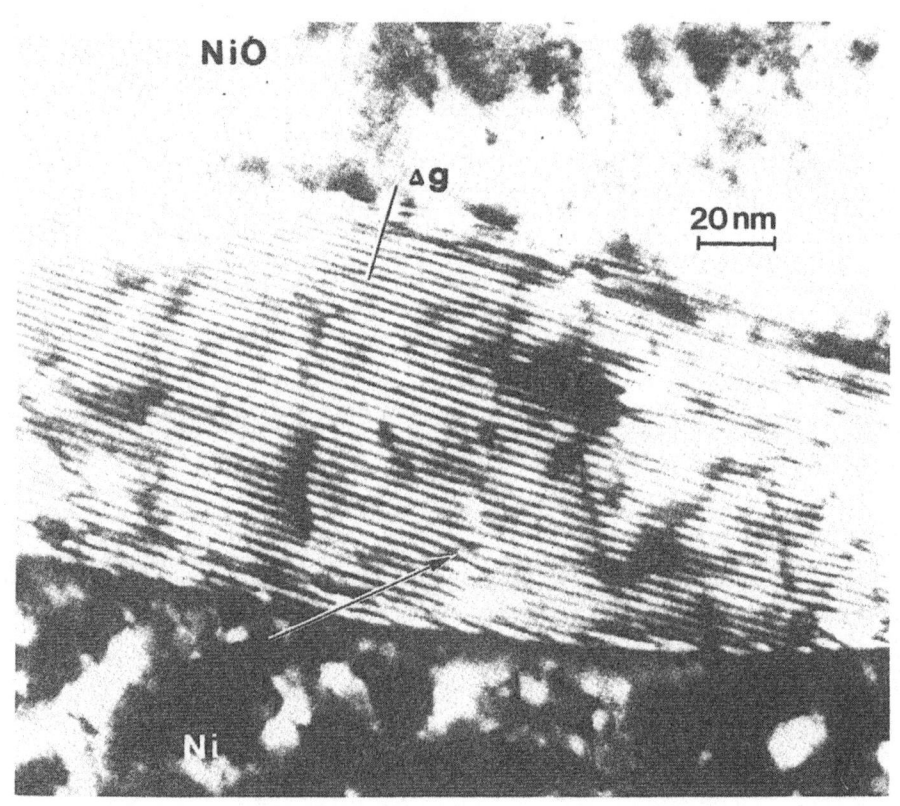

Figure 28. *Moiré fringe contrast appearing at an inclined interface be-tween NiO and Ni substrate, imaged in dynamical bright field conditions. The moiré fringes reveal the presence of an inter-face dislocation (arrowed).*

REFERENCES

1. A. V. Crewe, Chemica Scripta *14*, 17-20 (1978-79).
2. M. R. Pascucci, J. L. Hutchison and L. W. Hobbs, Radiation Effects *74*, 219-26 (1983).
3. P. G. Self, R. W. Glaisher and A. E. C. Spargo, Ultramicroscopy *18*, 49-62 (1985).
4. D. L. Phillips, A. H. Heuer and T. E. Mitchell, Phil. Mag. *A42*, 405-16.
5. D. J. H. Cockayne, J. R. Parsons and C. W. Hoelke, Phil. Mag. *24*, 139-50 (1971); D. J. H. Cockayne and R. Gronski, Phil. Mag. *44*, 159-75 (1981).
6. P. A. Doyle and P. S. Turner, Acta Cryst. *A24*, 390-97 (1968).
7. T. A. Nguyen and L. W. Hobbs, in *Electron Microscopy of Materials,* ed. W. Krakow, D. A. Smith and L. W. Hobbs, Proc. Mater. Res. Soc. *31*, 291-302 (1984).
8. C. Lebreton and L. W. Hobbs, Radiation Effects *74*, 227-36 (1983).
9. L. W. Hobbs, J. de Physique *37* [*C7*], 3-26 (1976).
10. F. W. Clinard, Jr., D. E. Peterson, D. L. Rohr and L. W. Hobbs, J. Nucl. Mater. *126*, 245-54 (1984).

BIBLIOGRAPHY

Recommended articles and texts for further reading are listed below.

P. W. Hawkes, *Electron Optics and Electron Microscopy*, Taylor & Francis, London (1972).

J. M. Cowley, *Diffraction Physics*, North-Holland, Amsterdam (1981).

J. C. H. Spence, *Experimental High Resolution Electron Microscopy*, Clarendon Press, Oxford (1981).

O. Saxton, *Computer Techniques for Image Processing in Electron Microscopy*, Academic Press, New York (1978).

P. B. Hirsch, A. Howie, R. B. Nicholson, D. W. Pashley and M. J. Whelan, *Electron Microscopy of Thin Crystals*, Butterworths, London (1967); 2nd edition, Robert E. Krieger Publishing Company, Huntington, NY (1977).

U. Valdrè and A. Zichichi, eds., *Electron Microscopy in Material Science*, Academic Press, New York (1971, 1975).

M. H. Loretto and R. E. Smallman, *Defect Analysis in Electron Microscopy*, Chapman & Hall: Halsted/Wiley, New York (1975).

L. W. Hobbs, Transmission Electron Microscopy of Defects Aggregates in Non-Metallic Crystalline Solids, in *Defects and their Structure in Non-Metallic Solids*, B. Henderson and A. E. Hughes, eds., Plenum New York, pp. 431-82 (1976).

S. Amelinckx, R. Gevers and J. Van Landuyt, eds., *Diffraction and Imaging Techniques in Material Science*, North-Holland, Amsterdam (1978).

G. Thomas and M. J. Goringe, *Transmission Electron Microscopy of Materials*, Wiley, New York (1979).

C. J. Humphreys, Scattering of Fast Electrons by Crystals, *Rep. Progr. Phys.* *42*, 1825-87 (1979).

J. J. Hren, J. I. Goldstein and D. C. Joy, eds., *Introduction to Analytical Electron Microscopy*, Plenum, New York (1979).

L. Reimer, *Transmission Electron Microscopy*, Springer-Verlag, New York (1984).

J. N. Chapman and A, J. Craven, eds., *Quantitative Electron Microscopy*, Scottish Universities Summer Schools in Physics, Edinburgh (1984).

NEUTRON SCATTERING FROM DEFECTS IN MATERIALS

Roger J. Stewart

J.J.Thomson Physical Laboratory

University of Reading

Reading, England

Introduction

The importance of neutron scattering in the study of defects in materials is that provides structural information about inhomogeneities ranging in scale from a few Angstoms up to defect regions with dimensions of the order of $5000\overset{\circ}{A}$ in bulk specimens. Such a wide range encompasses point defects (vacancies, interstitials and substitutional impurities) and their associated strain fields up to clusters of point defects (e.g. voids) or precipitates of second phase elements in an alloy or indeed any positional or compositional disorder. It is a complimentary technique to electron microscopy which provides very detailed information on the defect structure in real space over tiny localised volumes in the specimen. Neutron scattering provides a diffraction pattern of defect structure averaged over a volume of up to a few cubic centimeters. Of course a diffraction pattern cannot be unequivocally translated into real space, but this is balanced by the ability to investigate the defect structure throughout a large volume of a sample in a relatively short time (a few minutes on a high flux neutron source).

Neutron properties

Neutrons are uncharged particles with mass, $m = 1.00894$ amu (1 amu = 1.66×10^{-24} g), spin, $s=1/2$, and a magnetic moment of -1.913 nucl. mag.. The neutron spontaneosly decays ($n \rightarrow p + e^- + \nu$), but the half-life (11.7 mins) is large compared to the typical time during which a neutron interacts with matter and can thus be considered as stable for our purposes. In order to completely describe the state of a neutron its momentum ($\hbar k$) and its spin component with respect to an axis ($S_z=\pm\frac{1}{2}$) must be specified. However for unpolarised neutrons we usually describe them in terms of either their energy (E) or de Broglie wavelength (λ):

$$E = h^2|\underline{k}|^2/8\pi^2m, \quad \lambda = 2\pi/|\underline{k}| = h/mv$$

v is the neutron velocity. Fortunately a neutron which has an energy corresponding to room temperature (referred to as a thermal neutron) has a wavelength corresponding to the normal interatomic spacings in materials and is thus ideally suited to study both the static and dynamic structure of condensed matter. Neutrons with energies from 10^{-4} eV to 10 eV (30Å to 0.1Å in wavelength) are used in the study of condensed matter. In the study of defects only a small part of this wide range is employed, typically 4 to 12Å.

Interaction of neutrons with matter

Neutrons interact with matter in a variety of ways and because they are uncharged they can easily penetrate deep into a sample. The principal interaction with atoms is via the short range, strong nuclear interactions which are independent of the atomic number of the nucleus and vary in an apparently random way from nucleus to nucleus. The short range nature of the interaction (about 10^{-15}m) compared to the wavelength of the neutrons used in the study of condensed matter (about 10^{-7}m) means that the scattering from a single nucleus is spherically symmetric. The amplitude of the wave scattered by a single nucleus is denoted by b and is called the scattering length. The energy of the neutrons scattered by a crystal can either be the same as that of the incident neutron (elastic scattering) or can have gained or lost energy (inelastic scattering).

Atoms with unpaired electrons possess a magnetic moment which can interact with the magnetic moment of the neutron to produce scattering. This interaction is used extensively in the study of the magnetic structure of solids and can effect the scattering associated with inhomogeneities.

Instead of being scattered a neutron may be absorbed by a nucleus, which may then break up in one of several different ways, by the emmision of gammarays, or charged particles or indeed the emmision of 2 or 3 neutrons. As a result of these processes the nucleus is altered or in the case of fissile materials can even be split into two smaller nuclei.

The probability of elastic scattering or inelastic scattering or ab-

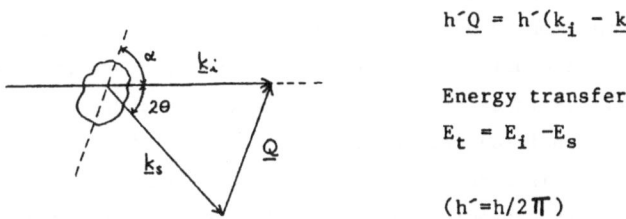

Momentum transfer

$h\llap{ˊ}\underline{Q} = h\llap{ˊ}(\underline{k}_i - \underline{k}_s)$

Energy transfer

$E_t = E_i - E_s$

$(h\llap{ˊ} = h/2\pi)$

Fig. 1. The scattering geometry. The wave vector $\underline{k}_i$ is the incident neutron direction and $\underline{k}_s$ is the scattered neutron direction.

sorption occuring in a given interaction will depend on the particular atomic nucleus and the energy of the incident neutron. The specific probability for a particular process to occur is quantitatively represented by a cross-section (σ). This can be envisaged as the effective area that a nucleus presents to the incident neutron and is measured in barns (1 barn = 10^{-24} cm^2). The absorption cross-section increases directly with increasing wavelength, but even so long wavelength neutrons suffer little attenuation in comparison to x-rays and sample thicknesses of the order of 1 cm are common. For scattering events the cross-section is often expressed as the probability of a neutron being scattered into unit solid angle and is referred to as the differential scattering cross-section ($d\sigma/d\Omega$), which has units of barns/steradian.

In an ideal scattering experiment the momentum of the incident and scattered neutrons would be determined. In order to investigate the properties of a sample we need to determine the energy and momentum transfer and the orientation of the sample, α , (figure 1). We will restrict our discussion to to elastic scattering, which gives a time averaged picture of the positions of the atoms in the sample under study. In fact in many experiments no energy discrimination is carried out on the scattered beam and thus the inelastic contribution is measured along with the elastic scattering to give the total differential scattering. For elastic scattering $E_t = 0$ and $|\underline{k}_i| = |\underline{k}_s| =$ $2\pi/\lambda$. The magnitude of the scattering vector $\underline{Q}$ is then given by $4\pi\sin\theta/\lambda$, written simply as Q in the rest of this chapter.

If we consider a sample with N_t atoms, we must take into account the fact that the scattering length, b, varies irregularly from isotope to isotope of the same atom, and, if the nuclear state has non-zero spin, b depends on the relative orientation of the neutron spin and the nuclear spin. Since the different isotopes of an element are randomly distributed in a sample their atomic positions are uncorrelated as are the orientations of their nuclear spins this leads to isotope and spin incoherence. The effects of this can be seen by expanding the expression for the differential cross-section (Lovesey 1984 equation 1.38)

$$\frac{d\overline{\sigma}}{d\Omega} = \sum_{nn'} \exp(i\underline{Q}.(\underline{R}_n - \underline{R}_{n'}))\langle b_n^* \cdot b_n \rangle \quad \ldots \ldots (1)$$

b_n is the scattering length of the isotope at the site represented by the position vector $\underline{R}_n$ and will in general depend on what nuclear spin is associated with the isotope. The quantity $\langle b_n^* \cdot b_n \rangle$ is the value of $b_n^* \cdot b_n$ averaged over random nuclear spin orientations and random isotope distributions. There is no correlation between the values of b_n and $b_{n'}$ if n and n' refer to different sites, thus

$$\langle b_n^* \cdot b_n \rangle = \langle b_{n'}^* \rangle \langle b_n \rangle = |\langle b \rangle|^2 \text{ if } n \neq n'$$

however if n = n', $\langle b_n^* \cdot b_n \rangle = \langle |b_n|^2 \rangle = \langle |b|^2 \rangle$

so that in general $\langle b_n^* \cdot b_n \rangle = |\langle b \rangle|^2 + \delta(n-n')(\langle |b|^2 \rangle - |\langle b \rangle|^2)$

δ is the Dirac delta function. Substituting this equation into (1) we obtain:

$$\frac{d\overline{\sigma}}{d\Omega} = N_t(\langle |b|^2 \rangle - |\langle b \rangle|^2) + |\langle b \rangle|^2 \left| \sum_n \exp(i\underline{Q}.\underline{R}_n) \right|^2$$

The first term is the incoherent cross-section and the second term the coherent cross-section clearly they are markedly different. The incoherent cross-section is independent of Q and the atomic positions and is completely isotropic (that is the incoherent scattering from a sample is the same in all directions). This is not completely true since thermal motions cause a decrease in the incoherent cross-section at high Q through the Debye-Waller factor (the same fall off also occurs for the coherent cross-section). It can be written in a more revealing form as $N_t \langle |b - \langle b \rangle|^2 \rangle$; thus it depends on the mean deviation of the scattering lengths from their average value. In contrast, the coherent cross-section depends on Q and the atomic positions and for certain orientations and scattering directions strong constructive

98

interference between the waves scattered from different nuclei will occur. The magnitude of the coherent cross-section depends on the average value of the scattering length squared.

Neglecting spin effects and using c_j for the fractional concentration of isotope j we can write:

$$\langle b \rangle = \sum_j c_j b_j \quad \text{and} \quad \langle |b|^2 \rangle = \sum_j c_j |b_j|^2.$$

Lovesey (1984 page 15) has shown that if spin effects are included then

$$\langle b \rangle = \sum_j [\ c_j((i_j+1)b_j^{(+)} + i_j b_j^{(-)})/(2i_j+1)\ ]$$

$$\langle |b|^2 \rangle = \sum_j [\ c_j((i_j+1)|b_j^{(+)}|^2 + i_j |b_j^{(-)}|^2)/(2i_j+1)\ ]$$

where i_j is the nuclear spin of isotope j. The scattering lengths $b^{(+)}$ and $b^{(-)}$ are associated with the two possible final states found ($i+\frac{1}{2}$ or $i-\frac{1}{2}$) when a neutron, with spin $\frac{1}{2}$ interacts with a nucleus of spin i.

Values for the cross-sections of elements are usually quoted after integration of the differential cross-section over 4π steradians. The total scattering cross-section, ($\sigma = 4\pi \langle |b|^2 \rangle$), the coherent part of the cross-section ($\sigma_c = 4\pi |\langle b \rangle|^2$ and the absorption cross-section (σ_a) have been compiled by Koester and Yelon (1982).

Neutron scattering from defects

The angular distribution of the neutrons scattered by a perfect crystal is determined by the interference of the neutron waves coherently scattered by the regular array of nuclei forming the crystal lattice. Any constructive interference gives rise to the well known Bragg peaks, the angular positions of which are related to the positions of the atoms by the Bragg relation $\lambda = 2d\sin\theta$. If one atom is removed from the lattice (i.e. a vacancy is created) the scattering will be changed since a proportion of the waves scattered from the remaining nuclei will not suffer destructive or constructive interference with the waves which would have been scattered by the missing atom. In practice the Bragg scattering from the crystal lattice makes this defect scattering very difficult to observe. However if the neutron wavelength is greater than twice the maximum interplanar spacing in the crystal ($2d_{max}$) there is no Bragg scattering and in principle the defect scattering can be observed. For most materials $2d_{max}$ lies in the range 4 to 8Å. The ease with which the defect scattering can be observed of course will depend on the concentration of the defects and how small the incoherent and absorption cross-sections are in the Q region of interest.

The differential defect scattering cross-section resulting from a vacancy is exactly the same as that for an isolated atom at the same position and has a magnitude of b^2. The coherent scattering amplitude for a vacancy in a lattice of A atoms is in fact $-b_A$ and that for an self interstitial $+b_A$ but the measured cross-section is proportional to the square of the coherent scattering amplitude and so the neutron scattering from vacancies or self interstitials is indistinguishable. This is in fact Babinet's principle in optics, which states that complementary objects produce the same diffraction effects (e.g. a screen pierced with holes or an array of discs of the same size as the holes and in the same geometrical arrangement). For M randomly positioned interstitials or vacancies the magnitude of the differential defect scattering cross-section is independent of Q with a value of Mb^2. Thus the defect concentration can be obtained directly from the cross-section. For substitutional impurities to be seen a difference in scattering length must exist between the substitutional atom and the atom it replaces.

Clusters of point defects (e.g voids) or agglomerations of substitutional impurities (e.g. precipitates) can be considered as particles made up of regularly spaced isotropic scattering centres for neutrons. Consider one particle of this type bathed in a neutron beam, all the scattering centres are then sources of scattered waves. When the scattering direction is the same as that of the incident beam, these scattered waves are all in phase (i.e. independent of the mutual disposition of the scattering centres). However, as the scattering angle increases, the difference in phase between the various scattered waves also increases. The amplitude of the resultant wave thus decreases with increasing angle because of the increasing destructive interference, it becomes zero when there are as many waves with phases between 0 and π as there are between π and 2π. This will occur when the scattering angle 2θ is of the order of λ/L, L being the average dimension of the particle. For n_p such particles, neglecting interparticle interference effects for the moment, the scattering would be n_p times that for a single particle. Thus the scattering from clusters of point defects or agglomerates of atoms in an otherwise perfect lattice varies with Q (or 2θ). From the experimentally determined Q dependence of the scattering, under ideal conditions, it is possible to determine the size, shape and composition of the inhomogeneity present.

100

Studies of the neutron scattering associated with defects began in the late 1950´s and continued through the 1960´s, but all these investigations were hampered by lack of intensity. The field has, however, developed substantially since the early 1970´s because of the availability of intense long wavelength neutron beams (for instance at the Institut Max von Laue-Paul Langevin in Grenoble) and/or the development of efficient large area position sensitive neutron detectors.

It is convenient to consider three regimes of defect scattering, although the division is to some extent arbitrary. Firstly, small angle scattering (scattering in the vicinity of the reciprical lattice point (000) produced by relatively large fluctuations in the scattering length density). Secondly, diffuse scattering (between Bragg peaks and between the lowest angle Bragg peak and the straight through beam, produced by strain around defects especially from the relatively large displacements in the first few shells of atoms around the defect). Finally, Huang scattering (close to Bragg peaks and difficult to separate quantitatively, it is produced by the long range part of the strain field around defects). The requirement to work near Bragg peaks in Huang scattering means that there is no general advantage in using neutrons and little work has been done. Thus Huang scattering will not be discussed in this chapter.

Small angle scattering

The term small angle neutron scattering (SANS) is used in materials science to cover the Q range $0 < Q < \pi/d$ where d is the interatomic distance in the material under investigation. In other words from the reciprocal lattice point (000) out to the first set of reciprocal lattice points. Now d is typically of the order of $3\overset{\circ}{A}$, therefore the upper limit for Q is $1\overset{\circ}{A}^{-1}$. This means that for $12\overset{\circ}{A}$ neutrons 2θ is 180° which is by no means a small angle. The terminology is a hang over from x-ray studies, where the wavelengths used are very much smaller (about $1\overset{\circ}{A}$), but the term SANS is extensively used.

Theoretical background to SANS

In this section the differential scattering cross-section arising from inhomogeneities distributed in an otherwise perfect lattice is derived and the various approximations which are employed in the analysis of SANS data are developed. SANS arises from fluctuations in the scattering length density $C(\underline{r})$ associated with a number density of nuclei $p(r)$. The amplitude of the wave scattered by an inhomogeneity can be written:

$$f(\underline{Q}) = \sum_j b_j \exp(-i\underline{Q}.\underline{R}_j) = \sum_j b_j \int \exp(-i\underline{Q}.\underline{r}) \, \delta(\underline{r} - \underline{R}_j) \, dr \quad \ldots\ldots 3$$

where $\underline{R}_j$ is the position vector of the jth nucleus in the inhomogeneity and b_j is the coherent scattering length of the jth nucleus. Defining the scattering length density $C(\underline{r}) = \sum_j b_j \, \delta(\underline{r} - \underline{R}_j)$, we can write:

$$f(\underline{Q}) = \int C(\underline{r}) \exp(-i\underline{Q}.\underline{r}) \, d\underline{r} = F\{C(\underline{r})\} \quad \ldots\ldots 4$$

$F\{C(\underline{r})\}$ is the Fourier transform of $C(\underline{r})$. The coherent differential scattering cross-section for one inhomogeneity is thus given by

$$\frac{d\sigma}{d\Omega} = f(\underline{Q})f^*(\underline{Q}) = |F\{C(\underline{r})\}|^2 \quad \ldots\ldots 5$$

To mathematically describe an inhomogeneity of mean scattering length density $\langle c_p \rangle$ in an otherwise homogeneous matrix of mean scattering length density $\langle c_m \rangle$ we introduce a shape function $g(\underline{r})$. $g(\underline{r})$ has a value of one inside the inhomogeneity and is zero outside the boundaries of the inhomogeneity. For a crystal containing n_p such inhomogeneities the scattering length density can be written as

$$C(\underline{r}) = \sum_{j=1}^{n_p} [\ \langle c_p \rangle g(\underline{r}-\underline{r}_j) + \langle c_m \rangle(1-g(\underline{r}-\underline{r}_j))\ ] \quad \ldots\ldots 6$$

where $\underline{r}_j$ is the position vector of the jth inhomogeneity. Substituting (6) into (5) the differential scattering cross-section is given by

$$\frac{d\sigma}{d\Omega} = |\sum_{j=1}^{n_p} [\ (\langle c_p \rangle - \langle c_m \rangle) \int g(\underline{r}-\underline{r}_j)\exp(-i\underline{Q}.\underline{r})dr + \langle c_m \rangle \int \exp(-i\underline{Q}.\underline{r})dr\ ]|^2 \quad \ldots 7$$

The second term describes the scattering associated with the Bragg peaks and can be omitted for wavelengths greater than $2d_{max}$, since the forward scattering is then the only contribution remaining from the whole lattice and it is experimentally inseparable from the transmitted neutron beam.

Now $F\{g(\underline{r}-\underline{r}_j)\} = F\{g(\underline{r})\}\exp(i\underline{Q}.\underline{r}_j)$ and thus equation (7) reduces to:

$$\frac{d\sigma}{d\Omega} = (\langle c_p \rangle - \langle c_m \rangle)^2 \ |\sum_{j=1}^{n_p} F\{g(\underline{r})\}\exp(i\underline{Q}.\underline{r}_j)|^2$$

$$= (\langle c_p \rangle - \langle c_m \rangle)^2 \ |F\{g(\underline{r})\}|^2 \sum_{i,j=1}^{n_p} \exp(i\underline{Q}.\underline{r}_{ij}) \quad \ldots\ldots 8$$

This equation shows that the differential scattering cross-section from inhomogeneities arises from three terms. The first the difference in scattering length density between the inhomogeneities and the crystal matrix ($\langle c_p \rangle - \langle c_m \rangle$) determines the magnitude of the scattering. The second term the square of the Fourier transform of the shape function determines the Q dependence of the cross-section. The final term describes the contribution to the cross-section arising from interference of the waves scattered by different inhomogeneities. For widely spaced (i.e. low concentrations) and randomly positioned inhomogeneities this so called interference function is equal to n_p. For convenience the interference function is defined as

$$I(Q) = (1/n_p) \sum_{i,j=i}^{n_p} \exp(i\underline{Q}.\underline{r}_{ij}) \quad \ldots \ldots 9$$

so that it is unity for low concentrations of inhomogeneities.

Let us consider the three terms in the differential cross-section (equ. 8) one at a time. The value of $\langle c_p \rangle$ is obtained from

$$\langle c_p \rangle = \sum_{j=1}^{N_p} b_j/V_p \quad \ldots \ldots 10$$

where V_p is the volume of the inhomogeneity and N_p is the number of atoms in the inhomogeneity. $\langle c_m \rangle$ is defined in a similar way. Consider an agglomerate of identical substitutional impurity atoms (B atoms) in a matrix of A atoms then

$$\langle c_p \rangle = (N_p/V_p)b_B = (1/v_p)b_B \quad \text{and} \quad \langle c_m \rangle = (1/v_m)b_A$$

where v_p is the volume occupied by one atom in the agglomerate and v_m the corresponding value in the matrix. For substitutional impurities $v_p = v_m$, thus

$$\langle c_p \rangle - \langle c_m \rangle = (1/v_m)(b_B - b_A) \quad \ldots \ldots 11$$

Hence the magnitude of the scattering, in this case, is directly proportional to the difference in the coherent scattering lengths of the atoms A and B.

The second term in equation (8) can be investigated using the theorem that the square of the Fourier transform of a function is equal to the Fourier transform of the auto-correlation function of that function.

$$|F\{g(\underline{r})\}|^2 = F\{ \int g(\underline{r}).g(\underline{r} - \underline{r}') \, d^3\underline{r} \} = F\{G(\underline{r})\} \quad \ldots \ldots 12$$

where $G(\underline{r})$ is the auto-correlation function of $g(\underline{r})$. For a spherical particle of radius R, $g(\underline{r}) = 1$ for $0 < |\underline{r}| \leqslant R$ and $g(\underline{r}) = 0$ for $|\underline{r}| > R$, it can be shown that:

$$G(\underline{r}) = 4\pi R^3.(1-0.75(r/R)+0.0875(r/R)^3)/3 \quad \text{for } |\underline{r}| \leqslant 2R$$

$$= 0 \quad \text{for } |\underline{r}| > 2R.$$

Fourier transforming we obtain

$$|F\{g(\underline{r})\}|^2 = [\ 4\pi R^3 \ (\sin QR - QR\cos QR)/(QR)^3\ ]^2 \quad ..... \ 13$$

It is illuminating to consider an ellipsoidal inhomogeneity. For one such inhomogeneity the Fourier transform of the shape function would produce on the detector plane a scattering pattern in the form of an ellipse, with its major axis at right angles to the major axis of the ellipsoid (figure 2). This is a well known feature of diffraction patterns, namely, that large distances in real space correspond to small distances in reciprocal space and small distances in real space to large distances in reciprocal space. Thus the scattering from large inhomogeneities will be concentrated at low Q, whereas small inhomogeneities produce Q dependent scattering at higher Q.

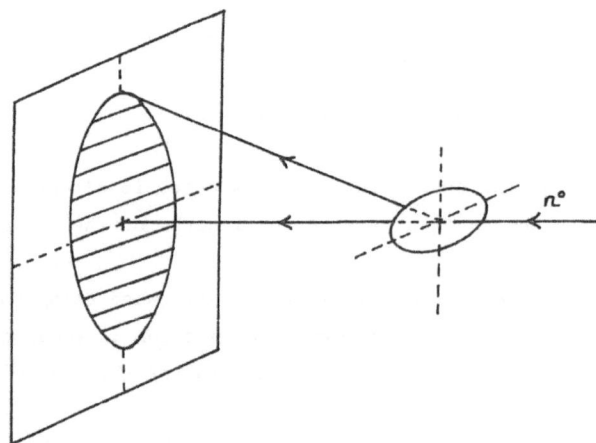

Fig 2. The form of the scattering from an oriented ellipsoid.

104

The functional form of the interference term (third term in equ.8) depends on the exact way the inhomogeneities are spatially distributed. Various models have been used to describe the spatial arrangement and the interference function can then be calculated. For n_p randomly distributed inhomogeneities Messoloras (1974) has defined a probability function $W(r)$ of finding an inhomogeneity at a distance r from another inhomogeneity, such that $W(r)=0$ for $r<D_p$ and $W(r)=1$ for $r \geqslant D_p$. D_p is the mean distance between inhomogeneities. The number of inhomogeneities in a crystal of volume V is given by:

$$n_p = 3V/(4\pi D_p^3) \quad \ldots \ldots \quad 14$$

Following James (1962) the interference function can be written as

$$I(Q) = 1 - (3/D_p^3) \int_0^\infty (1-W(r))r^2 (\sin Qr/Qr) dr$$

$$= 1 - 3(\sin QD_p - QD_p \cos QD_p)/(QD_p)^3 = 1 - Z(QD_p) \quad \ldots \ldots \quad 15$$

The function $Z(QD_p)$ has zeros for $QD_p = 4.492, 7.725, \ldots$, maxima at $QD_p = 0, 9.095, \ldots$, and minima at $QD_p = 5.763, 12.323, \ldots$.

The Q dependence of the differential cross-section in equation (8) is determined by the last two terms. $|F\{g(\underline{r})\}|^2$ decreases with increasing Q so that the cross-section becomes a maximum at the first maximum of the interference function, this is when $QD_p = 5.763$. Thus we can calculate a value for D_p from the position of this maxima. For values of $QD_p > 5.763$ the interference function $(1-Z(QD_p))$ has a value very close to unity and thus the Q dependence of the cross-section in this Q region is adequately described by $|F\{g(\underline{r})\}|^2$ alone providing the inhomogeneity concentration is not too high (<5%). This is an important result when it comes to trying to fit the experimental data using approximations for the differential scattering cross-section.

Other spatial models to describe the distribution of the inhomogeneities have been developed. Synecek(1962) has put forward a model in which the spatial probability distribution for the inhomogeneities has a Guassian form. In this case the peak in the cross-section occurs at a value of $QD_p=7.695$. Dusic et al (1985) used a spatial distribution function to describe inhomogeneities (voids) randomly arranged over the grain boundaries of an alloy and found that $QD_p = 6.539$. If the inhomogeneities were arranged on a super-lattice QD_p would have a value of 6.283 (i.e 2π).

Guinier approximation

The Guinier approximation (Guinier (1937)) is the most important of the various approximations which are used in the interpretation of SANS data. Thus all the assumptions which are implicit in the approximation will be outlined in this section. Returning to equation (8) we have

$$F\{g(\underline{r})\} = \int g(\underline{r})\exp(i\underline{Q}.\underline{r}) \; d^3\underline{r}$$

Now since $g(\underline{r})= 0$ outside the inhomogeneity we can write

$$F\{g(\underline{r})\} = \int_{V_p} \exp(i\underline{Q}.\underline{r}) \; d^3\underline{r} \quad .....16$$

where the integration is now made over the volume of the inhomogeneity. In order to evaluate this integral for an inhomogeneity of any shape we set up a co-ordinate system as shown in figure 3, with the incident beam direction along the z direction and the origin of the co-ordinates at the centre of gravity of the inhomogeneity. For small values of Q ($<0.2\overset{\circ}{A}^{-1}$) the scattering angles are small even for long wavelength neutrons, thus to a good approximation $\underline{Q}$ will be perpendicular to the z direction and we can choose the x-axis to be along $\underline{Q}$ (see figure 3). Therefore $\underline{Q}.\underline{r} = Qx$ and we may write

$$\int_{V_p} \exp(i\underline{Q}.\underline{r})d^3\underline{r} = \iiint_{V_p} \exp(iQx)dxdydz$$

$$= \int \exp(iQx)dx \iint dydz \;\; = \;\; \int \exp(iQx)A(x)dx$$

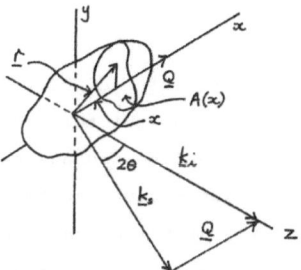

Fig. 3. Scattering geometry for the Guinier approximation

where A(x) is the cross-sectional area of the inhomogeneity perpendicular to the x-direction. For values of Qx less than the order of 1 we can expand the exponential, neglecting terms of power greater than 2 to obtain

$$\int A(x)dx + iQ \int QxA(x)dx - (Q^2/2) \int x^2A(x)dx + + +$$

The first term is just the volume of the inhomogeneity V_p and since the origin has been chosen to be the centre of gravity of the inhomogeneity the second term is zero. Hence we are left with

$$V_p - (Q^2/2) \int x^2A(x)dx \quad \ldots\ldots \ 17$$

In classical mechanics the integral $\int x^2A(x)dx$ is defined as the moment of inertia of the volume with respect to the y-z plane. Classical mechanics also defines the average inertial distance along x, R_x, such that the product $R_x^2 \cdot V_p$ gives the moment of inertia:

$$R_x^2 \cdot V_p = \int x^2A(x)dx \quad \ldots\ldots \ 18$$

Using this definition we can write equation (16) as

$$F\{g(\underline{r})\} = V_p(1-Q^2R_x^2/2) \quad \ldots\ldots \ 19$$

Now since it is implicit in our previous assumptions that QR_x is small we can write this as:

$$F\{g(\underline{r})\} = V_p \exp(-Q^2R_x^2/2) \quad \ldots\ldots \ 20$$

Thus the differential scattering cross-section for an inhomogeneity is

$$\frac{d\sigma}{d\Omega} = (\langle c_p \rangle - \langle c_m \rangle)^2 \cdot V_p^2 \exp(-Q^2R_x^2) \quad \ldots\ldots \ 21$$

This equation is valid for inhomogeneities of any shape, providing they have a common orientation. For inhomogeneities with random orientation a mean value for the cross-section will be measured at a particular scattering angle which, assuming I(Q) = 1 for the moment, is given by

$$\frac{d\sigma}{d\Omega} = n_p (\langle c_p \rangle - \langle c_m \rangle)^2 \cdot V_p^2 \exp(-Q^2\langle R_x^2 \rangle) \quad \ldots\ldots \ 22$$

107

$\langle R_x^2 \rangle$ is the average value of R_x^2, taking into account all possible orientations of the inhomogeneity with respect to the incident beam. We want to calculate a value for $\langle R_x^2 \rangle$, again we draw on classical mechanics and introduce the concept of the radius of gyration of a particle, R_g, with respect to its centre of gravity which is defined in an analogous way to equation (18):

$$V_p R_g^2 = \int r^2 dV = \iiint (x^2+y^2+z^2)dxdydz = V_p(R_x^2+R_y^2+R_z^2) \quad \ldots\ldots 23$$

R_x, R_y and R_z are the inertial distances with respect to the three co-ordinate planes. Thus $\langle R_g^2 \rangle = \langle R_x^2 \rangle + \langle R_y^2 \rangle + \langle R_z^2 \rangle$. But since $\langle R_x^2 \rangle$, $\langle R_y^2 \rangle$ and $\langle R_z^2 \rangle$ must be the same after any rotation we have that $\langle R_x^2 \rangle = \langle R_g^2 \rangle/3$, which we will write as $R_g^2/3$. Thus finally we have the so-called Guinier approximation for the cross-section

$$\frac{d\sigma}{d\Omega} = n_p \ (\langle c_p \rangle - \langle c_m \rangle)^2 \ V_p^2 \ . \ \exp((-Q^2 R_g^2)/3) = C_G \ \exp((-Q^2 R_g^2)/3) \quad \ldots 24$$

C_G is refered to as the Guinier constant. This approximation is valid for values of QR_g up to values of 1.2 - 2.0, depending on the shape of the inhomogeneities. A plot of $\log(d\sigma/d\Omega)$ against Q^2, the so-called Guinier plot (figure 4), has a slope of $(-R_g^2)/3$ and an intercept at $Q^2=0$ of $\log(C_G)$. Thus from the slope of such a plot we can determine the size of the inhomo-geneity if we know its shape. For a sphere of radius R_s, $V_p=4\pi R_s^3/3$ and $dV=4\pi r^2 dr$. Substituting these into equation (23) we have:

$$R_g^2 = (3/R_s^3) \int_0^{R_s} r^4 \ dr = (3/5)R_s^2$$

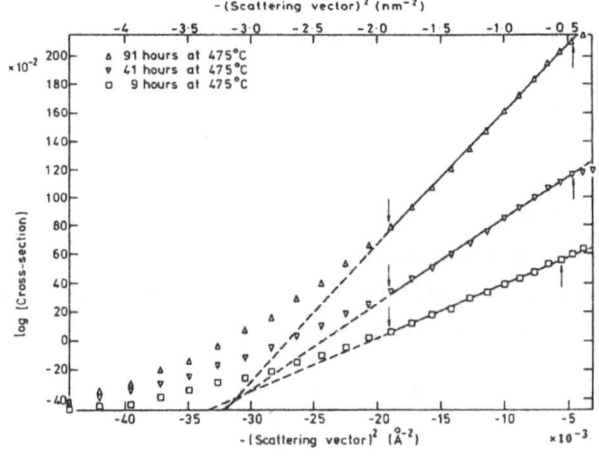

Fig. 4. Typical Guinier plots (arrows indicate Guinier range)

Guinier Constant

The Guinier constant contains information about the number of inhomogeneities, n_p, their volume, V_p, and their scattering length density difference with respect to the surrounding matrix, $<c_p>-<c_m>$. Consider an alloy made up of two elements A and B (concentration of A atoms > B atoms). Further assume there is an inhomogeneity (e.g. a precipitate) where the concentration of B atoms is m_p thus we have from (10)

$$c_p = (1/N_p v_p) \sum_{j=1}^{N_p} b_j = (1/N_p v_p)(m_p N_p b_B + (1-m_p)N_p b_A) \quad \ldots\ldots 25$$

If m_m represents the concentration of B atoms in the matrix outside the inhomogeneity then

$$c_m = (1/v_m)(m_m b_B + (1-m_m)b_A) \quad \ldots\ldots 26$$

If we assume for simplicity that the atomic volume is the same inside and outside the inhomogeneity (i.e. $v_p = v_m$) then

$$c_p - c_m = (1/v_m)(m_p - m_m)(b_A - b_B) \quad \ldots\ldots 27$$

Thus we can write the Guinier constant as follows

$$C_G = n_p V_p^2 (1/v_m)^2 (m_p - m_m)^2 (b_A - b_B)^2$$

$$= n_p N_p^2 (m_p - m_m)^2 (b_A - b_B)^2 \quad \ldots\ldots 28$$

since $N_p v_p = V_p$. Thus if the shape of the precipitate is known, V_p can be calculated from R_g. Hence, since b_A, b_B and v_m are known, we can obtain the product $n_p \cdot (m_p - m_m)$. If there is an interference peak we can obtain a value for n_p (from 14) and thus obtain a value for $(m_p - m_m)$. On the other hand if $(m_p - m_m)$ is known we can obtain a value for the number of inhomogeneities n_p even if there is no interference peak in the SANS data. In order to separate $(m_p - m_m)$ we need additional information. The total concentration of B atoms in the alloy can be obtained by some form of chemical analysis, say this is c_B. Thus from the conservation of the number of atoms we have $c_B N_t = m_p n_p N_p + m_m (N_t - n_p N_p)$ and hence we can determine m_p and m_m.

In experimental data the differential cross-section is usually quoted in barns/steradian/atom, $d\sigma/d\Omega$ divided by the total number of atoms bathed by the neutron beam. Thus C_G is divided by NV (number of atoms/cm^3 x volume).

Porod approximation

The Porod approximation for $|F\{g(r)\}|^2$ applies for QR_g values greater than about 2.5. It is not used as extensively as the Guinier approximation. The derivation of the approximation is rather involved and so will not be presented here, see Porod(1982). For randomly distributed and oriented inhomogeneities with a well defined internal surface the differential scattering cross-section can be approximated by:

$$\frac{d\sigma}{d\Omega} = n_p (\langle c_p - c_m \rangle)^2 \ 2\pi S_p Q^{-4} \quad \ldots\ldots \ 29$$

where S_p is the surface area of the inhomogeneity. The contribution from the interference term is constant in this Q region with a value of unity. Thus the cross-section falls off as $1/Q^4$ in the Porod region and is proportional to the surface area of the inhomogeneity. Using the definition for the Guinier constant we can write the Porod approximation as:

$$\frac{d\sigma}{d\Omega} = 2\pi C_G \cdot S_p / (v_p^2 \cdot Q^4) = C_p \cdot Q^{-4} \quad \ldots\ldots \ 30$$

Hence a plot of $d\sigma/d\Omega$ versus Q^{-4} should be linear at high Q values. However care has to be taken since the scattering power in this Q region is often very weak and incoherent scattering or other Q independent scattering from

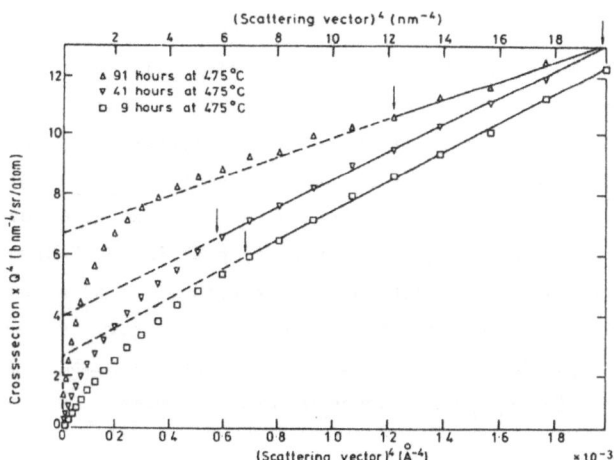

Fig. 5 Typical Porod plots

the sample can make the cross-section arising from the inhomogeneities difficult to obtain accurately. If this is the situation then a plot of $Q^4 \cdot d\overline{U}/d\Omega$ versus Q^4 can be made. The slope of this curve is then the Q independent scattering and the intercept at $Q^4 = 0$ gives the Porod constant, C_p (figure 5). Returning to equation (30) for a spherical inhomogeneity of radius R_p this becomes

$$\frac{d\overline{U}}{d\Omega} = (9/2)C_G/(R_p^4 \cdot Q^4) \quad \ldots\ldots \; 31$$

For spherical inhomogeneities the slope of the $d\overline{U}/d\Omega$ versus Q^{-4} plot would give a value for R_p providing C_G has been determined from a Guinier plot.

A distribution of inhomogeneity sizes is likely to be present in any sample and Baur and Gerold (1964) have shown that R_g^2 and R_p are equal to certain moment ratios of the size distribution:

$$R_g^2 = 3\langle R_s^8 \rangle / 5\langle R_s^6 \rangle \quad \text{and} \quad R_p = \langle R_s^3 \rangle / \langle R_s^2 \rangle$$

here R_s is the radius of any sphere and $\langle \ldots \rangle$ indicates the average over the size distribution. It is clear from these results that the Guinier approximation is weighted towards the larger sizes in the distribution and thus a value of R_s determined from R_g will usually be larger than R_p.

The Guinier and Porod plots shown in figures 4 and 5 are for an Fe-Cr-Al alloy with spherical Cr-rich precipitates. The spherical radii obtained from the Guinier plots for the three ageing times are 13.6Å, 17.2Å and 21.7Å. The corresponding values from the Porod plots are 13.9Å, 14.9Å and 17Å. These results clearly demonstrate the influence of a size distribution. The 9 hour ageing treatment corresponds to the very early stages of precipitation in this alloy when there might be expected to be little variation in the precipitate size and there is good agreement between the values of the radius. On the other hand the longer ageing treatments, when there is likely to be a distribution of precipitate sizes, show a marked difference.

Integrated cross-section

The integrated cross-section is given by $4\pi \int_0^\infty (d\overline{U}/d\Omega)Q^2 dQ$. For the simple model of a two phase system used in the derivation of equation (28) the integrated cross-section is given by:

$$4\pi \int_0^\infty (d\sigma/d\Omega)Q^2 dQ = (2\pi)^3 N^2 V(m_p - c_B)(c_B - m_m)(b_A - b_B)^2 \quad \ldots \ldots \text{ 32}$$

Unfortunately data is needed over a very wide Q range in order to determine the integrated cross-section. Attempts have been made using Porod´s approximation to extrapolate to infinity. But this can involve large uncertainties because of high incoherent scattering and/or scattering arising from isolated impurity atoms (Laue monatonic scattering) and isolated interstials (Q independent scattering) or indeed small agglomerates which give Q dependent scattering. Thus in practice the integrated cross-section has to be used with care if data has not been recorded over a wide Q range.

Neutron scattering equipment for defect studies

In order to investigate the scattering of neutrons by defects in a material four components are required. Firstly, a beam with a high intensity of long wavelength neutrons so that measurements can be made using neutrons of wavelength longer than the Bragg cut-off of the material. Secondly, a means of selecting a limited band of neutron wavelengths from the beam. Thirdly, a device for limiting the angular divergence of the beam; and, finally, a means of detecting the neutrons scattered by or transmitted through the material in a low background enviroment. Let us now consider each of these components.

Neutron beam production

Two types of neutron source are used in neutron scattering: steady-state reactors and accelerator based sources. Reactors produce a continous flux of high energy (about 2 MeV) neutrons as the result of the fission of uranium. Accelerator based sources are usually pulsed (typically 50 pulses/sec) and the neutrons are produced as a result of nuclear reactions induced by the impact of high energy electrons (typically 100 MeV) or protons (typically 800 MeV) with a target made from a high atomic number element. In the electron case, the electrons are decelerated very rapidly in the target and high energy gamma rays are produced as a result (bremsstralung). Some of these gamma rays then excite target nuclei which subsequently decay with the emission of a neutron. When high energy protons hit the target spallation occurs. As a result of spallation fragments of target nuclei are split off and some high energy neutrons are released. One proton can effect many nuclei and the neutron yield is high (typically 30 neutrons per proton). The high energy neutrons produced from either reactor or accelerator based sources have to be slowed down or moderated before they can be used in a neutron scattering experiment.

Neutron beams are brought out of a reactor pile (or the target station on a pulsed source) through holes running through the biological shield. The majority of the neutrons emerging from such beam holes have undergone a large number of collisions with atoms in the moderator surrounding the reactor core (or target on a pulsed source) and will tend to come into thermal equilibrium at the moderator temperature T. Such neutrons will have a distribution of velocities which follows a Maxwellian curve appropriate to the temperature T. The root mean square neutron velocity, v, characteristic of this temperature is given by $mv^2/2=3kT/2$ where k is the Boltzmann constant and m the neutron mass. The typical temperature of the moderator in a reactor is 310°K thus the root mean square neutron velocity is 2770 m sec^{-1} and the corresponding wavelength is 1.4Å. In a Maxwellian spectrum the long wavelength flux falls off as λ^{-5} and thus the flux at 6Å from a moderator at 310°K is about 0.1% of the peak flux at about 1Å. In order to increase the long wavelength flux the temperature of the moderator, or part of it in a reactor, must be reduced; thus shifting the peak flux in the Maxwellian distribution to longer wavelengths. It is impracticable to cool the whole moderator in a reactor because of the heat generated by the fission process, which in the case of the High Flux Beam Reactor (H.F.B.R.) at the I.L.L., Grenoble is 57 megawatts. However a small volume of the reactor moderator can be cooled to enhance the flux at long wavelengths. At the I.L.L. a 25 litre vessel containing liquid deuterium at 25°K is used which provides a very high intensity of neutrons with wavelengths of 5Å and longer.

Wavelength selection

Long wavelength neutrons have velocities which are relatively low, the velocity, v (m/sec) = 3958/λ (Å). Thus a limited band of velocities can be selected relatively easily using a mechanical velocity selector. A typical mechanical selector consists of a cylinder, made of material which absorbs neutrons (e.g. cadmium in the form of a Mg-10wt%Cd alloy), which has slots around its periphery. These slots are machined on a helical path in a direction parallel to the major axis of the cylinder. The principle of operation is to rotate the cylinder about its major axis at a specific speed thus allowing only those neutrons which travel along the slots without hitting the walls to be transmitted (see fig.6). The wavelength resolution obtained in this way is at best 3%, and typically 10%. The neutron transmission of a typical helical velocity selector is 30%.

On a pulsed source the various neutron wavelengths can be separated by time of flight. The time of flight of a neutron is 252.7λ μsec/m (wavelength in $\overset{\circ}{A}$), thus a $6\overset{\circ}{A}$ neutron takes 1500 μsec to cover one meter (equivalent to the cruising speed of Concorde M2.0). Flight time differences of a few microseconds can easily be measured and thus the achievable wavelength resolution is very good. Unfortunately after travelling a distance L, longer wavelength neutrons of pulse n are overtaken by shorter wavelength neutrons of pulse n+1 and thus choppers have to be used to limit the range of wavelengths available on a pulsed source so as to avoid pulse overlap problems. The maximum usable wavelength range $\Delta\lambda(\overset{\circ}{A})$ = prf(μsec)/252.7L(m) where prf is the pulse repetition time (2000 μsec on the S.N.S) and L is the length of the instrument from source to detector. The wavelengths transmitted by a chopper on a pulsed source is related to its position and phase relationship to the neutron pulse at the source. The phase relationship once set has to be accurately maintained. However this can be achieved and thus since a wide range of wavelengths can be simultaneously used on a pulsed beam a much wider Q range can be covered at one time than is possible on a non-pulsed beam. This is very important if the inhomogeneities are changing as a function of time and there is insufficient time to change the sample to detector distance or the incident wavelength to follow Q dependent changes.

Collimation

The simplest way to collimate a neutron beam is with a series of apertures cut in neutron absorbing material (cadmium, gadolinium or boron carbide). However if fine collimation is required or if there is insufficient distance available for apertures to work then Soller collimators are used. These are made up of a number of neutron absorbing blades set up parallel to the neutron beam direction and arranged so that they are perpendicular to the scattering plane. The overall transmission is determined by the thickness of the blades in relation to their spacing. Very thin blades made from stretched plastic coated with neutron absorbing gadolinium paint are available. A 10´ collimator made in this way has a transmission > 95%. If area detectors are used then two Soller collimators with blades at right angles can be used.

Detectors

Since neutrons are neutral particles all long wavelength neutron detectors are designed to detect the energetic charged secondary products arising from the absorption of neutrons by certain nuclei. The most important of which are:

$$^{10}B + n° = {}^{7}Li + {}^{4}He + 2.3MeV$$

$$^{3}He + n° = {}^{3}H + p + 0.77MeV$$

$$^{6}Li + n° = {}^{4}He + {}^{3}He + 4.79MeV$$

Gas counters, filled with ^{3}He gas (at up to pressures of 16 atm.) or BF_3 gas enriched in ^{10}B, in their simplest form consist of a cylinder of copper or stainless steel with a thin anode wire running axially along the cylinder. A voltage of a few kilovolts is maintained between the anode and the outer case of the detector. Charged particles emitted when a neutron is absorbed produce ionisation along their tracks and this charge is detected. The detectors are usually operated so that the electrons produced in the ionisation are accelerated and produce further ionisation. This results in a larger pulse, which is proportional to the initial ionisation, hence the name proportional detector. A 1 cm. thick 4 atm. ^{3}He detector records 70% of the $4\overset{\circ}{A}$ neutrons which pass through it, whereas a 3cm. thickness is necessary for a 2atm. $^{10}BF_3$ detector to achieve the same efficiency.

In scintillation detectors the neutron absorber is in solid form and is thus much denser than in gas detectors and so high detection efficiencies can be achieved with thicknesses of 1mm. ^{6}Li isotope distributed in a glass or plastic medium is widely used as the neutron absorber in scintillation detectors. The energetic charged particles produced on the absorption of a neutron liberate ion pairs along their tracks in the glass. The resulting electrons excite the active component of the glass Ce^{3+}, thus producing a flash of light which is detected by a photo-multiplier. Higher count rates are possible with scintillation detectors than with gas detectors because of a shorter dead-time. However, gamma rays are more difficult to discriminate from the neutrons in scintillation detectors.

Large position sensitive area detectors consisting of up to 16000 individual elements have come into prominence over the last decade or so and have made a major impact in defect studies. In parallel with this development came the utilisation of neutron guide tubes to transport neutron beams large distances from their source into low radiation background enviroments. When neutrons impinge on a flat surface at small glancing angles, u, total or mirror reflection can occur providing the incident angle is less than the critical angle, u_c. u_c is related to the relative refractive index for neutrons, n, between the two media involved by $cos\, u_c = n$. For nickel u_c is 6´ per $\overset{\circ}{A}$ and hence for a $10\overset{\circ}{A}$ neutron u_c is 1°. Thin layers of nickel

evaporated onto optically flat glass plates arranged in a box section with
the nickel plating on the internal surface are used to transport neutron
beams over large distances (typically up to 100m) with little attenuation of
the beam (few percent). Since the background radiation from the source falls
off as $1/r^2$ the use of neutron guides enables a neutron spectrometer to be
built in a low background enviroment which is crucial for defect studies.
Usually slightly curved neutron guide tubes are used so that there is no
direct line of sight from the source to the detector and high energy neutrons
and gamma rays in the beam, which are not reflected by the guide tube walls,
do not reach the spectrometer, but are absorbed in the shielding surrounding
the beam line. On pulsed sources the advantages of using guides has to be
balanced against the pulse overlap problems which are worsened by increasing
the overall length of an instrument.

The SANS instrument, D11, at the I.L.L. is an example of a spectrometer
employing many of the above features and it is used extensively in the study
of inhomogeneities in materials (see figure 6). The velocity selector is 60m
from the cold source and has a wavelength resolution of 8%. The beam
collimation is achieved using two apertures which can effectively be placed
2, 5, 10, 20 or 40m apart using a combination of straight neutron guide
tubes. The monochromatic collimated beam then impinges on the sample and the
scattered neutrons are detected on an position sensitive area $^{10}BF_3$ proportional detector consisting of 128x128 5x5mm. active elements. The detector
can be placed anywhere from 2 to 38m from the sample and thus the Q range
investigated in a particular experiment can be continuously varied from 0.3 -
0.001 $\overset{o}{A}^{-1}$. At the centre of the detector there is a beam stop.

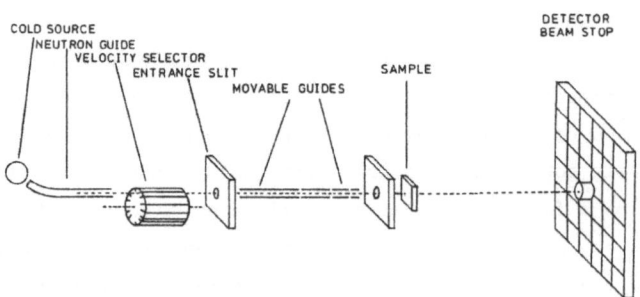

Fig. 6. Schematic diagram of the D11 SANS spectrometer

<u>Data analysis</u>

Consider a sample which is larger than the incident neutron beam as shown schematically in figure 7. The scattered neutron beam intensity, I_{EC}, measured on the detector when no sample is in the beam can be considered as arising from three different areas of the spectrometer A, B and C (see fig.7)

$$I_{EC} = I_A + I_B + I_C \quad \ldots\ldots 33$$

I_A is the contribution from neutrons scattered into the detector from the beam line prior to the sample position, I_B that from long wavelength neutrons transmitted through the sample position which are then scattered into the detector and I_C background neutrons picked up by the detector from all other sources. If a sample is now placed in the beam there will be an additional contribution from the neutrons scattered by the sample, I_D. However, the sample will transmit fewer neutrons thus the contribution from region B will be reduced. Let 1-T be the attenuation of the beam in passing through the sample, then the total intensity measured at the detector with the sample in the beam is

$$I_S = I_A + T.I_B + I_C + I_D \quad \ldots\ldots 34$$

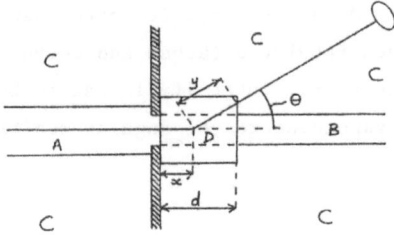

Fig. 7. Schematic diagram for data analysis calculations, showing regions giving rise to background scattering.

117

$T = \exp(-N\bar{\sigma}x)$ where N is the number of atoms per unit volume in the sample, $\bar{\sigma}$ the transmission cross-section of the sample and x the length of the sample through which the neutron beam passes. T can be obtained by a transmission measurement or from the known value of $\bar{\sigma}$. We wish to obtain I_D, but we cannot get it from measurements of I_{EC} and I_S alone. However, if we place a piece of cadmium directly in front of the sample all the long wavelength neutrons in the beam will effectively be absorbed by the cadmium. Thus the intensity on the detector in this case, I_{CD} would be

$$I_{CD} = I_A + I_C$$

I_D can then be separated as follows

$$I_D = I_S - I_{CD} - T \cdot (I_{EC} - I_{CD}) \quad \ldots\ldots 35$$

Of course some of the neutrons scattered by the sample will be absorbed by the sample itself thus an absorption correction has to be made to account for this using the known absorption cross-section of the sample. The exact value for the absorption correction, A, depends on the exact path the neutron follows in the sample before (x) and after (y) it is scattered which is related to the geometry of the sample and the scattering angles but can be evaluated using the integral:

$$A(\theta,\phi) = \int_V \exp(-\bar{\sigma}_T N(x+y)) dV/V \quad \ldots\ldots 36$$

V is the volume of the sample and ϕ is the angle around the beam line. For the typical geometry shown in figure 7 this integral becomes:

$$A(\theta,\phi) = \int_0^d \exp(-\bar{\sigma}_T N(x + (d-x)\sec\theta)) \pi R^2 dx / \pi R^2 d$$

$$= (\exp(-\bar{\sigma}_T N d \sec\theta) - \exp(-\bar{\sigma}_T N d))/(\bar{\sigma}_T N d(1-\sec\theta)) \quad \ldots 37$$

For $\theta=0$ this reduces to $A(0,\phi)=\exp(-\bar{\sigma}_T N d)$. For other sample geometries the absorption factors have been tabulated (Rouse and Cooper 1970). Thus to take account of absorption the measured intensity I_D has to be divided by the absorption correction corresponding to the angular position of the detector, $A(\theta,\phi)$.

A further correction is also necessary because the detector is not 100% efficient. This factor is taken into account by measuring the intensity of the beam scattered by a sample of well known scattering power. Vanadium is

118

usually used since the scattering is almost entirely incoherent (i.e. the same in all directions) and the incoherent cross-section is well known, with a value of 0.38 barns/steradian/atom. The measured intensity scattered by the vanadium crystal has to be corrected for absorption and the effects of parsitic contributions to the scattering in exactly the same way as required for the actual sample. The final equation for the cross-section of the sample in barns/steradian/atom is thus:

$$(d\bar{\sigma}/d\Omega)_S = [(I_D)_S/(N_S.A(\theta,\phi)_S)]/[(I_D)_V/(N_V.A(\theta,\phi)_V)] \times 0.38 \quad \ldots \ldots \; 38$$

The subscripts S and V refer to the sample and vanadium and N is the number of atoms bathed by the beam. Thus in order to obtain an accurate value for $(d\bar{\sigma}/d\Omega)_S$ between six and eight separate experiments need to be performed.

Inelastic scattering from the sample as a whole can effect the separation of the scattering associated with the inhomogeneities. However since the long wavelength neutrons used have energies less than 5 meV, only inelastic interactions in which the neutron gains energy from the crystal phonons are possible. Inelastic processes in which energy is lost by the neutrons to the crystal phonons require that energy and momentum must be conserved and this is not possible for neutrons with such low energies as a few meV. Thus the inelastic contribution can be reduced or indeed removed by cooling the sample to liquid nitrogen temperatures. If this is not possible, for instance if a sample is being studied as a function of temperature, then the separation of elastic and inelastic scattering events can be made using time of flight analysis. In many instances the defect scattering is obtained from the difference in scattering measured before and after a specific treatment (e.g. irradiation or application of heat). Thus the influence of inelastic scattering in many cases is effectively eliminated since any changes in the inelastic scattering as a result of the treatment are likely to be small.

The optimum sample thickness for maximum scattering intensity is given by $x_{op}=V/N_S\bar{\sigma}_T$ where $\bar{\sigma}_T$ is the total removal cross-section which includes absorption, incoherent scattering, inelastic scattering and any coherent scattering which is possible with the incident neutron wavelength. Thus for a material with $N_S/V=5\times10^{22}$ cm^{-3} and $\bar{\sigma}_T=20$ barns/atom, $x_{op}=1$ cm. For aluminium $x_{op}=22$ cm for 6Å neutrons whereas for nickel it is 5 mm. Multiple scattering can also be a problem in samples with a high inhomogeneity concentration. By this we mean that a neutron is scattered more than once in its path through the sample. This is very difficult to correct for and the best option is to avoid the problem by choosing a small enough sample.

Diffuse neutron scattering arising from lattice relaxation around defects

Consider a perfect crystal lattice containing N_t atoms into which n_p identical defect clusters have been randomly introduced. For generality we will consider each cluster to consist of N_I interstitials and N_V vacancies, and assume that N_R atoms are relaxed around each cluster. A relaxed atom and its vacated lattice site can be treated as a pseudo inter-stitial-vacancy pair. If an incident neutron beam of amplitude Ψ_i and wave vector $\underline{k}_i$ is incident on the crystal in a direction α with respect to the crystal axes; then the beam scattered by the defects at an angle 2θ has amplitude $\Psi_{fd}(\alpha,\theta)$ and wave vector $\underline{k}_f$. We describe the normal lattice positions of the host matrix with the position vector $\underline{R}_j$ and suppose that the position of each of the defect clusters is represented by $\underline{d}_m$ (defines cluster origin) and the position of each interstitial atom (or relaxed atom) in the cluster by $\underline{u}_{Is}$ (relative to cluster origin) and $\underline{u}_{Vt}$ the position of each vacancy (or relaxed atom original site) in the cluster. The scattered amplitude per unit solid angle is then, writing I for $N_I + N_R$ and V for $N_I + N_R$

$$\Psi_{fd}(\alpha,\theta) = \sum_{j=1}^{N_t} b\,\exp(-i\underline{Q}.\underline{R}_j) + \sum_{m=1}^{n_p}\sum_{s=1}^{I} b\,\exp(-i\underline{Q}.(\underline{d}_m+\underline{u}_{Is}))$$
$$+ \sum_{m=1}^{n_p}\sum_{t=1}^{V} (-b)\exp(-i\underline{Q}.(\underline{d}_m+\underline{u}_{Vt}))$$
$$= b(A + B - C) \quad\ldots\ldots 39$$

Note the scattering amplitude for a vacancy is $-b$. The scattered intensity is thus given by

$$\Psi_{fd}\,\Psi_{fd}^* = b^2(AA^*+AB^*-AC^*+BA^*+BB^*-BC^*-CA^*-CB^*+CC^*) \quad\ldots\ldots 40$$

This equation consists of a series of products of random walk summations. If one considers the situation beyond the Bragg cut-off, the first term (the coherent scattering from the unperturbed lattice) is zero. The coherent scattering from the lattice only undergoes constructive interference in the forward direction, which because of the crystal size is inseparable from the transmitted neutron beam. In addition AB^*, AC^*, BA^* and CA^* are each of the order of $n_p^{\frac{1}{2}}$, whereas BB^*, CC^*, BC^* and CB^* are each of the order of n_p. Now for the inhomogeneities which can be studied using neutron scattering n_p lies in the range 10^{10} to 10^{20} clusters/cm^3. Thus the terms of magnitude $n_p^{\frac{1}{2}}$ can be neglected in comparison with those of magnitude n_p. Hence we can rewrite equation (40) fully as:

$$\Psi_{fd}\Psi_{fd}{}^* = b^2 (BB^* + CC^* - BC^* - CB^*) = b^2 |\ B-C\ |^2$$

$$= b^2 |\sum_{m=1}^{n_p} \sum_{s=1}^{I} \exp(-i\underline{Q}.(\underline{d}_m+\underline{u}_{Is})) - \sum_{m=1}^{n_p} \sum_{t=1}^{V} \exp(-i\underline{Q}.(\underline{d}_m+\underline{u}_{Vt}))|^2$$

$$= b^2 |\sum_{m=1}^{n_p} \exp(-i\underline{Q}.\underline{d}_m) (\sum_{s=1}^{I} \exp(-i\underline{Q}.\underline{u}_{Is}) - \sum_{t=1}^{V} \exp(-i\underline{Q}.\underline{u}_{Vt}))|^2$$

$$= b^2 |\ E(F-G)\ |^2 = b^2 EE^* |\ F-G\ |^2 \ \ldots\ldots\ 41$$

The term EE^* consists of n_p^2 randomly oriented vectors, and thus by a random walk analysis, this sum is equal to n_p. Of course in practice the defect clusters are not arranged completely randomly throughout a crystal, but instead are randomly arranged on specific sites. Taking this into account the term EE^* becomes $n_p(n_d-n_p)/n_d$, where n_d is the total number of possible sites for a cluster in the sample. For most situations $n_d \gg n_p$ and thus this reduces to n_p.

The differential cross-section arising from defect clusters in a single crystal is obtained from equation (41) averaged over all the equivalent orientations of the cluster with respect to the neutron beam (represented in the following equation by $\langle\ldots\rangle$):

$$\frac{d\sigma}{d\Omega} = F_c (\sum_{s,s'=1}^{I} \langle\cos(\underline{Q}.(\underline{u}_{Is}-\underline{u}_{Is'}))\rangle + \sum_{t,t'=1}^{V} \langle\cos(\underline{Q}.(\underline{u}_{Vt}-\underline{u}_{Vt'}))\rangle$$

$$- 2 \times \sum_{s=1}^{I} \sum_{t=1}^{V} \langle\cos(\underline{Q}.(\underline{u}_{Is}-\underline{u}_{Vt}))\rangle) \ \ldots\ldots\ 42$$

where $F_c = b^2(n_p(n_d-n_p)/n_d)$. For polycrystalline material the cross-section must be averaged over all possible orientations of the cluster with respect to the neutron beam. Consider a typical term in equation (42), $\cos(\underline{Q}.\underline{R}) = \cos(QR\cos\alpha)$. Averaging over all possible orientations, α, this becomes:

$$(1/4\pi) \int_0^\pi \cos(QR\cos\alpha)\ 2\pi\sin\alpha\ d\alpha = \sin(QR)/QR$$

Hence the differential cross-section for defect clusters in a polycrystal is:

$$\frac{d\sigma}{d\Omega} = F_c [\sum_{s,s'=1}^{I} \sin(QRI_{ss'})/QRI_{ss'} + \sum_{t,t'=1}^{V} \sin(QRV_{tt'})/QRV_{tt'}$$

$$- 2 \times \sum_{s=1}^{I} \sum_{t=1}^{V} \sin(QRVI_{st})/QRVI_{st}\ ] \ \ldots\ 43$$

where $RI_{ss'} = |(\underline{u}_{Is}-\underline{u}_{Is'})|$ and similarly for $RV_{tt'}$ and RVI_{st}. At $Q=0$ the right hand side of this equation reduces to $(d\sigma/d\Omega) = F_c(N_I-N_V)^2$

121

and as Q becomes large it reduces to $d\overline{U}/d\Omega = F_c(2N_R+N_I+N_V)$. Thus at low Q the differential cross-section is independent of the relaxation associated with the defect cluster, whereas at high Q the relaxation does contribute.

The effects of relaxation on the differential scattering cross-section can be calculated using either equation (42) or (43), providing some assumptions are made to determine the positions of the atoms in the radial strain field produced by the defect. In an elastic continuum the amount of relaxation surrounding a point source of dilation is given by (Eshelby 1956),

$$\Delta R=a/R^2 \quad \ldots\ldots 44$$

where R is the distance from the point source of dilation and a is a constant determining the strength and sign of the dilation. The elastic continuum approach can be used to describe the relaxation after the first few nearest neighbour shells of atoms around the defect. Larkins and Stoneham (1971) and Clark et al (1971) have considered first shell displacements as large as 20% for point defects in diamond structured materials. Clark et al (1971) have also found that the displacements in the first two shells of atoms around a

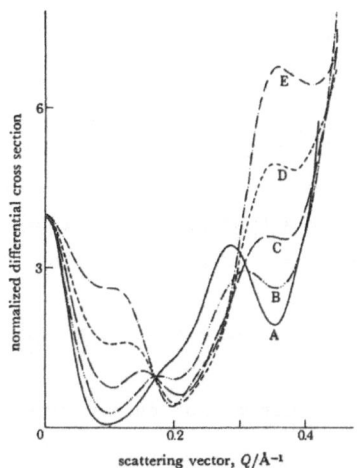

Fig. 8. Theoretical differential cross-section for divacancies in silicon. The first nearest neighbour (n.n.) atoms are relaxed 20% towards the divacancy. The next 4 n.n. shells are relaxed according to (44) with the exception of the 2nd. n.n., for which the relaxation is modified. The values of a in (44) for this are: a for A, $\frac{1}{2}$a for B, 0 for C, $-\frac{1}{2}$a for D and $-$a for E.

122

divacancy were of opposite sign. The discrimination that can be achieved in the determination of the details of the relaxation field associated with a defect are illustrated by the calculations of the differentiaal cross-section from various relaxation models associated with a divacancy in a single crystal of silicon. In these calculations the relaxations out to the 5th. nearest neighbour shell were included, involving a total of 54 atoms. The type of effects which may be observed are shown in figure 8. Here the first shell is relaxed inwards by 20% and the 3rd, 4th and 5th shells are given inward relaxations of amounts given by equation (44). The 2nd nearest neighbour shell is varied from inward to no relaxation to outward relaxation. Such opposite relaxations of the 2nd nearest neighbour shell of atoms surrounding a defect have been shown to occur in lattice calculations involving interatomic force simulations. The results shown in figure 8 show that the second shell relaxation considerably influences the differential cross-section for Q around 0.1, 0.28 and 0.35 $\overset{\circ}{A}^{-1}$.

Applications of neutron scattering in the study of inhomogeneities

In this section some representative examples of the application of neutron scattering to the study of inhomogeneities in materials will be discussed (see reviews by Kostorz(1979,1983), Schmatz(1973,1983), Mitchell and Stewart (1980) and the Annex to the Annual Report of the I.L.L.).

Irradiated materials

When crystals are irradiated with high energy particles, atoms are displaced from their normal positions. Various point defects (interstitials, vacancies) or small defect clusters (divacancies, di-interstitials, vacancy-impurity complexes, interstitial-impurity complexes) or large defect regions (voids) may be formed. At one extreme (e.g. bombardment with $\frac{1}{2}$ MeV electrons) the recoil energies of the primary displaced atoms are small (less than 100 eV) and only point defects are formed. At the other extreme (e.g. fission energy neutrons in a reactor), the recoil energies are about $\frac{1}{2}$ Mev and the energetic primary knock-on atoms produce many more (typically 1000) further displacements in coming to rest. Because of the penetration of the neutron, the resulting damage occurs throughout the material. Because of this wide size range long wavelength neutron scattering is ideally suited to the study of radiation induced defects.

Fission energy neutron damage in a GaAs single crystal has been studied by Gupta et al (1978). Their measurements of the neutron scattering from the

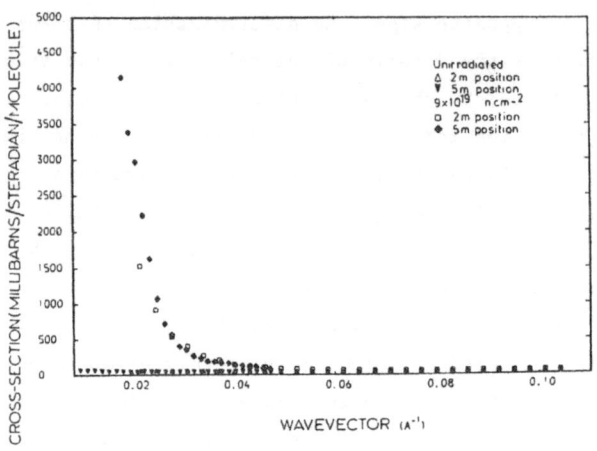

Fig. 9. The scattering from irradiated and unirradiated GaAs

irradiated sample and an unirradiated control sample are shown in figure 9. The scattering from the unirradiated sample was isotropic and with no Q dependence and therefore no observable inhomogeneities are present. In contrast there is pronounced small angle scattering in the irradiated sample associated with defect regions with a radius of gyration of 150 Å and the scattering was not isotropic about the incident beam direction when the sample was aligned with its <111> direction along the beam line (figure 10). The anisotropy in the scattering pattern is that of the Fourier transform of the shape function of the radiation induced inhomogeneity. Two models for the inhomogeneity were proposed: (i) ellipsoidal defect regions (major axis 516Å and minor axis 103Å) oriented in the 6 equivalent <100> directions; (ii) a

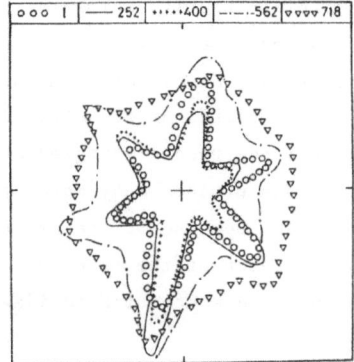

Fig. 10. Annealing behaviour of the anisotropic scattering from
irradiated GaAs. The diagram shows iso-intensity contours.

124

void surrounded by regions with a high defect concentration (giving rise to the oriented spikes) which could be rich in interstitials (probably in small agglomerates) which were ejected when a primary knock-on atom came to rest and created the void. Upon annealing to 718°C the spikes in the scattering pattern anneal out, this might be consistent with the interstitial rich regions annealing leaving the central void of model (ii). The radius of gyration at 718°C is 89Å.

Impurity atoms in crystals

Czochralski grown silicon single crystals are used extensively in the fabrication of VLSI devices. Such silicon is not chemically pure but commonly contains oxygen impurities at a level of about $10^{18}cm^{-3}$. Carbon, nitrogen and hydrogen are other common contaminants in silicon. These impurities can have either advantageous or deleterious effects in device fabrication depending on the thermal treatments to which the silicon is subjected during manufacture of a specific device. During the heating cycles necessary impurity atoms diffuse and agglomerate to form precipitates. A detailed understanding of the kinetic processes involved and the form and distribution of the precipitates is crucial to assure high device yields.

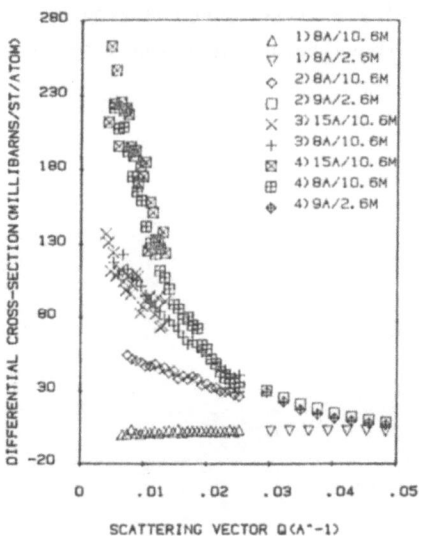

Fig.11. SANS from SiO_2 precipitates in silicon using different
wavelengths and sample to detector distances. 1) as grown
2) 48hrs at 750°C, 3) 96hrs at 750°C and 4) 431hrs at 750°C

A wide ranging study of oxygen precipitation in dislocation free silicon single crystals has been made using neutron scattering and infrared absorption measurements (Livingston et al 1984). On heating this material, oxygen atoms diffuse and aggregate to form precipitates of a second phase, which is believed to be a form of SiO_2. Although the total oxygen concentration is very low (few ppm) neutron scattering measurements of the precipitation in the temperature range 650°C to 850°C have been successfully made. Figure 11 shows the small angle scattering arising from the growth of precipitates at 750°C in a sample containing 7×10^{17} oxygen atoms/cm^3. The spectra shown are azimuthally averaged, because the scattering is very weak. The scattering from the untreated sample is flat with a cross-section close to zero, indicating that no appreciable agglomeration of impurities is present initially. There is no inter-precipitate interference peak because of the very low concentration of precipitates.

The Guinier plots of this data are shown in figure 12 and exhibit good linear regions at low Q. However the form of the cross-section at higher Q (evidence of a second Guinier region) indicates that the precipitates are not spherical. Cuboidal shaped SiO_2 precipitates have been observed in silicon using electron microscopy (Messoloras et al 1985) and thus a computer simulation was made of the expected scattering from cuboids oriented in the way observed in the electron microscope. The anisotropic scattering patterns so obtained were azimuthally averaged and the general form of these cross-sections were similar to those found experimentally. From a Guinier plot of

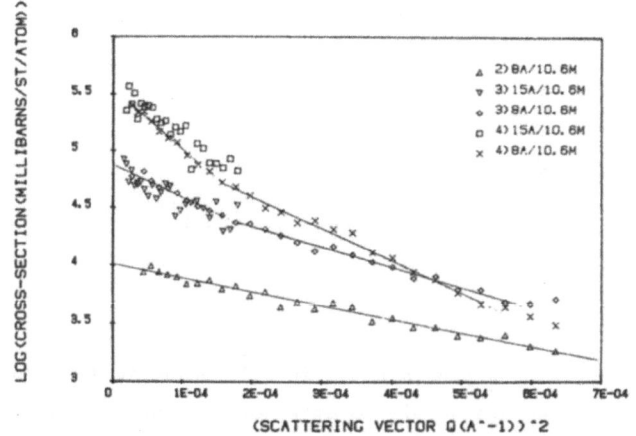

Fig.12. Guinier plots for the silicon data of fig.11

the computer simulated scattering for the oriented cuboids it was found that the radius of gyration obtained for the azimuthally averaged scattering was very close to the value of R_g derived assuming a random distribution of cuboidal precipitates; which is simply given by $R_g = L[(2w^2+1)/12]^{\frac{1}{2}}$ for a cuboid with sides of length L, wL and wL (w>1). Unfortunately a unique pair of values for L and w can not be found from R_g alone and the scattering in the Porod region is very weak so this can not be used.

Let us consider what information we can gain from the Guinier constant C_G in equation (24). For convenience the measured cross-sections are quoted in barns/steradian/atom $(1/VN)(d\bar{U}/d\Omega)$, where N is the number of atoms per unit volume and V the volume of the sample bathed by the neutron beam. Thus

$$C_G = n_p v_p^2 \ (\langle c_p \rangle - \langle c_m \rangle)/(NV) \ \ \ \dots \ 45$$

If we know $\langle c_p \rangle - \langle c_m \rangle$ then from the experimentally determined value of C_G we can determine the product $v_p^2 . n_p$. Now for a precipitate containing oxygen in a silicon matrix we have from equations (25) and (26) that

$$\langle c_p \rangle - \langle c_m \rangle = (1/v_p)(b_0 m_p - b_{Si}(1-m_p)) - (1/v_m)(b_0 m_m - b_{Si}(1-m_m)) \ \ \ \dots \ 46$$

Now b_0 and b_{Si} the scattering lengths of oxygen and silicon are known as is v_m the volume occupied by one atom in a silicon matrix. The precipitates are also known to be a form of SiO_2, thus $m_p = 2/3$. In addition the total oxygen concentration is only 10^{-5} and thus m_m can be taken to be zero. Thus we know all the terms on the right hand side of (46) except v_p, the volume per atom in the precipitate. This will depend on the form of SiO_2 and will vary from $11.4\overset{\circ}{A}^3$ for the high density phase coesite to $15.2\overset{\circ}{A}^3$ for amorphous SiO_2. Thus bearing in mind these assumptions a value for $\langle c_p \rangle - \langle c_m \rangle$ can be found and hence from C_G a value for $v_p^2 . n_p$. In order to separate these two terms we need additional information which we can get from infra-red absorption studies of the crystals. From the 9 μm absorption band the concentration of isolated oxygen atoms in the silicon matrix, m_i, can be obtained (Livingston et al 1984). Thus IR absorption data of the 9 μm band were taken before and after each heat treatment. The difference in the values of m_i, assuming that all the isolated oxygen atoms lost from the matrix form SiO_2 precipitates, give

$$m_i(t) - m_i(0) = (n_p/V).N_p.m_p = (n_p/V).(V_p/v_p).m_p \ \ \ \dots \ 47$$

127

where t refers to the heat treatment time. Since for SiO_2 precipitates m_p = 2/3 and a value for v_p can be assumed with little error, we can obtain a value for the product $n_p v_p / V$ from the IR data. Substituting for this into the the Guinier constant (45) we can determine V_p, from the experimental value for C_G, and hence n_p / V. Thus from the SANS data we have found R_g for the precipitates and from a combination of SANS and IR measurements the volume (V_p) of a precipitate and the precipitate number density n_p / V. Knowing V_p and R_g we can numerically determine the values of w and L for a cuboidal shaped precipitate since $V_p = w^2 L^3$ and $R_g^2 = L^2(2w^2+1)/12$. This is a good example of applying several techniques to reveal the form, number and distribution of the inhomogeneities in a crystal.

In the early stages of precipitation the SiO_2 precipitates in silicon are believed to be approximately spherical (one Guinier region in fig. 12). For spherical precipitates of radius R_s the rate of growth as a function of time, t, depends on the diffusion coefficient, D, in the silicon matrix of the element precipitating as follows:

$$R_s^2(t) - R_s^2(0) \propto 2Dt[(m_o - m_s)/(m_p - m_s)]$$

where m_o is the concentration of the precipitating element in the crystal before any thermal treatment and m_s is its concentration in the matrix close to the precipitate. Livingston et al (1984) found that the value of m_s was so small that it could be neglected, thus since at zero time there is no evidence for precipitation from the SANS data a plot of R_s^2 versus t should be linear. This was found and from the slope a value for D of 4.4×10^{-14} $cm^2 sec^{-1}$ was found for the 750°C results. This is in excellent agreement with values of D for oxygen in silicon found by other techniques and thus is confirmation that the precipitates contain oxygen.

Alloys

The mechanical properties of many modern alloys depend critically on the size and distribution of precipitate particles of a second phase. These in turn are affected by a variety of parameters such as the homogenisation temperature, ageing temperature and time, quench rate and temperature, impurity content and stress. Many neutron scattering studies of the growth and dissolution of precipitates in alloys have been made. In many cases when neighbouring elements in the periodic table are involved, neutron scattering can be used when x-ray techniques are extremely difficult because of poor contrast. This is because neighbouring elements can have significantly different neutron scattering lengths (e.g. AlMg, FeCr and NiFe alloys). Such experiments permit a full investigation of the kinetics of growth and dissolution of the precipitates to be made non-destructively and information concerning the influence of the various parameters outlined above can be obtained. In some cases where the changes in the precipitates as a result of some treatment is very fast (processes lasting a few minutes up to a few hours), the changes can be followed in situ in the neutron beam. Allen et al (1978) followed the growth and dissolution of precipitates in an Al-Zn alloy using a specially designed in beam furnace and were able to follow the dissolution even when the whole process lasted only 5 minutes. Miller et al (1978) and Dusic et al (1985) used a similar cell to carry out an in-beam study of the nucleation and growth of voids on grain boundaries up to the point of fracture in a nickel based superalloy subjected to temperature and stress. This is one of the advantages of using neutrons since, because of their low absorption by most materials, in beam furnaces, cryostats, pressure cells etc. are relatively easy to construct.

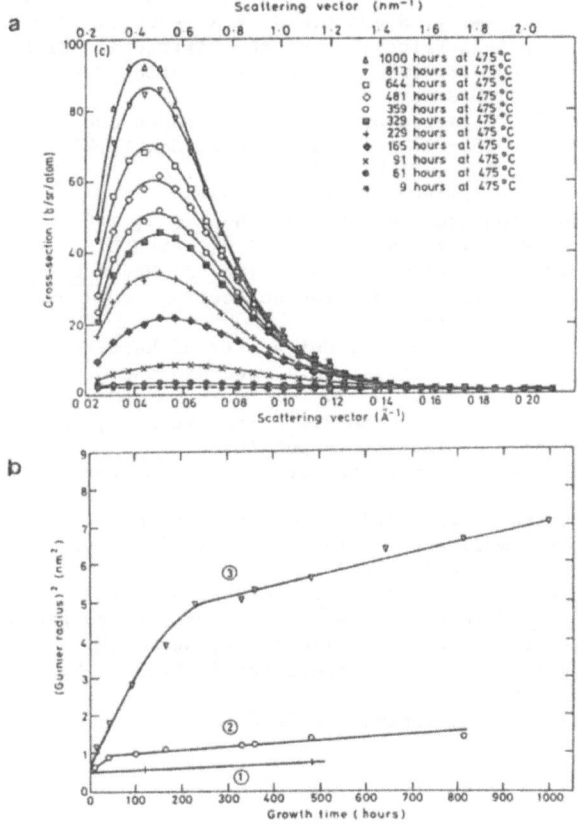

Fig.13 (a) SANS from an Fe-Cr-Al alloy aged at 475°C. (b) Cr-rich precipitate growth at 1) 350°C, 2) 400°C and 3) 475°C

The precipitation of a Cr-rich phase in an Fe-Cr-Al alloy containing small amounts of yttrium was followed using SANS (Messoloras et al 1984). The observed scattering at 475°C is shown in figure 13 together with the dependence of the average precipitate size (obtained from a Guinier plot) on time for various ageing temperatures. This is typical of the type of data that can be obtained. No incubation period was found before precipitation occured, however in an identical alloy without yttrium there was an incubation period. The dependence of the Guinier constant on ageing time is the most sensitive test for the presence of an incubation period.

References

Allen,D.R., Messoloras,S., Stewart,R.J. and Kostorz,G. 1978 J. Appl. Crystallogr. 11, 578

Baur,R. and Gerold,V. 1964 Acta. Metall. 12, 1449

Clark,C.D., Mitchell,E.W.J. and Stewart,R.J. 1971 Cryst. Lat. Defects 2, 105

Dusic,M., Messoloras,S. and Stewart,R.J. 1985 Met. Sc. & Techn. to be pub.

Eshelby,J.D. 1956 Solid State Physics 3, 79

Guinier,A. 1937 C.R. Hebd. Seance Acad. Sci., Paris 204, 1115

Gupta,S., Mitchell,E.W.J., Stewart,R.J. & Kostorz,G. 1978 Phil.Mag. A37, 227

James,R.W. 1962 Optical Principles of the Diffraction of X-rays, Wiley. p470

Koester,L. and Yelon,W.B. 1982 Neutron Diffraction Newsletter, Argonne Natl. Lab., IL 60439, U.S.A.

Kostorz,G. 1979 In "Treatise on materials science and technology" Ed. H. Herman, Volume 15, 227 (Academic Press, New York)

Kostorz,G. 1983 In "Physical Metallurgy" 3rd. edition. Editors R.W.Cahn & P.Haasen (North Holland, Amsterdam) p793

Larkins,F.P. and Stoneham,A.M. 1971 J. Phys. C 4, 143

Livingston,F.M., Messoloras,S., Newman,R.C., Pike,B.C., Stewart,R.J., Binns, M.J., Brown, W.P. and Wilkes,J.G. 1984 J. Phys. C 17, 6253

Lovesey,S.W. 1984 "Theory of neutron scattering from condensed matter", Vol 1 Clarendon Press, Oxford

Messoloras,S. 1974 Ph.D. Thesis, University of Reading, England

Messoloras,S., Pike,B.C., Stewart,R.J. and Windsor,C.G. 1984 Metal Science 18, 311

Messoloras,S., Kinder,S., Newman,R.C., Stewart,R.J., Bergholtz,W., Booker, R. and Hutchins,J. 1985 to be published

Miller,R.J.R., Messoloras,S., Stewart,R.J. and Kostorz 1978 J. Appl. Crystallogr. 11, 583

Mitchell,E.W.J. and Stewart,R.J. 1980 Phil. Trans. R. Soc. Lond. B290, 511

Porod,G. 1982 In "Small angle x-ray scattering" Editors O.Glatter & O.Kratky (Academic Press, London) p17

Rouse,K.D. and Cooper,M.J. 1970 Acta. Cryst. A26, 682

Schmatz,W. 1973 In "Treatise on materials science and technology" Ed. H. Herman, Volume 2, 105 (Academic Press, New York)

Schmatz,W. 1984 In "Methods of experimental physics" Vol 21, 147 (Academic Press, New York)

Synecek,V. 1962 J. Phys. Radium, 23, 828

ELASTIC NEUTRON DIFFRACTION AND DEFECT STRUCTURES

F.W. Beech

Department of Metallurgy and Materials Science
Imperial College London, London SW7 2BP
Department of Chemistry
University College London, London WC1H OAJ

INTRODUCTION

In this contribution to the school we wish to discuss the application of elastic neutron scattering techniques to the elucidation of defect structures. In particular we will address the problems of obtaining information on static defects in complex materials using powder diffraction and also the investigation of the dynamic defect structures found at high temperatures in the so-called superionically conducting materials.

DIFFRACTION AND STATIC DEFECTS: THE PROFILE REFINEMENT OF POWDER DATA

The name "profile refinement" is used to describe a particular technique for obtaining structural information from diffraction measurements on polycrystalline samples. The straightforward analysis of a powder diffraction pattern consists of associating individual peaks with particular Bragg reflections (indexing the pattern) and evaluating a diffraction intensity for each reflection by integrating the area under the peak in a plot of intensity versus scattering angle. Intensities thus obtained can be used in the usual way for structural determination. A problem arises when the resolution of the pattern is not fine enough to completely resolve all the Bragg peaks. In this case a "conventional" analysis requires that some arbitrary division be made of the intensity in overlapping peaks between the component Bragg reflections. The profile refinement technique, which was originally formulated by Rietveld (1969), addresses the problem by essentially forgetting about the individual Bragg reflections and refining the pattern as a whole. This is achieved by including each measured point of the diffraction pattern as an observation in a least squares fitting procedure.

The first requirement for a successful refinement is an accurate description for the shape of the peak created by the individual peak overlaps. For powder diffractometers, operating in constant wavelength

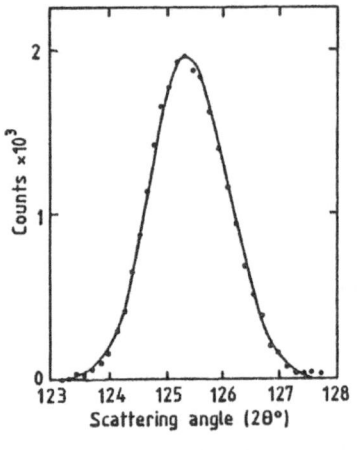

Fig 1a   A Gaussian peak shape
obtained using constant wavelength
neutrons
After Rietveld (1969)

a

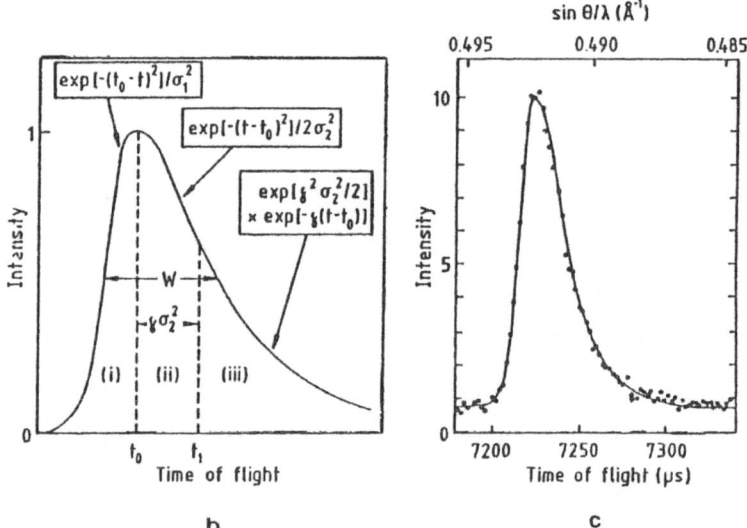

b   c

Fig 1b   Peak shape obtained using Pulsed Neutrons

Fig 1c   Fit to Ni 222 peak
          (b) and (c) after Cole and Windsor (1980)

(CW) mode, the peak shape depends on a number of factors besides the
crystallinity of the sample. The machine dependent effects include the
mosaic structure of the crystal monochromator and the geometry of the
defining collimators. The convolution of these factors produces a peak
shape which is almost exactly Gaussian (fig 1a). At very low and very
high scattering angles the Bragg peaks can show asymmetry which is a
function of the finite divergence of the incident beam. Under these
circumstances the simple form of the peak shape (H) has allowed its
dependence on the scattering angle to be analytically formulated in
terms of three parameters U, V and W. The Gaussian nature of the peaks
allows us to define simply the extent of an individual peak as 1.5H.
Since the tails of a Gaussian function fall rapidly with distance from

the maximum, little error is introduced via this assumption.

There are now several hybrid versions of the original Rietveld program in general circulation; they all however follow the same basic pattern. The program is split into two parts. The first part determines those reflections that can contribute to the intensity $Y_i(\text{calc})$ at any point i in the pattern in accordance with the 1.5H criterion. This part also corrects the observed intensities $Y_i(\text{obs})$ for the background which is estimated by interpolation of the background level in regions between the Bragg peaks. The second part of the program contains the main least squares routine which minimizes the function

$$M = \sum_i w_i [y_i(\text{obs}) - \frac{1}{C} y_i(\text{calc})]^2$$

where the summation is over all points in the pattern. The weighting factor (w) is taken to be inversely proportional to the variance of the quantity in the square brackets and C is an overall scale factor.

In essence the refinement consists of minimizing the sum of the squares of the weighted differences between the observed and calculated intensities for every point in the profile under the Bragg reflections. This is achieved by adjusting structural parameters (lattice parameters, atom co-ordinates and thermal parameters) and profile parameters (U, V, W and the diffractometer zero point). As is inferred in the title, the technique is a refinement process and not an ab initio method for structure determination. As such a good initial model for the structure is required for sensible starting values of the structural parameters. The initial values for the profile parameters would generally be typical values obtained in refinements of data collected on the diffractometer in use.

Several authors have suggested that the original formulation of the technique by Rietveld is not entirely satisfactory. In a sequence of papers Cooper and co-workers (1979) have discussed the point that it fails to recognise that the two groups of least-squares parameters are determined by different features of the diffraction pattern. The structural parameters depend on the integrated intensities of the reflections and not on their shape or positions, whereas the reverse is true for the profile parameters which are shape dependent. Cooper suggests that a refinement is satisfactory only if the goodness of fit is the same for the two groups of parameters. He suggests that this is most easily obtained if the profile parameters are refined first and then the structural parameters are refined in a separate second stage to the procedure.

Another deficiency is that the residuals associated with observations at different points on the same Bragg peak are correlated with one another but this effect is ignored during refinement. Clarke and Rollett (1982) have examined the effect of this correlation on the parameters of a few test cases including UO_2 and Al_2O_3. The correction for correlation produced increases in e.s.d's by up to a factor of two; the effect was most significant in the refinement of thermal parameters.

PULSED NEUTRON SOURCES AND POWDER DIFFRACTION

The installation of high resolution powder diffractometers on pulsed neutron sources, such as those at the Argonne Laboratory (U.S.A.) and the Rutherford Laboratory (U.K.), has greatly increased the quality of the diffraction data obtainable from a material. In particular the extension of high resolution into high Q ranges has resulted in the

collection of extensive data sets, containing reflections obtained at
low d spacings which are not accessible to CW diffractometers. The
extra information contained in reflections in this part of a diffraction
pattern enhances the accuracy obtainable on structural parameters.
These reflections also facilitate space group determination and thus it
may become possible, with judicious usage of synchrotron sources (X-rays
are much better for solving the phase problem) for ab initio powder
structure determinations to be performed.

It is our aim in this section to outline the procedures adopted
and the problems encountered in extending the profile technique to the
analysis of time of flight (tof) data. It should be recalled from the
lectures of Stewart (1985) that pulsed sources are created via the
interaction between an accelerated proton or electron beam and a heavy
metal (usually uranium) target. The resulting high energy neutrons are
moderated to epithermal and thermal energies before they are used for
diffraction. The moderation process creates a non-Gaussian and
distinctly asymmetric line profile. As we have already discussed, it is
essential that an accurate description of the peak shape is available if
a successful refinement is to be achieved. In principle this would be
possible using a minimum number of parameters if the full details of the
spectrometer were known. Unfortunately this level of detailed knowledge
does not exist and instead an empirical fitting approach has to be
adopted.

Cole and Windsor (1980) working with data obtained on the BSS on
the Harwell LINAC consider the lineshape, shown in fig 1b, to be made up
of three distinct sections. At the leading edge of the pulse (region 1
in fig 1b) the time of flight of the neutrons can be adequately
described in terms of a Gaussian function. The first part of the
trailing edge (region 2) also appears to be Gaussian but with a different
constant from the leading edge. The long remaining part of the trailing
edge (region 3) is a result of the slow decay of the thermalized neutron
pulse, which is represented by an exponential function.

Each region in fig 1b is determined by two parameters (intensity
and width) but since there is a smooth transition between the regions
it is relatively easy to inter-relate the parameters. The principal
parameters are σ_1 and σ_2, which determine the widths of the Gaussian
regions, and γ, the decay constant of the exponential region. The
three peak shape parameters exhibit a complex dependence on wavelength
(tof) whose empirical functions, displayed on fig 1b, require seven
coefficients to describe fully their behaviour (cf the three for CW
operation). In fig 1c we show the excellent level of fit, in this case
to the Ni 222 peak, obtainable from this description.

Currently the most widely used peak shape description is that due
to Von Dreele and co-workers (1982) which is programmed into the suite
of refinement programs for the GPPD on the IPNS at the Argonne
laboratory. The peak shape function is considered to be a convolution
of a much simpler moderator pulse shape than that of Cole and Windsor
and a Gaussian instrumental contribution. The moderator pulse is
described in terms of a rising exponential for the leading edge and a
different decaying exponential for the trailing edge. These two
processes are characterized by independent exponential rise and decay
constants which, with the width function that characterizes the
instrumental Gaussian, form the three profile shape parameters. Their
variation with tof can be described in six coefficients, although it
should be noted that in the refinements so far performed using this
model two of the parameters display a pronounced tendency to be zero.

TABLE 1 Comparison of structural information for Mg_2SiO_4 (after Langer *et al.*, 1981)

(a) Atomic Co-ordinates

X = parameters from X-ray single-crystal data. N = parameters from neutron single-crystal data. R = parameters from Rietveld refinement of neutron powder data.

		x	y	z
Mg(2)	X	0.9915(1)	0.2774(1)	0.25
	N	0.9914(3)	0.2773(2)	0.25
	R	0.9917(3)	0.2775(1)	0.25
	X	0.4265(1)	0.0930(1)	0.25
	N	0.4263(4)	0.0939(2)	0.25
	R	0.4268(3)	0.0940(1)	0.25
	X	0.7658(1)	0.0917(1)	0.25
	N	0.7657(3)	0.0915(2)	0.25
	R	0.7662(1)	0.0914(1)	0.25
	X	0.2218(1)	0.4471(1)	0.25
	N	0.2217(3)	0.4473(1)	0.25
	R	0.2211(3)	0.4472(1)	0.25
	X	0.2774(1)	0.1631(1)	0.0331(1)
	N	0.2776(2)	0.1629(1)	0.0330(1)
	R	0.2776(2)	0.1630(1)	0.0331(1)

(b) Thermal Parameters

Thermal parameters ($\times 10^5$) are of the form

$$\exp[-(\beta_{11}h^2 + \beta_{22}k^2 + \beta_{33}l^2 + 2\beta_{12}hk + 2\beta_{13}hl + 2\beta_{23}kl)].$$

		β_{11}	β_{22}	β_{33}	β_{12}	β_{13}	β_{23}
Mg(1)	X	394 (6)	132 (2)	254 (3)	-3 (2)	-42 (4)	-37 (2)
	N	191(48)	140(12)	253(12)	34(20)	-43(29)	-42(19)
	R	403(44)	144(11)	307(33)	17(19)	-20(30)	3(14)
Mg(2)	X	483 (7)	88 (2)	314 (4)	2 (2)		
	N	333(49)	106(12)	260(30)	11(21)	$\beta_{13} = \beta_{23} = 0$	
	R	348(43)	101(11)	307(32)	12(17)		
Si	X	241 (5)	78 (1)	208 (3)	4 (1)		
	N	211(58)	135(15)	158(36)	-3(26)	$\beta_{13} = \beta_{23} = 0$	
	R	-13(56)	81(11)	299(30)	-39(21)		
O(1)	X	266 (9)	117 (9)	302 (6)	4 (4)		
	N	284(45)	156(11)	244(25)	27(20)	$\beta_{12} = \beta_{23} = 0$	
	R	136(49)	107 (9)	344(25)	-25(18)		
O(2)	X	411(10)	77 (2)	324 (6)	-3 (4)		
	N	382(48)	98(11)	390(27)	-28(20)	$\beta_{12} = \beta_{23} = 0$	
	R	409(47)	77 (9)	276(26)	-9(18)		
O(3)	X	430 (7)	116 (2)	269 (4)	5 (3)	-20 (4)	50 (2)
	N	353(30)	174 (8)	232(19)	11(14)	-3(20)	48(12)
	R	273(37)	125 (7)	248(17)	88(12)	-54(21)	24 (8)

RELIABILITY OF THE PROFILE REFINEMENT TECHNIQUE

There have been a number of powder refinements performed on materials for which single crystal structural information is available. Thus we are in a position to assess the reliability of the structural parameters obtained from the technique.

Larger and co-workers (1981) using the IPNS facility have reported a comparison between powder diffraction experiments and single crystal experiments, using both X-rays and neutrons, on forsterite, Mg_2SiO_4. The positional parameters they report are reproduced in table 1a, and as can been seen the agreement in both absolute values and uncertainties between the various techniques is good. The thermal parameters, reproduced in table 1b, are also, with the notable exception of β_{11} for silicon, in good agreement with one another. There is no obvious reason for the errant value of β_{11} silicon, since in general thermal parameters are well handled by the technique. This is illustrated by the study of Albinati and co-workers (1980) who analysed the variation of the isotopic thermal parameters in UO_2 at five temperatures above ambient. They were able to apply both the Rietveld and integrated intensities methods in the data analysis and report no significant differences in the handling of thermal behaviour between the two techniques.

Thus while the analysis due to Rietveld is not thought to be a totally rigorous theoretical treatment of the problem, the consensus of opinion is that it is capable of producing high quality structural information and reasonable estimates of the corresponding standard deviations.

It has been extremely successful in extending the range of materials for which structural refinements from powder data can be attempted. We wish to end this section by highlighting some of the recent areas of application of the technique.

PHASE TRANSITIONS

Profile refinement has found wide application in the study of structural phase transitions as a function of both temperature and pressure, since it enables the small structural distortions and atomic displacements associated with such transitions to be determined with a high degree of accuracy. This ability has been fully exploited in the study of ferroelectric materials. This is especially true for those materials in which the transition causes twinning in single crystals. We refer the interested reader to the study of Sleight and co-workers (1979) on $BiVO_4$ as a representative example of this application of the technique. Also of note is the evolution of the structure of KD_2PO_4 with pressure performed by Nelmes (1980).

ZEOLITES

These aluminosilicate systems are classic materials for powder neutron diffraction experiments. This is because large single crystals of the materials are exceptionally difficult to prepare and also the neutron scattering lengths are sufficiently different to allow differentiation between silicon and aluminium ions (cf X-ray). The technique has been successfully applied to the investigation of the more symmetric zeolite framework structures and their cation distributions. Interested readers are referred to the recent papers of Cheetham and co-workers (1984).

A novel application recently reported by Fitch and co-workers (1985) is the localization of benzene molecules adsorbed into the cavities of zeolite Y. At low coverages (concentrations) of the organic they discovered that the benzene molecules were bound to the framework cations and located mainly in supercages in the structure. As the coverage was increased, van der Waals forces allowed the benzene molecules to occupy the less energetically favourable "window sites" formed between adjoining supercages.

The new generation of very high resolution diffractometers on intense sources should greatly facilitate the investigation of the large complex unit cells of the low symmetry zeolites. We expect a great deal of activity on these materials in the next few years.

NON-STOICHIOMETRY

The investigation of the accommodation of non-stoichiometry in a structure has been a central area of interest for the application of the technique. This is evidenced in the reference sections of the reviews of Cheetham and Taylor (1978) and Hewat (1979). The continuing power of the technique is demonstrated in the recent refinement of Bayerlein and co-workers (1984) of data taken on the IPNS for the pyrochlore structured materials $Pb_2Ru_2O_6._5$ and $PbTlNb_2O_6._5$. This study showed for the first time in this type of material, evidence for oxygen vacancy-anion ordering accompanied by A site cation displacements in response to the vacancies. The ordering was intimated by the measurement of weak reflections that could not be indexed in the expected cubic pyrochlore space group. The detection of the lowering of symmetry in these materials is rather a nice example of the subtle effects being revealed by the higher fluxes on pulsed sources.

SUPERIONICALLY CONDUCTING MATERIALS

The investigation of defect distributions in a lattice as a function of temperature has been fuelled by the interest in fast ion conducting materials. A fundamental difficulty in investigating these materials lies in the limited data sets that can be obtained from them. Those materials of most interest, for example the fluorites discussed by Chadwick (1985), tend to be cubic and with the intensity decrease associated with high temperature diffraction, we are unfortunately only measuring enough independent intensities to allow very simple modelling of the defect structure in the refinement.

In the next section we shall discuss fast ion conduction in a little more detail and outline the level of structural information that can be obtained from single crystal studies of these materials.

DIFFRACTION AND DYNAMIC DISORDER

As was discussed in the lectures of Corish (1985) one area of defect science that has received a considerable amount of attention in recent years is the generation of fast ion conduction in certain materials. They are characterized by the rapid diffusion of ions through the crystal lattice, the mobile ions moving through a framework formed by all the other ions in the structure. The ion jumps, from occupied to unoccupied positions, are thermally activated processes. The low level of excitation required to promote the jumps (typically $E_{act} \leqslant 0.5$ eV) can be visualised in terms of a low potential barrier between the positions. The linking of these individual potentials creates a low energy diffusion pathway in the lattice.

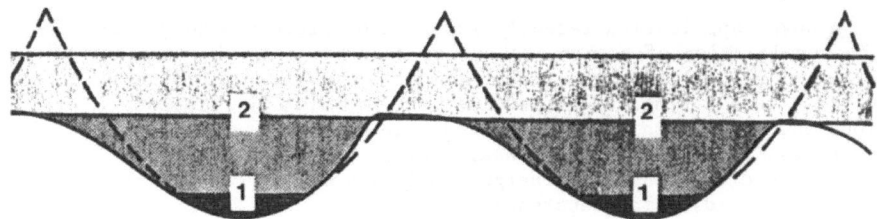

Fig 2   Schematic Visualisation of Diffusion Pathway
After Schulz (1982)

In fig 2 we show the generally visualised situation along the diffusion pathway. It is immediately apparent that the minima are broad and shallow, and it is a consequence of this phenomenon that the degree of certainty about the location of atomic positions in superionics falls rapidly with relatively small increases in temperature. The general arguments about this point can be easily comprehended by reference to fig 2. Clearly, only those ions with thermal energies below level 1 can be considered, without question, to occupy a defined position. Those between states 1 and 2 possess unusually large vibrational amplitudes. Once these exceed the radius of the ion then it becomes difficult to define uniquely a lattice position for the ion with any degree of confidence. Once an ion possesses a thermal energy larger than the highest potential barrier along the diffusion path i.e. energies about level 2, it has to be considered as being delocalized along the diffusion pathway.

As is discussed by Lechner (1983) molecular motions, such as rotations and proton tunnelling effects, are normally investigated by means of inelastic neutron scattering. In this section however, we would like to consider the kind of structural information that can be extracted from Bragg diffraction on superionic materials. In particular we are interested in extracting the averaged distribution of the mobile ions contained in the diffraction snapshot of the disorder. Bragg diffraction data from fast ion conductors are usually analysed within the space group of the averaged structure. Both dynamic and static disorder of the conducting ions are assumed then to be absorbed by the thermal and occupational parameters of the averaged structure. Thus we particularly want to discuss the choice of physically representative thermal parameters that allow the differentiation of the point defect formations from the large dynamic temperature effects inherent in superionic materials at high temperatures.

We have superimposed on fig 2 the general form of the potential generated from an harmonic temperature factor. It is apparent that the harmonic approach is only a reasonable approximation at points close to the lattice position. It is not able to reproduce the strong anharmonicity apparent in these systems. We do not intend to discuss the derivations and relative strengths of the current analytical descriptions for temperature factor equations incorporating anharmonic thermal motion. Interested readers are referred to the pioneering work of Willis and Pryor (1975). We would simply like to note that they are generally written in terms of series expansions and that only the Gram-Chalier expansion allows transformations from reciprocal space to real space to be performed without some level of approximation being involved. We should also like to draw attention to the point that the higher order coefficient in the series expansions are usually highly correlated. It is possible to give direct, qualitative interpretations,

to the low order coefficients. For example, the high site symmetry
displayed by fluorine ions in the cubic fluorite structure means that
anisotropic thermal behaviour can be examined fully by including terms
up to the third order in the expansion. This can be visualised as four
lobes, in the form of a tetrahedron, extending along the <111> directions
(Willis and Pryor 1975).

In less symmetric systems however, the higher terms in the series
expansions are required to describe fully the site thermal behaviour
(Johnson and Levy 1974). In order then to assess the physical
significance of the refined coefficients we require an interpretation
route that circumnavigates the coefficient correlation problem. An
approach pioneered by Schulz (1982) consists of fourier transforming
the coefficients into one real space quantity, the probability density
function (pdf). This can be envisaged as a statistical description of
the thermal vibrations of the atoms in the ensemble in as much as it
states the probability of a thermal vibration from an atomic position
into a neighbouring volume of the crystal. Summation of the individual
pdf's leads to the joint probability density function (PDF) and plots
of this function across a lattice plane give a simple pictorial
representation of the thermal behaviour of the system. A more detailed
understanding can be achieved according to Schulz if the PDF's are
further manipulated to create one particle atomic potentials. These
potentials can then be linked to construct an approximation to the
potential along the diffusion path. Hence information on atomic
positions, potential barriers and activation energies for jumps between
regular sites can in principle be obtained.

The literature in the last three years has seen an explosion in
the number of papers applying these concepts to the structural
investigation of a wide range of superionic materials. Chadwick (1985)
has already discussed the high temperature fluorites and we shall
illustrate the techniques by highlighting the analysis performed by
Schulz and co-workers (1979) on the lithium conducting solid electrolyte
Li_3N. As is discussed by Catlow (1985) this material is a fast and
anisotropic lithium conductor. Its structure, as is shown in fig 3,
is composed of (Li,N) layers interconnected by pure Li layers. Crystal
structure investigations as a function of temperature utilizing harmonic
temperature factors resulted in a decrease in the occupation
probabilities of the lithium sites in the (Li,N) plane, designated
Li(2) in fig 3, with temperature, as is schown schematically in fig 4a.
This effect is balanced by the emergence of an interstitial site
between two lithium sites. The incorporation of this site into the
refinement significantly reduced the R factor value. The obvious
interpretation of these results was that the lithium ion conductivity is
caused by jumps of lithium ions within the (Li,N) layer between regular
and interstitial sites. The observations accord with the observed
anisotropy in the conductivity but unfortunately the occupation of the
interstitial site results in unusually low Li-N bond lengths.

Re-evaluation of the data using anharmonic temperature factors
resulted in a significantly different picture. In this refinement the
lithium concentration at the Li(2) sites remained constant at full
occupancy and hence no interstitial site could be located. The
probability densities plotted in fig 4b clearly illustrate that the
Li(2) ions have large vibrational amplitudes along the Li(2)-Li(2)
connection lines. The potential evaluated in this case is plotted in
fig 4c. It has been interpreted in the following way; the Li(2) ions
occupy only their regular sites in the (Li,N) plane, no interstitial
sites are involved. Lithium ions with energies higher than the

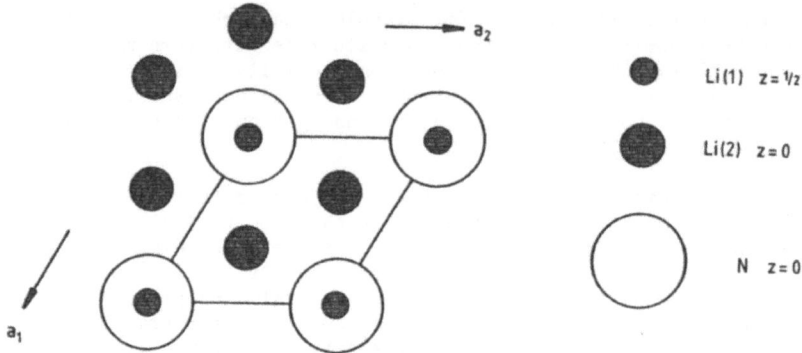

Fig 3   The Structure of Li₃N

potential barrier (heavy black line) are allowed to change site.
Therefore the potential barrier of fig 4c marks the activation energy
of a Li(2) ion for the jump between two regular positions in the
(Li,N) plane. From fig 4c a potential height of 0.3 eV can be read,
which is in excellent agreement with the activation energies from
transport measurements performed by Von Alpen and co-workers (1977)
which ranged from 0.27-0.29 eV.

It should be noted that this is the activation energy for a jump
between an occupied and unoccupied site, however, nominally there are
no vacancies on the regular sites in this model. This problem could be
explained if the vacancies existed in concentrations below the detection
limit of the technique, circa 5 mole percent for diffraction
experiments. This hypothesis would of course be consistent with the
molecular dynamics simulation of the material discussed in the
presentation of Catlow (1985).

SUMMARY

In the first section we discussed the level of structural
information that can be obtained from powder diffraction. In
particular we considered the situation where the data analysis is
complicated by the overlap of Bragg reflections. Under these
circumstances a conventional structural analysis would require an
arbitrary division of the total intensity into contributions from
component Bragg reflections.

The profile refinement technique addresses the problem by
including each measured point of the diffraction pattern as an
observation in a least squares fitting procedure. The least squares
minimization is achieved via the variation of two types of parameters
(a) structural and (b) profile. An accurate description of the first
parameters depends on a accurate determination of the integrated
intensities, while the profile parameters depend on the peak shape.
The simple Gaussian peak shape obtained on constant wavelength
diffractometers is undoubtedly one of the reasons why this technique
has enjoyed a large amount of success.

We then proceeded to discuss the extension of the technique to
data collected on pulsed neutron sources. The key to exploiting the
high resolutions of these diffractometers lies in correctly describing
the asymmetry of the peak shapes created by the neutron moderation

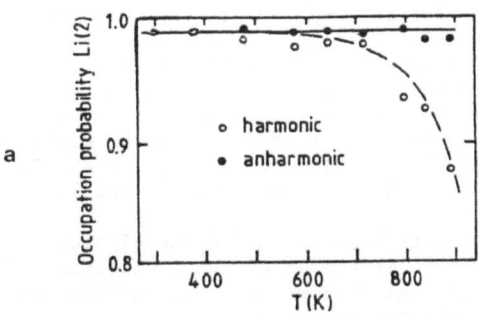

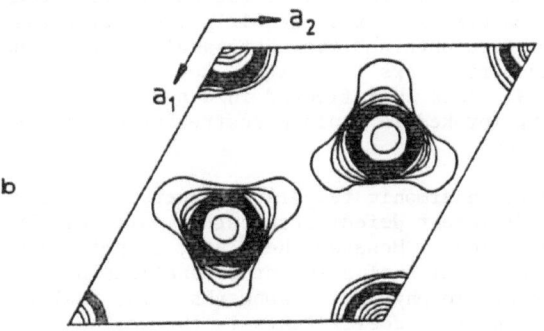

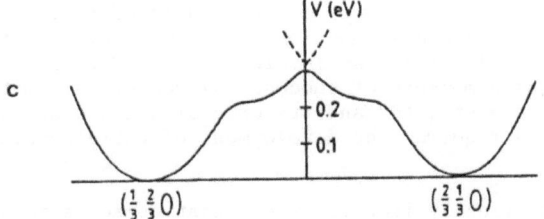

Fig 4a   Occupation probabilities vs Temperature for Li(2)

4b   Plot of Probability Density Function

4c   The generated one particle potential

After Schulz (1982).

processes. We continued the discussion by outlining the general arguments behind the current empirical fitting procedures. The key point of this discussion is that it is possible to extend successfully the technique to the analysis of time of flight data, although an increased number of profile parameters are required if the refinement is to account correctly for the peak shape. We ended this section of the presentation with a brief outline of the wide range of problems to which the technique has been successfully applied. We hope that we have shown that this very versatile technique is capable of producing high quality structural information on a diverse range of materials and end this section by noting for those readers who might like more detailed information about profile analysis, the excellent recent review of the subject by Albinati and Willis (1982).

In the second section we discussed the elucidation of the defect structures of superionic materials by the application of high temperature single crystal diffraction. We outlined the analysis problems created by the nature of the diffusion pathways of the mobile species. In order to extract reliable information on the averaged distribution of the nuclear density, it is necessary to describe correctly the anharmonic thermal vibrations experienced by the mobile ions in these materials. It is Schulz's contention that the macroscopic picture obtained from data analysis using harmonic temperature factors infers too large a degree of positional disorder on the mobile ions. This is reflected in a number of spurious interstitial positions being invoked to explain scattering density smeared out from assigned positions.

The use of anharmonic temperature factors can certainly lead to significantly different defect structures, particularly in terms of defect concentrations. However there are, as pointed out, problems in interpreting the results of a structure analysis using these temperature factors within simple physical frameworks. The qualitative pictures available for low order coefficients in the series expansions mean that highly symmetric structures are amenable to investigation. However, if this kind of work is to be extended successfully to less symmetric materials then one area where attention must be focussed is the interpretations of the higher order terms.

In this respect the approach of Schulz of fourier transforming the coefficients into one function and then working in terms of one particle potentials offers an interesting way forward. Despite the undoubtedly large measure of success enjoyed by this method it is still relatively unsophisticated and its extension for usage on highly complex centres requires the development of correspondingly complex potentials.

We would finally like to stress that defect structures are complex phenomena and as such the results obtained from one technique should be allied to the findings of other techniques if a balanced picture of the situation is to arise.

It is hoped that this presentation has been complementary to the lectures of Stewart (1985) and that as such the reader has grasped the wide range of properties that can be probed via neutron scattering. Obviously in the allotted time it has not been possible to cover every topic that we would have liked to have done. One obvious omission is the use of diffuse scattering to probe short range ordering of defects. This technique basically provides information on the defect arrangement in a lattice as seen from a particular site in the interior of the

142

crystal. This subject is covered rather nicely at an introductory
level in an article by Fender (1973).

REFERENCES

Albinati A. and Willis B.T.M. (1982) J. Appl. Cryst. 15 361-374.
Albinati A., Cooper M.J., Rouse K.D., Thomas M.W. and Willis B.T.M.
    (1980) Acta Cryst. A36 265-270.
Bayerlein R.A., Horowitz H.S., Longo J.M., Leonowicz M.E., Jorgensen
    J.D. and Rotella F.J. (1984) J.S.S.C. 51 253-265.
Catlow C.R.A. (1985) This Publication.
Chadwick A.V. (1985) This Publication.
Cheetham A.K. and Taylor J.C. (1978) J.S.S.C. 21 253.
Cheetham A.K., Eddy M.M. and Thomas J.M. (1984) J. Chem. Soc. Chem. Comm.
    1337-1340.
Clarke C.P. and Rollett J.S. (1982) J. Appl. Cryst. 15 361-374.
Cole I. and Windsor C.G. (1980) Nuclear Instruments and Methods 171
    107-113.
Cooper M.J., Sakata M. and Rouse K.D. (1979) 167-187 in "Accuracy in
    Powder Diffraction" N.B.S. Special Publication 567 eds Block S.
    and Hubbard C.R.
Corish J. (1985) This Publication.
Fender B.E.F. (1973) 250-270 in "Chemical Applications of Thermal
    Neutron Scattering" ed Willis B.T.M. Oxford University Press.
Fitch A.N. (1985) Personal Communication.
Hewat A. (1979) 111-143 in "Accuracy in Powder Diffraction" N.B.S.
    Special Publication 567 eds Block S. and Hubbard C.R.
Johnson C.K. and Levy H.A. (1974) 313-336 in "International tables for
    X-ray Crystallography" vol 4 Kynoch Press Birmingham England.
Larger G.A., Ross F.K., Rotella F.J. and Jorgensen J.D. (1981) J. Appl.
    Cryst. 14 137-139.
Lechner R.E. (1981) 169-226 in "Mass Transport in Solids" Proceedings
    of NATO ASI Lannion 28 June - 11 July 1981 ed Bénière F. and
    Catlow C.R.A. Plenum London.
Nelmes R.J. (1980) Ferroelectrics 24 237-245.
Rietveld H.M. (1969) J. Appl. Cryst. 2 65-71.
Schulz H. and Zucker U. (1979) 495-499 in "Fast Ion Transport in Solids:
    electrodes and electrolytes" eds Vashista P., Mundy J.N. and
    Shenoy G.K. North Holland.
Schulz H. (1982) Ann. Rev. Mat. Sci. 12 351-376.
Sleight A.W., Chen H.Y. and Ferretti A. (1979) Mat. Res. Bull. 14
    1571-1579.
Stewart R.J. (1985) This Publication.
Von Alpen U., Rabenau A. and Talet G.H. (1977) Appl. Phys. Lett. 30
    621-623.
Von Dreele R.B., Jorgensen J.D. and Windsor C.G. (1982) J. Appl. Cryst.
    15 581-589.
Willis, B.T.M. and Pryor A.W. (1975) "Thermal Vibrations in
    Crystallography" Cambridge University Press.

APPLICATION OF THE POSITRON ANNIHILATION TECHNIQUE IN STUDIES OF

DEFECTS IN SOLIDS

Morten Eldrup

Metallurgy Department
Risø National Laboratory
DK-4000 Roskilde, Denmark

ABSTRACT

The basic principles of positron annihilation physics are discussed and the four most important experimental techniques are described (i.e. the positron lifetime, the angular correlation, the Doppler broadening, and the low--energy-positron beam techniques). Several examples are discussed, in particular for metals and molecular crystals, which illustrate the sensitivity of the positron annihilation techniques to vacancy type defects. For example it is shown how information can be obtained about vacancy formation energies, vacancy migration and clustering, vacancy-impurity interactions, densities of rare gasses in bubbles in metals, and defect density profiles in near-surface regions.

1. INTRODUCTION

The positron was discovered by Anderson in 1932[1], although its existence had already been predicted by Dirac[2] a few years earlier. However, it was not until about a year later that Blackett and Occhialini[3] demonstrated that the predicted particle was identical to the observed one. The exciting account of the discovery of the positron can be found in Ref. 4.
As predicted by Dirac's relativistic quantum theory[2] the positron is the antiparticle to the electron. This means that the positron has the same mass and spin as the electron, but has the opposite charge, viz. one positive elementary charge. Furthermore, if a positron is surrounded by one or more electrons the positron may annihilate with one of the electrons, i.e. both particles disappear and their masses are transformed into energy which is emitted as γ-quanta, normally two or three. The properties of these γ-quanta, such as their energies, emission directions, and time of emission which can all be measured, provide useful information about the behaviour of the annihilating positron-electron pair and consequently about the material in which the positrons annihilate. Very briefly, this is the principle of the Positron Annihilation Technique (PAT, also referred to as Positron Annihilation Spectroscopy, PAS, or simply PA).
Positron Annihilation research is a very wide field as can be judged from the proceedings of the most recent international conferences[5-7] and other recent books.[8,9] To illustrate this breath let us mention a few other recent reviews. They describe works concerning fundamental properties of positronium, including tests of quantum electrodynamics,[10,11] positron-atom

scattering in gases,[12,13] positron and positronium chemistry,[14-17] solid state physics, including both perfect and defected materials,[8,9,18-21] surface physics,[22-24] and medical physics.[25]

In the present paper we shall concentrate on a discussion of the possibilities to study defects in solids, primarily metals and molecular crystals. First, however, we shall discuss some of the basic physical principles of positron annihilation and then describe the most important experimental techniques used in positron annihilation studies.

2. POSITRONS IN SOLIDS

In conventional positron annihilation experiments the positrons are injected into a solid with a mean energy of the order of 200 keV. They slow down to thermal energies in about 10^{-12}-10^{-11} sec (1-10 psec) by ionisation and excitation of the solid. During this time they penetrate a distance of 10-1000 μm depending on the density of the solid (the penetration depth is roughly inversely proportional to the density). Hence, the positrons probe bulk material in such experiments. In recent years a rapid development of a new technique has taken place, viz. of low-energy-positron beams.[22-24] With such beams it is possible to inject monoenergetic positrons with variable energy and hence variable penetration depth into a sample and thus study surface and near-surface properties as will be illustrated in Section 4.

Annihilation of Positrons

The characteristics of positron-electron annihilation are derived from quantum electrodynamics.[26] Certain spin selection rules determine the number of γ-quanta emitted at annihilation. For a free positron in a medium, annihilation with emission of 2 γ-quanta is most probable. It occurs when the annihilating positron and electron have antiparallel spins. The cross-section for annihilation with emission of 3 γ-quanta (for parallel positron and electron spins) is 1/379 of the 2γ cross-section. Annihilations with emission of 1, or more than 3 quanta have even smaller probability. Hence, for annihilation of free positrons, only 2γ-emission is important.[27,28] (The 3γ annihilation is important though when positron-electron bound states with parallel spins, like ortho-positronium, are formed. See below).

The two annihilation quanta have a total energy of $E = 2m_o c^2 = 2 \times 511$ keV (m_o is the electron or positron rest mass, c the velocity of light) and are emitted in almost opposite directions as sketched in Fig. 1. The angle Θ is usually very small, typically < 20 milliradian (mrad) (20 mrad $\simeq$ 1°).

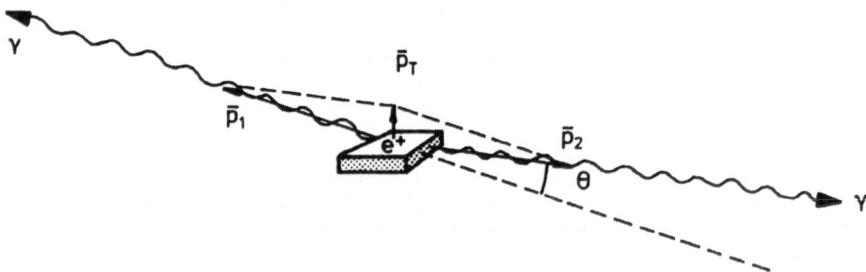

Fig. 1. Emission of two annihilation γ-quanta in almost opposite directions. The vectors represent the momentum of the two quanta ($\bar{p}_1$ and $\bar{p}_2$) and the total momentum of the pair ($\bar{p}_T$). The angle Θ is usually very small, typically Θ<1°.

In the center-of-mass frame for the annihilating electron-positron pair the two γ-quanta are emitted in exactly opposite directions because of momentum conservation, and both have an energy of $m_o c^2$. Typically the annihilating pair has a total kinetic energy of a few eV so that in the laboratory frame the two γ's are not emitted in exactly opposite directions (Fig. 1), and their energies deviate slightly from $m_o c^2$ (see further in Sec. 3). In the independent-particle approximation the distribution of the total momentum of γ-quanta, $\bar{p}_T$, is given by the annihilation rate[20,29]:

$$\Gamma(\bar{p}_T) \propto \sum_j \left| \int \exp(-i\bar{p}_T \cdot \bar{r}) \psi^+(\bar{r}) \psi_j^-(\bar{r}) d\bar{r} \right|^2 \qquad (1)$$

where ψ^+ and ψ_j^- are the wavefunctions for the positron and the j'th electron state, respectively, both taken at the same $\bar{r}$ value. Roughly speaking Eq. (1) expresses the momentum conservation in the annihilation process, i.e. the momentum distribution of the γ's reflects the momentum distribution of the annihilating electron-positron pair. It can be determined by measuring the distribution of angles Θ, since in a good approximation we have:

$$\Theta = p_T / m_o c \qquad (2)$$

In many cases, in particular in metals, the positron momentum is low compared to the electron momentum. Hence, a measured momentum distribution is essentially that of the electrons. Such measurements are being used to determine the electronic structure of metals and alloys.[18,29,30]

On integrating Eq. (1) over all $\bar{p}_T$ we obtain the total annihilation rate[20] which is (including the proportionality constant):

$$\lambda = \pi r_o^2 c \int \rho^-(\bar{r}) \rho^+(\bar{r}) d\bar{r} \qquad (3)$$

where r_o is the classical electron radius ($= e^2/m_o c^2$) and ρ^+ and ρ^- are the positron and total electron density of $\bar{r}$, respectively. From Eq. (3) we see that the positron mean lifetime (normally just referred to as the lifetime) $\tau = \lambda^{-1}$ is a measure of the electron density sampled by the positron.

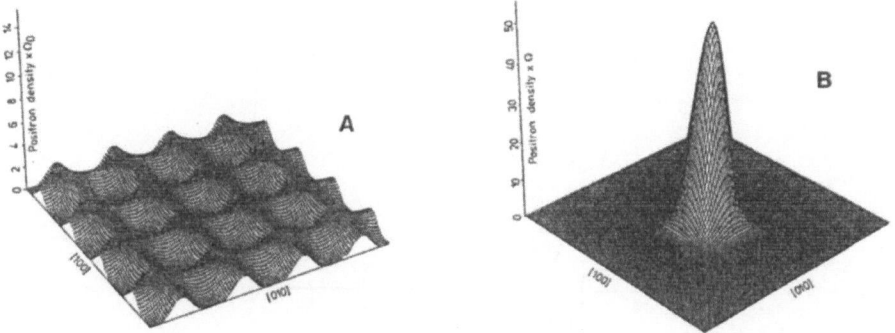

Fig. 2. Calculated positron densities in tungsten. A) The positron in the perfect lattice in a Bloch function state. The density vanishes at the cores of the positive ions and has maxima in the interstices. B) The positron trapped in a vacancy where it is strongly localized. From Ref. 32.

Positrons in Metals

Generally, positron behaviour in metals is simpler than in insulators, so let us first consider the metal case. In a perfect metal lattice the positron will, after slowing down, diffuse around, normally in a delocalized, Bloch-function, state. Positron lifetimes in metals are 100-400 psec, depending upon the metal.[31] During its lifetime the positron typically diffuses about 200 nm. Figure 2A illustrates the positron density in a perfect lattice.

If the metal contains defects such as vacancies, vacancy clusters and dislocations, i.e. regions of less than average density, positrons may become trapped, i.e. localized at these defects. This is because the positron is repelled by the positively charged ion cores. Hence, structural defects which represent missing (or a reduced density of) ions will provide attractive potentials for positrons (see e.g. ref. 33 for a more detailed description of the trapping potential). Figure 2B illustrates the density of a positron trapped at a vacancy. Trapped in such a defect the positron will experience a lower electron density than in the bulk material and its lifetime is therefore increased (Eq. (3)). Furthermore, the average conduction electron momentum at the defect is lower than in the bulk and the positron overlaps less with the high-momentum core electrons. Both these effects lead to a narrower total-momentum distribution for the annihilation quanta (Eq. (1)). These changes in annihilation characteristics for defect-trapped positrons (which can be measured; see next section) are the basis for the now well established use of PAT to metal defect studies.

It should particularly be noted that positrons are not trapped by interstitials or small clusters of these. Thus, the positron is specifically sensitive to vacancy type defects (low density regions).

The annihilation characteristics depend on the type of defect in which the positron is trapped. Hence it is possible to a large extent to differentiate between different types of defects, and from the rates with which the defects trap positrons defect concentrations can be derived.

For example, for positrons trapped in vacancy clusters their lifetime will generally increase with cluster size as illustrated by the calculated curves[34] shown in Fig. 3. In fact, theoretical calculations can to-day reliably predict lifetimes for positrons trapped at defect clusters of varying size, configuration and impurity content.[35-37] This will be further illustrated in Sec. 4.

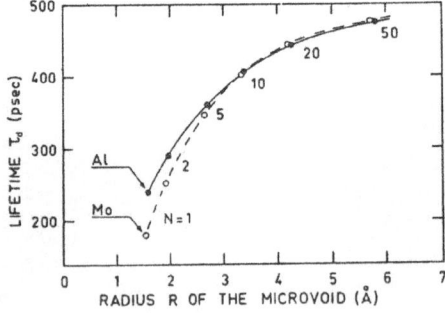

Fig. 3. The increase of the lifetime of a positron trapped in a 3-dimensional vacancy cluster (a void) as a function of cluster size, calculated in a spherical jellium model. For large clusters the lifetime saturates at about 500 psec. From Ref. 34.

Positronium in Insulators

In most insulators a fraction of the injected positrons will form a bound state with an electron, the so-called positronium atom (Ps), a light analog to the hydrogen atom, before annihilation. Two ground states of Ps exist. The singlet or para-Ps state, in which the positron and electron spins are antiparallel, has in vacuum an intrinsic lifetime against 2γ-decay of 125 psec. The triplet or ortho-Ps state has parallel particle spins and therefore, in vacuum, it can only decay by 3γ-annihilation with the much longer mean lifetime of 142 nsec.[10,11,38]

The Ps formation process in condensed matter has been a matter of strong debate.[16,17,38-41] However, there are both theoretical and experimental arguments that a certain (significant) fraction of the Ps yield (if not all) is a result of the so-called spur processes. In the spur process which leads to Ps formation the thermalized positron combines with one of the thermalized excess electrons created during the slowing down of the positron. This reaction competes with positron and electron reactions with other species (e.g. positive ions, radicals, or surrounding molecules). As a result, only a certain fraction of the positrons form Ps. Hence, in many insulators three different "positron particles" exist, viz. the "free" positron, the para-Ps, and the ortho-Ps.

The characteristics of Ps in a condensed material are changed compared to those for Ps in vacuum because of the overlap of the surrounding electrons with the Ps atom.[38] Two effects of this overlap are particularly important. One effect is the so-called pick-off process by which the positron in ortho-Ps annihilates with one of the surrounding electrons which has its spin opposite to that of the positron. In this process two γ-quanta are emitted and the ortho-Ps lifetime is strongly reduced from the 142 nsec vacuum value to a few nsec or less. The exact magnitude of the lifetime is determined by the amount of overlap between the positron in ortho-Ps and the surrounding electrons (similar to Eq. (3)). The ortho-Ps lifetime is thus a measure of the electron density sampled by the Ps atom.

The momentum distribution $\Gamma(\bar{p})_T)$ arising from the pick-off process is relatively wide (typically of the order of 10 mrad, see Section 5) as is the distribution from "free" positrons, since in both cases annihilation takes place with electrons of rather high momentum (equivalent to a kinetic energy of 5-10 eV). Annihilation of para-Ps on the other hand normally gives rise to a rather narrow momentum distribution. This is so, because Ps often is thermalized before annihilation. The total Ps momentum is therefore low (equivalent to a kinetic energy of $\sim$ 0.02 eV), and this is reflected in $\Gamma(\bar{p}_T)$ for para-Ps since para-Ps mainly annihilates intrinsically.

The second important effect of the Ps overlap with the surrounding electrons in condensed materials is due to the (repulsive) exchange interaction between these electrons and the electron in Ps. The Ps atom will therefore tend to occupy regions of less than average electron density in a material. Such regions may for example be vacancies or clusters of vacancies, and for ortho-Ps, trapping in these regions will lead to a longer lifetime because of a reduction in the pick-off rate resulting from the lower electron density. The localization of Ps results in an increase of its zero-point energy (and momentum) which again leads to an observable widening of the momentum distribution from para-Ps (see Section 5).

3. EXPERIMENTAL TECHNIQUES

For a number of years three experimental techniques have been dominating in PAT experiments, viz. the lifetime, the angular correlation, and the Doppler broadening techniques. Recently a rapid development of so-called low--energy-positron beams has taken place. We shall briefly describe these four techniques below. They are all based on nuclear spectroscopy by which the emitted γ-quanta (or in certain cases the positron itself) are detected. More detailed discussions of the techniques may be found in Refs. 23,24,31,42,43.

For all four techniques the sources of the positrons are in most cases radioactive isotopes such as ^{58}Co, ^{64}Cu, ^{68}Ge and ^{22}Na which emit positrons (e$^+$ or β$^+$). For lifetime measurements ^{22}Na is particularly useful, because a 1.28 MeV γ-quantum is emitted essentialy simultaneously with the positron, thus signaling the "birth" of the positron. This makes it possible to measure the positron lifetime as we shall now see.

Positron Lifetime Technique

The principle of a positron lifetime spectrometer is shown in Fig. 4. The source is often made by drying a droplet of a ^{22}NaCl aqueous solution

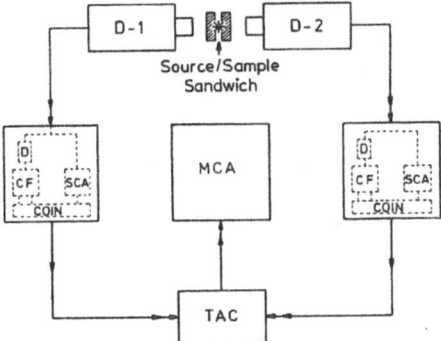

Fig. 4. Schematic diagram of a simple, modern lifetime spectrometer. The source is sandwiched between two samples. D-1 and D-2 are two detectors. DISCR are simultaneous time and energy discriminators as indicated inside the boxes, TAC is a time-to-amplitude converter, and MCA a multi-channel analyzer. From Ref. 44.

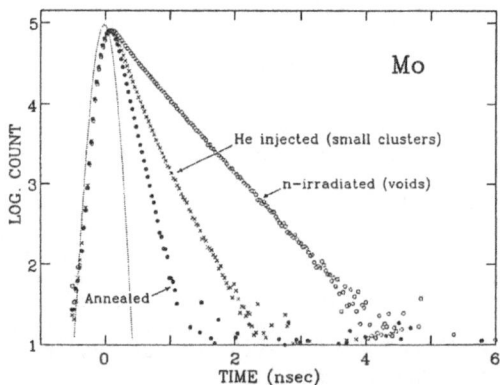

Fig. 5. Positron lifetime spectra for three molybdenum samples after subtraction of a constant background. The slopes on the right-hand side of the spectra equals the positron annihilation rates (inverse lifetime). A clear difference is seen between the lifetimes of positrons annihilating in bulk Mo ("Annealed"), trapped in small vacancy clusters ("He injected"), and trapped in large voids ("n-irradiated"). The dotted curve is the time resulution curve of the lifetime spectrometer.

150

on a thin foil ($\sim$ 1 mg/cm^2) or directly on the sample. Typical source strengths are 1-50 μCi. The source is then sandwiched between two pieces of the sample which are thick enough to ensure that all positrons are stopped in the sample. One of the two detectors (including the connected electronics) detects the 1.28 MeV γ-quantum, i.e. signals the emission of the positron, while the other detector detects one of the 511 keV annihilation γ-quanta, i.e. signals the disappearance of the positron. The time difference between the two detector signals, i.e. the lifetime of the positron, is measured by the Time-to-Amplitude Converter, and this information is stored in the multi-channel analyzer (MCA). Typically 10^5-10^7 annihilating positrons are recorded and the MCA produces a histogram showing the distribution of positron life-times, normally referred to as the lifetime spectrum. Fig. 5 shows examples of such spectra (see also Section 5).

   Analysis of lifetime spectra. If the positrons all annihilate from the same state (e.g. bulk Mo) the lifetime spectrum consists of only one decaying exponential (although somewhat broadened by the finite resolution of the spectrometer, see Fig. 5). However, positrons may after slowing down exist in different states (e.g. free e$^+$ and Ps) or a transition from one state to another (e.g. trapping into defects) may take place after thermalization. In such cases positrons will annihilate with different rates determined by their surroundings, and a lifetime spectrum will contain more than one component:

$$N(t) = \left[\sum_i A_i \exp(-\lambda_i t) \right] * P(t) \qquad (4)$$

where $*$ symbolizes convolution of the ideal spectrum with the spectrometer time resolution function, $P(t)$. The two longlived spectra in Fig. 5 consist of more than one component. In a lifetime spectrum the physics information is contained in the values of λ_i and A_i (the latter more often represented by the intensity, $I_i = (A_i/\lambda_i)/\sum_i(A_i/\lambda_i)$, the normalized area under the i'th component). Computer programs exist which fit the model function, Eq. (4), to measured spectra (see e.g. Ref. 45 and refs. therein) and in this way extract the I_i's and λ_i's.

   From Fig. 5 it is obvious that the spectrometer time resolution has roughly the same magnitude as the shorter positron lifetimes in the spectra. Therefore, in order to extract the maximum amount of information from the spectra it is necessary that the resolution curve is stable and well-defined. This puts rather strong requirements on the stability of the lifetime spec-trometer. Another problem in the analysis of lifetime spectra is to make cor-rections for the positrons with annihilate in the source material (or sur-rounding foils). This correction normally amounts to a few percent of the total intensity. However, an improper correction may have a rather strong influence on the values of the fitted lifetimes and intensities. Finally, it is not necessarily obvious how many lifetime components are contained in a measured spectrum. An important parameter in the analysis is therefore the number of components used. Justification of the choice of a certain number can in many cases be obtained from previous knowledge about the sample, from the consistency of the obtained results (see Sections 4 and 5), and partly from the goodness of the fit. The data analysis problems are further discus-sed e.g. in Refs. 20,27,31,43-45.

Angular Correlation Measurements

   Figure 6 shows the principle of a conventional angular correlation ap-paratus. It is used to measure the angular (and hence momentum) distribution of the two 511 keV annihilation quanta (see Section 2). The source is typi-cally of a strength of 1-50 mCi. It shines positrons down on the sample. The directions of the two emitted photons are defined by collimators which are narrow slits in lead blocks. Typically the sample to detector distance is a couple of meters, and practical angular resolutions will be 0.5-1 mrad.

The slits closest to the detectors are long slits perpendicular to the plane of the drawing, and the detectors long scintillation crystals mounted just behind the slits, each of them coupled to one or more photomultiplier tubes. One of the detectors (including the collimator) is fixed, while the other one can move to cover angles Θ in the range of about -30 mrad to 30 mrad.

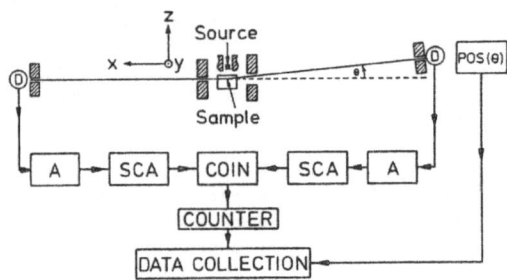

Fig. 6. Schematic diagram of a (1-Dimensional) angular correlation apparatus. The source is external to the sample. D represents a long scintillation detector perpendicular to the drawing, A an amplifier, SCA a single channel analyser, and COIN a coincidence unit. The DATA COLLECTION will often be a minicomputer. The rectangular hatched areas are lead collimators in the form of long slits perpendicular to the drawing. POS(Θ) measures the angle Θ. From Ref. 44.

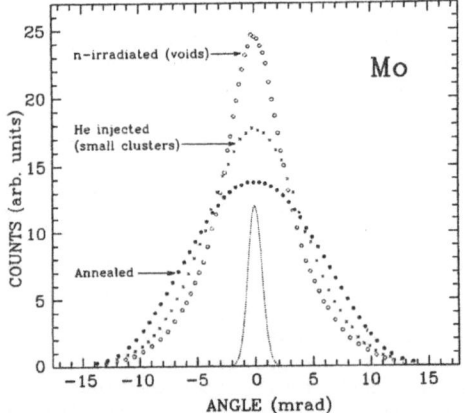

Fig. 7. Angular correlation (ACAR) curves for three molybdenum samples (equivalent to those used for Fig. 5). A clear narrowing of the curves is seen when positrons are trapped in small vacancy clusters ("He injected") or in large voids ("n-irradiated"). This is because annihilation in these cases takes place with electrons of lower momentum than it does in bulk Mo ("Annealed"). The curves are normalized to the same area. The dotted curve shows approximately the angular resolution of the apparatus.

152

The following electronics count the number of pairs of 511 keV annihilation quanta in a given time as a function of the angle Θ. The data are stored in the data collection unit which is often a minicomputer, that also controls the measurement (e.g. measuring time for each Θ, change of Θ, maybe change of sample temperature etc.). The resulting distribution of angles between the annihilation quanta is called an angular correlation or ACAR (Angular Correlation of Annihilation Radiation) curve. Typical measuring times for a whole curve are from a few hours to a day. Such curves are shown in Fig. 7.

The ACAR apparatus shown in Fig. 6 is a so-called 1-dimensional machine, since it only measures one component of the momentum, viz. the z-component, p_z, but integrates over p_x and p_y, i.e. $N_{1D}(p_z) = \iint N_{3D}(\bar{p}_T)dp_x dp_y$. Here N_{3D} is the experimental 3-dimensional momentum distribution, equivalent to the theoretical $\Gamma(\bar{p}_T)$ in Eq. (1). In recent years, however, a development of so-called 2-dimensional ACAR machines has taken place.[42,46] In these, γ-detectors which are position sensitive in two dimensions are used. Thus, the above integration is only over p_x and a 2-dimensional momentum distribution is obtained, viz.: $N_{2D}(p_y, p_z) = \int N_{3D}(\bar{p}_T)dp_x$. Examples of such curves are shown in Section 4.

Analysis of ACAR curves. ACAR curves contain information about the electronic structure in the surroundings of the positron (Eq. (1)). This is more detailed information than contained in lifetime spectra in which each positron state is characterized by only one number (the lifetime, τ). On the other hand it may be difficult to extract the additional information, since in most cases the functional forms of the curves are not known (as they are for lifetime spectra). This is particularly true when positrons annihilate from several different states, thus giving rise to a composite ACAR curve.

In order to separate the various components of such a composite curve it is often necessary to have independent knowledge from other measurements of the shapes of at least some of the components which may be present. By using flexible computer programs[45] one can fit the measured curve by a sum of components of different shapes (one for each positron state) and in this way obtain shapes and/or intensities of unknown components. This approach is further discussed in e.g. Refs. 45 and 47. A simpler way of analysis is to calculate one or more "shape parameters", e.g. the normalized height of the curve for Θ = 0. This does not make use of the whole curve, but may be sufficient in many applications, e.g. in defect studies, although in unfavorable cases important features of the curves may be overlooked.[47] Similar ways of analysis are often used for Doppler broadening spectra and will be further illustrated below.

Recently a very careful and detailed analysis of 2-dimensional ACAR curves has been carried out in a study of thermally generated defects in aluminium.[48] In that work, measurements were compared with theoretical calculations of the expected curves for positrons in bulk Al and trapped in mono- and divacancies. In order to take full advantage of the potential of PAT for defect spectroscopy it is important that such detailed comparisons between theory and experiments are carried out.[43,48] The results will be illustrated in Section 4.

Doppler Broadening Measurements

If the momentum of the annihilating pair has a component, p_x, in the direction of the emitted γ-quanta, the frequencies of these quanta and hence their energies will be Doppler shifted.[8] The energy shift is:

$$\Delta E = cp_x/2 , \tag{5}$$

one quantum having the energy $E_0 + \Delta E$, the other one $E_0 - \Delta E$, where $E_0 = m_o c^2$ = 511 keV. Equation (5) shows that ΔE is proportional to p_x. Hence, by measuring the distribution of ΔE one obtains the distribution of one component of the momentum, just like in a 1-dimensional ACAR measurement. Such a measurement can be done with a solid-state detector (intrinsic germanium (Ge) or lithium drifted germanium (Ge(Li))) of good resolution (1.1-1.5 keV FWHM at 511 keV). A "Doppler broadening spectrometer" is shown schematically in Fig. 8. The signal from the detector is amplified and then recorded by the MCA. High stability of the system is required and the MCA is therefore normally connected to a digital stabilizer. On expansion of the energy scale in the region around 511 keV, spectra like the ones shown in Fig. 9 are obtained.

By comparison with Fig. 7 it is seen that the resolution in Fig. 9 is much poorer than in Fig. 7. Since 1 keV is equivalent to 3.914 mrad the resolution curve of a Doppler broadening spectrometer is wider than equivalent to 4 mrad. This is at least five times broader than the resolution of a typical angular correlation apparatus. Advantages of the Doppler compared to the angular correlation technique are that only weak sources (the same as for lifetime measurements) are needed, and that the experimental set-up is simpler, e.g. there are no stringent requirements on the position of the detector with respect to the source/sample arrangement, and all the – relatively few – units of the set-up are commercially avaiable. Normally a higher countrate is easily obtained so that a spectrum can be measured in a couple of hours or less (at the sacrifice of resolution the same can be obtained in angular correlation measurements, though).

Analysis of Doppler broadening spectra. Such spectra can be analysed in the same ways as discussed above for angular correlation curves. However, in most cases a simple analysis is preferred which characterises the spectrum by only one or two parameters. These are often called S and W. S is the area under the spectrum in a narrow energy range symmetrical around 511 keV (A, hatched in Fig. 9), while W is the sum of the areas under the "wings" of the spectrum, i.e. in two regions symmetrical around 511 keV ($B_1 + B_2$, hatched in Fig. 9), both S and W being normalized to the total area of the spectrum. Instead of S, sometimes H is used which is defined as the normalized peak height of the spectrum.

Thus, the narrower the spectrum is the larger is S and the smaller is W. If a narrowing is a result of a transition of a fraction of the positrons from one state with a broad spectrum (e.g. in the bulk metal) to another state with a narrower spectrum (e.g. trapped in a defect) a useful parameter to extract is $R = \Delta S/\Delta W$, where ΔS and ΔW are the changes in S and W. It is easy to see that R depends only on the spectra for the two states, not on the fraction of positrons that make the transition. R can therefore be used to identify the annihilation sites for the positrons.[49]

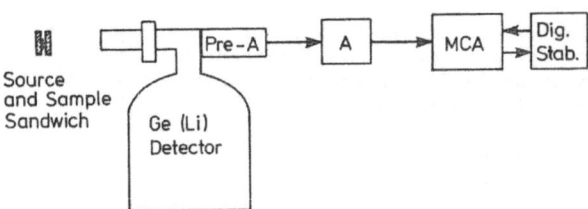

Fig. 8. Schematic diagram of a Doppler broadening spectrometer. The source is sandwiched between the two samples. Ge(Li) represents the solid-state detector cooled by liquid nitrogen. PRE-A and A are pre- and main-amplifier, and MCA a multi-channel analyzer which is often connected to a digital stabilizer (Dig. Stab.).

Analysis in terms of one or more parameters like H, S, W and R are also frequently used for ACAR curves.

More detailed discussions of the Doppler broadening technique can be found in Refs. 31, 43.

Low-Energy Positron Beams

In recent years beams of mono-energetic positrons with variable energy have been developed.[22-24,50] The principle of such a beam is sketched in Fig. 10. Positrons from a radioactive source (here [58]Co) are injected into a single-crystal metal moderator. (Instead of a radioactive isotope as the source, positrons can also be produced through pair-creation by the bremsstrahlung generated when high-energy electrons from an accelerator hit a target.[51]) In the moderator the positrons slow down, and a few of them thermalize so close to the moderator surface that they may diffuse back to it before they annihilate. Having reached the surface the positrons have a certain probability of being re-emitted with an energy equal to the negativ value of the workfunction of the positron in the metal. This is of the order of an electron-Volt. The mono-energetic beam of positrons from the moderator can then be accelerated to the desired energy and injected into the sample (right hand side of Fig. 10). The mean penetration depth of the positrons increases with energy, so by varying the positron incident energy it is possible to probe different depths below the surface. After slowing down the positron may undergo different reactions as indicated for a metal sample in the figure. The positron may annihilate while diffusing in the solid or it may get trapped at a defect before annihilation. It may also reach the surface (as mentioned above for the moderator), where it may be ejected either as a free positron or in a Ps state, or it may become trapped in a surface state. If the sample contains a high concentration of defects that trap positrons, only a small fraction of the injected positrons will return to the surface compared to the case where no defects are present.

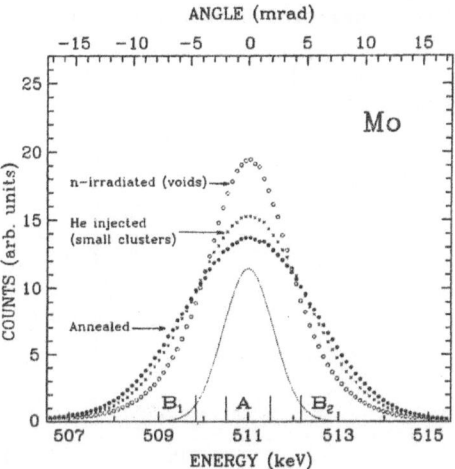

Fig. 9. Doppler broadening spectra for three molybdenum samples (equivalent to those used for Figs. 5 and 7). All three curves are normalized to the same area. The dotted curve is the resolution curve of the spectrometer. For comparison with Fig. 6 a Θ-scale is shown at the top which is equivalent to the energy scale at the bottom. The hatched regions show typical integration regions for determination of S and W parameters.

Thus by measuring this fraction, either by a measurement of the number of emitted positrons (using a so-called channeltron to detect them) or of the number of emitted ortho-Ps atoms (by detecting the ortho-Ps 3γ-decay) (Fig. 10), the concentration of defects in the sample can be estimated.[23,24,50] This will be further illustrated in Section 4.

4. DEFECTS IN METALS

As discussed in Section 2 positrons may become trapped in defects in metals, notably vacancies and clusters of these and other regions of less than average density. In the following a few selected examples of this will be shown. First, however, a simple model, the so-called trapping model, will be described. This model is often used to extract quantitative information about defects from PAT data.[27,33,52] In its simplest version the model assumes that positrons (or Ps atoms) initially are in the bulk of the material where they annihilate with the rate $\lambda_b = \tau_b^{-1}$. From the bulk they may become trapped in one type of defects with a time independent trapping rate κ. For a trapped positron the annihilation rate is $\lambda_d = \tau_d^{-1}$. With these assumptions we have the following rate equations for the disappearance of positrons in the bulk and in the trap:

$$\frac{dP_b(t)}{dt} = - \lambda_b P_b - \kappa P_b \qquad (6)$$

$$\frac{dP_d(t)}{dt} = - \lambda_d P_d + \kappa P_b \qquad (7)$$

where P_b and P_d are the probabilities that the positron is in the bulk and in the trap, respectively. The solutions of the equations result in a lifetime spectrum having two exponential components with the lifetimes:

$$\tau_1 = (\lambda_b + \kappa)^{-1} \quad , \quad \tau_2 = \lambda_d^{-1} \qquad (8)$$

and relative intensities:

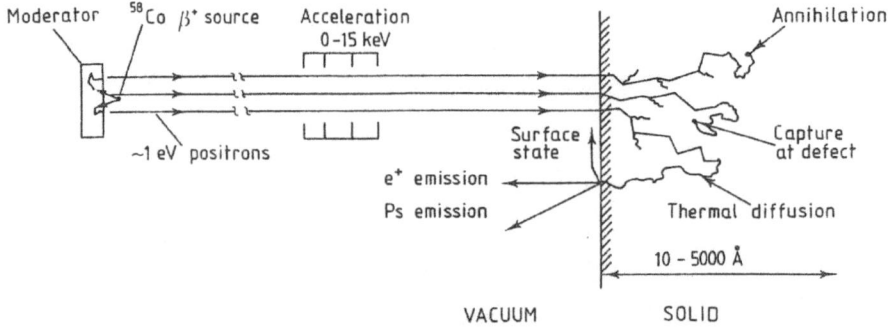

Fig. 10. The principle of a slow-positron beam. Positrons from a radioactive source (^{58}Co) are slowed down to thermal energies in the moderator. A small fraction is emitted from the moderator surface and is accelerated to be injected into the solid sample. Here different possible processes may take place as indicated in the figure. From Ref. 50.

$$I_1 = 1 - I_2 \quad , \quad I_2 = \kappa/(\lambda_b - \lambda_d + \kappa) \tag{9}$$

κ can be calculated from experimental I_2 values (Eq. (9)) and inserted into Eq. (8). If the τ_1 values calculated this way are in agreement with the measured τ_1 values, the experimental data are consistent with the trapping model.

The mean lifetime is given by

$$\bar{\tau} = I_1\tau_1 + I_2\tau_2 = \tau_b(1 + \kappa\tau_d)/(1 + \kappa\tau_b) \tag{10}$$

In an ACAR or Doppler broadening curve the (narrower) component due to trapped positrons will have the intensity:

$$I_N = \kappa/(\lambda_b + \kappa) \tag{11}$$

More generally, the parameters defined for momentum measurements in Section 3, viz. H, S and W as well as $\bar{\tau}$ (Eq. (10)) all vary with κ in the same way:

$$F = (\lambda_b F_b + \kappa F_d)/(\lambda_b + \kappa) \tag{12}$$

where F_b and F_d are the characteristic parameter values in the bulk and in the defect, respectively. For increasing trapping rate F changes in a sigmoidal way from F_b at low κ to F_d at high κ values. From a measured value of F the trapping rate can be determined (Eq. (12))

$$\kappa = \lambda_b(F-F_b)/(F_d-F) \tag{13}$$

As discussed in e.g. Refs. 27,52,53 the above equations can be expanded to describe more complicated trapping situations which occur, if e.g. more than one type of positron trap exists and/or detrapping can take place. An additional complication can be that for short times κ is no longer time independent.[53] Normally it is a good assumption that the trapping rate is proportional to the defect concentration, C, i.e.

$$\kappa = \mu C \tag{14}$$

where μ is called the specific trapping rate for the particular type of defect. For vacancies μ is of the order of 10^{15} sec^{-1} (Ref. 33), so changes in lifetimes and intensities (Eqs. (8)-(10)) can be detected for relative vacancy concentrations as low as 10^{-7}-10^{-6}. For vacancy clusters the specific trapping rate increases with cluster size, so that concentrations of large voids can be detected which are 2-3 orders of magnitude lower than for vacancies.[33]

Vacancies in Equilibrium

The equilibrium concentration of vacancies is given by[54,55]

$$C_V = \exp(S_V^F/k) \exp(- H_V^F/kT) \tag{15}$$

where S_V^F and H_V^F are the vacancy formation entropy and enthalpy, respectively, k is Boltzmann's constant, and T the absolute temperature. With increasing temperature C_V increases and so does the positron trapping rate (Eq. (14)). The mean lifetime and the momentum parameters will therefore increase in a sigmoidal way (Eq. (12)). This is illustrated in Fig. 11A by the variation of the ACAR curve peak height, H, for several metals.[56] At low temperatures the vacancy concentration is too low to cause any appreciable trapping of positrons and H changes only little with T. However, at about 0.6 T_m (Refs. 54,55) (T_m is the melting temperature) the vacancy concentration has reached

Table 1. Some vacancy formation enthalpies from PAT compared to results obtained from quenching studies. From Ref. 57.

	H_V^F (eV)	
	PAT	Quenching
Al	0.66	0.66
Ag	1.11	1.10
Au	0.97	0.94
Cu	1.31	1.30
Ni	1.8	1.6
Mo	3.0	3.2
W	3.8	3.6

10^{-7}-10^{-6} and trapping of positrons is seen as an increase of H. H continues to increase with increasing vacancy concentration until at temperatures close to the melting point vacancy concentrations of 10^{-4}-10^{-3} are reached and essentially all positrons are trapped before annihilation. Hence, a saturation of the measured parameter is observed (Fig. 11A). Using Eq. (13) the trapping rate can be obtained when inserting H for F, and from an Arrhenius plot of κ (Fig. 11B) the vacancy formation enthalpy can be derived.

Studies similar to the ones of Fig. 11 have been carried out on a number of metals.[54,55,57] It has also been shown that by careful measurements and analysis it is possible to separate two components in the lifetime spectra in agreement with the trapping model[58,59] (Eqs. (8) and (9)). These measurements yield the most precise or the only values of the vacancy formation enthalpy for many metals.[54,55,57] Some of the values are listed in Table 1 and compared with results from quenching studies.[55,57]

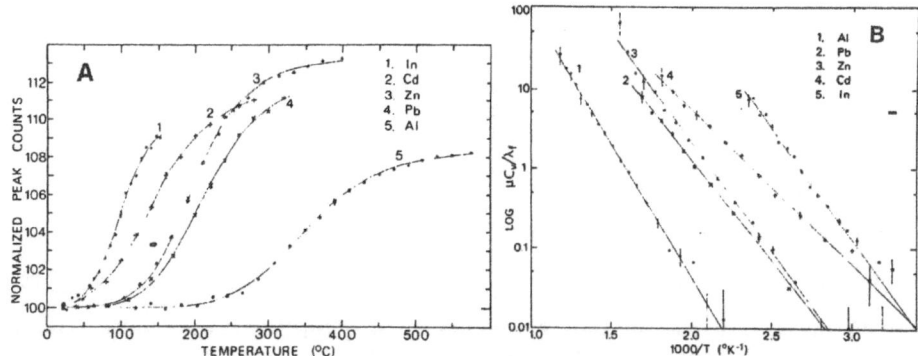

Fig. 11. A) The peak height (H) of angular correlation curves as functions of temperature for various metals. H has been normalized to 100 at low temperatures. B) The normalized trapping rate ($\kappa=\mu C_V$) as a function of inverse temperature (Arrhenius plot) as derived from the data of Fig. 11A by use of Eq. (13). The slopes of the straight lines that are fitted to the points equal the vacancy formation enthalpy (Eq. (15)). From Ref. 56.

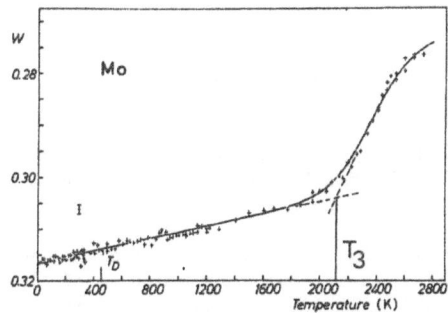

Fig. 12. The W parameter as function of temperature for Mo. The full curve is a fit of the trapping model (Eq. (12)). T_3 is defined as the on-set temperature for positron trapping in vacancies. From Refs. 55 and 60.

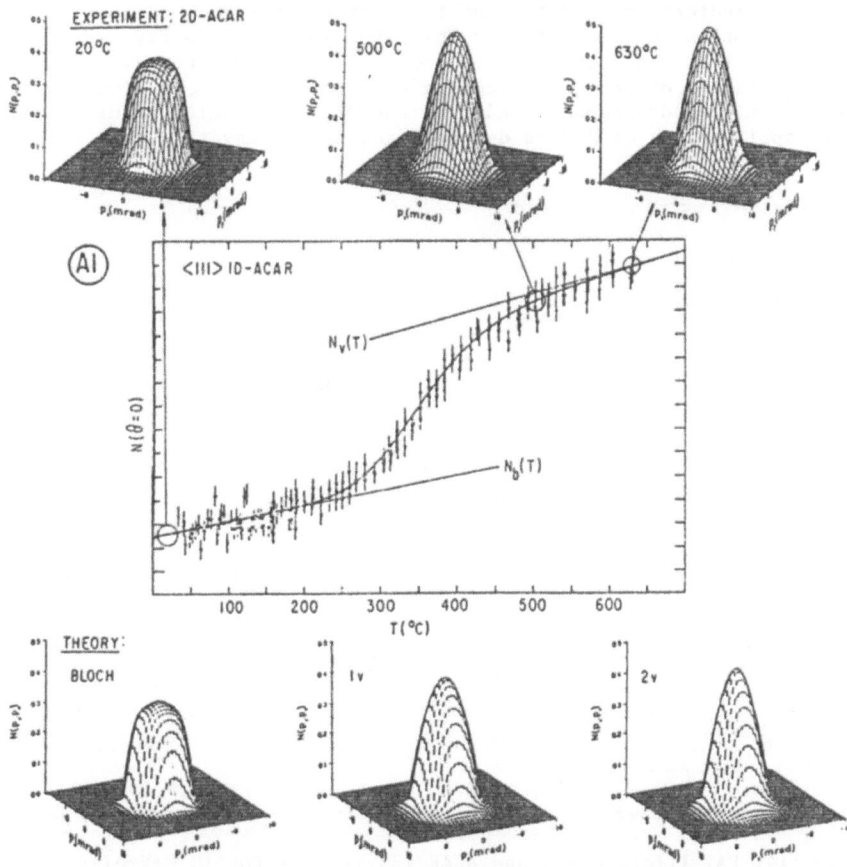

Fig. 13. Experimental and theoretical 2-dimensional ACAR curves for aluminium. The experimental curves are measured at temperatures at which positrons are not trapped (20°C), are trapped mainly by monovacancies (500°C), and by mono- and possibly divacancies (630°C) as illustrated on the sigmoidal curve in the center (obtained by measurements of 1D-ACAR). The theoretical curves are for positrons in the bulk (Bloch state), the monovacancy-trapped, and the divacancy-trapped states. From Refs. 48,57.

159

A relatively simple way to extract H_V^F from positron measurements is based on the approximate empirical relationship:

$$H_V^F = A \times kT_3 \qquad (16)$$

where T_3 is the on-set temperature for positron trapping (see Fig. 12) and A is a constant equal to 17 for bcc and 13.5 for fcc and hcp mtals.[54,60] Eq. (16) is particularly useful for those metals in which saturation trapping is not reached before melting.[54,55,60]

As seen in Fig. 12, the W parameter changes linearly with temperature below T_3. This is ascribed to thermal expansion of the lattice (but may have a more complex nature[61]). Similarly, the curves in Figs. 11A and 12 are not completely saturated close to the melting point. These temperature variations (of F_b and F_d, Eq. (12)) have to be taken into account in the analysis of the measurements. It has been discussed[48,60,62,63] whether the variation at high temperature could, at least in part, be ascribed to divacancy formation. A detailed comparison of 2-D angular correlation curves with theoretical predictions at different temperatures has been carried out to solve this problem for aluminium.[48] This is illustrated in Fig. 13. The conclusion is that at high temperatures, apart from monovacancies, an appreciable concentration of positron-trapping defects, most likely divacancies, is present.[48] This study seems to indicate the direction which future similarly detailed applications of PAT to defect spectroscopy may take.

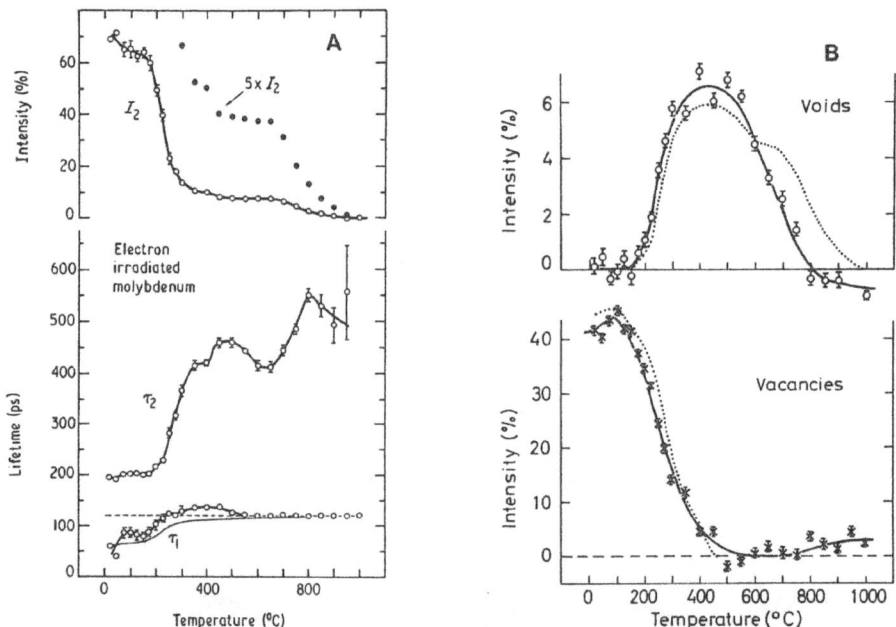

Fig. 14. PAT lifetime (A) and ACAR (B) results for 10 MeV-electron irradiated molybdenum as functions of annealing temperature. In A) the τ_2, I_2 component is due to positrons trapped in vacancies and/or voids. The dashed line is the positron lifetime in annealed Mo. In B) are shown the relative intensities of the vacancy and void components extracted from ACAR curves. The dotted curves show the expected intensities as derived from the lifetime data by a "trapping model" analysis with two traps. From Ref. 47.

Vacancy formation in alloys has also been studied by PAT. In dilute alloys the vacancy-solute binding enthalpy can be determined by measuring the increase in vacancy concentration due to this binding. However, because small differences between curves have to be evaluated it is rather difficult to do this with a reasonable precision, although it can be done. [57,64]

Non-equilibrium Vacancies and Vacancy Clustering

In the following we shall illustrate how the specific sensitivity of positrons to vacancies and clusters of these can be used to follow, during annealing, the migration and clustering of vacancies created at low temperature by irradiation.

Fig. 14A shows results for molybdenum electron irradiated at room temperature.[47] The irradiation creates Frenkel pairs, and about half of the positrons now become trapped in the vacancies with a lifetime of $\tau_2 \simeq 200$ psec ($\tau_b \simeq 120$ psec). On annealing τ_2 starts to increase at about 200°C. The increase shows that in this temperature range (the so-called stage III) 3-dimensional vacancy clusters (voids) are formed (compare with Fig. 4), i.e. vacancies are migrating in stage III. Simultaneously the intensity of the longlived component, I_2, and hence the trapping rate decrease strongly, partly because the trapping rate per vacancy decreases with increasing cluster size, partly because some of the vacancies cluster into two-dimensional vacancy loops rather than three-dimensional voids. Positron trapping in such loops in Mo gives rise to a lifetime in the range 120-180 psec, probably about 155 psec.[65] However, the main reason for the decrease of I_2 is that a large fraction of the migrating vacancies recombine with interstitials produced during the irradiation. The vacancy – interstitial recombination in stage III was well known for many metals, but for a number of them a long debate has taken place about which of the two defects was migrating in this temperature range.[47,66] For molybdenum the results of Fig. 14A clearly show that vacancies migrate in stage III.[47]

By detailed analysis of the ACAR curves measured as a function of annealing temperature it was possible to separate the characteristic shapes of the components due to positrons in the bulk, trapped in vacancies, and trapped in voids, respectively.[47] They were very similar to the three shapes shown in Fig. 7. The intensities of the components for vacancies and voids are shown in Fig. 14B (with the intensity for the bulk component they add up to 100%). It is seen how the "vacancy intensity" (i.e. the fraction of positrons trapped in vacancies) decreases essentially to zero in stage III while the "void intensity" increases. At the highest temperatures the voids anneal out. The dotted curves are the intensities calculated from the lifetime data using the trapping model for two traps. The good agreement shows the consistency of the lifetime and ACAR measurements.[47] After annealing to 900°C a very low density of ~ 30Å diameter voids could be seen by electron microscopy. This clearly illustrates the important property of PAT, viz. that it is possible to follow the evolution of 3-dimensional vacancy clusters from single vacancies up to a size visible by electron microscopy.

Impurity-Defect Interaction

If a metal contains impurities, these impurities may interact with the defects in the metal and thereby change the characteristic behaviour of the defects, e.g. their annealing behaviour. In the following, examples will be given which illustrate how this can be seen by PAT measurements.

Nitrogen in molybdenum. Figur 15 shows lifetime results for annealing of electron irradiated pure and nitrogen doped molybdenum.[67] The behaviour for the pure metal is the same as shown in Fig. 14, i.e. the lifetime increase starting at about 210°C shows the agglomeration of vacancies into clusters. Doping with nitrogen and subsequent irradiation gives rise to a slightly shorter lifetime on average in the vacancies, because some of them

161

have trapped a nitrogen atom, which increases the electron density in these vacancies. At about 160°C τ_2 decreases further. This is believed to be due to migration of nitrogen,[67] since after the migration more vacancies (probably all) will contain nitrogen. In the stage III temperature interval no changes are observed in τ_2 or I_2, i.e. no clustering or loss of vacancies occur. This means that the nitrogen-vacancy binding prevents free migration of va-

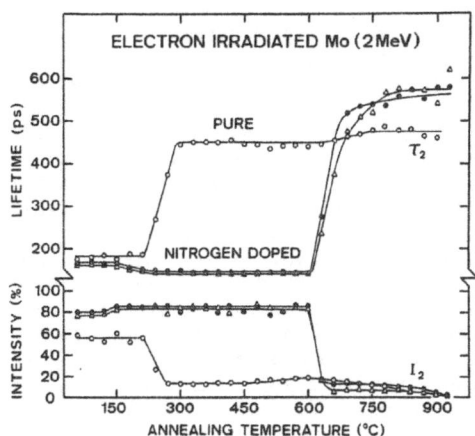

Fig. 15. The lifetime of positrons trapped in defects and the associated intensity as functions of annealing temperature for electron irradiated pure and nitrogen doped molybdenum. From Ref. 67.

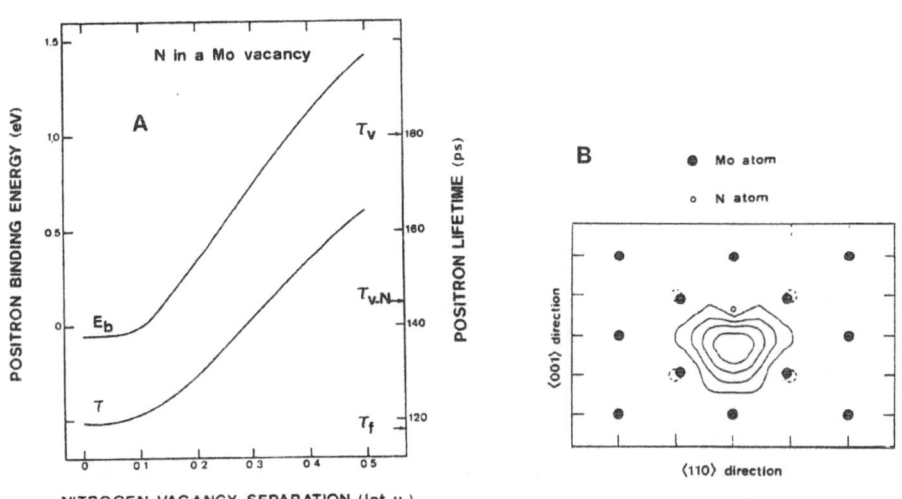

Fig. 16. A) shows the calculated lifetime (τ) and binding energy (E_b) for a positron trapped in a Mo vacancy containing a nitrogen atom as a function of the position of the nitrogen. Experimental lifetimes are shown by arrows on the right hand axis. In B) is shown the calculated wavefunction for a positron trapped in a nitrogen decorated Mo vacancy, the nitrogen position being determined by τ_{V-N}. From Ref. 37.

162

cancies. Not until 650°C is a sharp change of both τ_2 and I_2 observed which shows that clustering takes place, probably after dissociation of the vacancy-nitrogen pair.[67] At higher temperatures τ_2 is longer than in the pure sample. Since this cannot be due to a larger size (the lifetime saturates at about 500 psec in pure vacancy clusters, see Fig. 3) it must be ascribed to nitrogen trapped in the vacancy clusters[67], a result also found in other studies.[65] The lifetime of the positron trapped in a nitrogen-vacancy defect (τ_{V-N} = 145 psec) can be used to estimate the position of the nitrogen atom in the vacancy[37] as illustrated in Fig. 16. The experimental value of τ_{V-N} is equivalent to a displacement of the nitrogen of about 0.3 lattice units from the centre of the vacancy along the <100> direction, a position shown in the right-hand part of the figure.

Figures 15 and 16 clearly illustrate the possibilities of studying gas--defect interaction, in particular when experiments are supplemented by theoretical calculations. Other recent examples of this can be found in e.g. Refs. 7,37,65,68.

Gas bubbles in metals. If gas is present in metals which are subject to irradiation, gas filled cavities, so-called bubbles, may be formed. It is of great interest to monitor the dynamics of such bubbles and obtain information about the gas density in the bubbles.[69] In particular the influence of helium has been the subject of many investigations.[69] Since positrons become trapped in bubbles, PAT studies can contribute to this area of research. Let us illustrate this by some lifetime results for copper containing a high concentration (~ 3 at %) of krypton.[70] (Fig. 17). Up to an annealing temperature of about 300°C the lifetime of the positrons (which are all trapped in Kr-

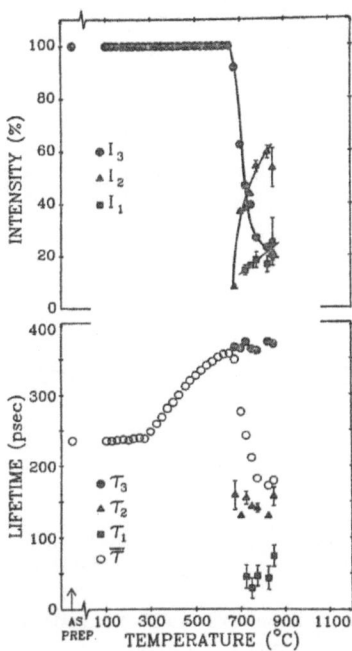

Fig. 17. Positron lifetime parameters as function of annealing temperature for Cu containing ~ 3 at % Kr. For temperatures at and above 675°C, where more than one component is resolved, the mean lifetime (open circles) is given by $\bar{\tau} = \Sigma \tau_i I_i$. From Ref. 70.

163

-filled bubbles) is about 240 psec. This reduction of τ compared to the roughly 500 psec expected for empty voids (of ~ 30Å diameter[70], Fig. 4) is a result of the annihilation with the electrons of the Kr atoms. The higher the Kr density is in the bubbles, the smaller is τ. Hence, from the magnitude of τ it is possible to calculate the Kr density. This was done in Ref. 37 resulting in $\rho_{Kr} = 3.0 \times 10^{22}$ cm^{-3}. Electron diffraction later revealed that the Krypton in the bubbles was crystalline at room temperature, the measured lattice parameter being equivalent to a density of 2.9×10^{22} cm^{-3} (Ref. 71), i.e. confirming the PAT estimate. Above 300°C (at which temperature the Krypton melts[71]) the lifetime increases (Fig. 17), thus showing a decreasing Krypton density in the bubbles, probably as a result of bubble coalescence.[70] At about 700°C a strong decrease of the trapping rate into bubbles takes place, as evidenced by the behaviour of I_3. At the same time a major loss of Kr takes place, i.e. a large fraction of the bubbles coalesce to form open channels to the surface.

The application of PAT to yield information about bubbles and voids is being developed these years with the aim of giving data for cavity sizes and densities as well as gas densities in the bubbles. In this respect, two advantages of PAT compared to electron microscopy are that PAT is non-destructive and that submicroscopic cavities can be detected. A clear disadvantage is that PAT is an averaging technique, i.e. it does not produce e.g. detailed size distributions of cavities and it may in some cases be difficult to separate components arising from different types of defects.

Carbon in iron. As a last example of impurity-defect interactions let us look at some results which very clearly show the interaction of vacancies and vacancy clusters in iron with carbon impurities.[72,73]

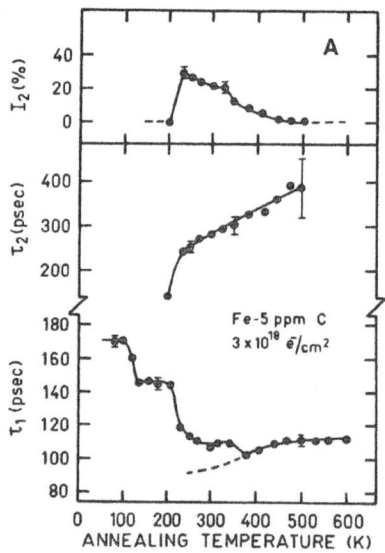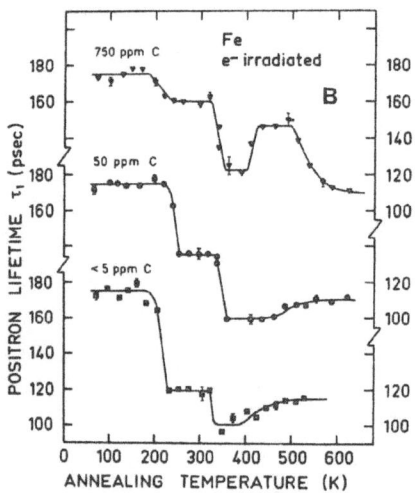

Fig. 18. Positron lifetime parameters as functions of annealing temperature for iron containing carbon impurities after electron irradiation (3 MeV) below 77K. τ_1 is the mean lifetime for positrons annihilating in the bulk, in vacancies and/or vacancy-impurity complexes. τ_2 is due to vacancy clusters. A) shows all parameters for nominally pure iron; B) compares the short lifetimes for 3 different carbon contents, and for higher electron doses than in A). From Ref. 72.

Figure 18 shows the annealing behaviour of low temperature electron irradiated iron.[72] After irradiation only one lifetime component of τ_1 = 175 psec, due to positron trapping in mono-vacancies, is observed. On annealing to 120K τ_1 decreases for the low-dose sample (Fig. 18A) because interstitials created during irradiation become mobile and recombine with some of the vacancies. In the high-dose samples (Fig. 18B) no drop of τ_1 is seen at 120K, since even if about half of the vacancies disappear there is still a sufficient concentration left to trap all positrons before annihilation. At 220K τ_1 drops further and a long-lived component (τ_2, I_2) appears for all samples, signaling the migration and clustering of vacancies (state III). The definitive determination of this temperature was an important conclusion, since the vacancy migration temperature had been the subject of much controversy over a long period.[72] Between 220K and 320K τ_1 is constant for all samples, but the level increases with carbon concentration (Fig. 18B), i.e. more vacancies survive stage III annealing at high carbon concentrations. This is explained by trapping of the migrating vacancies by the immobile carbon atoms, the vacancy – carbon pairs retaining the ability to trap positrons. These have a lifetime of 160 psec when trapped in the vacancy-carbon pairs (determined as the highest τ_1 value in this temperature interval, for 750 ppm C). At about 350K migration of interstitial carbon is known to occur.[72] This results in a clear decrease of τ_1, believed to be due to further decoration of the carbon-vacancy pairs with migrating carbon atoms which results in disappearance of the ability of the decorated vacancies to trap positrons. Finally, at the highest temperatures total recovery of the samples takes place (the details for the 750 ppm sample though not being fully understood). A detailed discussion of the data analysis is given in Ref. 72. The vacancy and vacancy cluster concentrations are calculated as functions of annealing temperature from the lifetime data using an extended version of the trapping model, Eqs. (6)-(14).

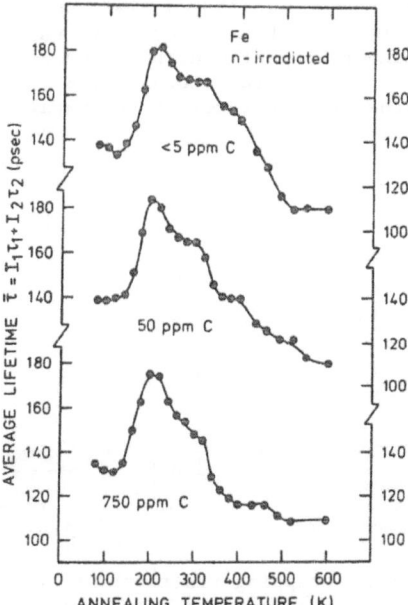

Fig. 19. Average positron lifetime $\bar{\tau}$ as a function of annealing temperature for iron with three different carbon contents which has been neutron irradiated at 77K to a dose of 4×10^{16} fast neutrons per cm^2. From Ref. 73.

Figure 19 shows the annealing behaviour of iron neutron irradiated at 77K for three different carbon concentrations.[73] The average positron lifetime increases in all three cases at about 170K i.e. below the temperature (220K) for free vacancy migration. A more detailed analysis of the data shows that the long lifetime τ_2 increases from about 200 psec to 300 psec at ~ 170K. This behaviour shows that vacancy agglomerates present already after irradiation in the collision cascades grow in this temperature region. This phenomenon is not sensitive to the carbon content and is ascribed to short-range correlated vacancy migration within the collision cascade. Because of the high local concentration of vacancies the vacancy migration is observed at lower temperatures than after electron irradiation, and the carbon content has only a small effect. Only at around 350K an effect of carbon concentration is clearly observed. At this temperature – as for electron irradiated iron (Fig. 18) – a decrease of τ is seen, the magnitude of which depends on carbon doping. As above, this decrease is a result of free carbon migration.

As mentioned above, the lifetime of a positron trapped in a vacancy-carbon pair was estimated to be 160 psec.[72] Calculations similar to the ones leading to Fig. 16 show that this lifetime can be accounted for if the carbon atom is in an asymmetrical position in the vacancy, displaced about 0.5 lattice constants.[35] This is in good agreement with other estimates.[35]

In this section a few examples have been shown of how conventional PAT can give information about defects in metals, both in equilibrium and non-equilibrium and about their interaction with impurities. A number of other examples can be found e.g. in Refs. 5-7,54,74.

Depth Profiling of Defects

Let us finally, before leaving metals, show a recent result which illustrates the possibilities of estimating depth distributions of defects close to a surface by the new low-energy-positron-beam technique.[50,75]

As briefly mentioned above (Section 3), low-energy positrons injected into a metal may diffuse back to the surface where they have a certain probability to escape. The higher the incident positron energy is, the smaller is the probability of the positron to return to the surface, because it may annihilate before it reaches the surface. If the metal contains defects,

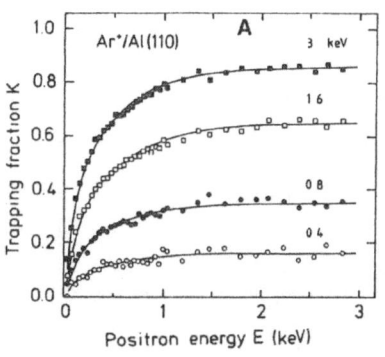
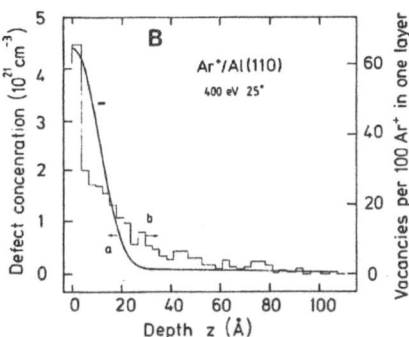

Fig. 20. A) Experimental trapping fractions, K, as function of positron incident energy for an Al surface after argon-ion bombardment. Ion energies are indicated, their incident angle being 75°. The curves are best fits of the theory as described in the text. B) Defect concentration profiles derived from a) experiments and b) Monte Carlo simulations. From Ref. 75.

trapping of the diffusing positrons into these defects further decreases the probability of the positrons to reach the surface compared to the perfect metal. A convenient measure of this effect is the so-called trapping fraction[50,75]

$$K(E) = (J_b(E) - J_d(E))/J_b(E) \qquad (17)$$

where $J_i(E)$ is the probability for a positron which has been injected with energy E of diffusing back to the surface in a perfect sample (i=b) or in a sample containing defects (i=d). Figure 20A shows K(E) for an Al surface, damaged by Ar-sputtering. K increases at low E and approaches saturation for E > 1-1.5 keV. At these energies the positrons are implanted beyond the defected overlayer. Hence, during their diffusion back towards the surface these positrons have the same probability (independent of E) of getting trapped in the damaged layer. Thus, the value of K for E > 1.5 keV depends on the total defect density, while the shape of K(E) contains information about the defect distribution.[50,75] In order to extract this information one has to solve the one-dimensional diffusion equation for the positron with inclusion of annihilation and of trapping into defects of a certain depth distribution, C(z), and taking into account the positron penetration profile. From the solution, $J_d(E)$ and hence K(E) are derived and C(z) is then varied to obtain a good fit to the data of Fig. 20A.[50,75] Figure 20B shows a defect concentration profile obtained this way by parameterizing C(z) as a Gaussian followed by an exponential tail and on using a specific trapping rate $\mu = 5 \times 10^{14}$ sec^{-1} (Eq. (14)).[50,75] The detailed shape of the derived profile may depend on the way in which C(z) is parameterized. However, the main features remain, i.e. the narrow peak close to the surface and the long tail extending to about 100Å, in agreement with the simulation shown in Fig. 20B.

Earlier work[76] has also demonstrated that it is possible to detect the near-surface damage created by He-irradiation of crystalline and glassy metals. Furthermore, apart from probing defects near the surface, low-energy

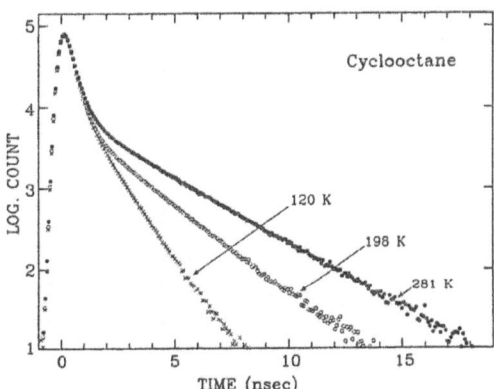

Fig. 21. Positron lifetime spectra for the plastic crystal cyclooctane. The short-lived part of the spectrum is due to para-Ps and to positrons not forming Ps. The longlived part is due to ortho-Ps. At 120 K, in the dense brittle phase, the ortho-Ps lifetime τ_{o-Ps} = 1.0 nsec. At 198 K, just above the phase transition to the less dense plastic phase τ_{o-Ps} = 1.8 nsec, and at 281 K, i.e. close to the melting point, τ_{o-Ps} = 2.7 nsec because Ps becomes trapped in thermally generated vacancies. From Ref. 78.

positron beams have been used in studies of the surface itself, e.g. by low energy positron diffraction[23] (LEPD, which offers certain advantages over the conventional LEED) and re-emitted-positron energy-loss spectroscopy (REPELS).[77] Very active research is going on these years to clarify the interactions of positrons with e.g. defects and impurities on metal surfaces.

5. DEFECTS IN MOLECULAR CRYSTALS

In molecular crystals the lattice positions are occupied by molecules (rather than atoms as they are in metals). A vacancy in such a crystal is a missing molecule on one of the lattice positions.

In most molecular solids a fraction of the injected positrons forms Ps, as discussed in Section 2. Because Ps is attracted to regions of lower-than--average density it may become trapped in vacancy-type defects in molecular crystals. The trapping leads to an increase of the ortho-Ps lifetime as illustrated in Fig. 21 and will normally also be reflected in the measured momentum distribution as shown below for ice. Like in Section 4 we discuss first equilibrium vacancies and then vacancies in non-equilibrium and their clustering.

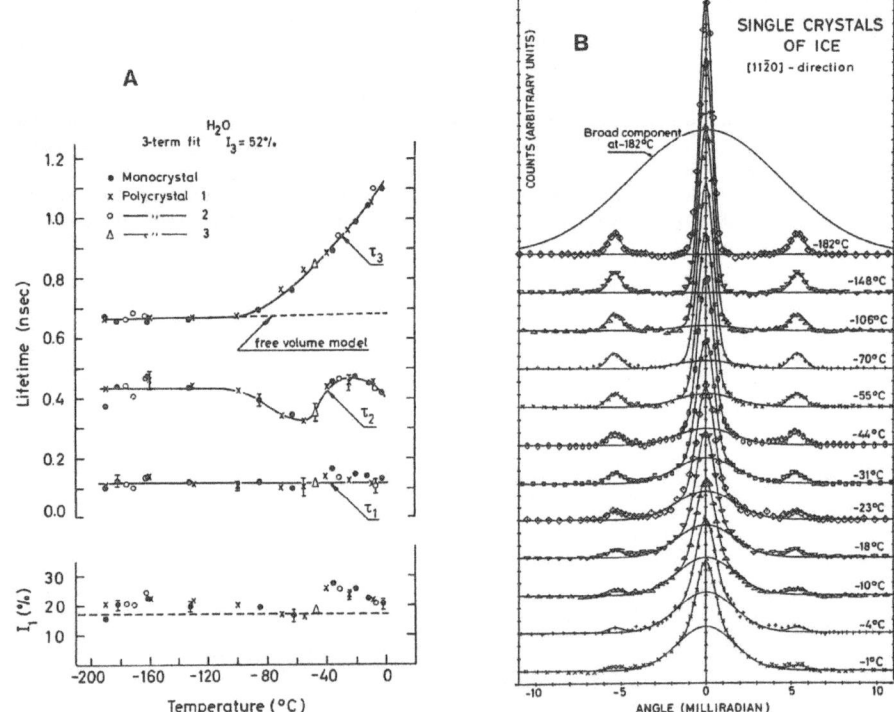

Fig. 22. Positron lifetime (A) and ACAR (B) results for ice as functions of temperature. In B) the broad component shown at the top is due to ortho-Ps pick-off and free positron annihilation. This component has been subtracted from all curves, leaving only the para-Ps peaks. The increase of the ortho-Ps lifetime, τ_3, the decrease of the narrow (< 1 mrad) ACAR para-Ps peaks and the simultaneous growth of a middle-broad (~ 3.6 mrad) component above -100°C are attributed to Ps trapping in thermally generated vacancies. From Refs. 79,80.

Vacancies in Equilibrium

The transition of Ps from being in a delocalized state in a perfect crystalline material at low temperatures to being in a localized, vacancy-trapped, state at higher temperatures is clearly illustrated by the results for crystalline ice in Fig. 22. At low temperatures, the ortho-Ps lifetime $\tau_3 \approx 0.7$ nsec, but above $-100°C$ it increases much stronger than expected from the lattice thermal expansion ("free volume model", Fig. 22A). This increase is ascribed to Ps trapping in thermally generated vacancies[79,80] in analogy with the curves in Fig. 11.

As briefly explained in the following, the narrow peaks (FWHM < 1 mrad) found at low temperatures in ACAR curves (Fig. 22B) and the positions of the "side-peaks" give evidence that Ps is in a delocalized (Bloch function) state.[80,81] (Similar peaks are also observed in quartz and a number of ionic crystals[82]).

Localization of Ps leads to an increase of its zero-point energy (and momentum, cf. Heisenbergs uncertainty principle). For localization in a harmonic potential, for example, we have the relationship[38]

$$X_{1/2} \times \Theta_{1/2} = 10.7 \text{ mrad Å} \tag{18}$$

Here $X_{1/2}$ and $\theta_{1/2}$ are the full widths at half maxima of the Ps spatial distribution and the para-Ps ACAR component, respectively. Hence, narrow components of $\Theta_{1/2} < 1$ mrad show that the Ps wavefunction extends at least 10-15Å, i.e. over several unit cells. Furthermore, the side-peaks are positioned exactly at angles equivalent to reciprocal lattice vectors of the ice lattice.[80,81] This shows that the Ps wavefunction is influenced by the periodicity of the lattice, i.e. must be delocalized. More quantitatively: If a Bloch-function is inserted for the Ps centre-of-mass wavefunction in Eq. (1) (slightly modified to take into account that one electron is bound to the positron) the resulting $\Gamma(\bar{p}_T)$ has peaks centered at reciprocal lattice vectors $\bar{g}$, i.e. for $\bar{p}_T = \hbar\bar{g}$ (Refs. 38,80,81). With increasing temperature above $-100°C$ the intensities of the narrow peaks decrease and a somewhat broader ($\Theta_{1/2} \sim 3.6$ mrad) component emerges, ascribed to para-Ps trapped (and therefore localized) at thermally generated vacancies.[80,81] From Eq. (18) we find $X_{1/2} \approx 3.0$Å, in good agreement with the size of an ice vacancy. On using the trapping model (Eqs. (8)-(14)) it is possible to estimate the vacancy formation energy from these data.[80] From the non-equilibrium measurements, to be discussed below, the vacancy migration energy can be extracted.[80,83] Both these energies are estimated to be 0.3 eV (which entails an equilibrium vacancy concentration in ice close to the melting point of the order 10^{-6} rather than the earlier estimate of about 10^{-10}, Refs. 80,83).

Measurements on a number of the so-called plastic crystals demonstrate the possibilities of extracting vacancy formation energies for some of these crystals, too.[78,84,85] Fig. 21 showed characteristic lifetime spectra for one of these crystals, and in Fig. 23A are shown the average ortho-Ps lifetime, τ_o, and its intensity, I_o, for another crystal, succinonitrile.[86] The transformation from the low temperature brittle phase to the high temperature plastic phase is clearly defined and hysteresis is seen. Since the brittle phase is more dense than the plastic phase, τ_o is shorter in the former one. Like for ice the sigmoidal increase of τ_o with temperature in the plastic phase is attributed to Ps trapping in thermally generated vacancies the concentration of which increases with temperature (Eq. (15), compare also Fig. 11). Close to the melting point all ortho-Ps becomes trapped in vacancies where they have a lifetime of 2.45 nsec (Fig. 23A). On using this vacancy lifetime in the data analysis it is possible to separate two longlived components[86], one for ortho-Ps in the bulk and the other one for vacancy-trapped ortho-Ps. By applying the trapping model (Eqs. (8)-(9)) on the two ortho-Ps components, the trapping rate, κ, is derived. At the same time the calculated bulk lifetime (Eq. (8)) is found to be in good agreement with the measurements, thus demonstrating the consistency of the trapping model with the data.[86] Fig. 23B shows κ plotted in an Arrhenius plot which leads to an activation energy of 0.36 eV. Hence, on the assumption that the vacancy concentration is proportional to κ, like in metals (Eq. (14)), the vacancy formation enthalpy (Eq. (15)) is H_V^F = 0.36 eV. It is surprising that this value is only half the expected enthalpy, viz. the sublimation energy.[86,87] Similarly, for other soft plastic crystals (having a rather low entropy of fusion[87]) the H_V^F values as determined by PAT are lower than the sublimation energy,[78,84,85] while for harder plastic crystals (high entropy of fusion) H_V^F is found in good agreement with the sublimation energy.[84] The "anomalous" H_V^F values for the low-entropy-of-fusion materials are not yet understood.

From measurements on a number of crystals it has been possible to establish a useful correlation between the vacancy volume and the lifetime of ortho-Ps trapped in the vacancy.[85] This is shown in Fig. 24. The figure also includes data for liquids in which Ps normally creates a cavity (called a Ps bubble) because of the repulsion between Ps and the molecules. Hence, on using the correlation in Fig. 24 it is possible to obtain information about sizes of microcavities in simple molecular substances by measurements of the lifetime of ortho-Ps trapped in these cavities.

Finally in this section, it should be pointed out that in a number of molecular crystals, in particular ordinary brittle crystals, it is not possible to derive H_V^F, because the sigmoidal lifetime increase (Fig. 23A) is not observed. This is probably so because Ps becomes trapped at other defects than vacancies (or maybe self-traped in the lattice). The nature of these defects is not known at present.[84,85]

Non-equilibrium Vacancies and Vacancy Clustering

In metals it is relatively easy to create vacancies at low temperatures by irradiation (see Section 4). In molecular crystals a main effect of ir-

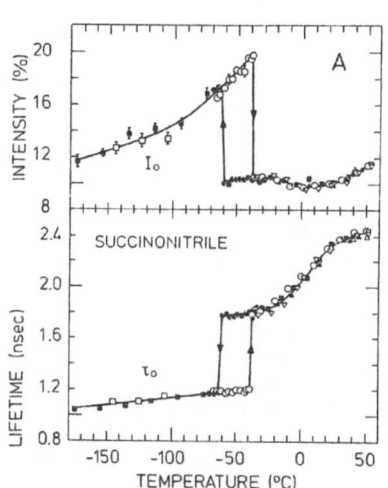 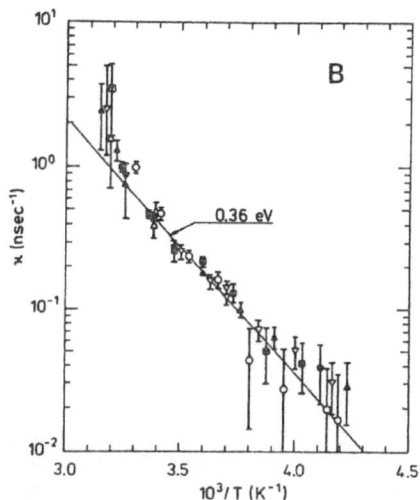

Fig. 23. Positron lifetime results for the plastic crystal succinonitrile. A) The average ortho-Ps lifetime, τ_o, and its intensity, I_o, as functions of temperature. B) The trapping rate, κ, for ortho-Ps as deduced from the trapping model vs. the inverse absolute temperature in the plastic phase. The slope of the straight line is equivalent to an activation energy of 0.36 eV. From Ref. 86.

radiation will be dissociation of some of the molecules. This dissociation may lead to the formation of molecular vacancies, but in addition to the creation of other molecular species, radicals, and ions. The defect population is therefore considerably more complicated than in metals. This may influence both the Ps formation process as well as subsequent Ps reactions (chemical reactions and trapping), and the separation of the various processes in the measured data may be very difficult. However, in the relatively simple molecular crystal, ice, it has been possible to observe the formation of vacancies during low-temperature γ-irradiation and vacancy clustering during annealing.[83]

Another way to create vacancies in ice at low temperatures has been found by PAT measurements.[83] It consists of a doping the ice with hydrogen fluoride (HF) in parts per million (ppm) concentration followed by a quenching of the crystal from close to the melting point down to low temperature. The results of the subsequent annealing are shown in Fig. 25. Two ortho-Ps components are resolved, one associated with ortho-Ps trapped in vacancies ($\tau_3 = 1.2$ nsec) and a longer one (τ_4, I_4) due to clusters of vacancies. At low temperatures all ortho-Ps ($\approx 54\%$) becomes trapped in vacancies ($I_3 \approx 54\%$, $I_4 \approx 0\%$). At about 100K the intensity I_4 of a longer-lived component ($\tau_4 \approx 2.5$ nsec) starts to increase. This is due to vacancy migration which leads to the formation of vacancy clusters, each containing 3–4 vacancies as judged from the lifetime value (Fig. 24). Above 115K τ_4 shows a strong increase as a result of cluster growth, either by growth of the larger clusters at the expense of the smaller ones, or because the already formed clusters migrate and coalesece. During this clustering process vacancies are lost, and with increasing cluster size the effective trapping rate per vacancy decreases. Hence, at the highest temperatures I_4 decreases to zero (compare Fig. 14). In the temperature range above 100K the intensity I_3 due to ortho-Ps trapped in monovacancies decreases because of the loss of monovacancies to clusters, until the thermal equilibrium value is reached at about 150K.[83] From the vacancy migration temperature of about 100K the vacancy migration energy was estimated at 0.3 eV as mentioned above.[83]

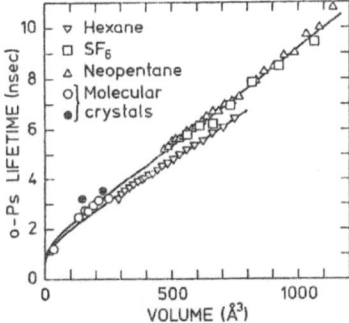

Fig. 24. The relationship between the ortho-Ps lifetime in a cavity and the volume of the cavity. O,●: ortho-Ps lifetimes in various molecular crystals close to the melting point as a function of molecular volume ($\approx$ vacancy volume); ●: crystals with high divacancy concentrations; Δ, □: lifetimes in bubbles in liquids as a function of bubble volume. The curves represent a model calculation fitted to the data for the liquids. From Ref. 85.

6. OTHER DEFECTS AND MATERIALS

In the previous two sections only point defects and their 3-dimensional agglomerates in metals and molecular crystals were discussed, primarily because these defects and substances provide the simplest examples to illustrate the potential of PAT for the study of defects in solids. However, other defects and other materials have also been investigated by PAT. In general there are clear effects of defects on the annihilation characteristics, although in many cases the defect population is rather complicated and the PAT results therefore difficult to interprete in detail. In such cases in particular, PAT should be used in combination with other techniques in defect studies. In the following a few references are given to such work, not treated above.

When metals are deformed, primarily dislocations are generated. Positrons are trapped at defects in deformed metals, either in the dislocations (maybe being localized at jogs on the dislocations), in point defects associated with the dislocations, or in individual point defects created during the deformation. A number of annealing studies of deformed metals have been performed by PAT. More details can be found in e.g. Refs. 57,74,88.

In Section 4 vacancies in dilute alloys were briefly mentioned. However, also in concentrated alloys it seems possible to obtain information about the effective vacancy formation energy.[57,89] A number of studies of precipi-

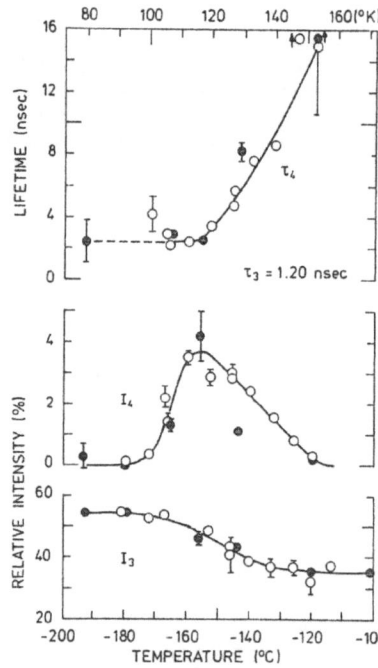

Fig. 25. Annealing of 25 ppm HF-doped ice crystals after quenching to liquid nitrogen temperature. Lifetimes and intensities of the two long-lived components (trapped ortho-Ps) are shown for two different samples. The shorter liftime, τ_3 = 1.2 nsec, is due to ortho-Ps trapped in vacancies. The increase of I_4 is associated with migration of vacancies which form vacancy clusters, the increase of τ_4 with the growth of the size of these clusters. From Ref. 83.

172

tation phenomena in concentrated alloys have also been carried out (see e.g. Ref. 90).

In amorphous alloys the positron behaviour in as-prepared samples is strongly influenced by the presence of inherent defects and inhomogeneities, since the positrons become trapped in these. Therefore the observed effects of additional defect creation by e.g. irradiation or deformation are smaller than for crystalline metals, but the defect recovery on annealing, for example, can in most cases easily be followed by PAT.[91,92]

Semiconductors have not nearly been studied by PAT to the same extent as metals. In spite of this, several papers have demonstrated that positrons are sensitive to vacancy type defects in semiconductors. This is for example exemplified by studies of defects in both crystalline[93-95] and amorphous[96] Si, by which e.g. a relationship between positron lifetime and vacancy cluster size was established[94] and the vacancy formation enthalpy determined.[95]

In ionic crystals a number of different types of defects trap positrons. In particular trapping by F-centers is well established. In an F-center the positron forms Ps with the trapped electron. A detailed discussion of positron studies of ionic crystals can be found in Ref. 97.

7. CONCLUSION

In the present paper the positron annihilation technique was described, both some basic physics principles and the experimental techniques. Emphasis was placed on the sensitivity of the techniques to defects in solids, and this sensitivity was illustrated by a number of examples, in particular for metals and molecular crystals. For metals it was pointed out that positrons are sensitive to vacancy type defects, the main sensitivity being for sizes from monovacancies up to clusters containing 50-100 vacancies and for relative concentrations of 10^{-4}-10^{-7} and below, depending upon the size, thus covering the range from the smallest defects up to sizes visible by electron microscopes. It was exemplified that vacancy - impurity interactions can be observed and near-surface defects detected (by low-energy-positron beams). Similarly, it was shown that vacancy type defects can be investigated in molecular solids.

It is hoped that the paper has clearly illustrated why the positron annihilation technique has now become a well-established technique for studies of defects, especially in metals, but at the same time a technique which is still being developed.

ACKNOWLEDGEMENTS

The author wants to thank O.E. Mogensen for many years of inspiring collaboration in positron annihilation research, K.O. Jensen for critically reading the manuscript, and N.J. Pedersen for technical assistance.

REFERENCES

1. C. D. Andersen, The Apparent Existence of Easily Deflectable Positives, Science 76: 238 (1932), and The Positive Electron, Phys. Rev. 43: 491(1933).
2. P. A. M. Dirac, On the Annihilation of Electrons and Protons, Proc. Camb. Phil. Soc. 26:361 (1929-30), and A Theory of Electrons and Protons, Proc. Roy. Soc. A126:360 (1930).
3. P. M. S. Blackett and G. P. S. Occhialini, Some Photographs of the Tracks of Penetrating Radiation, Proc. Roy. Soc. A139:699 (1933).
4. N. R. Hanson, "The Concept of the Positron", Cambridge University Press, Cambridge (1963).

5. R. R. Hasiguti and K. Fujiwara, Eds., "Positron Annihilation" (Proc. 5'th Int. Conf. on Positron Annihilation), Japan Institute of Metals, Sendai, Japan (1979).

6. P. G. Coleman, S. C. Sharma, and L. M. Diana, Eds., "Positron Annihilation" (Proc. 6'th Int. Conf. on Positron Annihilation), North-Holland, Amsterdam (1982).

7. P. C. Jain, R. M. Singru, K. P. Gopinathan, Eds., "Positron Annihilation" (Proc. 7'th Int. Conf. on Positron Annihilation), World Scientific, Singapore (1985).

8. P. Hautojärvi, Ed. "Positrons in Solids" (Tropics in Current Physics), Springer, Berlin (1979).

9. W. Brandt and A. Dupasquier, Eds., "Positron Solid-State Physics", (Proc. Int. School of Physics, "Enrico Fermi", 1981), North-Holland, Amsterdam (1983).

10. S. Berko and H. N. Pendleton, Positronium, Ann. Rev. Nucl. Part. Sci. 30:543 (1980).

11. A. Rich, Recent experimental advances in positronium research, Rev. Mod. Phys. 53:127 (1981).

12. T. C. Griffith and G. R. Heyland, Experimental aspects of the study of the interaction of low-energy positrons with gases, Phys. Rep. 39C:169 (1978).

13. M. Charlton, Experimental studies of positrons scattering in gases, Rep. Prog. Phys. 48:737 (1985).

14. H. J. Ache, Ed. "Positronium and Muonium Chemistry", Adv. Chem. Ser. 175, Am. Chem. Soc., Washington D.C. (1979).

15. B. Lévay, Chemical structure studies with positrons and mesons, At. Energy Rev. 17:413 (1979).

16. O. E. Mogensen, Positronium formation in condensed matter and high--density gases, in: "Positron Annihilation", Ref. 6, p. 763 (1982).

17. F. M. Jacobsen, Positronium formation in gases and liquids, in: "Positron Scattering in Gases", J. W. Humbertson and M. R. C. McDowell, Eds., Plenum, New York (1984), p. 85.

18. S. Berko, Fermi surface studies in disordered alloys: Positron annihilation experiments, in: "Electrons in Disordered Metals and at Metallic Surfaces", P. Phariseau, B. L. Gyorffy, and L. Scheire, Eds., Plenum, New York (1979) p. 239.

19. R. W. Siegel, Positron annihilation spectroscopy, Ann. Rev. Mat. Sci, 10:393 (1980).

20. R. N. West, Positrons as solid state probes, in: "Nuclear Physics Methods in Materials Research", K. Bethge, H. Baumann, H. Jex, and F. Rauch, Eds., Vieweg, Braunschweig (1980) p. 234.

21. P. Hautojärvi, Positron annihilation studies of vacancy-type defects, Hyperfine Int. 15/16:357 (1983).

22. A. P. Mills, Jr., Studying surfaces of solids using slow positron beams, Comments Solid State Phys. 10:173 (1982).

23. A. P. Mills, Jr., Experimentation with low-energy-positron beams, in: "Positron Solid-State Physics", Ref. 9, p. 432 (1983).

24. K. G. Lynn, Slow positrons in the study of surface and near-surface defects, in: "Positron Solid-State Physics", Ref. 9, p. 609 (1983).

25. S. E. Derenzo, T. F. Budinger, R. H. Huesman, and J. L. Cahoon, Dynamic positron emission tomography in man using small bismuth germanate crystals, in: "Positron Annihilation", Ref. 6, p. 935 (1982).

26. A. I. Akhiezer and V. B. Berestetskii, "Quantum Electrodynamics", Interscience Publishers, New York (1965).

27. R. N. West, Positron studies of condensed matter, Adv. Phys. 22:263 (1974).

28. V. I. Goldanskii, Physical chemistry of the positron and positronium, At. Energy Rev. 6:1 (1968).

29. S. Berko, Momentum density and Fermi-surface measurements in metals by positron annihilation, in: "Positron Solid-State Physics", Ref. 9, p. 64.

30. P. E. Mijnarends, Electron momentum densities in metals and alloys, in: "Positrons in Solids", Ref. 8, p. 25 (1979) and, Momentum density in metals and alloys: Theory, in: "Positron Solid-State Physics", Ref. 9, p. 146 (1983).

31. I. K. Mackenzie, Experimental methods of annihilation time and energy spectrometry, in: "Positron Solid-State Physics", Ref. 9, p. 196 (1983).

32. R. P. Gupta, R. W. Siegel, Positron trapping and annihilation at vacancies in bcc refractory metals, J. Phys. F. 10:L7 (1980).

33. R. M. Nieminen and M. J. Manninen, Positrons in imperfect solids: Theory, in: "Positrons in Solids", Ref. 8, p. 145 (1979).

34. P. Hautojärvi, J. Heiniö, M. Manninen, and R. Nieminen, The effect of microvoid size on positron annihilation characteristics and residual resistivity in metals, Phil. Mag. 35:973 (1977).

35. M. J. Puska and R. M. Nieminen, Carbon-vacancy interaction in α iron: interpetation of positron annihilation results, J. Phys. F. 12:L211 (1982), and, Defect spectroscopy with positrons: a general calculational method, J. Phys. F. 13:333 (1983).

36. C. Corbel, M. Puska, and R. M. Nieminen, Computed positron lifetimes in vacancies and vacancy-iron clusters in gold, Rad. Effects 79:305 (1983).

37. H. E. Hansen, R. M. Nieminen, and M. J. Puska, Computational analysis of positron experiments, J. Phys. F. 14:1299 (1984).

38. A. Dupasquier, Positronium like systems in solids, in: "Positron Solid-State Physics", Ref. 9, p. 510 (1983).

39. O. E. Mogensen, The spur reaction model of positronium formation, J. Chem. Phys. 60:998 (1974), and, Effect of an external electric field on the positronium formation in the positron spur, Appl. Phys. 6:315 (1975).

40. M. Eldrup, A. Vehanen, P. J. Schultz, and K. G. Lynn, Positronium formation and diffusion in a molecular solid studied with variable energy positrons, Phys. Rev. Lett. 51:2007 (1983); ibid 53:954 (1984); and, Positronium formation and diffusion in crystalline and amorphous ice using a variable-energy positron beam, Phys. Rev. B. 32:7048 (1985).

41. O. E. Mogensen, Comment on: Positronium formation and......, submitted to Phys. Rev. B.

42. S. Berko, Two-dimensional angular correlation of annihilation radiation experiments, in: "Positron Annihilation", Ref. 5, p. 65 (1979).

43. L. C. Smedskjaer and M. J. Fluss, Experimental methods of positron annihilation for the study of defects in metals, in: "Methods of Experimental Physics", Vol. 21:77-145 (1983), J. N. Mundy, S. J. Rothmann, M. J. Fluss, and L. C. Smedskjaer, Eds., Academic Press, New York (1983).

44. F. M. Jacobsen, A positron lifetime study of properties of light particles in liquids, Risø Report 433 (1981).

45. P. Kirkegaard, M. Eldrup, O. E. Mogensen, and N. J. Pedersen, Program system for analysing positron lifetime spectra and angular correlation curves, Comput. Phys. Commun. 23:307 (1981).

46. A. A. Manuel, L. Oberli, T. Jarlborg, R. Sachot, P. Descouts, and M. Peter, Progress in 2-D angular correlation of positron annihilation using high-density proportional chambers, in: "Positron Annihilation", Ref. 6, p. 281 (1982).

47. M. Eldrup, O. E. Mogensen, and J. H. Evans, A positron annihilation study of the annealing of electron irradiated molybdenum, J. Phys. F. 6:499 (1976).

48. M. J. Fluss, S. Berko, B. Chakraborty, K. R. Hoffmann, P. Lippel, and R. W. Siegel, Positron annihilation spectroscopy of the equilibrium vacancy ensemble in aluminium, J. Phys. F. 14:2831 (1984).

49. S. Mantl and W. Triftshäuser, Defect annealing studies on metals by positron annihilation and electrical resistivity measurements, Phys. Rev. B. 17:1645 (1978).

50. J. Mäkinen, A. Vehanen, P. Hautojärvi, H. Huomo, and J. Lahtinen, Measurements of vacancy-type defect distribution in argon-sputtered Al (110) studied with variable-energy positrons, Submitted to Surf. Science for publication.

51. R. H. Howell, M. J. Fluss, I. J. Rosenberg, and P. Meyer, Low-energy, high-intensity positron beam experiments with a Linac, Nucl. Instr. Meth. B. 10/11:373 (1985).

52. R. N. West, Positron studies of lattice defects in metals, in: "Positrons in Solids", Ref. 8, p. 89 (1979).

53. A. Seeger, The study of defects in crystals by positron annihilation, Appl. Phys. 4:183 (1974).

54. K. Maier, Defects in thermal equilibrium: Positron annihilation and other methods, in: "Positron Solid-State Physics", Ref. 9, p. 265 (1983).

55. H. E. Schaefer, Thermal equilibrium studies of vacancies in metals by positron annihilation, in: "Positron Annilation", Ref. 6, p. 369 (1982).

56. B. T. A. McKee, W. Triftshäuser, A. T. Stewart, Vacancy-formation energies in metals from positron annihilation, Phys. Rev. Lett. 28:358 (1972).

57. R. W. Siegel, Positron annihilation spectroscopy of defects in metals - an assessment, in: "Positron Annihilation", Ref. 6, p. 351 (1982).

58. C.-K. Hu, S. Berko, G. R. Gruzalski, W. K. Warburton, Positron annihilation lifetime and Doppler profile studies in Pb, Pb(Tl), and Pb(Cd), in: "Positron Annihilation", Ref. 5, p. 231 (1979).

59. W. Lühr-Tanck, Th. Kurschat, Th. Hehenkamp, High resolution positron-lifetime study in silver, Phys. Rev. B. 31:6994 (1985).

60. K. Maier, M. Peo, B. Saile, H. E. Schaefer, A. Seeger, High-temperature positron annihilation and vacancy formation in refractory metals, Phil. Mag. A 40:701 (1979).

61. L. C. Smedskjaer, Positron prevacancy effects in pure annealed metals, in: "Positron Solid-State Physics", Ref. 9, p. 597 (1983).

62. G. M. Hood and R. J. Schultz, Positron annihilation and vacancy formation in Al, J. Phys. F. 10:545 (1980).

63. M. J. Fluss, S. Berko, B. Chakraborty, P. Lippel, R. W. Siegel, A monovacancy divacancy model interpretation of positron annihilation measurements in aluminium, J. Phys. F. 14:2855 (1984).

64. L. C. Smedskjaer, M. J. Fluss, D. G. Legnini, M. K. Chason, R. W. Siegel, Positron annihilation measurements of vacancy formation in Ni and Ni(Ge), in: "Positron Annihilation", Ref. 6, p. 526 (1982).

65. H. E. Hansen, B. Nielsen, K. Petersen, Annealing of high energy nitrogen and oxygen radiation damage in molybdenum studied by positrons, Rad. Effects 77:1 (1983).

66. A. Seeger, The interpretation of radiation damage in metals, and W. Schilling, P. Ehrhart, K. Sonnenberg, Interpretation of defect reactions in irradiated metals by the one interstitial model, both in: "Fundamental Aspects of Radiation Damage in Metals", M. T. Robinson, F. W. Yong Jr., Eds., US-ERDA CONF-751006, Oak Ridge, Tennessee (1975), pp. 493 and 470.

67. B. Nielsen, A. van Veen, L. M. Caspers, H. A. Filius, H. E. Hansen, K. Petersen, The interaction between nitrogen and defects in Mo studied by the positron annihilation technique, in: "Positron Annihilation", Ref. 6, p. 438 (1982).

68. P. Hautojärvi, H. Huomo, M. Puska, A. Vehanen, Vacancy recovery and vacancy-hydrogen interaction in niobium and tantalum studied by positrons, Phys. Rev. B. 32:4326 (1985).

69. H. Ullmaier, Ed., Proceedings of the "International Symposium on Fundamental Aspects of Helium in Metals", Rad. Effects 78:1-426 (1983).

70. K. O. Jensen, M. Eldrup, J. H. Evans, Positron annihilation studies of copper and nickel containing high concentrations of krypton, in:

"Positron Annihilation", Ref. 7 (1985) and M. Eldrup, J. H. Evans, A positron annihilation study of copper containing a high concentration of krypton, J. Phys. F. 12:1265 (1982).

71. J. H. Evans and D. J. Mazey, Evidence for solid krypton bubbles in copper, nickel and gold at 293K, J. Phys. F. 15:L1 (1985).

72. A. Vehanen, P. Hautojärvi, J. Johansson, J. Yli-Kauppila, P. Moser, Vacancies and carbon impurities in α-iron: Electron irradiation, Phys. Rev. B. 25:762 (1982).

73. P. Hautojärvi, L. Pöllänen, A. Vehanen, J. Yli-Kauppila, Vacancies and carbon impurities in α-iron: Neutron irradiation, J. Nucl. Mat. 114:250 (1983).

74. K. Petersen, Studies of nonequilibrium defects in metals, in: "Positron Solid-State Physics", Ref. 9, p. 298 (1983).

75. A. Vehanen, J. Mäkinen, P. Hautojärvi, H. Huomo, J. Lahtinen, R. M. Nieminen, S. Valkealahti, Near-surface defect profiling with slow positrons: Argon-sputtered Al(110), Phys. Rev. B. 32:7561 (1985).

76. W. Triftshäuser and G. Kögel, Defect structures below the surface in metals investigated by monoenergetic positrons, Phys. Rev. Lett. 48:1741 (1982).

77. D. A. Fischer, K. G. Lynn, W. E. Frieze, Reemitted-positron energy-loss spectroscopy: A novel probe of adsorbate vibrational levels, Phys. Rev. Lett. 50:1149 (1983).

78. J. Bruce, J. N. Sherwood, N. J. Pedersen, M. Eldrup, A positron annihilation study of the plastic crystal cyclooctane, in: "Positron Annihilation", Ref. 7, p. 181 (1985).

79. M. Eldrup, O. Mogensen, G. Trumpy, Positron lifetimes in pure and doped ice and in water, J. Chem. Phys. 57:495 (1972).

80. O. E. Mogensen and M. Eldrup, Vacancies in pure ice studied by positron annihilation techniques, J. Glaciology 21:85 (1978), and Positronium Bloch function and trapping of positronium in vacancies in ice, Risø Rep. No. 366 (1977).

81. O. E. Mogensen, G. Kvajić, M. Eldrup, M. Milosević-Kvajić, Angular correlation of annihilation photons in ice single crystals, Phys. Rev. B 4:71 (1971).

82. K. Fujiwara, Motion of positronium in some insulating crystals, in: "Positron Annihilation", Ref. 6, p. 615 (1982).

83. M. Eldrup, Vacancy migration and void formation in γ-irradiated ice, J. Chem. Phys. 64:5283 (1976), and M. Eldrup, O. E. Mogensen, J. H. Bilgram, Vacancies in HF-doped and in irradiated ice by positron annihilation techniques, J. Glaciology 21:101 (1978).

84. D. Lightbody, J. N. Sherwood, M. Eldrup, Vacancy formation energies in plastic crystals using positron annihilation techniques, Mol. Cryst. Liq. Cryst. 96:197 (1983).

85. M. Eldrup, On positron studies of molecular crystals, in: "Positron Annihilation", Ref. 6, p. 753 (1982).

86. M. Eldrup, N. J. Pedersen, J. N. Sherwood, Positron annihilation study of defects in succinonitrile, Phys. Rev. Lett. 43:1407 (1979).

87. J. N. Sherwood, Ed., "The Plastically Crystalline State", Wiley, Chichester (1979).

88. J. G. Byrne, A review of positron studies of the annealing of the cold worked state, Met. Trans. A 10A:791 (1979) and J. G. Byrne, Dislocation studies with positrons, in: "Dislocations in Solids", Vol. 6, F. R. N. Nabarro, Ed., North-Holland, Amsterdam (1983) p. 265.

89. H. Fukushima and M. Doyama, The formation energies of a vacancy in pure Cu, Cu-Si, Cu-Ga, and Cu-γMn solid solutions by positron annihilation, J. Phys. F. 6:677 (1976).

90. G. Dlubek, O. Brümmer, P. Hautojärvi, J. Yli-Kauppila, A positron study of age-hardenable Al-Zn-Mg alloys, Phil. Mag. A 44:239 (1981), and R. Krause, G. Dlubek, G. Wendrock, Structural changes during post--ageing of an Al-Zn (15%) alloy at 100°C studied by positron annihilation, small angle X-ray scattering and microhardness measure-

ments, <u>Cryst</u>. <u>Res</u>. <u>Technol</u>. 20:1495 (1985) and refs. therein.

91. N. Shiotani, Positron studies of amorphous alloys, <u>in</u>: "Positron Anni-
     hilation", Ref. 6, p. 561 (1982).

92. P. Hautojärvi and J. Yli-Kauppila, Positron annihilation in amorphous
     metals, <u>Nucl</u>. <u>Instr</u>. <u>Meth</u>. 199:75 (1982).

93. S. Dannefaer, N. Fruensgaard, S. Kupca, B. Hogg, D. Kerr, A positron
     study of plastic deformation of silicon, <u>Can</u>. <u>J</u>. <u>Phys</u>. 61:451 (1983)
     and references therein.

94. W. Fuhs, U. Holzhauer, S. Mantl, F. W. Richter, R. Sturm, Annihilation
     of positrons in electron-irradiated Silicon crystals, <u>Phys</u>. <u>Stat</u>.
     <u>Sol</u>. <u>(b)</u> 89:69 (1978).

95. S. Dannefaer, P. Mascher, D. Kerr, Monovacancy formation enthalpy in
     silicon (preprint).

96. S. Dannefaer, D. Kerr, B. G. Hogg, A study of defects in amorphous
     silicon films, <u>J</u>. <u>Appl</u>. <u>Phys</u>. 54:155 (1983).

97. A. Dupasquier, Positrons in ionic solids, <u>in</u>: "Positrons in Solids",
     Ref. 8, p. 197 (1979).

X-RAY TOPOGRAPHY AND RELATED TECHNIQUES

USING SYNCHROTRON RADIATION

Michèle Sauvage-Simkin*

Laboratoire de Minéralogie-Cristallographie, associé au
CNRS et aux Universités Paris VI et Paris VII
4 Place Jussieu - 75230 Paris Cedex 05, France

INTRODUCTION

X-Ray Topography (XRT) is a non-destructive imaging technique based
on the difference in reflecting power between perfect and imperfect crystal regions. When coupled with proper detectors, it provides a two-
dimensional map of the defect content in a single crystal. However, X-rays
are almost exclusively sensitive to the strain field associated with a
particular imperfection and hence, only those defects producing strains
larger than the minimum detectable limit (now 10^{-8}) over areas broader
than the spatial resolution (about 1μm) may ever be detected by this
method. Moreover, these figures are utmost performances which cannot be
achieved on standard set-ups as will be developed in the following. When
compared to Transmission Electron Microscopy (TEM)[1], XRT is then able to
reveal non destructively the same type of defects : dislocations, stacking faults, planar boundaries, precipitates or impurity induced strains
with a much better sensitivity but a poorer spatial resolution. XRT is
thus restricted to study single crystals with a low density of imperfections. Contrary to TEM, it will be mostly suitable to investigate the
initial stages of processes like microplasticity, phase transformations,
since the field of view is broad enough (about a few cm^2) to image the
whole sample submitted to an applied stress or a thermal treatment,
specially with the new synchrotron radiation (SR) topography cameras.
The advent of this new powerful X-ray source has significantely enlarged
the field of application of XRT which suffered in the past from the long
exposures preventing real time experiments on evolving systems. A factor
of about 100 in the intensity delivered by the SR sources compared to
laboratory generators, together with the advances in X-ray video detectors enable now to follow the transformations taking place in single
crystals under controlled external stimulations. Examples dealing with
growth processes from the melt or in the solid state, phase transformations, plastic deformation will be described in the following.

TEM remains of course the only technique to image defects in the
Angström range such as dislocation cores or interface structures. Moreover, the large variety of electron excited secondary processes provide

---

* and LURE, CNRS et Université Paris-Sud, 91405 Orsay, France

additionnal information on the defect induced modifications of the physical properties. To complete this short introduction on defect imaging by diffraction techniques, one should say a few words about neutron topography (NT). At present the spatial resolution is of a few ten μm and exposures last several hours. Nevertheless, there are some questions to which NT only is able to provide an answer, for instance problems connected with the spatial configuration of heli-magnetic domains[2].

People interested in the visualization of imperfections in single crystals are then provided with a variety of techniques, mostly complementary for the scale of observation and the type of defects to which they respond.

The present paper will now focus on X-ray imaging and diffractometric methods, with a particular emphasis on the progresses linked to SR sources since the mid-seventies. The first part will recall some basic X-ray diffraction principles which may appear somewhat tedious but are necessary to understand most applications. In a second section, the specific properties of SR sources will be described together with indications on the available XRT stations at the various SR facilities. The last section will provide examples of recent applications. Since this review covers a rather broad field, the reader is referred to more specialized books for additionnal information[3,4,5] ; the present article should be taken more as an introductory guide to the subject that an exhaustive compilation.

BASIC PRINCIPLES OF X-RAY PROPAGATION

As everybody knows, the periodic crystal lattice is able to diffract radiations with wavelength λ in the Angström range as long as the incidence angle θ_B satisfies Bragg's law :

$$2d \sin \theta_B = n\lambda \qquad\qquad n \text{ integer} \qquad\qquad (1)$$

where d is the spacing of the family of lattice planes. However, this very simple geometrical statement does not suffice to account for the exact interaction of X-ray waves with the presently available single crystals. The formalism of dynamical theory has to be used.

Plane-Wave - Perfect Crystal

The interaction of a plane wave, characterized by a vacuum wave vector $\vec{K}_{ov}$, with a perfect crystal defined by a triply periodic dielectric susceptibility $\chi(\vec{r})$ is governed by Maxwell's equations. Hence the electric displacement $\vec{D}(r)$ in the medium should satisfy :

$$\Delta \vec{D} + \text{curl. curl } (1+\chi) \vec{D} = 0 \qquad\qquad (2)$$

where the susceptibility $\chi(r)$ can be expanded in a Fourier series

$$\chi(r) = \sum_h \chi_h e^{-2\pi i \vec{h}.\vec{r}} \qquad\qquad (3)$$

the infinite summation extending over all reciprocal lattice vectors $\vec{h}$. Following Laue's treatment[6], it can be shown that solutions of (2) appear as linear combinations of Bloch waves of the form :

$$\vec{D}(r) = \sum_{j} \vec{D}_j(r) = \sum_{j} \sum_{h} \vec{D}_{hj} \; e^{-2 \, i\pi \vec{K}_{hj} \cdot \vec{r}} \qquad (4)$$

All wave vectors $\vec{K}_{hj}$ pertaining to a given Bloch wave are deduced from one of them, say $\vec{K}_{oj}$ by adding a reciprocal lattice vector $\vec{h}$:

$$\vec{K}_{hj} = \vec{K}_{oj} + \vec{h} \qquad (5)$$

$\vec{K}_{oj}$ is usually selected as the one related to $\vec{K}_{ov}$ by the condition of continuity of the tangential component. Equation (5) thus defines a bundle of vectors having their origin on the nodes of the reciprocal lattice and a common extremity P_j referred to as the tie-point of the solution. Introducing expansion (4) into equation (2) leads to an infinite system of equations for the unknown amplitudes $\vec{D}_{hj}$. It comes out that only those waves for which $|K_{hj}|$ is close to $|K_{oj}| \sim n/\lambda$, n being the refractive index for X-rays, have a non vanishing amplitude. The normal situation for X-rays is the so-called two-beam case where one of the $\vec{D}_{hj}$ (h $\neq$ 0) is important.

Let us introduce now a geometrical support to facilitate the understanding. Classically, one uses the Ewald sphere of radius $1/\lambda$ going through the origin 0 of the reciprocal lattice and whose center L_a is such that OL_a is identical to $\vec{K}_{ov}$. Here we will instead consider the two spheres of radium $1/\lambda$ centered on nodes 0 and H, intersecting in L_a (fig. 1). The extremity M of the actual vacuum wave vector $\vec{K}_{ov}$ lies somewhere on the "O" sphere. Hence, by applying the boundary condition, the extremity P_j of the crystal wave vector $\vec{K}_{oj}$ is obtained on the inner "O" sphere with radius n/λ. The corresponding $\vec{K}_{hj}$ is then readily derived ($\vec{K}_{hj} = \overrightarrow{HP_j}$). It is clear that for an arbitrary location of M, $|K_{hj}|$ is very different from n/λ and only a refracted wave $\vec{D}_{oj}$ propagates in the medium. On the contrary, when M

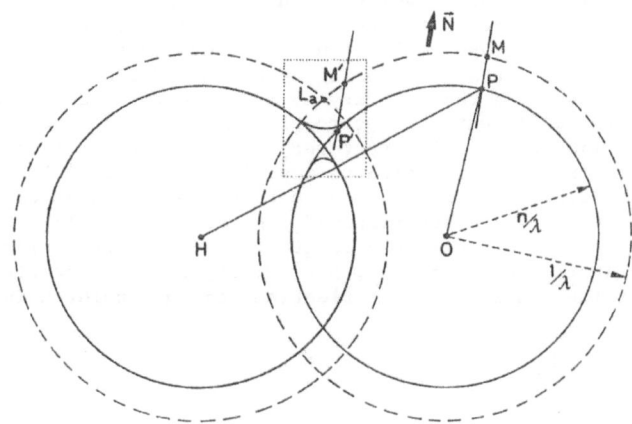

Fig. 1 Deriving the crystal wave vectors in the two-beam case.
$\vec{n}$, normal to the crystal surface. Boxed area enlarged in Fig. 2.

lies in a close vicinity of L_a (see M' in fig. 1), both $|K_{oj}|$ and $|K_{hj}|$ are near to n/λ and a reflected wave also exists in the crystal. In that case, the simple representation sketched in fig. 1 is no longer valid and the actual loci of tie-points P_j are branches of a hyperbola, merging tangentially into the inner "O" and "H" spheres of fig. 1. A blow-up of this region is presented in fig. 2. ; in the narrow range of existence of the reflected wave, the asymptotic circles may be replaced by their tangent T_o and T_h. Fig. 2 represents the well known dispersion surface. When M coincides with L_a, the incident ray fulfills exactly the geometrical Bragg's law (equation 1). Hence $\overline{L_aM}$ indicates the departure from Bragg incidence $\Delta\theta$ since $\overline{L_aM} = \Delta\theta/\lambda$.

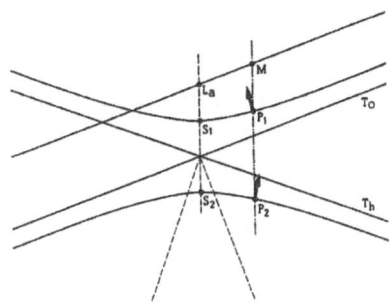

Fig. 2 The dispersion surface in the symmetrical transmission case : surface normal $MP_{1,2}$ perpendicular to the reciprocal vector $\overrightarrow{OH}$.

Applying the boundary condition produces two tie-points P_1 and P_2 on either branch. Hence two Bloch waves propagate in the crystal :

$$\begin{cases} \vec{D}_1 = \vec{D}_{o1}\, e^{-2\pi i\, \overrightarrow{OP}_1 \cdot \vec{r}} + \vec{D}_{h1}\, e^{-2\pi i\, \overrightarrow{HP}_1 \cdot \vec{r}} \\ \vec{D}_2 = \vec{D}_{o2}\, e^{-2\pi i\, OP_2 \cdot r} + D_{h2}\, e^{-2\pi i\, \overrightarrow{HP}_2 \cdot \vec{r}} \end{cases} \tag{6}$$

$\vec{D}_1$ and $\vec{D}_2$, referred to as wavefields, travel along the normals to the dispersion surface at their respective tie-points (see bold arrows in fig. 2). Both waves building up a wavefield are coherent and interfere which results in a beat of the amplitude D_j with a spatial periodicity equal to the lattice plane spacing $(1/|\overrightarrow{HP}_j - \overrightarrow{OP}_j| = 1/|\vec{h}| = d)$. Two systems of standing waves with the lattice periodicity are then present in the crystal under Bragg reflection. Moreover, these two systems have opposite phases, the nodes being on maxima of electronic density for wavefield 1 whereas the contrary holds for wavefield 2 ; this accounts for the different absorption of the two wavefields under Bragg condition, in simple crystal structures, known as the Borrmann effect and has found a novel application to impurity and adatoms location as will be shown later. Interferences also take place between wavefields 1 and 2 whenever they are superimposed. The periodicity $1/|\overrightarrow{P_1P_2}|$ is then much larger; a characteristic length for X-ray diffraction is the so-called extinction distance, Λ_o, periodicity of this interference phenomenon at the center of the reflection domain : point M in L_a, in fig. 2.

$$\Lambda_o = \frac{1}{|\overline{S_1S_2}|} \tag{7}$$

$|\overline{S_1S_2}|$ being of the order of $10^{-5} - 10^{-6}\,\overset{\circ}{A}{}^{-1}$, Λ_o ranges from a few micrometers to a few ten micrometers. The dynamical theory predictions which have been exposed in this section only hold for single crystals perfect enough over a length much larger than Λ_o, where they have actually been confirmed. For crystals either very small or too severely

distorted, the kinematical theory which ignores the dispersion surface
may as well be used. Both theories converge to the same result for
the observable reflected intensity in the limit of small crystals.

The reflecting power R_h, averaged over the oscillations, depends
sharply on the departure from exact Bragg incidence $\Delta\theta$. Rocking curves
corresponding to the two possible geometries of transmission (fig. 3a)
and reflection (fig. 3b) are presented. In the transmission or Laüe case,
the overall shape is gaussian, centered at $\Delta\theta = 0$, with a maximum
reflectivity close to 0.5, the remaining intensity is found in the re-
fracted waves D_{oj}. In the reflection or Bragg case, a reflectivity
close to 1 is observed in the so-called total reflection domain, which
is no longer centered at $\Delta\theta = 0$: a shift $\Delta\theta_R$ is observed which depends
on the wavelength and on the angle α between the reflecting planes and
the crystal surface ($\alpha = \pi/2$ and $\alpha = 0$ are referred to as the "symmetrical" Laüe and Bragg cases, respectively). This refraction effect can
be efficiently used to eliminate harmonics from the output beam of X-ray
Monochromators for Synchrotron Radiation[7]. It is clear from fig. 3 that
intensity is observed in the reflected beam within a finite angular range

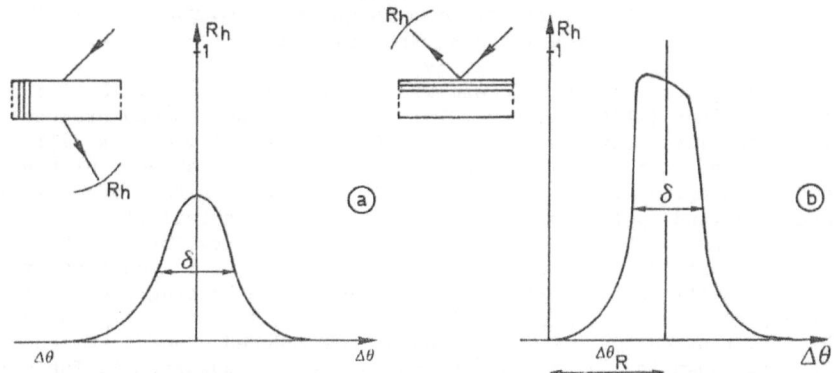

Fig. 3   Rocking curves in the symmetrical Laüe case (a) and
Bragg case (b). Thick crystals, low absorption.

characterized by the full width at half maximum (FWHM) δ of the rocking
curve. For a given Bragg reflection, δ depends on λ and α. Typical
values range from a fraction of an arc second to a few ten seconds. It
is easy to understand that any local distortion of the crystal leading
to a variation of lattice parameter Δd or (and) to a rotation of the
lattice plane $\Delta\phi$ will change the local value of $\Delta\theta$ and induce an intensity variation which can be detected in plane-wave imaging methods.

When the crystal thickness becomes of the order of Λ_o, averaging
over the oscillations is no longer justified. Since it has found important applications in the characterization of thin epitaxial layers,
the rocking curve expected in the symmetrical Bragg case for a thin
perfect crystal with thickness equal to $\Lambda_o/3$ is presented in fig. 4.
The spacing of the minima in the oscillating wings is inversely proportional to the crystal thickness and provides an accurate way to
estimate this parameter. When compared to the thick crystal rocking
curve (fig. 3b), the peak reflectivity is lower whereas the FWHM δ is
larger, both trends being more pronounced as the ratio t/Λ_o decreases.

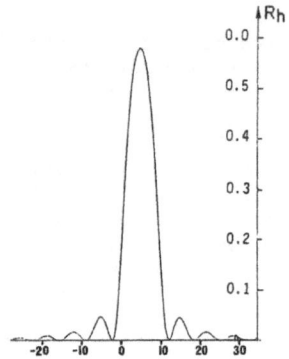

Fig. 4   Calculated rocking curve for a self-supporting thin crystal with $t = \Lambda_o/3$. Symmetrical Bragg case.

Non Plane-Wave - Perfect Crystals

The results exposed above were based on the assumption that the incident wave could be defined by a unique $\vec{K}_{ov}$ implying that both the wavelength and angle spreads were vanishingly small. Such waves are not spontaneously delivered by any X-ray source : a laboratory generator gives a divergent beam within the natural bandwith of characteristic lines ($\Delta\lambda/\lambda \sim$ a few 10^{-4}) whereas SR sources may produce highly parallel beams but with an infinite spectral width as will be shown in the next section. As a consequence, in the absence of beam conditionning, a single crystal will be irradiated with a beam divergent enough or (and) polychromatic enough for the whole reflection domain to be excited. The spherical wave theory, developed by N. Kato[8] has to be used to predict the X-ray propagation but it won't be exposed here. It suffices to stress that the observed reflecting power will correspond to the integration of the curves presented in fig. 3 and 4. The resulting curves, as a function of the reduced crystal thickness t/Λ_o are shown in fig. 5 a and b, for negligible absorption. The dotted line corresponds to the kinematical integrated intensity, simply proportional to the crystal thickness, both results are identical in the very thin crystal limit ($t/\Lambda_o \leqslant 0.2$). Severely distorted areas with large strain gradients behave kinematically whereas the host perfect crystal responds dynamically, here lies the origin of contrast in imaging techniques recording an integrated intensity such as the Lang[9] and Berg-Barrett[10] methods in the laboratory or White Beam Topography (WBT)

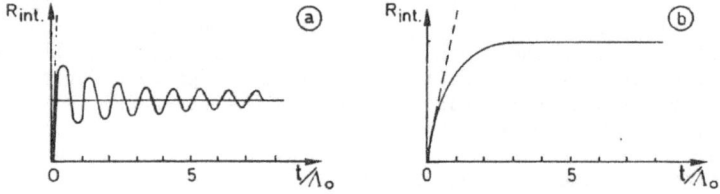

Fig. 5   Integrated intensity in the Laue (a) and Bragg (b) cases.

and Broad Band Monochromatic Topograph with SR sources. An overall smooth rotation or d spacing variation, within the incident beam divergence or spectral width, would not produce a detectable contrast with such methods.

Imperfect Crystals

Dealing with perfect crystals in the preceding sections, a few hints of how the exposed schemes would be altered by crystal imperfections have been given. As mentioned before, a defect is only active through the strain introduced in the host lattice. For most defects of interest, a displacement field $u(\vec{r})$ can be defined outside of the core region. The resulting additionnal variation of departure from Bragg incidence $\delta(\Delta\theta)$ reads[11]

$$\delta(\Delta\theta) = \frac{\lambda}{\sin 2\theta} \frac{\partial}{\partial s_h} (\vec{h}.\vec{u}) \tag{8}$$

where s_h is the coordinate along the reflected direction. Plane wave methods which are particularly sensitive to variations of $\Delta\theta$ are then able to map the strains (spatial derivatives of $u(\vec{r})$) around the defects. It is obvious from formula (8) that a defect with a displacement field normal to the diffraction vector $\vec{h}$ won't be detected on this particular reflection. Looking for the extinction of defect images for several Bragg reflections is precisely the best way to identify Burgers vectors or fault vectors both in XRT and TEM methods.

As was outlined previously, imaging methods based on integrated intensity do not detect constant or very slowly varying strains, fairly large strain gradients are required to produce a noticeable contrast in the images. An important parameter is then a quantity β, proportionnal to the second derivative of the displacement :

$$\beta \propto \frac{\partial^2}{\partial s_o \partial s_h} (\vec{h}.\vec{u}) \tag{9}$$

where s_o is the coordinate along the transmitted direction. Numerical techniques have been elaborated to solve the differential equations governing X-ray propagation in distorted crystals, based on formalisms developed independently by Takagi[12,13] and Taupin[14]. Simulated rocking curves or simulated images (see Authier and Epelboin in ref. 3) can be reconstructed with a model strain field and compared to the experimental data as will be shown later on in this review.

SYNCHROTRON RADIATION TOPOGRAPHIC TECHNIQUES

Synchrotron Radiation Sources (see H. Winick in ref. 5)

Synchrotron radiation is emitted by a charged particle submitted to a centrifugal acceleration. Formula 10 gives the power radiated on a circular trajectory of radius ρ by a particle of rest mass m_o, charge e, kinetic energy E and velocity βc

$$P = \frac{2e^2 c}{3\rho^2} \beta^4 (\frac{E}{m_o c^2})^4 \tag{10}$$

The emitted electromagnetic spectrum is continuous with a maximum close to a critical wavelength λ_c given by :

$$\lambda_c = 5.59 \frac{\rho}{E^3} \tag{11}$$

Table 1:Characteristics of some X-ray SR Sources

source		λ_c (Å)	N(1.54 Å) Ph/sec/mrad/.1%BW	vert. size (mm)	horiz. size (mm)	diverg. (mrad)
Euopean SR Facility	BM	0.9	10^{13}	0.23	0.21	0.018
ESRF	W	0.5	3.10^{14}	0.09	0.14	0.036
ADONE (I)	W	4.3	$3.4.10^{12}$	0.55	3.2	0.18
SRS (UK)	BM	4.	3.10^{12}	0.53	6.21	0.12
	W	0.9	$1.2.10^{13}$	0.39	12.2	0.12
DCI (F)		3.6	$2.4.10^{12}$	2.44	6.25	0.14
DORIS (FRG)	BM	0.55	$4.4.10^{12}$	1.5	3.	0.15
	W	2.3	$1.3.10^{14}$	0.9	3.5	0.08
CHESS (US)	BM	1.	4.10^{12}	2.3	3.3	0.15
	W	0.4	3.10^{13}	2.8	4.4	0.12
NSLS (US)		2.4	8.10^{12}	0.23	0.58	0.023
SPEAR (US)	BM	2.7	3.10^{12}	0.64	2.	0.05
	W1	1.	$4.5.10^{13}$	0.35	10.6	0.07
	W2	1.7	$2.5.10^{14}$	0.35	10.6	0.07
PHOTON FACTORY (J)	BM	3.0	3.10^{12}	1.4	4.5	0.32
	W	0.7	$2.5.10^{13}$	1.6	4.4	0.42
VEPP3 (USSR)		3.	$1.5.10^{12}$	0.23	1.7	0.05

BM:bending magnet;W :wiggler

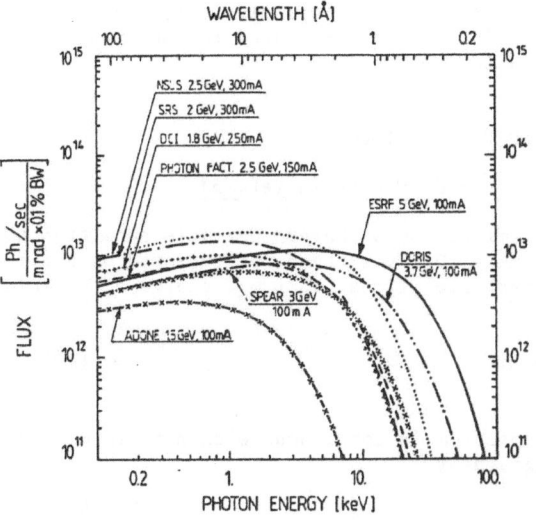

Fig.6 Photon fluxes from planned and existing X-ray SR sources.

with ρ expressed in meters and E in GeV. Formulae 10 and 11 show that electrons or positrons accelerated in rings with radii of a few meters, up to kinetic energies in the GeV range will be very efficient hard X-ray sources. In table 1 are listed the relevant features of the present and future storage rings where SR facilities are operating or planned. Fig. 6 shows the emitted spectrum for some of them. SR sources present many advantages for X-ray topography and related techniques :

- high brilliance which means high photon fluxes per unit area of the source and per unit solid angle
- white spectrum
- narrow divergence (in the 10^{-4} rad. range)
- spatially extended beam on the sample
- nearly pure linear polarization
- pulsed time structure in the MH_z range

These properties have governed the design of the experimental set-ups dedicated to XRT which will be described now.

White Beam Topography (see J. Miltat in ref. 4)

It is well known since the discovery of X-rays in 1912 that a single crystal immersed in a white beam diffracts selected wavelengths λ's satisfying the Bragg condition (1) for different sets of properly oriented lattice planes (the so-called Laue diagram). Due to the parallelism and two dimensional extension of the SR beam, each of these spots is

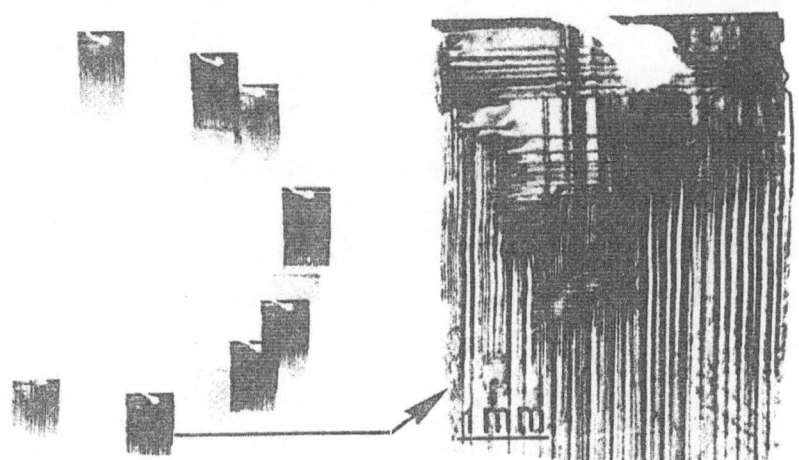

Fig. 7   White beam topography of $Gd_2(MoO_4)_3$ crystals. Full pattern and enlarged spot showing ferroelectric domain walls. (courtesy of C. Malgrange and B. Capelle).

actually an X-ray topograph which can be recorded on a fine grain emulsion with exposures of a few seconds compared to several minutes with the best X-ray generators and more likely hours with the standard ones (fig. 7). The spatial resolution r of the images is controlled by the angular divergence $\Delta\theta_o$ impinging on one point of the sample and by the sample detector distance D according to the simple formula :

$$r = D \times \Delta\theta_o \qquad (12)$$

$\Delta\theta_O$ depends on the so-called emittance of the source, a function of the electron bunch size and trajectories at the extraction point. For most sources $\Delta\theta_O$ is comprised between 10^{-4} and 10^{-5} rad. and is usually much smaller in the vertical plane than in the horizontal plane. Keeping r below 5 µm still allows sample-detector distances of about 10 cm which enables to surround the sample with rather cumbersome equipments often necessary for temperature conditionning, controlled atmospheres or magnetic studies.

Very efficient multipurpose white beam cameras have been set on line at the SRS (Daresbury - UK)[15] and at the Photon Factory (Tsukuba, Japan)[16]. The latter, presented in fig. 8, is equipped with two X-ray video detectors

Fig. 8 The WBT camera on line at the Photon Factory (after ref. 16)

provided with the necessary degrees of freedom to track the desired diffraction spot. A fluorescent screen where the whole Laue pattern is formed is viewed by a standard TV camera and compared to a computer generated pattern, the sample can then be brought in the proper orientation and the X-ray sensing cameras set on the selected spots. The required accuracy of angular positioning is a fraction of a degree, compared to a few seconds in laboratory set-ups working with characteristic lines. Moreover, crystals distorted by stresses or thermal gradients, crystals with subgrains, still give homogeneous images, each area extracting the proper wavelength from the incident white beam. In addition, crystals with uncontrollable orientation because grown in-situ always give useful diffraction spots.

These unique advantages enable to understand why WBT coupled to video detectors has permitted significant advances in the fields of crystal growth, phase transitions and plastic deformation studied in real time.

Video detectors. Standard direct X-ray video detectors have spatial resolutions at best in the 25 µm range. However, 6 µm resolution has been achieved with a very elaborate saticon tube[17]. For indirect detectors where an X-ray phosphor provides an intermediary image, optical magnification can be used to compensate for the poor resolution of the TV tube.

However, the resulting enlarged TV image is often too noisy and on-line image processors have to be included in the set-up[18].

Monochromatic Synchrotron Radiation Topography

Despite its obvious advantages, WBT cannot always be used for several reasons among which one can quote :

- radiation damage
- heat damage
- high fluorescence background from the sample or its immediate surroundings
- unavoidable parasitic scattering
- problems with the superposition of several harmonics in each spot (λ, $\lambda/2$ ..., λ/n)

Simple monochromators delivering rather broad bands in the 10^{-4} range both in angle and wavelength spread can be designed. However, if one wishes to know the exact diffraction conditions, each couple formed by the monochromator and the sample has to be studied carefully. The easiest way to obtain this information is the Du Mond diagram approach[19] which is simply the graph $\lambda = f(\theta)$ taking into account the finite width δ of the reflection domain. For the crystal monochromator, this graph enables to estimate the spectral and angular distribution of the beam extracted from the white SR spectrum once the overall divergence $\Delta\theta_0$ received by the monochromator is known (fig. 9)

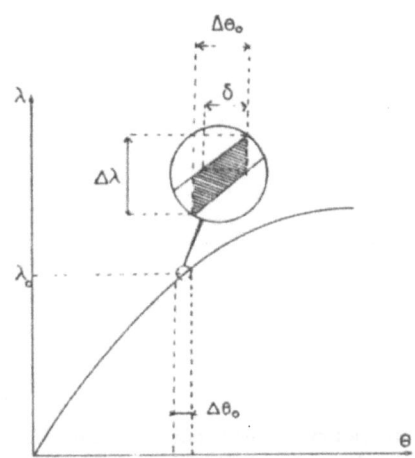

Fig. 9 Du Mond diagram for a single crystal in a white divergent beam : the shaded area in the blow-up gives the FWHM λ and θ spreads of the output beam

Such diagrams may also be drawn for multiple crystal settings with a proper choice of origin and sense of the θ axes. Two situations have to be distinguished, the so-called (+, +) setting displaid in fig. 10a and the (+,-) setting represented in fig. 10b. A particular case is the (+,-) parallellel setting where the two reflections are identical (fig. 10b, lower blow-up). The intensity output is proportionnal to the shaded overlap of the individual reflection bands. It appears that, except for the parallel case, the whole reflection domain of the sample is usually excited and one is dealing here with an integrated intensity method. However, the spectral and angular bandwith of the monochromator is limited (fig. 9) and severely distorted crystals may not be fully imaged. Fig. 11 shows misfit dislocation in a complex heterostructure where the built-in strains induce a curvature of radius R $\sim$ 20 m. Consequently, a

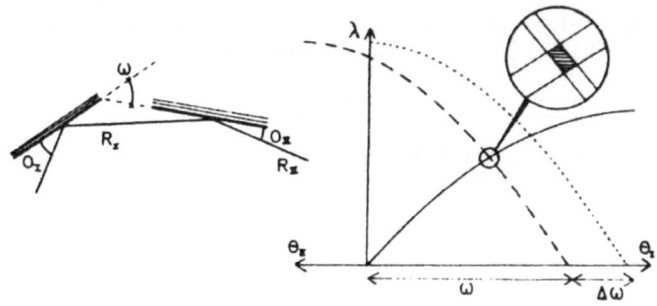

Fig. 10a   Du Mond diagramm for a two crystal set up in the (+,+)
geometry. Full curve : monochromator ; dashed curve :
sample when the angle between the reflecting planes
is ω ; dotted curve : sample after a rotation Δω.

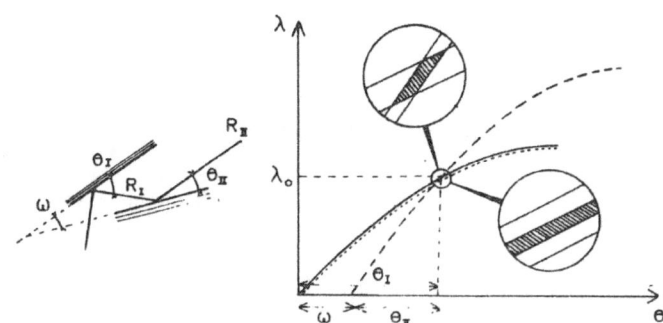

Fig. 10b   Du Mond diagram in the (+,-) geometry ; full curve : monochromator ; dashed curve and upper blow-up : non parallel
case ; dotted curve and lower blow-up : parallel setting.

stationnary topograph recorded in a non parallel (+,-) set-up shows only
a band-shape image (fig. 11a), with strong contrasts whereas a full image,
with a reduced dynamics, is obtained by rocking the sample in front of
the monochromatic beam (fig. 11b). Due to this possible partial imaging,

190

it is of primary importance, in the case of evolving phenomena, to follow the sample image on a TV monitor in order to be sure that the area of interest is not out of contrast.

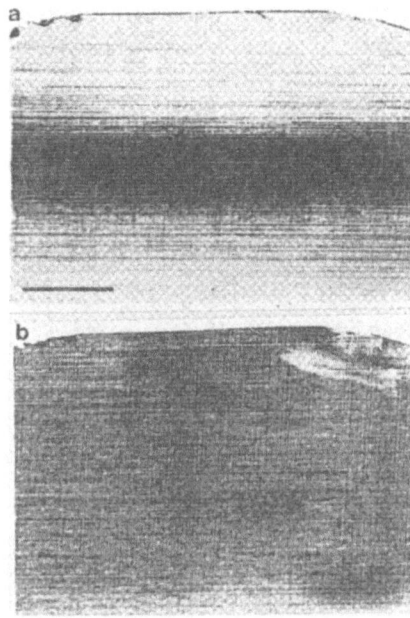

Fig. 11a   Static SR Topograph.
Monochromator : Ge 220,
$\lambda = 1.4$ Å. Available
divergence on the   sample
(GaAs 400) in a non
parallel (+,-) geometry :
14".

b   Fully integrated SR.
Topograph.
Scale mark : 2 mm.

As was mentionned above, in the parallel (+,-) setting, only a fraction of the reflection domain of the sample is excited. One is then getting close to a plane wave situation and indeed, in the laboratory, "plane waves" are simulated by such two crystal geometries where the intrinsic width of the first crystal is reduced by an asymmetric cut ($\theta-\alpha \sim 1°$). However, a matched perfect monochromator crystal has to be tailored for each Bragg reflection of each sample. The brilliance of SR sources has made feasible the design of beam conditionners delivering quasi plane waves. Fig. 12 gives the principle of a triple reflection silicon monolithic monochromator designed by H. Hashizume for the LURE-DCI (Orsay-France) two-axis spectrometer[20].

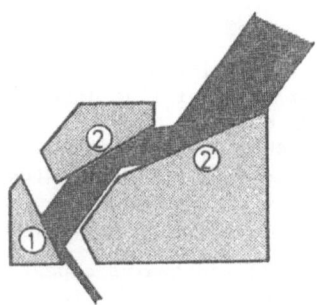

Fig. 12   LURE-DCI plane-wave,
harmonic-free monochromator :
$\lambda = 1.2378$ Å $\Delta\lambda/\lambda = 7 \cdot 10^{-6}$,
angular divergence :
$2 \cdot 10^{-6}$ rad.

Strains in the $10^{-5} - 10^{-6}$ range have been imaged and measured with this beam conditionner in a variety of samples, thus relaxing the constraint of a matched first crystal. However, narrow band pass monolithic

191

monochromators are not tunable. Separate crystals devices enabling to overcome this drawback have thus been operated in various SR facilities[21].

EXAMPLES OF APPLICATIONS

Since earlier reviews may be found in references 3, 4 and 5, the emphasis will be mostly put here on recent applications.

Crystal Growth

In situ crystal growth has been followed in two very different situations by WBT.

Solid state growth. Continuing the series of experiments on recrystallization in ultra high vacuum environment[22], C. Jourdan and J. Gastaldi have been able to study the structure of moving grain boundaries in Aluminium[23]. Long range periodic structures with periodicity in the 1-50 μm range have be found in boundaries close to the {110} and {100} crystallographic orientation of the recrystallized grain ; no such structures are ever observed in {111} oriented boundaries (fig. 13).

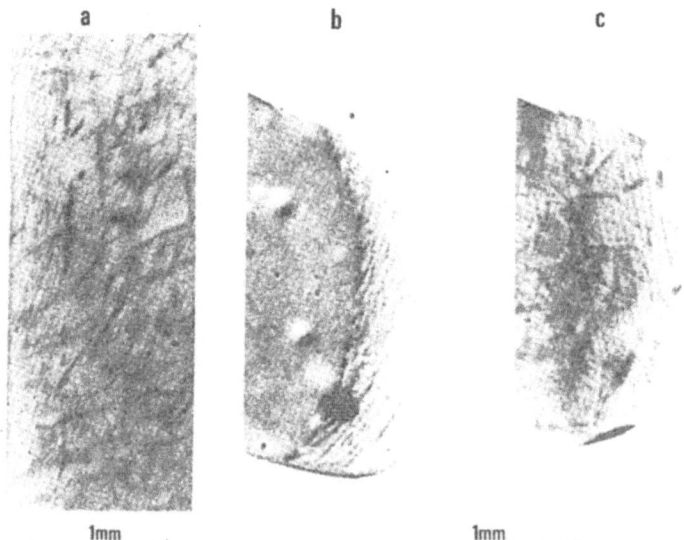

Fig. 13   WBT of moving grain boundaries in aluminium
          a) {110} left boundary - regular array of line defects
          b) {100} right boundary - coarse defects
          c) {111} right boundary - no defect, wedge Pendellösung
             fringes (courtesy of J. Gastaldi and C. Jourdan, after[23])

At the beginning of growth, the recrystallized grains appear as facetted, nearly strain-free polyhedra, bounded by low index faces, as can be found in solution or vapor growth (fig. 14). The crystal perfection is assessed by the presence of straight wedge shape fringes due to the Pendellösung interference. When present, stresses induced by the surrounding matrix are revealed by a fringe curvature[24].

Fig. 14
WBT-thickness fringes in a
quasi perfect recrystallized Aluminium grain.
Larger dimension : about
1.5 mm.

Growth from the melt. In an other type of experiment, recently performed at the Photon Factory, growth from the melt of a tin single crystal
has been followed in-situ by WBT[25] with a high resolution TV detector
(fig. 15). A comparison between single and polycrystal melt-growth behaviour could be made and the efficiency of the different grown-in dislocations for growth rate modification was estimated.

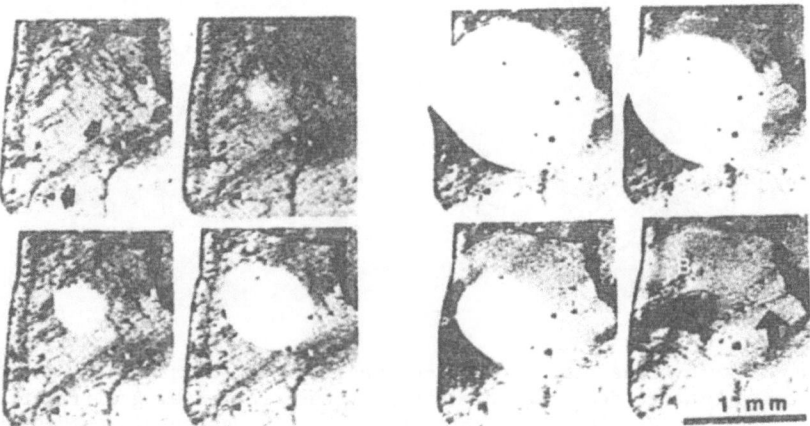

Fig. 15   Melt-growth sequence of a tin crystal. Improvement
of crystal quality except for a dislocation bundle B.

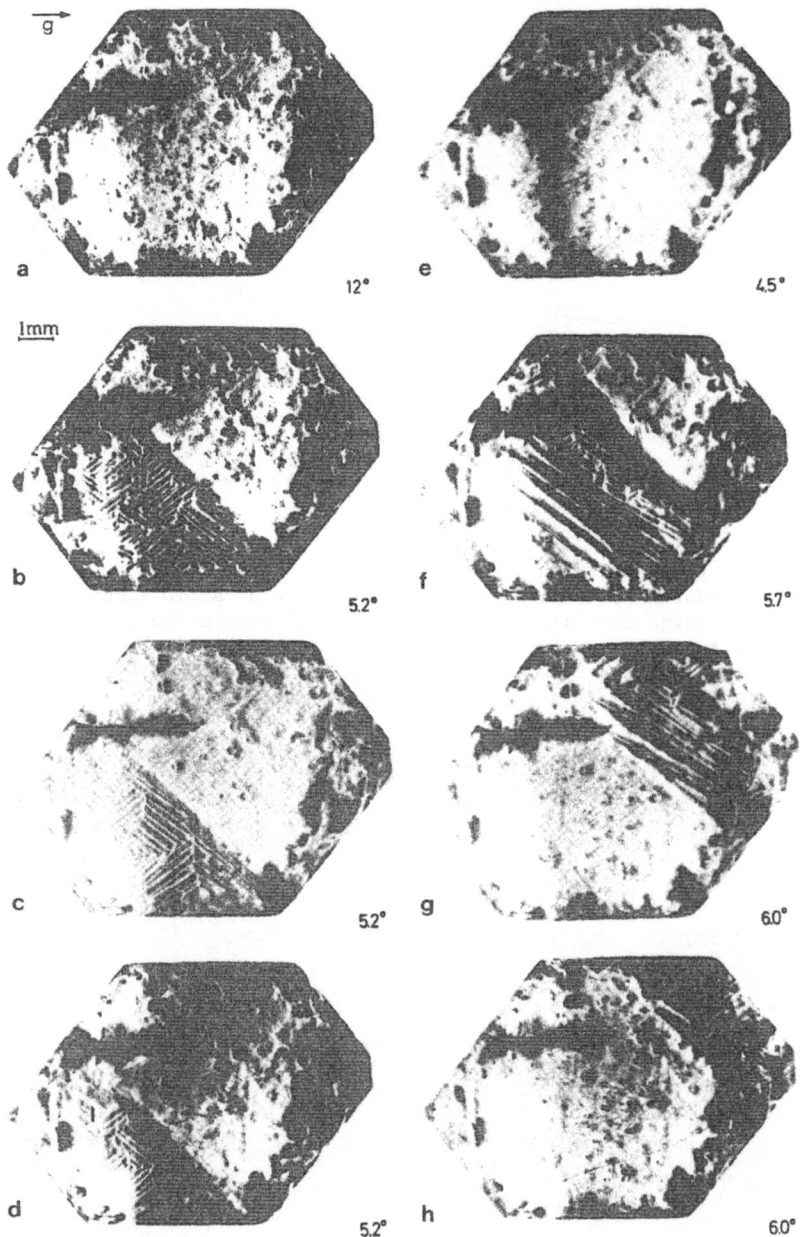

Fig. 16   Lock-in transition of a gel-grown TMA_2ZnCl_4 crystal. a → d
temperature decrease, e → h temperature increase. g diffraction
vector. IC : incommensurate phase, F : ferroelectric commensurate
phase, S mixed region with the "chevron-like" pattern (Courtesy
of M.Ribet et al., after[31]).

Phase Transformations

The major interest of imaging techniques in that field is to provide information on the morphology of the phase front and to reveal possible interactions with the preexisting lattice defects : dislocations, growth sector boundaries ... Moreover, for all studies involving phenomena taking place in a very narrow temperature range, the ability, provided by SR sources, to follow the transformation in real time and to record photographic plates within very short exposures, has significantely relaxed the stability constraints on the experimental set-ups.

When starting with a single crystal, one should distinguish between conservative and non-conservative phase transformations. In the first case, the coherence of the single crystal is preserved and only slight modifications of the structure take place. In the second case, a polycrystal is produced after the transformation. Examples of both situations have been studied by SR topography.

Quartz. The $\alpha \to \beta$ transition occuring at 846K has been known for years. It has been shown that the pattern formed by α_1, α_2 twins in the low temperature phase α was influenced by the presence of growth sector boundaries[26]. Actually, this transition is more complex since an incommensurate phase, intermediate between α and β has been discovered by neutron[27] and X-ray[28] diffraction. WBT performed both on a high power generator and at the Photon Factory has enabled to identify the six modulation vectors, $\vec{q}$, and to follow their temperature behaviour. A domain pattern, connected with single $\vec{q}$ regions in the incommensurate phase, has been revealed in the topographic images[29]. Similar studies on quartz and berlinite, $AlPO_4$, isomorphic to quartz, are in progress at LURE-DCI[30].

TMA_2ZnCl_4. Tetramethylammonium tetrachlorozincate crystals show a sequence of incommensurate and commensurate phases in the 0-25°C temperature range. Since the sample is very sensitive to radiation damage, the SR topographic study[31] has been performed with a Ge 220 monochromator, in the (+,+) setting, on the LURE-DCI two axis spectrometer ($\lambda \sim 0.6$ Å). A sequence of topographs recorded during a reversible lock-in transition between an incommensurate phase IC and a ferroelectric commensurate phase F is displaid in Fig. 16. The quasi-periodic "chevron-like" pattern appearing in the IC phase within an 0.5°C temperature interval is interpreted as the outcrop on the sample surface of arrays of "deperiodisation" lines[32]. Such lines are assumed to separate IC and F areas or two different IC regions finely intertwined. This is the first topographic evidence of a macroperiodicity related to a structure existing on a much finer scale in these rather sophisticated materials. Such contrasts are only visible in nearly strain-free crystals as those grown in gel-trapped solutions.

Titanium. Here is the case of a non conservative transition. The $\alpha \to \beta$ transformation taking place at 870°C has been followed in-situ under ultra high vacuum at LURE-DCI. A confirmation of the shear model introduced to interpret the martensitic type of the transition has been obtained by WBT[33]. More recently the process of growing large single crystals both in the α and β phase has been controlled and understood[34]. Since the newly formed crystals have not a simple orientation relationship with the matrix, it is of primary importance to work here with WBT.

Polytypic Structures. Although this is not a phase transition but more likely a phase coexistence problem, these studies will be mentioned here. WBT coupled to a computer-aided indexing of the patterns has enabled to characterize the highly complex structure of SiC polytypes[35]. An example of composite sample reflections is shown in Fig. 17, taken at the SRS (Daresbury, UK).

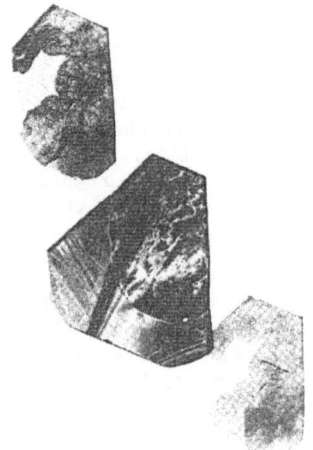

Fig. 17 WBT reconstructed puzzle of
a mixed polytype SiC crystal
about 6mm across. (Courtesy
of G.R.Fisher and P.Barnes,
after[35]).

Plastic Deformation

Since WBT enables to obtain homogeneous images even in highly strained samples, this field of application has been explored from the very
beginning of SR topography experimentation. Reviews of the major results
up to 1983 may be found in references 3, 4 and 5. The two examples presented in the following are dealing with semiconductor plasticity ;
however, the recent thorough study of slip nucleation in Fe-3.5 % Si bcc
single crystals[36] extending a previous work[37] ought to be mentioned too.

Plastic relaxation at a crack tip. A crack propagation may be stopped by the plastic relaxation of the large stresses concentrated at the
tip. The expansion kinetics of the plastic zone under a constant stress,
in silicon crystals has been followed as a function of both the temperature and applied stress[38]. Above 923 K, the extension of the plastic zone
is found to reach a saturation value wich depends on the stress as shown
in Fig. 18.

Transmission of plastic deformation through a grain boundary. This
question is of primary importance to understand the mechanical behaviour
of polycrystalline materials. A model system is provided by silicon bicrystals where the crystallographic type of the boundary and its orientation with respect to the slip systems activated by a given applied stress,
may be selected at will[39]. In Fig. 19 are represented two possible processes by which plastic deformation may propagate through the boundary :
dislocations, generated on purpose in one grain, may either actually cross
the boundary whenever the slip systems to which they belong intersect
along their Burgers vector at the boundary (fig. 19a) or pile-up and
induce stresses intense enough to activate new slip systems in the other
grain (fig. 19b).

Although WBT would seem the appropriate technique, these two experiments on silicon plasticity have been performed with a Ge 220 monochromator, in the (+,+) setting, in order to avoid a severe diffusion and
fluorescence background from the sample surroundings.

196

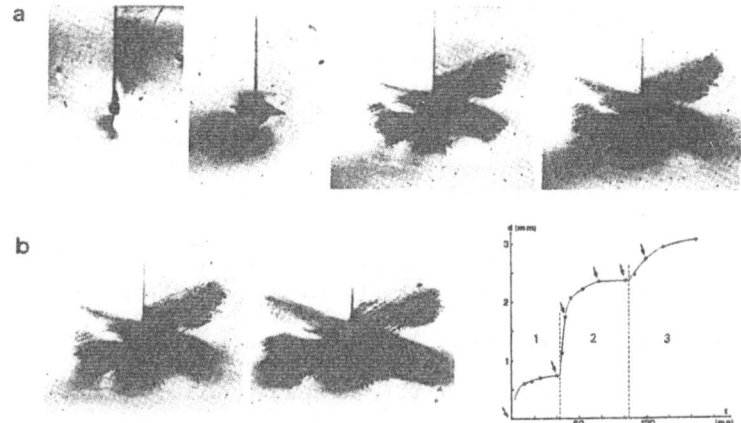

Fig. 18   Crack tip plasticity in silicon ; kinetics of the plastic
           zone extension under a step increase of the applied stress.
           a) 220 SR topographs b) size of the plastic zone versus time
           (courtesy of G.Michot and A.George, after[38]).

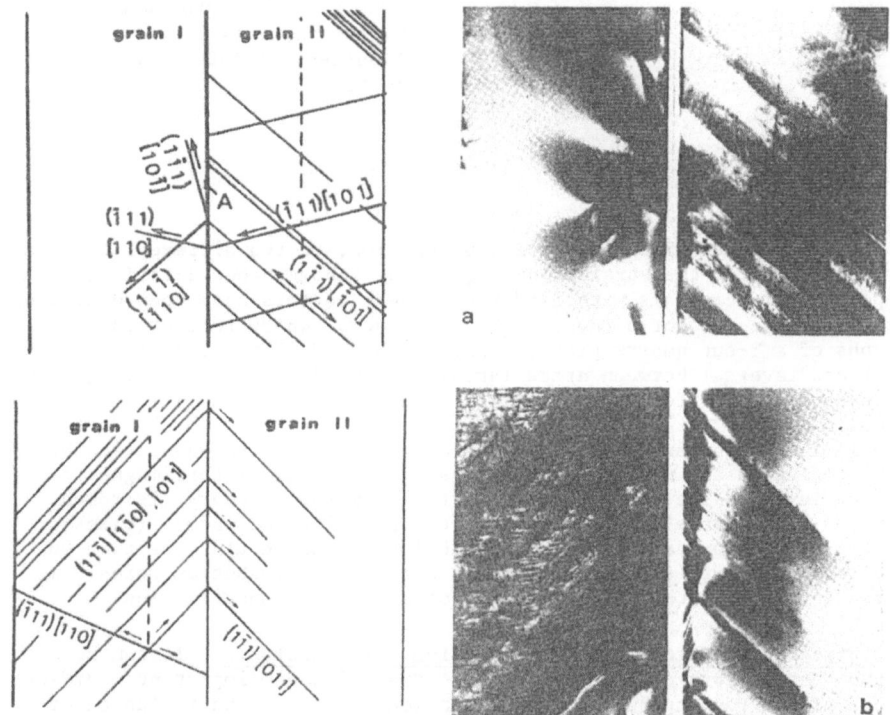

Fig. 19 Transmission of Plastic Deformation in silicon Σ = 9 bicrystals.
        a) Dislocations generated in grain I at a scratch (dashed line):
        the boundary filters one type of dislocations.
        b) Dislocations generated in grain II : secondary generation in
        grain I under the pile-up stresses (courtesy of A.Jacques et
        al. after[39]).

197

Quantitative Strain Measurements

As outlined previously, X-ray topography and diffractometry can be applied not only to image lattice defects but also to measure the associated strains. Plane wave (PW) techniques ought to be used for this purpose. Some applications to the estimate of either grown-in strains, such as those associated with impurity segregation or to strains introduced on purpose by various treatments (implantation, stacking of mismatched layers) will be presented.

Surface strain mapping by PW reflection topography. By taking a series of topographs at regular angular intervals along the rocking curve of a crystal, it is possible to measure the local strain with respect to a given area selected as a reference. The measurements correspond to average values over the penetration length of the radiation (1 - 50 μm). The method has been applied at LURE-DCI, with the triple reflection

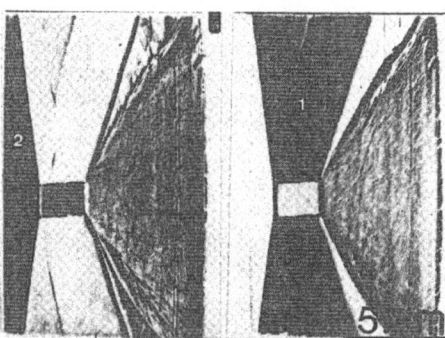

Fig. 20 PW reflection topographs On both wings of a quartz Rocking Curve. Zone 1 : "Z" sector. Zone 2 : "$\bar{X}$" sector with a higher impurity concentration. (Courtesy of A.Zarka and T. Soledad).

monochromator described in a previous section, to two different classes of problems. First the strains induced by the inhomgeneous distribution of impurities in crystals with different growth sectors have been evaluated (see reference 40 for a review). Fig. 20 shows two PW reflection topographs of a Y-cut quartz plate, separated by a 3.2" rotation ; the complete contrast reversal between areas labelled 1 and 2 corresponds to a relative parameter difference $(a_2-a_1)/\bar{a}$ equal to 7.10^{-6} (see [41]).

In the second example, strains are generated at the boundaries between implanted and non-implanted areas in an epitaxial ferrimagnetic garnet layer[42]. The perfect crystal, far from the boundary, is taken here as a reference. The analysis of the position and nature (black or white) of the boundary image as a function of the angular position on the rocking curve (fig. 21), provides a strain distribution which compares favorably with the predictions of a boundary modelisation.

Strain depth profiling by PW rocking curve analysis. It can be shown that a depth dependent state of strain induces important modifications of the rocking curve. Such situations are found in implanted or shallow diffused samples where the strain is a continuous function of depth but also in multiple heterostructures epitaxially grown on a possibly mismatched substrate, where the normal strain might be a step-like function of depth. The technique consists in introducing a model strain in the Takagi-Taupin equations until a good fit is obtained between the experimental and calculated rocking curves[43]. Although a very efficient work can be performed with laboratory two-crystal spectrometers in the (+,-) parallel setting, SR radiation plane wave facilities enable to take data with a larger variety of wavelengths and Bragg reflections in all kinds of samples.

198

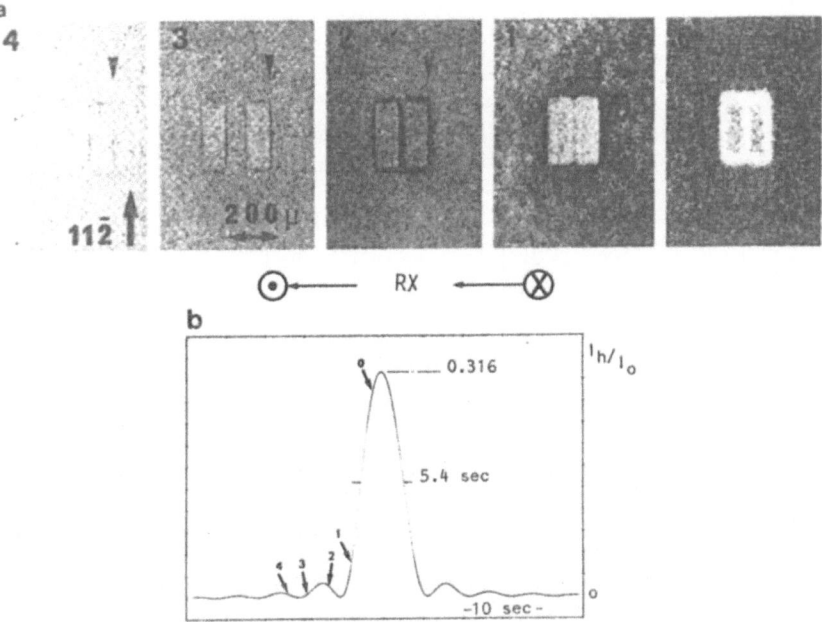

Fig. 21   a)   Series of PW topographs. Contrast variations of implantation
               boundaries at various departure from Bragg incidence in the
               matrix

          b)   Location of the topographs on a calculated profile for the this
               epilayer.   (Courtesy of J. Miltat, after [42]).

For example, the strain profile introduced by a uniform implantation
of 220 keV Ne ions ($2.10^{14}/cm^2$) in a ferrimagnetic garnet layer could
be accurately reconstructed by J.Miltat[42].

In the case of epitaxial heterostructures, the simplest case consists
in a single thin layer grown on a thick substrate with a relative mismatch $\Delta a/a$ between the two taking place abruptly at the interface. One
could expect a resulting rocking curve built with the succession along
the angle axis of the profiles shown in Fig. 3b and 4 shifted by a
quantity $\Delta\theta = -(\Delta a/a) \tan \theta$. Actually, interferences take place and
modify the fine structure of the thin crystal rocking curve. When one
compares Fig. 4 which is the calculated rocking curve of a self-supporting
2 μm thick $Ga_{0.7}Al_{0.3}As$ layer with the composite profile obtained
when the same layer is abruptly grown of a thick GaAs substrate
($\Delta a/a \sim 8.10^{-4}$) one notices an asymmetry appearing between the secondary
maxima on both sides of the epilayer peak (fig. 22a).   When the strain

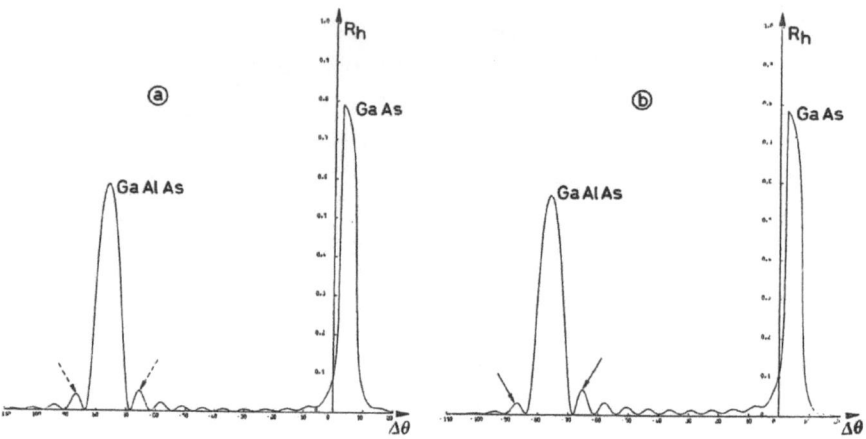

Fig. 22   Computed profiles for a $Ga_{0.7}Al_{0.3}As/GaAs$ heterojunction
           Layer thickness : $\Lambda_o/3$, $\Delta a/a = 8\ 10^{-4}$.

   a) abrupt junction - slight asymmetry between the secondary
      maxima (dotted arrows)
   b) smooth junction spread over 2000 Å = larger asymmetry
      (full arrows)

at the interface is allowed to vary smoothly over an intermediate region
of thickness Δ_z the asymmetry increases as shown in Fig. 22b where a
transition zone of about 2000 Å has been introduced between the same
epilayer and substrate for the computation. Semiconductor and garnet
heterojunctions have been evaluated by this method[44]. An optimized pro-
gramming technique using both analytical and numerical methods has been
developed[45] in order to establish clearly the depth sensitivity of the
method which depends in each case on several parameters (layer thickness,
$\Delta a/a$ ...), and is at best of a few ten angströms.

    SR plane wave rocking curves of more complex systems like semi-
conductor multiple heterojunctions[46] or strained layer superlattices[47]
have also been recorded and compared to simulated profiles. It has then
been possible to establish non destructively the depth distribution of
strain in these materials of primary importance in modern technology.

    Time resolved strain evaluation. To complete this overview of strain
characterization, two different experiments taking advantage of the time
structure of the SR sources should be quoted.

- a depth profiling of the strain and hence of the temperature raise
induced in silicon by laser annealing has been obtained by recording
PW rocking curves at different delay times with respect to the laser pulse
(in the 10 ns range) at CHESS (Cornell-USA)[48].

- stroboscopic WBT at the beam frequency has enabled to image the strains
produced by acoustic waves in the MH_z frequency range. Bulk standing waves
were observed in quartz[49] whereas the strains associated with travelling
surface acoustic waves have been visualized in $LiNbO_3$ samples both at
the SRS (Daresbury, UK)[50] and Hasylab (Hamburg - FRG)[51]. In the latter
case, a careful analysis of the contrast prooved that it came almost
entirely from the surface plane curvature, alternately convex and concave,
since it had the full periodicity of the acoustic wave (Fig. 23). An
additionnal contrast interpreted in terms of strain induced wavefield
curvature has also been discussed[52].

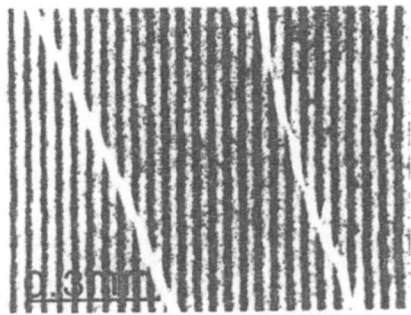

Fig. 23 WB stroboscopic reflection topograph of a 35,4 MH$_z$ surface acoustic wave in LiNbO$_3$. The twisted shadow of a 25μm wire is an evidence for the focusing and defocusing effect of the surface undulations (Courtesy of H.Cerva and W.Graeff after[51]).

MONOLAYER ASSESSMENT BY SR X-RAY METHODS

In the last section, X-ray diffractometry and topography were shown to provide information on the shallow strain distribution in single crystals, with a depth sensitivity of a few ten angströms in the best cases. However, special techniques can be thought of which are able to characterize the sample, on the monolayer scale, either at the surface or at a buried interface.

Standing Wave Experiments

The existence of standing wavefields with the lattice planes periodicity in a crystal under Bragg reflection was demonstrated in the first section of this paper (equation 6). In the reflection geometry (see fig. 3b), it can be shown that this standing wavefield is made to scan the actual atomic plane network by rocking the crystal accross the reflection domain. An impurity atom can then be accurately located, within a few hundreths of lattice spacing, by a synchronous detection of its fluorescence emission when recording the host crystal rocking curve[53]. The method has been applied to the registration of a fractional monolayer of adatoms deposited on the silicon surface[54], a problem also studied by standing waves produced in an X-ray interferometer at Hasylab (Hamburg, FRG)[55]. As an example, fig. 24 shows experimental data for Bi implanted silicon as a function of the dose together with the calculated best fit : it comes out that the substitutional fraction of Bi atoms varies from about 1 at low doses to about 0.4 at high doses[56].

This technique is developing rapidly in the various SR facilities, ultra high vacuum set-ups are in projects.

Surface Diffraction

A novel technique, namely the grazing incidence diffraction[57], has been introduced to determine the crystallographic structure of surfaces either clean or topped with adsorbates. When the incidence angle is kept below the critical value for total external reflection (a few ten minutes for hard X-rays since the refraction index is about $1-10^{-5}$), the radiation does not penetrate the material over more than a few ten angströms. Bragg reflections can then be recorded which originate only from the surface layers. In particular, in the case of reconstruction, the surface structure has a symmetry lower than the bulk, and fractionnal reflections, existing only for the topmost layers can be collected and used to perform a classical structure factor analysis leading to the surface atoms positions. The parasitic bulk contribution either as a diffused background

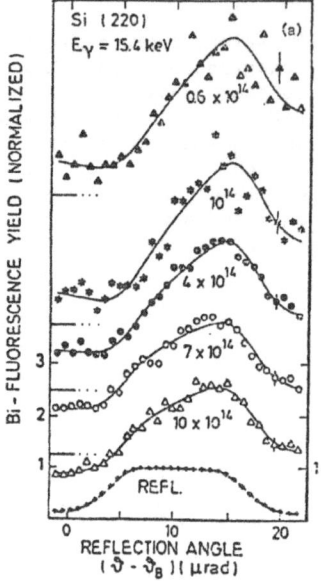

Fig. 24 Bi fluorescene as a function of the dose in implanted silicon. A random distribution would give a curve identical to the reflection profile. The intensity of the right maximum measures the substitional fraction (Courtesy of N. Hertel et al., after[56])

or as Bragg peak tails is considerably reduced. Clean surface studies must be performed in ultra high vacuum with all the necessary surface characterization surrounding equipments. The most convincing results have been obtained on the reconstructed 111 (2 x 2) surface of InSb[58]. In-situ grazing incidence ultra high vacuum chambers are under operation[59] or about to operate[60] in different SR facilities.

CONCLUSION

The aim of this paper was to give an overview of the increased potentialities brought to X-ray imaging and diffractometric techniques by the advent of SR sources. The associated improvement of X-ray video detectors has enabled significant steps forward in the field of real time experiments, specially for phase transformations in the recent years. A refinement of numerical computer simulations has on the other hand, supported the experimental efforts to go to higher sensitivities with sophisticated plane wave methods. The next generation of optimized SR sources will reinforce the trend towards the detection of very weak effects : surface studies and possibly surface magnetic diffraction, nuclear diffraction .

ACKNOWLEDGMENTS

The author wishes to express her gratitude towards her colleagues and friends from the SR facilities all around the world for having provided her with their latest results.

202

REFERENCES

1. L.W. Hobbs, this volume
2. J. Baruchel, S.B. Palmer and M. Schlenker, J. Physique, 42 : 1279 (1981)
3. "Applications of X-ray Topographic Methods to Materials Science". Ed. S. Weissmann, F. Balibar and J.F. Petroff - Plenum N.Y. (1984)
4. "Characterization of Crystal Growth Defects by X-Ray Methods" Ed. B.K. Tanner and D.K. Bowen - Plenum N.Y. (1980)
5. "Synchrotron Radiation Research". Ed. H. Winick and S. Doniach - Plenum N.Y. (1980)
6. M. Von Laue "Roëntgenstrahlinterferenzen" 3rd ed. Akademische Verlags Gesellschaft - Frankfürt (1961)
7. H. Hashizume, J. Appl. Cryst., 16 : 420 (1983)
8. N. Kato, Acta Cryst., 13 : 349 (1960)
9. A.R. Lang, Acta Cryst., 12 : 249 (1959)
10. J.B. Newkirk, Trans. AIME, 215 : 483 (1959)
11. A. Authier, J. Physique, 27 : 57 (1966)
12. S. Takagi, Acta Cryst., 15 : 311 (1962)
13. S. Takagi, J. Phys. Soc. Jap., 26 : 1239 (1969)
14. D. Taupin, Bull. Soc. Fr. Miner. Crist., 87 : 469 (1964)
15. D.K. Bowen, G.F. Clark, S.T. Davies, J.R.S. Nicholson, K.J.Roberts, J.N. Scherwood and B.K. Tanner, Nucl. Instr. Methods 208 : 277 (1983)
16. S. Suzuki, M. Ando, K. Hayakawa, O. Nittono, H. Hashizume, S. Kishino and K. Kohara, Nucl. Instr. Methods, 227 : 584 (1984)
17. J. Chikawa, F. Sato, T. Kawamura, T. Kuriyama, T. Yamashita and N. Goto, in "X-ray Instrumentation for the Photon Factory". - Eds S. Hosoya, Y. Iitaka and H. Hashizume - KTK/Reidel Pub. Co. Tokyo (1984)
18. B.K. Tanner, G.F. Clark, P.A. Goddard, D.K. Bowen, S.T. Davies and O.P. Aleshko-Ozhevsky, Nucl. Instr. Methods, 208 : 713 (1983
19. J.W.M. Du Mond, Phys. Rev., 52 : 871 (1937)
20. J.F. Petroff, M. Sauvage, P. Riglet and H. Hashizume, Phil. Mag. A 42 : 319 (1980)
21. D.E. Sayers, S.M. Heald, M.A. Pick, J.I. Budnick, E.A. Stern and J. Wong, Nucl. Instr. Methods 208 : 631 (1983)
22. J. Gastaldi, C. Jourdan, R. Marzo, C. Allasia and J.N. Jullien J. Appl. Cryst., 15 : 391 (1982)
23. J. Gastaldi and C. Jourdan, Phil. Mag. A, 50 : 319 (1984)
24. C. Jourdan and J. Gastaldi, Proceedings of the Conference on "Grain Boundary Structure and Related Phenomena". Minakami Spa, Japan, Nov. (1985)
25. O. Nittono, T. Ogawa, S.K. Gong and S. Nagakura, Jpn J. Appl. Phys. 23 : L 723 (1984)
26. A. Zarka, J. Appl. Cryst., 16 : 354 (1983)
27. G. Dolino and J.P. Bachheimer, Solid State Short Commun., 45 : 259 (1983)
28. K. Gouhara, Y.H. Li and N. Kato, J. Phys. Soc. Jpn, 52 : 3697 (1983)
29. K. Gouhara and N. Kato, J. Phys. Soc. Jpn, 53 : 2177 (1984)

30. A. Zarka and B. Capelle, extended abstracts of the IXth European Crystallographic Conference, Torino (Italy) Sept. 1985
31. M. Ribet, S. Gits-Leon, F. Lefaucheux and M.C. Robert, Ferroelectrics, in press
32. V. Janovec, Physics Letters, 99A : 384 (1983)
33. C. Jourdan and J. Gastaldi, J. Appl. Cryst., 16 : 646 (1983)
34. C. Jourdan and J. Gastaldi, private communication
35. G.R. Fisher and P. Barnes, J. Appl. Cryst., 17 : 231 (1984)
36. M. Dudley, J. Miltat and D.K. Bowen, Phil. Mag. A, 50 : 487 (1984)

37. J. Miltat and D.K. Bowen, J. Physique, 40 : 389 (1979)
38. G. Michot and A. George, Scripta Metallurgica, 16 : 519 (1982)
39. A. Jacques, E. Marlière, J.P. Michel and A. George, in "Dislocations in Solids" ed. H. Suzuki, T. Ninomiya, K. Sumino and S. Takeuchi ; University of Tokyo Press (1985), p. 655
40. M.C. Robert and F. Lefaucheux, J. Cryst. Growth, 65 : 637 (1983)
41. A. Zarka, Liu Lin and M. Sauvage, J. Cryst. Growth, 62 : 409 (1983)
42. J. Miltat, IEEE Trans. Mag., Mag. 20 : 1114 (1984)
43. J. Burgeat and D. Taupin, Acta Cryst., A24 : 99 (1968)
44. M. Sauvage-Simkin and J.F. Petroff, in ref. 3, 421
45. S. Ben Soussan, D.E.A. unpublished report Paris 1984
46. M.J. Hill, B.K. Tanner, M.A.G. Halliwell and M.H. Lyons, J. Appl.Cryst., in press
47. M. Sauvage, P. Voisin, C. Delalande, P. Etienne and P. Delescluse, Proceedings of the Conference on Modulated Structures - Tokyo, Sept. 1985 - to be published in Surface Science
48. B.C. Larson, C.W. White, T.S. Noggle and D.M. Mills, Phys. Rev. Letters, 48 : 337 (1982)
49. C.C. Gluer, W. Graeff and H. Möller, Nucl. Instr. Methods, 208 : 701 (1983)
50. R.W. Whatmore, P.A. Goddard, B.K. Tanner and G.F. Clark, Nature, 299 : 44 (1982)
51. H. Cerva and W. Graeff, Phys. Stat. Sol., a82 : 35 (1984)
52. H. Cerva and W. Graeff, Phys. Stat. Sol., a87 : 507 (1985)
53. S.K. Andersen, J.A. Golovchenko and G. Mair, Phys. Rev. Letters, 37 : 1141 (1976)
54. J.A. Golovchenko, J.R. Patel, D.R. Kaplan, P.L. Cowan and M.J. Bedzyk, Phys. Rev. Letters, 49 : 560 (1982)
55. G. Materlik, A. Frahm and M.J. Bedzyk, Phys. Rev. Letters, 52 : 441 (1984)
56. N. Hertel, G. Materlik and J. Zegenhagen, Z. Phys. B, 58 :   (1985)
57. W.C. Marra, P. Eisenberger and A.Y. Cho, J. Appl. Phys., 50 : 6927 (1979)
58. J. Bohr, R. Feidenhans'l, M. Nielsen, M. Toney, R.L. Johnson, I.K. Robinson, Phys. Rev. Letters, 54 : 1275 (1985)
59. S. Brennan and P. Eisenberger, Nucl. Instr. Methods, 222 : 164 (1984)
60. P. Fuoss and I.K. Robinson, Nucl. Instr. Methods, 222 : 171 (1984)

SPECTROSCOPIC STUDIES OF DEFECTS IN

IONIC AND SEMI-IONIC SOLIDS

Johann-Martin Spaeth

University of Paderborn
Fachbereich Physik
Warburger Str. 100A, 4790 Paderborn, FRG

1. INTRODUCTION

The investigation of point defects in solids is of increasing importance for that part of solid state physics which often is referred to as
'materials science'. A small concentration of impurities or native
defects can dominate many important bulk properties like electrical
conductivity and optical and mechanical properties. For the development
of materials for specific applications it is decisive to understand the
defect structure. In 'simple' hosts like the alkali halides, the properties of 'model' defects like the F centre or atomic hydrogen on various
sites have been studied in great detail and are rather well understood.
The experimental tools were mostly spectroscopic techniques, which were
to a great deal developed alongside with these studies. Currently the
interest has focussed more on the application of these techniques to more
complicated systems like hosts of a more complicated atomic structure, of
semi-ionic or covalent nature and more complicated defects, be it due to
the nature of the impurity, aggregation of impurities or of impurities
and intrinsic defects, low symmetry etc.

In the last decade spectroscopy has greatly profited from the advancement of experimental techniques, especially from the use of microcomputers
in experiments. Now the investigation of complicated systems, which are
of practical importance,can be attempted. Examples are III-V semiconductor materials for microelectronics, the development of laser-active material (color centre lasers), radiation damage in material withstanding high
temperatures.

In the series of 3 lectures it will not be possible to cover all
current important spectroscopic techniques. Methods of defect identification, that is the determination of the 'atomic' and electronic structure, will be discussed predominantly, as well as the direct correlation
of optical and structural defect properties. The most powerful method is
the magnetic resonance spectroscopy, especially when applying the technique of multiple resonances, like electron nuclear double resonance (ENDOR),
optically detected electron spin resonance (ODESR) and optically detected
ENDOR (ODENDOR). The aim of the lectures is to introduce to these methods
and to demonstrate their actual use. The examples given are mostly taken
from the work of the Paderborn group as a matter of convenience.

2. OPTICAL ABSORPTION AND EMISSION

In ionic and semi-ionic crystals many defects have localised energy states within the band gap and possess optical absorptions with energies below the band gap energy. Fig. 1 shows this schematically for the transition from the defect ground state to the first excited state. The absorption is then observed within the 'optical window' of the crystal, where no band-band transitions ($\hbar\omega > E_g$) and no transitions due to lattice vibrations occur. In the alkali halides this window ranges from about 6-10 eV to about 0.1 eV.

The correlation of such a defect-induced optical absorption with a particular defect is usually a difficult task. This is demonstrated, e.g. by the long history of the research on color centres in alkali halides (1). Assignments are mostly attempted by variation of the impurity concentration or the irradiation time, if it is an intrinsic defect like a vacancy, electron or hole centre as consequence of radiation damage. However, this is not always unambiguous. Several defects can be produced simultaneously with overlapping absorption bands. Often there is one characteristic lumincescence band. The measurement of its excitation spectrum can be used to identify the associated absorption bands provided no other defect luminescence bands overlap it strongly.

In ionic and semi-ionic crystals most defect states are sensitive to the positions of the nearby atoms or ions, so that the form of the absorption and the form and the energy position of the emission depend on the vibrations of the surrounding ions. This is conveniently discussed in the approximation of the configuration coordinate diagram (CC-diagram), in which the lattice vibrations are represented by a single localised mode with a configuration coordinate Q and linear coupling is assumed for the vibrational energy, resulting in a parabolic energy curve for both the ground and excited states (see Fig. 2). They have the form

$$E = 1/2 \ KQ^2 \qquad\qquad 2.1$$

where K is a force constant. Absorption and emission are vertical transitions (Born-Oppenheimer approximation). After the electrical dipole transition of the absorption has occurred, the electron distribution is changed and accordingly the lattice will relax and adjust to a new equilibrium position represented by $Q_o + \Delta Q$ in Fig. 2. From the vibrational ground state of the excited state parabola the emission takes place into a ground state configuration, which has to relax back into the original configuration. The relaxation process involves phonon emission which is

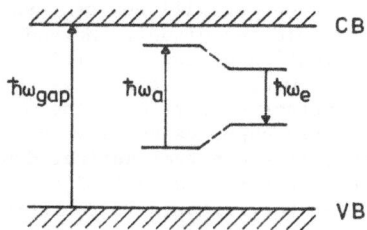

Fig. 1. Schematic representation of optical intracenter absorption and emission.

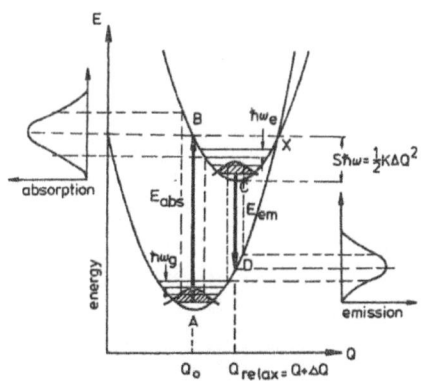

Fig. 2. CC-diagram of optical absorption and emission.

the reason for the observed Stokes shift ΔE between the energies of absorption and emission ($\Delta E = E_{abs} - E_{em}$). ΔQ is directly related to the strength of the electron phonon coupling, which is characterised by the Huang-Rhys factor S

$$S \cdot \hbar\omega = 1/2 \; K \cdot \Delta Q^2 \qquad\qquad 2.2$$

where ω is the vibrational frequency (assumed to be equal in ground and excited states in the simplest approximation).

For large electron phonon coupling (S >10) the shapes of the absorption and emission bands are approximately Gaussian with a half-width proportional to $\hbar\omega \cdot \sqrt{S}$. This is indicated in Fig. 2. Typical half-widths are several tenths of an eV. As an example the lowest energy absorption of laser-active $F_A(Tl^+)$ centres and their emission bands in several alkali halides are shown in Fig. 3 (2). A discussion of the structure of these centres and their optical properties will be presented in sec.7.

If the coupling is weak (S < 1) both ground and excited state parabolae are hardly displaced against each other and the dominant feature is the 'zero phonon line' (ZPL). The transition involves no vibrational energies. Typical examples are transitions within the $4f^n$ configuration in rare earth ions in solids. The 4f configuration is well shielded (3).

In intermediate coupling (1<S<6) the ZPL is resolved, but the multiphonon structure is the dominant feature of the spectrum. The intensity of the ZPL relative to the whole band is given by exp (-S). It becomes undetectible for high values of S. An example for the absorption of a ZPL and a phonon replica structure is given in Fig. 4 for F_3-centres in LiF (4). For further details and the influence of the Jahn-Teller effect on the shape of the bands see (5,6).

When a ZPL can be measured, the application of uniaxial stress and magnetic and electrical fields can cause a splitting of the ZPL, from which structural information such as defect symmetry can be derived. (see e.g. Ref. 1, chapters 5 and 6 and further references therein).

Defects having a strong electron phonon coupling may not show a luminescence or only a very weak one. In the simple framework of the CC-diagram a nonradiative de-excitation occurs, when the excited state energy reached in a Frank-Condon absorption transition (point B in Fig. 2) lies

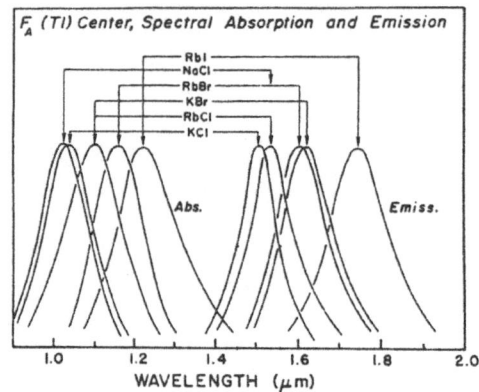

Fig. 3. Lowest energy absorption and emission of $F_A(Tl^+)$
centres in several alkali halides (After (2)).

above the intersection of ground and excited state potential energy curves
(point C in Fig. 2). In this case the system can be de-excited through
the intersection directly into the ground state under phonon emission.
In the approximation of linear coupling and equal vibration frequencies
for ground and excited state this occurs when $E_{em} < 1/2 \, E_{abs}$ (Dexter-
Klick-Russell-rule) (7,8).

3. ELECTRON SPIN RESONANCE (ESR)

Paramagnetic defects can be identified by means of electron spin
resonance (ESR). Defect identification is based on the resolution of the
hyperfine (hf) interactions, that is the interaction between the magnetic
moment of the unpaired electron (or hole) and the magnetic moment of a
central (impurity) nucleus or the magnetic moments of the nuclei of the
surrounding lattice atoms. The latter interaction is mostly called ligand
hyperfine or superhyperfine (shf) interaction. In ionic crystals ESR is,
of course, restricted to defects which are paramagnetic in their ground
state or which can be excited into paramagnetic excited states as in the
case of many electron systems (S > 1/2). In semiconductors, there is often
the possibility to 'make' the defects paramagnetic by raising or lowering
the Fermi energy E_F. This can be achieved by additional doping of shallow
donors or acceptors and is currently applied to the study of so-called deep
level defects, which have levels near the middle of band gap.

Fig. 5a shows schematically the basic ESR experiment. A magnetic
dipole transition is induced between the Zeeman levels for $m_s = \pm 1/2$ if
the resonance condition

$$\hbar\omega = g\beta_e B_o \qquad\qquad 3.1$$

is fulfilled. g is the electronic g-factor, β_e the Bohr magneton.
Usually one varies the magnetic field and keeps ω constant (Fig. 5b). For
fields around 0.3 T one uses microwave irradiation of $\sim$ 10 GHz (X-band) and
for fields around 0.8 T radiation of $\sim$24 GHz (K-band). Upon resonance
the microwave absorption is measured. (In order to be able to use lock-in
techniques for sensitivity enhancement one measures the derivative of the
absorption as consequence of magnetic field modulation) (9).

208

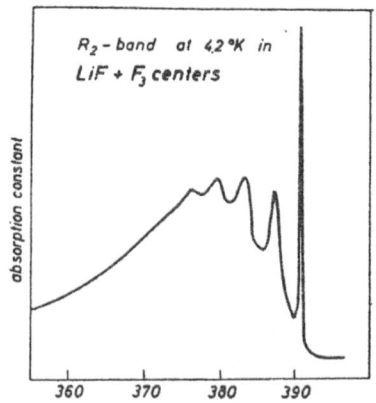

Fig. 4. Absorption spectrum of the R_2-band of F_3-centres
in LiF. After (4).

The experiment of Fig. 5 does not give much information on the defect structure although deviations of the g-factor from the value of the free electron (g_e = 2.0023) contain some relevant information (9,10). Defect identification is based on the splitting of such a resonance line due to hf or shf interactions. Fig. 6 shows the level diagram for the simple case of the hf splitting due to a nucleus with I = 1/2. According to the ESR selection rules (9)

$$\Delta m_s = \pm 1, \quad \Delta m_I = 0 \qquad\qquad 3.2$$

there are now two ESR transitions split by the hf interaction constant $A(\theta)$. The resonances are described by the Spin Hamiltonian

$$\hat{H} = g\beta_e \vec{B}_0 \hat{\vec{S}} + \hat{\vec{I}} \tilde{A} \hat{\vec{S}} - g_I \beta_n \vec{B}_0 \hat{\vec{I}} \qquad\qquad 3.3$$

where β_n is the nuclear magneton, g_I the nuclear g-factor and $\tilde{A}$ the hf tensor. In perturbation theory of first order one obtains for the

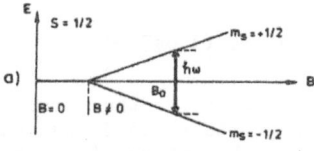

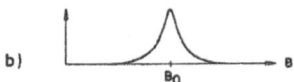

Fig. 5. Schematic representation of the basic ESR experiment.

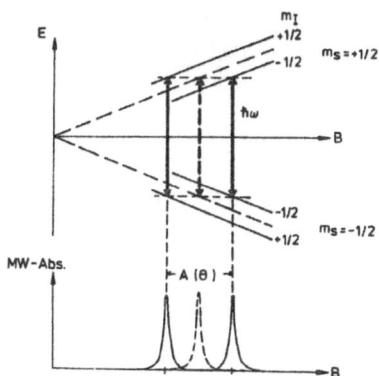

Fig. 6. Level diagram and ESR for a hyperfine interaction with
a nucleus of I = 1/2.

resonance field positions

$$h\nu_{ESR} = g\beta_e B_{res} + m_I\, A(\theta) \qquad\qquad 3.4$$

Since the nuclear spin quantum number m_I has the two values $m_I = \pm\, 1/2$
and ν_{ESR} = constant, there are two resonance fields B_{res1} and B_{res2}
separated by $A(\theta)/g\beta_e$.

Fig. 7 shows how a hf interaction can be used for defect identification for the example of Te$^+$ defects in Si (11). 92% of the Te isotopes
are diamagnetic and give rise to the central ESR line near 3500 G (measured
in X-band) corresponding to $g \approx 2$. The two magnetic isotopes ^{125}Te
(7% abundant) and ^{123}Te (0.9% abundant) both have I = 1/2 and according to
(3.4) doublet splitting. The ESR line intensities follow the isotope
abundance and the relative splittings are in the ratio of the respective
nuclear moments (see below). Therefore the defect is unambiguously identified as being due to Te impurities. However, no further structure due
to shf interactions is resolved. Therefore, the site of Te$^+$ in the lattice
cannot be determined from ESR. Te$^+$ has an isotropic hf interaction. In
general, the hf interaction is anisotropic and the spectrum is angular
dependent as indicated in 3.4. Again, in first order and for axial symmetry
of the defect the resonance fields are given by

$$h\nu_{ESR} = g\beta_e B_{res} + m_I\, (a + b\,(3\cos^2\theta-1)) \qquad\qquad 3.5$$

where a is the isotropic hf constant, b the anisotropic hf constant.
θ is the angle between the magnetic field B_0 and the principal axis z
of the hf tensor. In general, the interaction constants a and b and b'
are related to the principal values of the hf tensor by

$$A = (a \cdot \underset{\sim}{\tilde{1}} + \tilde{B}) \qquad\qquad 3.6$$

$$b = \frac{1}{2}\, B_{zz} \qquad\qquad 3.7$$

$$b' = \frac{1}{2}\, (B_{xx} - B_{yy}) \qquad\qquad 3.8$$

where b' describes the deviation from axial symmetry.

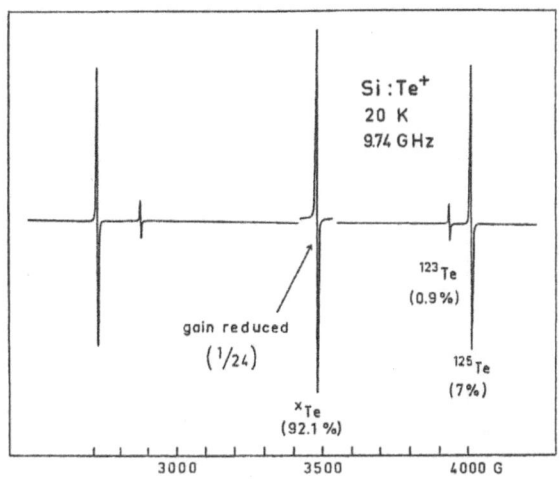

Fig. 7. ESR spectrum of Te$^+$ centres in Silicon. After (11).

Fig. 8a shows the ESR lines of Tl^0(1) centres in KCl (and some other simultaneously produced defects). Again, since both Tl isotopes ^{203}Tl and ^{205}Tl have I = 1/2, the spectra consist of doublets. The splitting is, however, angular dependent. The angular dependence in Fig. 8b for rotation of the magnetic field in a (100)-plane shows this as well as the distribution of the centres in the crystal with respect to their axes. They have axial symmetry about a (100) direction. Tl^0(1) centres are Tl^0 atoms next to an anion vacancy, while Tl^0(2) centres have two anion vacancies next the atom (12). The angular dependence and isotope distribution leads here already to a structural model. ESR cannot reveal, however, whether Tl^0 resides precisely on a cation lattice site or is relaxed towards the vacancy, and it cannot either be said whether the vacancy is filled with another impurity since the angular dependence reflects only the tetragonal symmetry of the defect and not further shf structure is resolved.

Fig. 8a shows also another feature of conventional ESR spectroscopy. Since the ionising radiation producing these centres creates many defects simultaneously, it is not easy to unambiguously assign all the ESR lines to particular centres and follow their angular dependence. It will be shown below that with optical detection of the ESR each defect can be measured selectively, which greatly facilitates the analysis.

Fig. 9 shows the ESR spectrum of atomic hydrogen on interstitial sites, cation and anion vacancy sites in KCl (13). The central hf splitting with the proton (I = 1/2) is in all three cases practically the same. Only for the interstitial site a shf interaction with nearest neighbours (13) is resolved. The substitutional sites cannot be inferred from the ESR spectrum. They could be established only by resolving the shf interactions with ENDOR experiments.

The hf interaction constants are determined by the electronic wave function of the defects and the nuclear moments of the nuclei. In a simple one particle approximation the isotropic constant a is given by

$$a_1 = \frac{2\mu o}{3} g\beta_e \, g_I \, \beta_n \, |\psi(r_1)|^2 \qquad\qquad 3.9$$

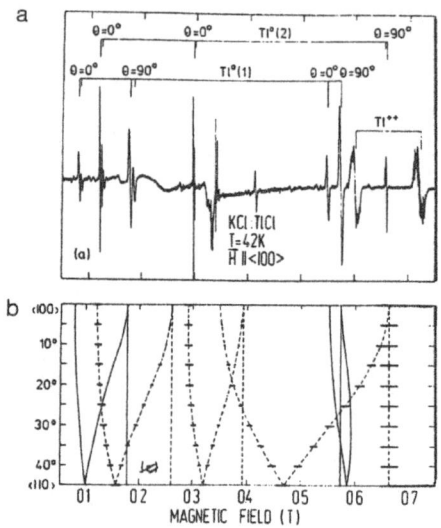

Fig. 8. (a) ESR spectrum of $Tl^o(1)$ and $Tl^o(2)$ centres in KCl:Tl
for $B_o||[100]$ measured at 42 K. After (12).
(b) The calculated angular variation of the spectra in
a (100) plane. Straight lines are due to $Tl^o(1)$,
broken lines are due to $Tl^o(2)$ centres. The horizontal
bars indicate the transition probabilities for $Tl^o(2)$.
After (12).

where $\psi(r)$ is the wave function of the defect, $\psi(r_1)$ its amplitude at the
site r_1 of a particular nucleus. The anisotropic tensor elements are
given by

$$B_{ik} = \frac{\mu_o}{4\pi} \, g\beta_e g_I \beta_n \int (\frac{3}{r^5} x_i x_k - \frac{1}{r^3} \delta_{ik}) \, |\psi(\vec{r})|^2 \, dV \qquad 3.10$$

$\vec{r}$ means the radius vector from the nuclear site of concern (origin) where
the origin is spared in the integral of equ. (3.10) (10).

Thus the hf constants are proportional to g_I and therefore the interaction constants of different isotopes must be in the ratio of their
respective g_I factors. This is used for their identification.

4. ELECTRON NUCLEAR DOUBLE RESONANCE (ENDOR)

The information needed for determining the defect structure is in its
shf structure. The most direct way to measure it would be by measuring the
nuclear magnetic resonance (NMR) of the neighbouring nuclei having an shf
interaction with the unpaired defect electron. An NMR measurement is not
possible since its sensitivity is too low for the defect concentrations in
question, which are usually of the order of 10^{-6} or lower. But by detecting
the NMR through the desaturation of a partially saturated ESR transition,
the sensitivity can be enhanced by several orders of magnitude. This
electron nuclear double resonance (ENDOR) experiment was originally
introduced by G. Feher (14). The experiments described and explained below
use the stationary ENDOR method introduced by H. Seidel (15). The stationary ENDOR is the more widely applicable technique.

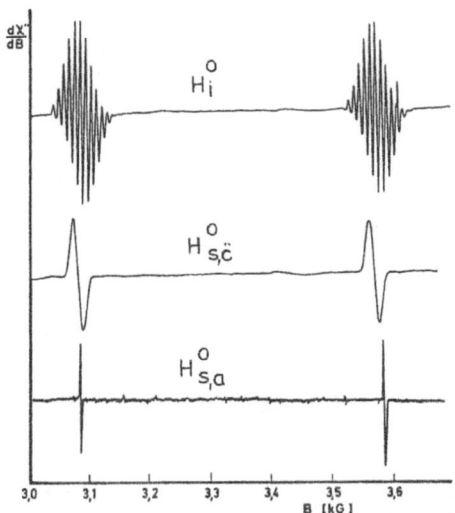

Fig. 9. The ESR spectra of three atomic hydrogen centres in KCl.
The hydrogen atoms occupy interstitial sites (H_i^o-centres),
anion vacancy sites ($H_{s,a}^o$-centres) and cation vacancy
sites ($H_{s,c}^o$-centres). After (13).

Fig. 10 shows for $S = 1/2$ and one nucleus with $I = 3/2$ schematically
the allowed ESR transitions with resolved shf structure of the ESR. In
an ENDOR experiment one of the allowed ESR transitions is partially saturated, that is one chooses the microwave power high enough, so that the
transition probability $W_{MW} \propto \gamma B_1^2$ is of the order or larger than the spin
lattice relaxation rate $W_{REL} \propto 1/T_1$. γ is the gyromagnetic ratio of the
electron, B_1 the microwave field amplitude, T_1 the spin lattice relaxation
time. If that is the case, then the spin population of the levels connected by the microwave transition deviates from the Boltzmann equilibrium
distribution. If $W_{MW} \gg W_{REL}$, then these levels become equally populated.
This results in a decrease of the observable microwave absorption, since

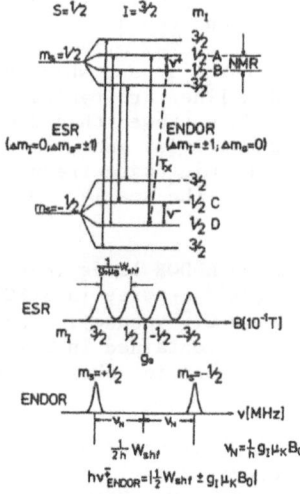

Fig. 10. Level diagram to explain the ENDOR mechanism (see text).

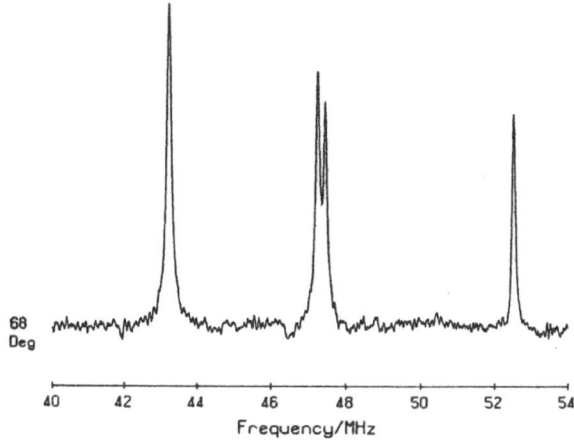

Fig. 11. ENDOR lines of the nearest ^{19}F neighbours of $F(Cl^-)$
centres in SrFCl.

microwave absorption and emission probabilities are equal. The levels
not connected by the microwave transitions are not affected. Therefore,
e.g. the level population $m_S = +1/2$, $m_I = +1/2$ is now inverted with
respect to the level $m_S = +1/2$, $m_I = -1/2$ (see Fig. 10). If these two
levels are connected by an NMR transition, the level populations can be
equalised, which results in a population decrease of the level $m_S = +1/2$,
$m_I = +1/2$ leading to a desaturation of the (partially) saturated transition
$m_S = -1/2$, $m_I = +1/2$ to $m_S = +1/2$, $m_I = +1/2$. This desaturation is monitored. It occurs for two NMR frequencies in the example of Fig. 10, since
the NMR frequencies for $m_S = +1/2$ and $m_S = -1/2$ are different (see below).
Thus, each nucleus gives rise to 2 ENDOR lines (for $S = 1/2$). A cross-
relaxation T_x (see Fig.10) allows the stationary observation of the
desaturation (Stationary ENDOR (15)). If several nuclei with the same
or similar shf interactions are coupled to the unpaired electron, then the
ESR pattern becomes complicated and the shf structure is usually not
resolved any more. In ENDOR all nuclei with the same interaction give
rise only to two (for $S- 1/2$) ENDOR lines, which greatly enhances the
resolution. ENDOR lines as NMR lines are typically 10-100 kHz wide, about
two or three orders of magnitude narrower than the inhomogeneous ESR lines.
Thus, in ENDOR one uses the sensitivity enhancement due to a quantum shift
from frequencies of $\sim$ MHz to the microwave frequencies of $\sim$ GHz and the
increased resolution power due to the smaller NMR line width and the reduction of the number of lines.

Fig. 11 shows as an example ENDOR lines (ESR desaturation) of the
4 nearest ^{19}F neighbours of $F(Cl^-)$ centres in SrFCl (for $m_S = -1/2$), where
the unpaired electron occupies a Cl^- vacancy. The structure model of BaFCl,
which has the same structure, is contained in Fig. 21. A similar pattern
is measured for $m_S = +1/2$ at frequencies shifted by $2\nu_n$ (^{19}F) to lower
frequencies.

If $I > 1/2$, in addition a quadrupole interaction can occur, which is
the interaction between the electric field gradient at the site of the
nucleus and its quadrupole moment. The ENDOR spectra are described by
the following Spin Hamiltonian (for a one electron system, that is without
fine structure)

214

$$\underset{\sim}{H} = g\beta_e \vec{B}_o \vec{\underset{\sim}{S}} + \sum_1 (\vec{\underset{\sim}{S}} \; \tilde{A}_1 \; \vec{\underset{\sim}{I}}_1 - g_I \; \beta_n \; \vec{B}_o \; \vec{\underset{\sim}{I}}_1 + \vec{\underset{\sim}{I}}_1 \tilde{Q}_1 \vec{\underset{\sim}{I}}_1) \qquad 4.1$$

The sum runs over all nuclei interacting with the unpaired electron. For simplicity it is assumed that g is isotropic (which is generally not the case). Q is the traceless quadrupole interaction tensor with the elements

$$Q_{ik} = \frac{eQ}{2I(2I-1)} \quad \frac{d^2V}{dx_i dx_k}\bigg|_{r=0} \qquad 4.2$$

Q is the quadrupole moment, V the electrical potential. The spectra are usually analysed in terms of the quadrupole interaction constants:

$$q = \frac{1}{2} Q_{zz}, \quad q' = \frac{1}{2} (Q_{xx} - Q_{yy}) \qquad 4.3$$

The selection rule for ENDOR transitions is:

$$\Delta m_s = 0, \qquad \Delta m_I = \pm 1 \qquad 4.4$$

If the shf and quadrupole interaction is small compared to the electron Zeeman term then the quantisation of the electron spin is not influenced by these interactions and the nuclei are independent of each other. They can be treated separately and the sum in equ. (4.1) can be omitted. In perturbation theory of first order, that is with the conditions

$$|B_{ik}|, \quad |Q_{ik}| \quad << \quad |a \pm \frac{1}{m_s} \; g_I \beta_n B_o| \qquad 4.5$$

$$\nu^{\pm}_{ENDOR} = |\frac{1}{h} m_s \; W_{shf} \mp \nu_n \pm \frac{1}{h} m_q \; W_Q| \qquad 4.6$$

with the following abbreviations:

$$W_{shf} = a + b \, (3\cos^2 \theta - 1) + b' \sin^2 \theta \cos 2\delta \qquad 4.7$$

$$W_Q = 3 \; \{q(3\cos^2 \theta' - 1) + q' \sin^2 \theta' \cos 2\delta'\} \qquad 4.8$$

θ, δ and θ', δ' are the polar angles of B_o in the principal shf and quadrupole axis system, respectively.

$$\nu_n = \frac{1}{h} g_I \; \beta_n \; B_o \qquad 4.9$$

ν_n is the Larmor frequency of a free nucleus in the magnetic field B_o.

$$m_q = \frac{m_I + m_{I'}}{2} \qquad 4.10$$

m_q is the average between the two nuclear spin quantum numbers, which are connected by the transition.

215

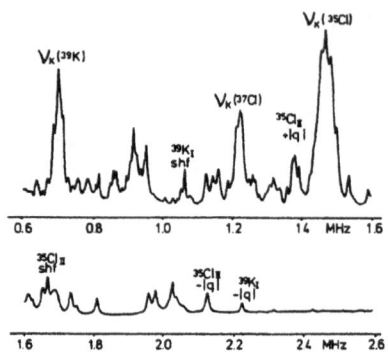

Fig. 12.   ENDOR spectrum of $H_{s,a}^{o}$-centres in KCl.   The high field
           ESR line was saturated (B = 3514 G).   T = 40 K, B_o nearly
           || to [100].   After (16).

        If there is no quadrupole interaction and S = 1/2, then each inter-
acting nucleus gives a pair of lines according to equ. (4.6) ('hf ENDOR-
lines').   The lines are separated by $2\nu_n$ if 1/2 $W_{shf} > h\nu_n$ and by W_{shf} if
$h\nu_n > \frac{1}{2} \cdot W_{shf}$.   If there is a quadrupole splitting, then each 'hf'−ENDOR
line is split into a characteristic multiplet, e.g. for I = 3/2 into a
triplet.   However, this splitting is easily recognised only, if $W_Q << W_{shf}$.
If these interactions are of the same order, then the first order approximation leading to equ. (4.6 - 4.8) breaks down.   The assignment of particular ENDOR lines to hf or quadrupole transitions becomes difficult.

        The chemical nature of a nucleus responsible for a particular hf
ENDOR line can usually be determined by measuring the shift of the ENDOR
line position when varying the magnetic field B_o through the ESR line.
According to equ. (4.6) and (4.9) the line position varies with ν_n which
is proportional to g_I and therefore characteristic for a particular
nucleus.   Furthermore, in the presence of several magnetic isotopes one
can use the repeated appearance of ENDOR lines with a frequency ratio
equal to the ratio of nuclear g_I factors.

        In Fig. 12 as an example an ENDOR spectrum is shown of atomic hydrogen
on anion vacancy sites ($H_{s,a}^{o}$-centres) in KCl, where several nuclei are
identified (ν_K in Fig. 12 = ν_n, the notation I and II stands for first
and second neighbour shell).   The quadrupole transitions are not recognised
easily without a detailed analysis (16).

        In order to determine the defect structure and the interaction para-
meters the dependence of the ENDOR line positions upon variation of the
magnetic field with respect to the crystal orientation must be measured
and analysed.   This is the major problem in an ENDOR analysis and the
essential tool for the determination of the defect structure.

        Fig. 13 a-c show such an angular dependence for a cubic crystal, such
as an alkali halide, calculated according to equ. (4.6 and 4.7) for the
first three neighbour shells of a defect on a lattice site.   The patterns
are characteristic for (100), (110) and (111) 'symmetry' of the neighbour
nuclei.   For each m_s-value such a pattern is observed.   From the number of
such patterns according to (4.6) one can infer the electron spin of the
defect and thus often its charge state.

        Each nucleus has its own principal axis system for the shf and quadru-
pole tensors.   Often, their orientation in a crystal is determined by

216

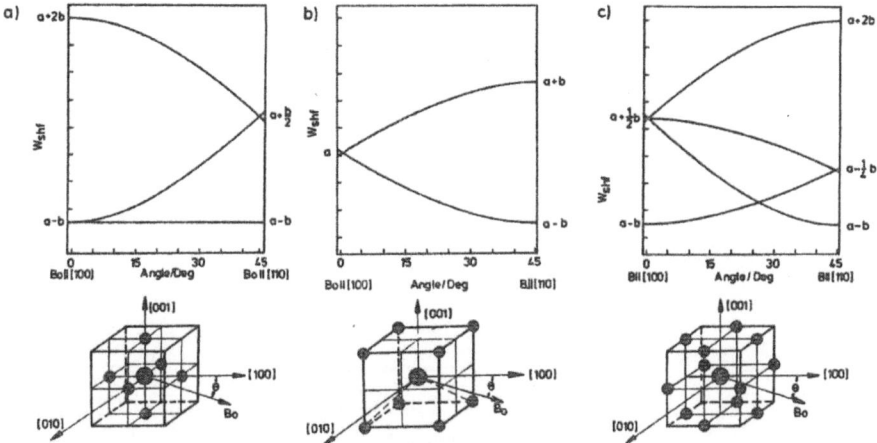

Fig. 13. Calculated ENDOR angular dependence for a defect on
a cubic substitutional site. (a) (100)-neighbours,
(b) (111)-neighbours, (c) (110)-neighbours.

symmetry. Otherwise, they must be determined from the analysis of the
angular dependence of the ENDOR spectra. If the defect centre (impurity)
and the respective nucleus are in a mirror plane, then two principal axes
must be in this mirror plane. If the connection line between the nucleus
and the centre is a threefold or higher symmetry axis, then the tensor is
axially symmetric with its axis in this symmetry axis.

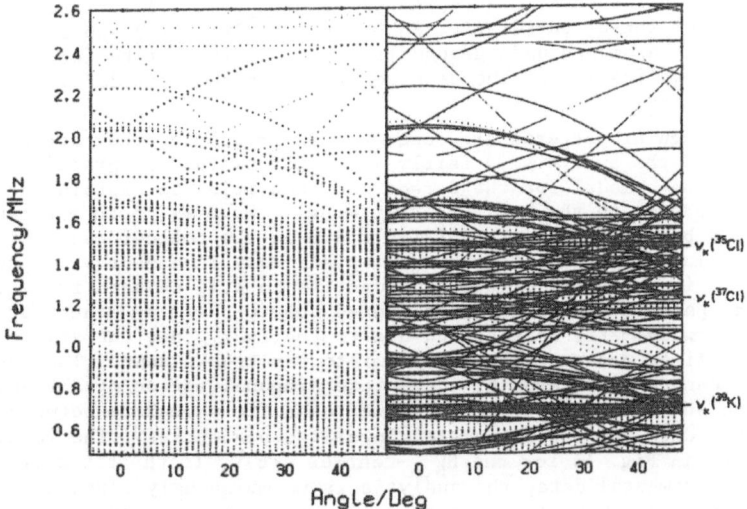

Fig. 14. Angular dependence of the ENDOR lines of H_S^0,a-centres in
KCl. The crystal was rotated in a (100)-plane between [100]
(0°) and [110] (45°). Left: experimental results. Right:
calculated angular dependence after determination of the Spin
Hamiltonian parameters for 4 neighbour shells in comparison
to the experimental results. Sfter (16).

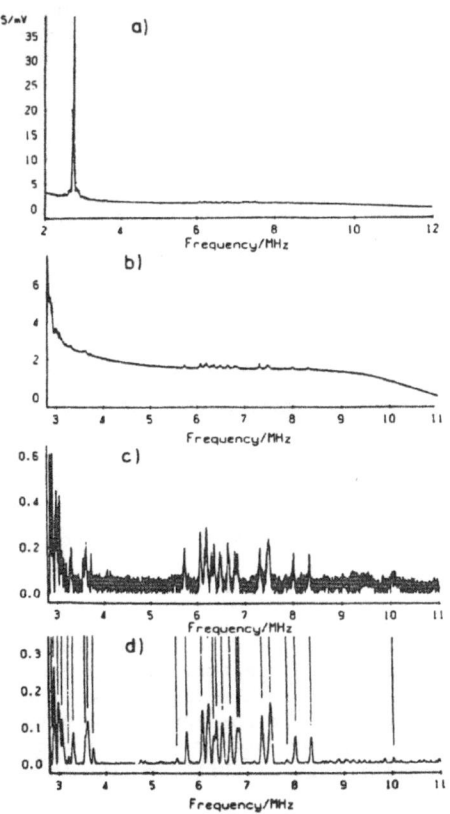

Fig. 15. Digital data processing for an ENDOR spectrum of Fe_i^o-centres in Si. (a) spectrum as measured, (b) subtraction of the strong distant ENDOR at ν_n (^{29}Si), (c) subtraction of a smooth underground line, (d) digital filtering and application of the peak search algorithm (for details see text).

If the angular patterns are separated in frequency, they are easily recognised and the analysis is fairly straightforward. In practice, one usually assumes a model for the defect and then tries to explain the ENDOR angular dependence on the basis of that model. However, both experiment and analysis become often very difficult due to the following complications. One is that many angular patterns overlap strongly in a narrow frequency. range. This is the case for the $H_{s,a}^o$-centres in KCl, as seen from Fig. 14. The angular pattern can become complicated due to the many centre orientations of low symmetry defects and due to the fact, that the simple solution of first order of the Spin Hamiltonian breaks down. The angular dependence can be calculated by a complete diagonalisation of the Spin Hamiltonian and the analysis is performed as a fit procedure between the experimental data and the calculated angular dependencies. The result of this is seen in Fig. 14 for the $H_{s,a}^o$-centres (16). If the fit agrees with the experimental data, the analysis is unambiguously correct. There are many more experimental points than parameters to be extracted from them. The number of ENDOR lines can be very high. As an example, the Ni^{3+} centre in GaP may be mentioned, where for each field orientation over 600 ENDOR lines were observed (17). Because of this, with conventional ENDOR spectrometers only comparatively simple problems could be solved. In recent years considerable progress was made by setting up computer controlled ENDOR spectrometers and by using computers for the data processing and analysis of the spectra.

218

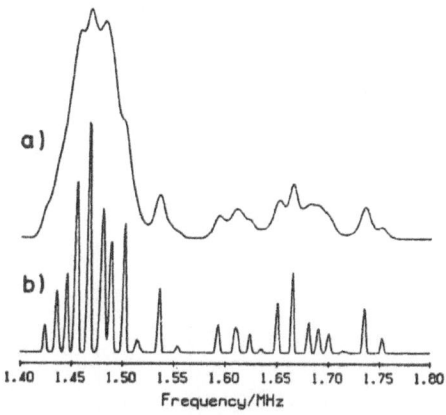

Fig. 16. Deconvolution of spectra to enhance the resolution
of ENDOR, (a) section of the spectrum of Fig. 14,
(b) after application of the deconvolution program.
After (19).

The angular dependence must be measured in small angular steps and due to
the low signal to noise repeatedly many times. In the computer controlled
ENDOR spectrometer the computer controls the following experimental parameters: rf (NMR frequencies), magnetic field, crystal orientation, sample
temperature and cavity matching. Thus, the angular dependence can be
measured automatically, also as a function of temperature. A full angular dependence can take up to 2-3 weeks of continuous measurement. The
ENDOR lines are stored in the computer and with a specially developed
software (see below) their positions can be determined automatically and
a computer plot of the angular dependence can be made. Fig. 14 is such
a plot.

The application of digital methods to the processing of the experimental data is summarized in Fig. 15 for the example of interstitial Fe^o
centres in Si (18). Fig. 15a shows the ENDOR lines of interest between
6 and 8 MHz, which are very weak compared to the 'distant ENDOR' due to
the free ^{29}Si nuclei and to a strong underground signal. The latter and
the distant ENDOR must be subtracted in order to deal better with the lines
of interest. With a special algorithm the underground is subtracted. It
does not assume a particular form of the underground. It 'eliminates' the
sharp peaks from the rest, which then is subtracted from the total spectrum (19). The resulting spectrum is that of Fig. 15c, which contains a
number of ENDOR lines and, of course, noise. The signal to noise ratio
of ENDOR spectra is usually not too good. The ENDOR effect in solid state
defects is mostly below 1% of the ESR signal. Low defect concentration
and the limited ENDOR effect are the major reasons for the poor signal to
noise ratio. Here the use of digital filtering has proved to be very
advantageous. The major ideas behind this are the following: what is
wanted is a smoothing of the spectra without disturbing the signals.
Conventional RC-filters have a poor trade-off between noise reduction,
signal distortion and speed of measurement. A simple digital method would
be a running average algorithm, which replaces a data point by the average
of the original data point and its unfiltered left and right neighbour
points. This symmetric average over (2 N + 1) data points produces the
classic noise reduction of any averaging process of uncorrelated data.
This idea can be improved by assinging different weights to the neighbouring data points.

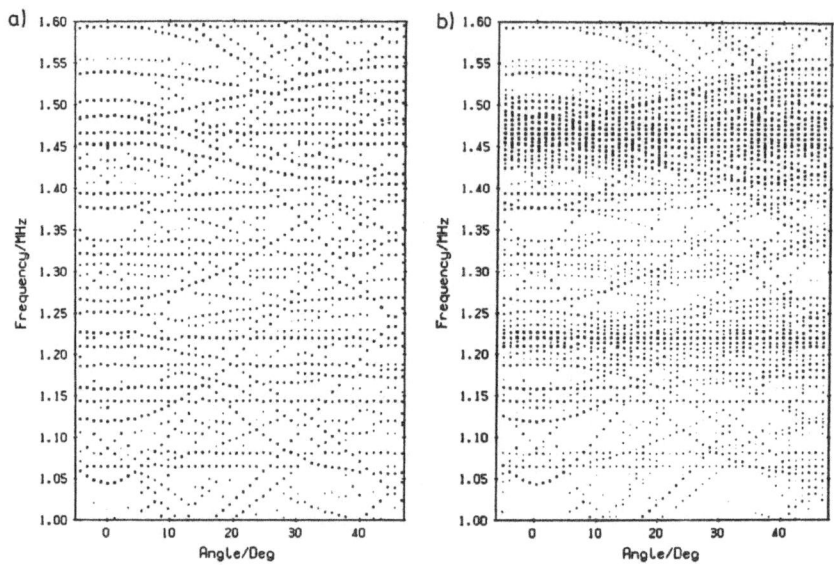

Fig. 17. (a) Section of the ENDOR angular dependence of Fig. 14 ($H_{s,a}^o$-centres in KCl), (b) the same section after application of the deconvolution program. After (19).

$$f[K] = \sum_{i = -N}^{N} a[i] \cdot Y[K - i] \qquad 4.11$$

Here Y denotes the unfiltered and f the filtered data points. The weights a[i] describe the digital filter used. For a simple running average one has

$$a[i] = {}^{1}/(2N + 1) \qquad 4.12$$

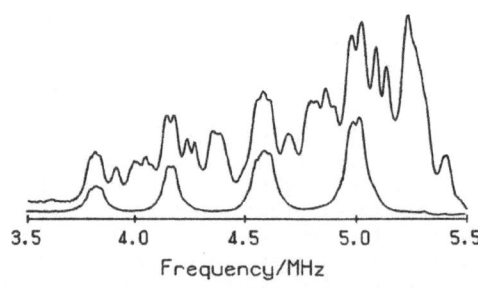

Fig. 18. Lower trace: section of the ENDOR spectrum of F centres in KCl:F⁻ prior to conversion to $F_H(F^-)$ centres. The lines are due to second shell ^{37}Cl and ^{35}Cl nuclei. Upper trace: the same section after conversion of about 50% F centres into $F_H(F^-)$ centres. After (26).

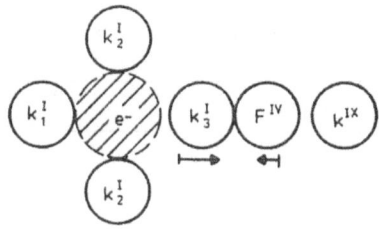

Fig. 19. Model of $F_H(F^-)$ centres in KCl as derived from the
ENDOR analysis.

If one requests, that the filtering process conserves additive constants, linear slopes, and parabolic peaks, area, first, second and third
moments and minimises noise under these constraints, it can be shown that
the weight function

$$a[i] = 3 \ ((3N^2 + 3N -1)-5i^2)/((2N-1)\cdot(2N+1)(2N+3)) \qquad 4.13$$

is optimal for any line shape (20). Such a filter is called DISPO filter
(digital smoothing with polynominal coefficient) (21). Compared to an RC-
filter it typically decreases the signal distortion by a factor of 20. For
small signal distortion (1%) and equal scan speed it reduces the noise by
an additional factor of 5 compared to the RC-filter (22,23).

Fig. 15d shows the application of such a filter to the spectrum of
Fig. 15c. The line positions are determined by a special peak search
algorithm, in which the 2nd derivative of the smoothed spectrum is calculated. A peak is, where this has a local minimum (19). This is exact for
symmetrical ENDOR lines of Gaussian or Lorentzian shape. The measurement
must make sure that the line shape remains symmetrical for this to be
applied. Therefore, for this kind of ENDOR spectroscopy, the method of
'transient ENDOR' originally introduced by G. Feher (14) is not applied,
since there the line shapes are nonsymmetrical.

When too many ENDOR lines overlap, the application of the peak search
algorithm is not sufficient. An improvement can be reached by applying
a deconvolution algorithm, which decomposes the spectra in an iterative
process, since the exact shape and width of the single ENDOR lines are not
known beforehand.

Fig. 16a shows a section of the spectrum of Fig. 12. After the
application of the deconvolution procedure one obtains Fig. 16b. In
Fig. 17a and 17b the ENDOR angular dependence of $H^o_{s,a}$-centres in KCl with
and without application of the deconvolution is compared. Clearly visible
is in the section between 1.45 - 1.50 MHz and 1.20 - 1.25 MHz the advantage of this procedure. The angular dependence can be followed much better
and can be analysed. The analysis yields as the site for the H^o atom the
anion vacancy, the values of the shf and quadrupole interaction constants
given in table 1 as an example for typical ENDOR results. The Spin Hamiltonian had to be diagonalised numerically (16).

Table 1. Shf and quadrupole constants of $H^o_{s,a}$-centres in KCl (in kHz), (T = 40 K). After (16)

shell	constant	$H^o_{s,a}$
$^{39}K_I$	a	253
	b	219
	q	198
$^{35}Cl_{II}$	a	57
	b	312
	b'	-3
	q	-88
	q'	-94
$^{35}Cl_{IV}$	a	37
	b	54
	q	±45
$^{39}K_V$	a	4
	b	11
	b'	≈0
	φ_B	$26.0^o \pm 0.2^o$
	q	±39
	q'	±17
	φ_Q	$13.5^o \pm 0.2^o$

The interaction constants are uncertain to ±1 kHz.
φ_B, $\varphi_Q = \sphericalangle(z, [100])$.

Not always it is possible to unambiguously determine the defect structure by the analysis of the ENDOR angular dependence. In crystals with
high symmetry as silicon or III-V semiconductors, like GaP, with a diamond
and zinc blende structure, respectively, there is a particular difficulty.
With respect to the substitutional site and the interstitial site with T_d
symmetry, there are the same kind of neighbour shells with the same occupancy with (100), (110) or (111) symmetry. Only the sequence is different.
However, from the experiments alone one can only determine the symmetry
type of a neighbour nucleus and not say which distance it has from the
centre. This would require a knowledge of the wave function Ψ or at least how
it varies with distance from the defect centre. If this was known, an
estimate of the interaction parameters would allow a shell assignment of
the measured nuclei according to equ. (3.9), (3.10) and (4.2). If this
is not known, then one cannot distinguish between the two sites. This
difficulty came up recently when investigating Te^+ and Se^+ centres in Si
(24, 25) and Ni^{3+} centres in GaP (17). In ionic solids one can usually
assume that $\psi(r)$ falls off monotonically with distance and therefore
difficulties in assigning neighbour nuclei usually do not occur. This is
generally not the case in semiconductors, where oscillations of the wave
functions can occur, also for deep level defects.

Photochemical reactions of point defects can also be studied by ENDOR.
As an example the conversion of F centres in KCl into $F_H(F^-)$ centres by
bleaching the optical F centre absorption band in KCl doped with F at a
temperature where the anion vacancies are mobile is shown in Fig. 18. In
the lower trace a section of the second shell Cl ENDOR spectrum of the F
centres prior to the conversion is shown, above the same section after
conversion of about 50% of the F centres into $F_H(F^-)$ centres. New ENDOR
lines appear between the original ones. In ESR the new centres cannot be

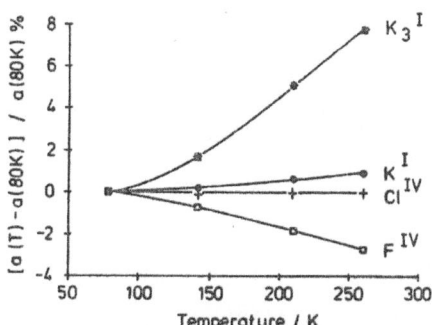

Fig. 20. Temperature dependence of the isotropic shf constant a
of K_3^I and F^{IV} neighbours of $F_H(F^-)$ centres compared to
the corresponding temperature dependence of a of K^I
and Cl^{IV} neighbours of F centres.

distinguished from the F centres. Fig. 19 shows the structure model
derived from the analysis of the $F_H(F^-)$ spectrum (26). F^- occupies a 4th
shell position and not a 2nd shell position as may have been assumed.
A separation of the ENDOR spectra can be difficult if the 'old' spectrum
is only partially converted and overlaps strongly the 'new' one. There
are, however, methods to separate these ENDOR spectra, as will be shown
in the following sections.

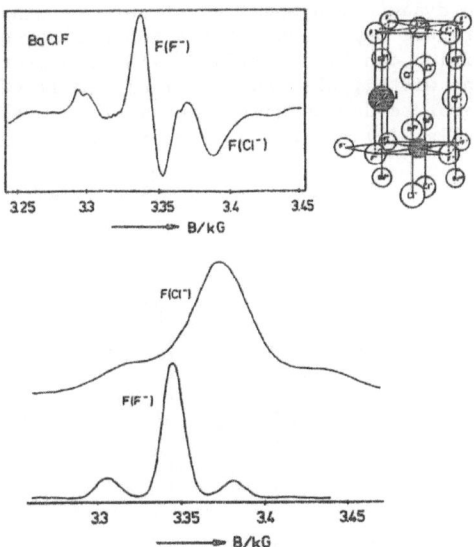

Fig. 21. Model of the two F centres in BaFCl (F(Cl$^-$) and F(F$^-$)
centres), their superimposed ESR spectra (derivative)
and the ENDOR-induced ESR spectra of both centres measured
using a ^{19}F ENDOR line of each centre. After (32).

223

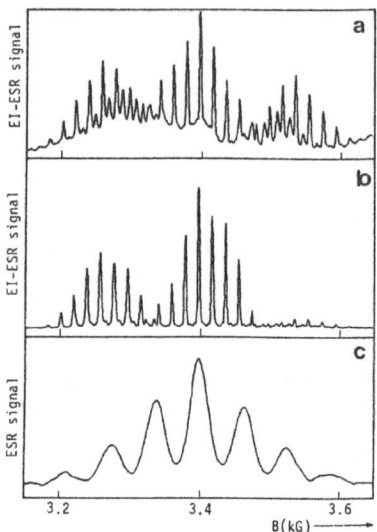

Fig. 22. ESR and ENDOR-induced ESR spectrum of X-irradiated $KMgF_3$
doped with Fe^{2+}, (a) integrated ESR spectrum after
X-irradiation at room temperature, $B_0 \| [100]$, (b) ENDOR-
induced ESR spectrum for the ^{19}F ENDOR line at 42.5 MHz.
The spectrum is due to Fe^{3+}, (c) ENDOR-induced ESR spectrum
for the ^{19}F ENDOR line at 21.0 MHz. The spectrum is due to
F centres. After (34).

With ENDOR also dynamical properties of defects can be studied by
measuring the temperature dependence of the ENDOR line positions. Since
each ENDOR line position can be determined with high precision (to 1-10 kHz,
depending on the line width), also comparatively small effects can be seen.
In Fig. 20 the temperature dependence of the isotropic shf interaction
constant of K_3^I (see Fig. 19) and F^{IV} neighbours of $F_H(F^-)$ centres

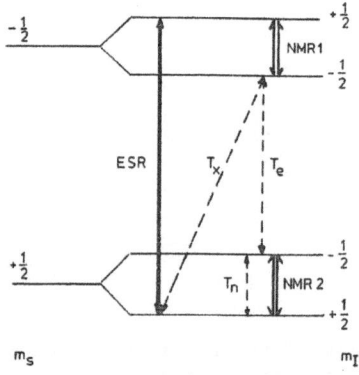

Fig. 23. Level scheme to explain the special tripel resonance
experiment (DOUBLE ENDOR) (see text).

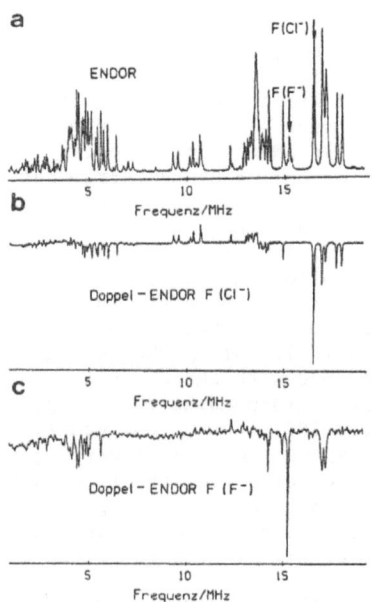

Fig. 24. (a) Part of the ENDOR spectrum of F(Cl⁻) and F(F⁻) centres simultaneously present in BaFCl, (b) ENDOR spectrum for setting one ENDOR frequency to an ENDOR line of F(Cl⁻) centres (see mark in Fig. 24a), (c) DOUBLE ENDOR spectrum for setting one ENDOR frequency to an ENDOR line of F(F⁻) centres (see mark in Fig. 24a). After (37).

and of the corresponding K^I and Cl^{IV} neighbours in F centres is plotted between 77 and 300 K (27). The strong temperature dependence of K_3^I of $F_H(F^-)$ reflects that K^I, which is displaced by 11% towards F^-, has more room compared to K^I in the F centre and experiences a large local mode vibration, while F^{IV} is knocked further away through these vibrations.

Another example is the strong temperature dependence of the shf interaction observed for atomic hydrogen centres on anion or cation sites (16, 28,29), which have local vibrational modes with very high amplitudes due to the light hydrogen mass.

In crystals with structural phase transitions the lattice changes its symmetry at T_c. This can be 'seen' by a paramagnetic probe, which therefore can be used to study such transitions. A recent example is the investigation of Mn^{2+} in $RbCdF_3$, which goes at 124 K from a cubic to a tetragonal phase. With ENDOR it was possible to study the order parameter as a function of temperature, which is directly reflected in the splitting of the ENDOR lines with temperature upon going through T_c for suitable field orientations. It turned out that the order parameter measured is smaller than the intrinsic value, since the lattice relaxes around the probe and this 'decouples' it from the lattice somewhat (30,31).

5. ENDOR-INDUCED ESR

Overlapping ESR spectra of different defects cause overlapping ENDOR spectra. It can be very difficult if not impossible to analyse their angular dependence, especially if many ENDOR lines occur in a narrow

frequence range. Furthermore, there may be weaker ESR spectra buried
under stronger ones having observable ENDOR lines, which can be analysed,
but the ESR spectrum cannot be measured. The two type of F centres possible
in BaFCl, e.g. are produced simultaneously. Their ESR spectra strongly
overlap (see Fig. 21). In such a case one can measure a kind of excitation spectrum of a particular ENDOR line belonging to one defect, which
gives an image of the corresponding ESR spectrum belonging to the same
defect (ENDOR-induced ESR-spectrum). This can be seen from Fig. 10. The
ENDOR transitions can be measured by setting B_O to either of the 4 shf
ESR transitions for m_I = 3/2, 1/2, -1/2 and -3/2, thus the ENDOR line
intensity measured should follow the ESR line pattern in the middle of
Fig. 10. According to equ. (4.6), however, the frequency of an ENDOR line
depends on ν_n and therefore on B_O. Because of this one must 'correct' the
ENDOR frequency for the variation of B_O when going through the ESR-spectrum. This can easily be done with the computer controlled spectrometer
in other ways provided the corresponding nuclear g_I factor is known. In
measuring the ENDOR-induced ESR spectrum one monitors the ENDOR line
intensity of a particular ENDOR line, while varying the magnetic field
through the ESR spectrum and correcting the frequency position according to
equ. (4.6). The resulting spectrum is an image of the (integrated) ESR
spectrum of that defect, to which the ENDOR line (nucleus) belongs. In
this way the ESR spectrum of different defects can be separated.

Fig. 21 shows this for the 2 F centres in BaFCl. The two ENDOR-
induced ESR spectra were measured using ^{19}F ENDOR lines of both centres.
For S = 1/2 and no quadrupole interaction experienced by the nucleus taken
for the measurement, the ENDOR-induced ESR spectrum corresponds to the true
ESR line shape, if the cross relaxation does not depend on m_I and is the
dominating electron spin relaxation process. This was observed in several
cases (32).

If the transition used is a quadrupole ENDOR line then not the true
ESR lines are measured with the ENDOR-induced ESR spectrum. The deviations can, however, be minimised, if one choses an ENDOR line with a very
small interaction, that is near the Larmor frequency ν_n of the nucleus.
For details see (32,33).

For systems with S > 1/2 there is usually a fine structure interaction splitting of the electron Zeeman levels. Then in ENDOR-induced ESR
one measures only some of the possible ESR transitions. This is demonstrated for Fe^{3+} centres in $KMgF_3$, which were produced by X-irradiation
of $KMgF_3$ doped with Fe^{2+}. In Fig. 22a the integrated ESR spectrum of the
5 ESR transitions is reproduced with partly resolved shf structure with
six nearest ^{19}F neighbours. Fig. 22b shows the ENDOR-induced ESR spectrum
measured with a ^{19}F ENDOR line at 42.5 MHz displaying 2 of the five Fe^{3+}
ESR transitions, while Fig. 22c shows the ENDOR-induced ESR spectrum
measured with a ^{19}F ENDOR line at 21.0 MHz. Clearly, another ESR spectrum
appears. It was buried under the Fe^{3+} spectrum beyond recognition and
turned out to be due to simultaneously produced F centres (34). A detailed
analysis also allows the determination of the relative signs of quadrupole
interaction constants, shf interaction constants and fine structure
constants (32,33).

6. DOUBLE ENDOR

Although with ENDOR-induced ESR experiments each ENDOR line can be
'labelled' to a particular defect in case of the simultaneous presence of
several defects, this can be a tedious task, especially if one has to
follow a complicated angular dependence. Such a situation is usually
encountered in radiation damage studies. Therefore, a method is called

for with which the ENDOR spectra of different defects can be measured
separately. This can be done by measuring a triple resonance, in which
2 NMR frequencies are applied simultaneously together with the microwaves.
The simplest case, the so called 'special triple resonance' (special DOUBLE
ENDOR) is schematically shown in Fig. 23 for the simple case of $S = 1/2$,
$I = 1/2$. If stationary ENDOR is measured for the transition 'NMR1' between
$m_s = -1/2$, $m_I = 1/2$ and $-1/2$, then the signal height is determined by the
ESR transition probability (that is by B_1^2), the NMR1 transition probability
(that is B_{rf}^2 (NMR1)) and the cross relaxation time T_x. It is assumed (and
a condition for the experiment), that $T_x > T_e$, the electron spin lattice
relaxation time, due to the comparatively long nuclear spin lattice
relaxation time T_n. If then a second rf frequency is applied between the
levels $m_s = 1/2$, $m_I = 1/2$ and $-1/2$ (NMR2) then T_n is effectively shortened
by this transition and therefore T_x is also shortened, which results in an
enhancement of the monitored ENDOR signal at the frequency NMR1. In the
experiment one irradiates with the fixed ENDOR frequency NMR1, monitors
the ENDOR line intensity of the line at NMR1, while sweeping the second
rf frequency. When the transition NMR2 is induced the ENDOR line intensity NMR1 increases. The increase is the DOUBLE ENDOR signal and is
deteced with a double lock-in technique.

In the stationary DOUBLE ENDOR spectrum positive and negative signals
are observed (35,36). Negative signals occur if the second NMR frequency
is induced between nuclear states belonging to the same m_s quantum number.
Fig. 24 shows the DOUBLE ENDOR spectrum for the two F centres in BaFCl.
In Fig. 24a the ENDOR lines of both F centres are superimposed, a full
analysis was not possible. In Fig. 24b the fixed ENDOR frequency NMR1
was set to one ENDOR line belonging to F(Cl$^-$) centres and NMR2 was swept
between 1 and 19 MHz and the DOUBLE ENDOR effect was recorded. In
Fig. 24c the analogous experiment was made for an F(F$^-$) ENDOR line. Both
DOUBLE ENDOR spectra show only lines due to the F(Cl$^-$) or F(F$^-$) centres
alone. Especially around 5 MHz both centres have many ENDOR lines, which
otherwise could not have been separated (37).

DOUBLE ENDOR is also very important to analyse low symmetry defects.
The defects are distributed over several orientations in the crystal. The
ESR and ENDOR spectra of these orientations overlap. In a sense each
defect orientation is equivalent to a new defect species. With DOUBLE
ENDOR the spectra of one particular defect orientation can be measured
separately, which greatly facilitates the analysis or makes it at all
possible. In a recent investigation of O$^-$ centres in α-Al$_2$O$_3$, which had
very low symmetry (that is 'no' symmetry) a definite assignment of the
quadrupole ENDOR lines to their corresponding 'hf' ENDOR lines was only
possible after one particular centre orientation could be measured
separately (38).

7. OPTICALLY DETECTED MAGNETIC ELECTRON SPIN RESONANCE (ODESR)

There are several ways to detect ESR optically. Which of the techniques is applied depends on the system studied and on the kind of question
one wants to answer. Compared to 'conventional' ESR the optical detection,
if possible, has several advantages with respect to the identification of
defects:

    (i)   higher sensitivity
    (ii) higher selectivity
   (iii) direct correlation between optical and structural
         properties
    (iv) possibility to investigate optically excited states.

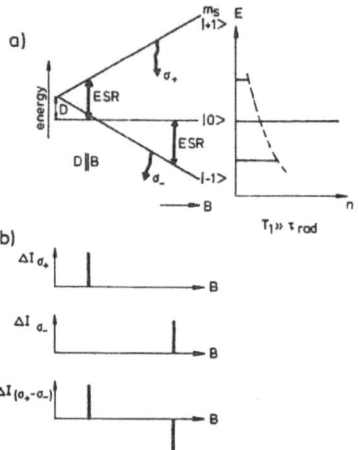

Fig. 25. Level scheme of triplet states to illustrate the optical
detection of ESR via a microwave-induced change in the
luminescence intensity or via the magnetic circular polarisation of the emission (MCPE). It is assumed that
$B_0 || D_{zz}$, $T_1 \gg \tau_{rad}$.

The sensitivity enhancement can be up to $10^4 - 10^5$, compared to the
conventional ESR, depending on the method used and the physical parameters
of the system. The high selectivity is based on the possibility to use
single absorption or emission bands for the detection. The bands of
different defects usually do not overlap entirely and hence, these defects
can be measured separately. A correlation of optical properties with
structural property is based on the same argument. Relaxed excited states
of sufficient radiative lifetimes can be studied. The condition is, that
the microwave transition rate is of the same order or bigger than the
reciprocal radiation lifetime. If this condition is not fulfilled, during
the lifetime of the state no occupation changes of the Zeeman levels can
be induced by the microwaves. ESR spectra of excited states with lifetimes
of the order of microseconds have been detected several times. The actual
technical limit, which is mainly given by the microwave amplitude to be
applied at low temperature, is probably at lifetimes of the order of
100 n sec.

The optical detection of ESR is based on either the observation of a
microwave-induced intensity change in luminescence or on the observation
of microwave-induced changes of the degree of polarisation of either
absorption or emission. A condition is that the Zeeman levels involved
are occupied differently and that the radiative transitions differ in
polarisation properties. It should be noted that the Zeeman splitting
of the levels is only of the order of 10^{-4} eV, while the phonon broadening
of the levels due to electron phonon coupling is 3 orders of magnitude
bigger (see section 2). Fortunately, the polarisation of the optical
absorption or emission transitions is usually only little affected by
the phonons, so that microwave-induced population changes within the
Zeeman levels can be observed.

The examples and methods discussed below are chosen mainly with
respect to defect identification. Not mentioned, e.g. are interesting
studies like radiative/nonradiative transitions or electron transfer
in excited states for which ODMR is an excellent tool, nor the study of
relaxed excited states if the defect model is known from ground state ESR.

228

Defects with 2 valence electrons, e.g. in a ns^2 configuration, often have relaxed excited triplet states, in which the two electrons have parallel spins and $S = 1$. The optical absorption leads first to a singlet excited state with $S = 0$, which then relaxes into the triplet state by intersystem crossing. The two electrons possess then a fine structure interaction (e.g. due to dipole-dipole interaction), so that the Spin Hamiltonian of equ. (3.3) obtains an additional term

$$H_{FS} = \vec{S} \, \tilde{D} \, \vec{S} \qquad\qquad 7.1$$

In perturbation theory of first order the energy is given for an axially symmetric fine structure tensor and B_0 parallel to its principal axis (then $D_{zz} = D$)

$$E_{FS} = D \left[m_s^2 - \frac{1}{3} \cdot S (S + 1) \right] \qquad\qquad 7.2$$

Fig. 25 shows the level scheme as a function of the magnetic field of the Spin Hamiltonian containing the term 7.1 and the electron Zeeman interaction. The three levels are all occupied from a higher singlet state. Radiative transitions from the triplet states into the singlet ground states are forbidden. However, the states $|+1\rangle$ and $|-1\rangle$ can mix with excited singlet states via the spin orbit interaction and therefore a finite radiative transitions probability into the ground state is observed for these levels, while $|0\rangle$ cannot decay by radiative transitions. The radiative lifetime of the $|+1\rangle$ and $|-1\rangle$ levels depends on the size of the spin orbit interaction and the energy separation of excited singlet states from the triplet state.

Due to the radiative decay of the $|+1\rangle$ and $|-1\rangle$ levels, these levels are less populated in a stationary state compared to the level $|0\rangle$, where population is accumulated (see Fig. 25). It is assumed that the spin lattice relaxation time T_1 is large compared to the radiative lifetime.

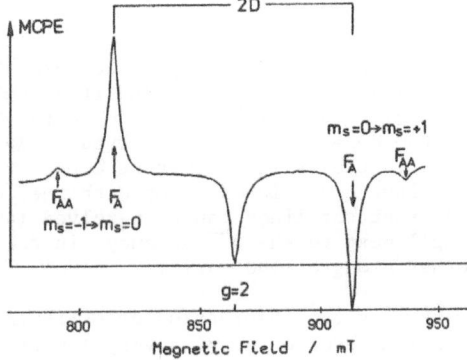

Fig. 26. Magnetic circular polarisation of the emission (MCPE) recorded at $\lambda > 700$ nm under cw-microwave irradiation (24 GHz) as a function of magnetic field ($B_0 \parallel [100]$ in CaO containing F_A(Mg) and F_{AA}(Mg) centres. T = 1.6 K. After (41).

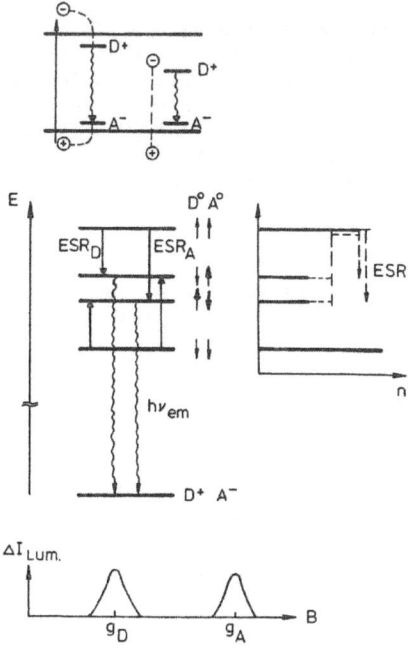

Fig. 27. Schematic representation of donor-acceptor recombination
in semiconductors and of the mechanism to observe ESR as
a change in intensity of the donor-acceptor recombination
luminescence.

Therefore the microwave transtion $|0\rangle \to |+1\rangle$ and $|-1\rangle \to |0\rangle$ shift popula-
tion into the radiative levels, upon which the luminescence intensity is
enhanced. This can easily be observed. On the other hand, one can also
observe the magnetic circular polarisation of the emission ($I_{\sigma+} - I_{\sigma-}$)
(MCPE), where for the low field transition an enhancement, for the high
field transition a decrease is observed. (See Fig.25b, where $D > 0$ is
assumed).

F centres in CaO, where an O^{--} vacancy is occupied by two electrons,
do have such relaxed excited triplet states and their ODESR was observed
in such a way (39,40). Fig. 26 shows as an example the ESR spectrum of
F_A and F_{AA} centres in CaO doped with Mg^{++} observed at $\lambda > 700$ nm (1). Both
centres differ slightly in their fine structure constant D. Measurements
with higher resolution show a shf interaction with the 10% abundant ^{25}Mg
($I = 5/2$). Six equidistant shf lines can be resolved (40,41). In the F_A
centre there is one Mg^{++} next to the O^{--} vacancy, in the F_{AA} centre two
Mg^{++} opposite each other along a (100) axis.

In semicoductor physics the observation of ODESR in the donor-
acceptor recombination luminescence plays a very important role for defect
identification. The fundamental process is illustrated in Fig. 27. Upon
an optical band-band transition an electron-hole pair is created. The
electron is captured by an ionised donor D^+ and the hole by a negatively
charged acceptor A^- (semi-insulating or p-type semiconductor) according
to

$$D^+ + A^- + e^- + h^+ \to D^0 + A^0 \qquad\qquad 7.3$$

230

Both D^O and A^O are paramagnetic. The unpaired electron at the donor and the unpaired hole at the acceptor can recombine under emission of a luminescence radiation

$$D^O + A^O \rightarrow D^+ + A^- + h\nu \qquad\qquad 7.4$$

The level scheme of D^O and A^O in a magnetic field is shown in Fig. 27. The two spins are assumed to be very weakly coupled (42). The energy positions of the levels are therefore determined only by their respective g-factors and the magnetic field. Since from the two triplet states a radiative decay into the singlet ground state is forbidden, only from the two singlet states, where the two spins are antiparallel, a luminescence is observed. Therefore, in a stationary state, the population of the two triplet states is higher compared to the two singlet states, similarly as discussed above for the state $|0>$ in the highly coupled triplet system.

Upon the microwave transitions indicated in Fig. 27 the recombination luminescence is increased and in principle one can observe the ESR lines of both the donor and acceptor provided their g-factors are different enough. Usually, one observes only the donor resonance since the p-type hole states of the acceptors experience a dynamical Jahn-Teller effect with the consequence that the resonance is very difficult to observe. It can be observed upon application of uniaxial stress if this is sufficiently large to suppress the dynamical Jahn-Teller effect.

With this technique recently the ODESR of anion antisite defects in GaP:Zn (p-type) could be observed. Upon band-band excitation two strongly overlapping luminescence bands can be observed, one peaking at 0.95 eV, the other at 1.20 eV. In the 0.95 eV luminescence a doublet ODESR spectrum is observed, which is due to a P atom on a Ga site ('antisite') with one unpaired electron. The doublet splitting is due to the hf interaction with the ^{31}P nucleus $(I = 1/2)$ (43,44). Using a specific modulation technique also the shf interaction with 4 nearest ^{31}P-neighbours could be resolved from which the structure model could be inferred (43). The acceptor resonance could not be seen.

Fig. 28 shows the ODESR spectrum measured in the 1.20 eV luminescence, which is also due to an antisite defect, however, in a triplet state and

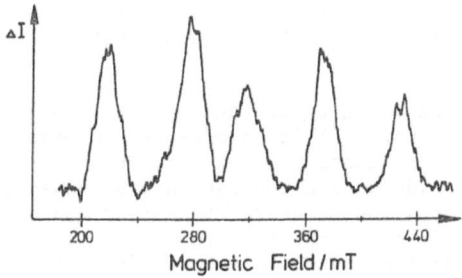

Magnetic Field / mT

Fig. 28. X-band ODESR spectrum observed as a microwave-induced change of the donor-acceptor recombination luminescence in GaP:Zn (excitation wavelength 514 nm). The spectrum shows the donor PP_3Y_P in a triplet state and the acceptor (central line) which is probably Fe_i^{3+}. $B_o || [111]$. After (45).

231

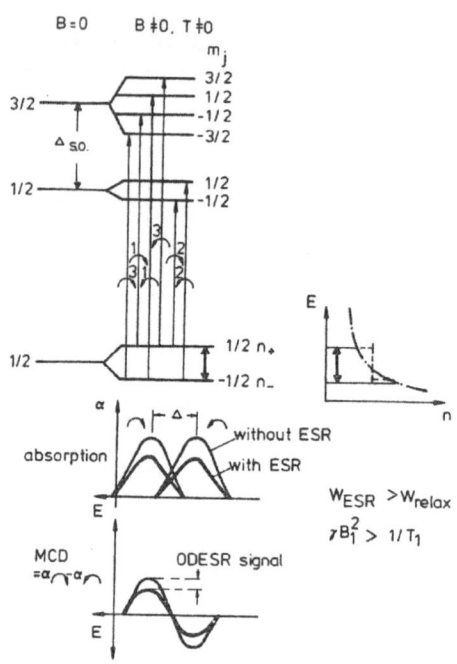

Fig. 29. Simple model to explain the magnetic circular dichroism (MCD) of the absorption and its microwave-induced decrease to detect the ESR transitions.

which also contains the ESR line of the acceptor taking part in the luminescence process. The two low and high field lines are due to the antisite, the splitting between the two lines of each doublet is due to the ^{31}P hf interaction, while the separation of the centres of the two doublets (between approx. 240 and 400 mT) corresponds to the separation of the lines in Fig. 26 and is due to the fine structure interaction. It was concluded that the structure model is a PP_3Yp defect with an unknown atom on one of the 4 nearest neighbour P sites and that the acceptor may be a nearby interstitial Fe^{3+} impurity (45). The level of the defect is at $E_v + 1.5$ eV. The PP_3Yp defect can be excited into its triplet state by direct excitation from the valence band with sub-band gap light, a possibility also indicated schematically in Fig. 27. This excitation enhances the luminescence, since the defects can be excited in the bulk of the crystal and not only in a thin surface layer, as is usually the case for band-band transitions due to the high absorption constant. When measuring the ODESR via recombination luminescence one common difficulty is, that there is a distance distribution between donors and acceptors and that the exchange interaction between them can be large and varying according to distances. It is especially large for shallow donors and acceptors. This results in a broadening of the ESR lines. Therefore, in most cases the donor resonances show no hf or shf structure and a defect identification can only be based on the g-factors. There are many resonances known, where the defect could not be identified. The broadening effect is less important for 'deep level' defects. One way to overcome this difficulty would be to measure the ODESR spectrum with time resolution. It was shown for shallow In donors in ZnO that the resolution can be greatly enhanced by exciting the luminescence with pulsed light and by taking only the long lifetime tails of the luminescence for the ODESR measurement, which comes from the distant donor-acceptor pairs for which the exchange interaction is only small (46). Little work was done so far with time resolved ODESR in solid state defects.

232

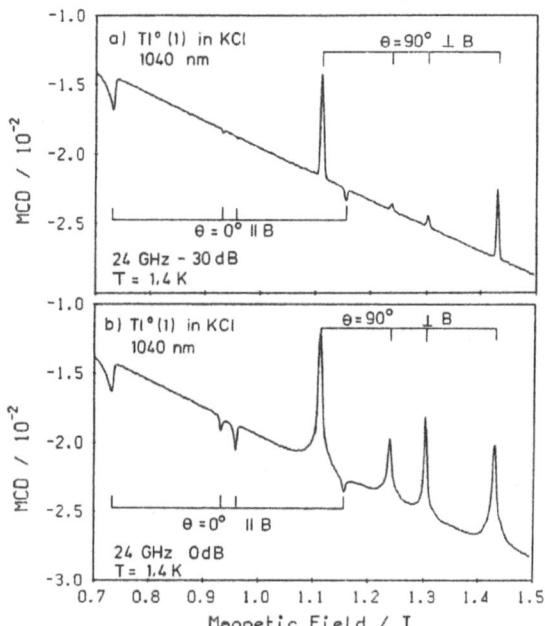

Fig. 30. (a) ODESR spectrum of $Tl^0(1)$ centres in KCl detected as microwave-induced change of the MCD of the absorption band at 1040 nm, measured at low microwave power (24 GHz, T = 1.5 K), (b) the same for high microwave power. After (48).

The investigation of a ground state ESR with optical detection has recently become of particular interest. The reasons are the enhanced sensitivity, its selectivity and the possibility to directly correlate optical bands with ESR spectra and thus with the defect structure. The experiment is based on the observation of the microwave-induced change of the magnetic circular dichroism (MCD) of an absorption band. In Fig.29 it is schematically indicated for an 'atomic' optical transition s → p how the method works.

The allowed optical absorption transition for right and left circular polarised light are indicated along with the relative matrix elements. The absorptions for both polarisations are split in energy by the spin orbit splitting of the excited state. If this splitting is small compared to the phonon width of the absorption bands the measurement of the MCD as a function of photon energy yields a derivative structure.

The MCD is defined as

$$MCD = \frac{d}{4} \ (\alpha_+ - \alpha_-) \qquad\qquad 7.5$$

where d is the crystal thickness, α_+ is the absorption constant for right and α_- for left circular polarised light. If the absorption comes from a Kramers doublet with S = 1/2, one obtains (47)

$$MCD = \frac{1}{2} \ \alpha_o \ d \ \frac{\sigma_+ - \sigma_-}{\sigma_+ + \sigma_-} \ \frac{n_- - n_+}{n_- + n_+} \qquad\qquad 7.6$$

α_o is the absorption constant for unpolarised, light σ_+ and σ_- are the cross sections for right and left polarised light, respectively, and n_+ and n_-

233

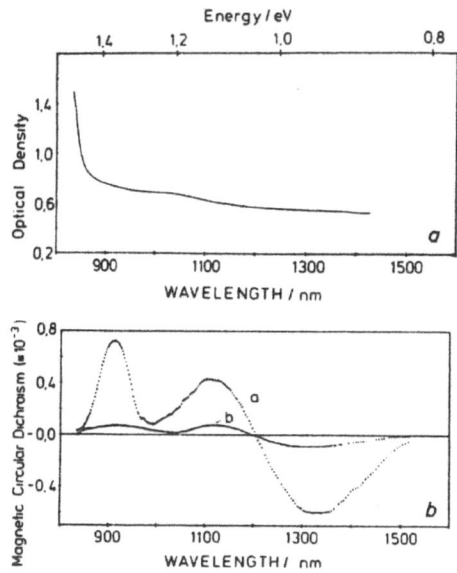

Fig. 31. (a) Optical absorption of LEC-grown undoped 'as-grown' semi-
         insulating GaAs, (b) magnetic circular dichroism of the
         absorption (MCD) of the same crystal. (B = 2 T, T = 4.2 K)
         (Curve a). Curve b is the 'MCD-tagged by ESR'. After (49).

are the occupation numbers for the $m_s = \pm 1/2$ states. The MCD is thus proportional to the occupation difference $n_+ - n_-$ and the cross section
difference $\sigma_+ - \sigma_-$.

The occupation difference $n_+ - n_-$ can be decreased by a microwave
transition, provided the transition rate is of the order or larger than
the spin lattice relaxation rate $1/T_1$. Such an ESR transition thus results
in a decrease of the MCD, which is monitored (see Fig. 29). One measures
the MCD as a function of the magnetic field under microwave irradiation.
The decrease due to the resonance is observed on the variation of the MCD
as a function of the magnetic field according to

$$MCD \quad \propto \quad \tanh \quad (g\beta_e B/2kT) \qquad\qquad 7.7$$

For low temperatures and the usual field variation this is practically
a linear function of B. As an example Fig. 30 shows the ODESR spectrum
measured in the absorption band at 1040 nm of the laseractive thallium
centres in KCl. (48) (see sections 2 and 3). This absorption has a paramagnetic MCD as expected. It is negative. From Fig. 30 one sees immediately that not a single ESR line appears near $g \approx 2$ as would be expected
for an $F_A(Tl^+)$ centre, but altogether 8 transitions. Comparison with the
$Tl^0(1)$ centres ESR data (12) shows that the ODESR spectrum can be described
exactly by their data (calculated for K-band, 24 GHz). Thus the laser-
active centres are indeed identical with the $Tl^0(1)$ centres previously
investigated by conventional ESR (48). Here a direct correlation with the
optical absorption band at 1040 nm was possible. The ODESR spectrum
contains allowed transitions of centres with their axes parallel and
perpendicular to the field orientation (indicated in Fig. 30) as well as
'forbidden' transitions, in which also the Tl nuclear quantum numbers
change. Their intensity decreases with decreasing microwave power

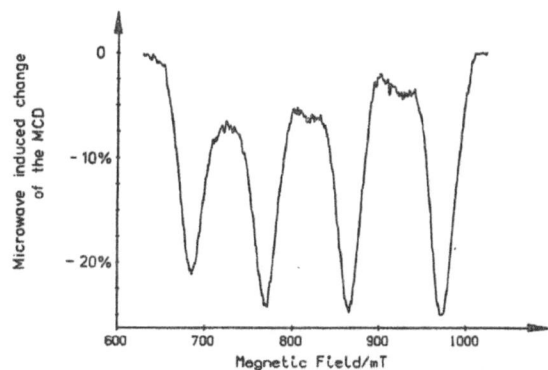

Fig. 32. ODESR spectrum of anion antisite defects $AsAs_4$ in s.i. undoped GaAs 'as-grown'. $(B_0 || [100]$, T = 1.4 K, ν_{ESR}=24 GHz, measured at λ = 1350 nm). After (49).

(see Fig. 30a and 30b). The centre axis is, as mentioned before, the connection line between the Tl^0 atom and the anion vacancy. (For details see Ref. 48).

Fig. 31 shows the optical absorption and the MCD of 'as-grown' undoped GaAs, which was grown with the LEC (liquid encapsulated Czochralski) method. In the absorption below the gap energy of 1.52 eV only a very weak band at 1.18 eV due to the so called EL2-defects is detectable, whereas the MCD measurement reveals the existence of further intracentre absorption bands due to paramagnetic defects. In the direct measurement the absorption is not detectable. The sensitivity enhancement when measuring the MCD is a consequence of the applied form of modulation spectroscopy (49).

When measuring the ODESR in this MCD, e.g. at 1350 nm, the spectrum of Fig. 32 is observed, which is due to $AsAs_4$ antisite defects, that is due to As atoms on a Ga site analogous to the PP_4 defects in GaP discussed above. The splitting into 4 lines is due to the hf interaction with the central ^{75}As nucleus (I = 3/2). The ESR of antisite defects in as-grown material is very weak, the signal to noise ratio is only of the order of 2-4 (concentration approx. 10^{16} cm^{-3}). Here the signal to noise ratio is better by about 2 orders of magnitude. It should be mentioned, however, that the antisite defects observed in ODESR have a T_1 of several seconds at 1.4 K, while those observed in conventional ESR have a very short T_1 ($\sim 10^{-5}$ sec.) (50). Therefore, both species are different although the spectra look the same and at present it cannot be said, which is the concentration of those measured in ODESR. Possibly, the signal to noise gain is even higher than 2 orders of magnitude. With present techniques the MCD can be measured down to $2-5 \cdot 10^{-5}$ for an optical density of one.

8. EXCITATION SPECTROSCOPY OF ODESR

The optical detection of ESR spectra opens up the possibility to observe the ESR lines upon variation of the optical wavelength. This can be particularly useful for the determination of all the optical transitions belonging to one particular defect even if these transitions are buried under optical absorption (or emissions) of simultaneously present different defects.

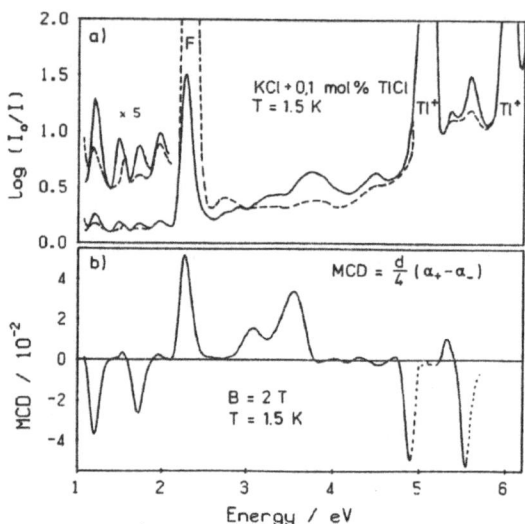

Fig. 33. (a) Absorption spectrum of KCl:Tl after X-irradiation at -40 °C (dashed line) and after F centre bleaching (solid line), (b) magnetic circular dichroism (MCD) of the bleached sample. After (48).

Fig. 33 shows again for the example of the $Tl^o(1)$ centres in KCl the optical absorption spectrum and the MCD spectrum measured after producing the $Tl^o(1)$ centres with X-irradiation at -40 °C, upon which many other defects, many of them Tl-related, are also created. In order to identify the absorption bands belonging to the $Tl^o(1)$ centres, one can monitor the ODESR of one particular ESR line (e.g. of the parallel or perpendicular spectrum for $B_o || (100)$), and change the wavelength of the light inducing the optical absorption transitions. Since the level scheme of Fig. 29 does not specify a particular excited state, the ODESR is always observed, if a spin orbit split excited state is reached. Thus, the excitation spectrum of the ODESR lines reveals all those optical absorptions, which belong to one particular centre (or centre orientation). Fig. 34 shows this for parallel and perpendicular $Tl^o(1)$ centres in KCl. Altogether 8 optical $Tl^o(1)$ absorptions could be detected in this way (48,51). This kind of excitation spectrum was called 'MCD-tagged by ESR' (48). A Tl-dimer centre with very similar optical properties as the $Tl^o(1)$ centre but with a completely different ESR spectrum could be separated and analysed with this method too. In the case of the analogous $Ga^o(1)$ and $In^o(1)$ centres the positions of the absorption bands could only be detected in this way, since they were buried in shoulders of other much stronger absorptions (53) and had no emissions, which in principle could also be used to detect other absorption bands by measuring the excitation spectrum of the emission. In the case of the $Tl^o(1)$ centres this was also not possible since the mentioned Tl-dimer centres emitted at the same wavelength as the $Tl^o(1)$ centres.

Another kind of excitation spectroscopy can be performed in semiconductors. The aim is to unambiguously determine the level of a defect in the gap, the structure of which is known from ESR. The usual methods like DLTS (deep level transient spectroscopy) or related methods can determine the energy position of the level, but not identify the corresponding defect. The method is based on a suitable position of the Fermi level such, that the level in question is not occupied. It is then

236

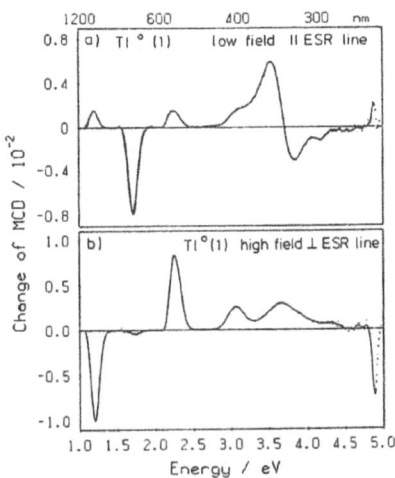

Fig. 34. (a) Magnetic circular dichroism (MCD) at 1.4 K of the
'parallel' Tl⁰(1) centres in KCl as tagged by the corresponding
low field ESR line, (b) the same for 'perpendicular' Tl⁰(1)
centres tagged by the high field ESR line. After (48).

occupied with a second light beam from the valence band, while the first
light beam is used to measure the ODESR via, e.g. a microwave-induced
decrease of the MCD.

A recent example is the investigation of the antisite defects $AsAs_4$
in p-type GaAs. Before occupying the D^+/D^{++} level with electrons from the
valence band no MCD can be measured. In Fig. 35 the onset of the MCD is
plotted as a function of the wavelength of the exciting light ('second
beam'). As from 0.52 eV onwards the MCD can be detected and also the ESR
spectrum. From 0.74 eV onwards the MCD decreases again, since the paramagnetic $AsAs_4$ defect captures a second electron and becomes diamagnetic
(D^0/D^+ level) (54). The method can be characterised as 'photo ODESR'
measurement in analogy to similar experiments performed earlier in conventional ESR (55).

9. OPTICALLY DETECTED ENDOR

As mentioned in the section on ESR for a structure determination of
defects it is desirable to resolve the shf structure and thus to measure
ENDOR. This is in principle also possible with optical detection and thus
to combine its advantages with the higher resolution of shf interactions,
which is characteristic for ENDOR. However, only very few experiments of
this kind of tripel resonance were reported for solid state defects. One
'self-ENDOR' experiment, that is ENDOR with the central nucleus, was
reported on shallow In donors in ZnO, where the tripel resonance was monitored via the donor-acceptor recombination luminescence (46). Another
experiment uses the microwave-induced decrease of the MCD and is schematically explained in Fig. 36 for the case of S = 1/2, a central nucleus of
I_c = 3/2 and one ligand nucleus of I_1 = 3/2, which is the (simplified)
scheme for the $AsAs_4$ antisite defects in GaAs. In the experiment one sets
the magnetic field onto a particular position in the ODESR line, e.g. into

237

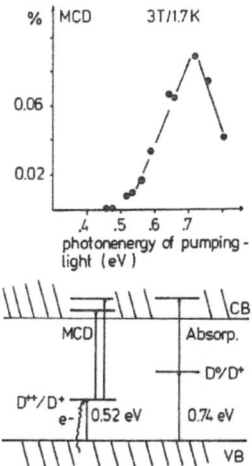

Fig. 35. Excitation spectrum of the ODESR lines of anion antisite
AsAs₄ defects as measured by photo-ODESR and the resulting
level position for the two charge states of the defect.

the flank. The microwave then induces transitions between a particular
combination of $m_{I,c}$ and $m_{I,1}$ states. Thus, only the population difference
between these states is diminished (equalised in the extreme case). Therefore, only a subset of spin packets contributes to the decrease of the MCD,
which is monitored. The microwave is 'sharp' enough not to hit more than
this subset of m_I values if spin diffusion etc. is neglected. If then an
NMR frequency is simultaneously applied, which combines the $m_{I,1}$ states
with its neighbouring ones, then more spin packets are shifted into the
microwave pump channel and contribute to the decrease in MCD. Thus, upon
such an NMR resonance the decrease of the MCD is enhanced. This is monitored for the detection of this tripel resonance.

Fig. 37 shows the ODENDOR spectrum of AsAs₄ defects in semi-insulating
'as-grown' GaAs/Cr. In Fig. 37a the ENDOR lines are due to the nearest
⁷⁵As neighbours, in Fig. 37b due to the second nearest ⁷⁵As neighbours
(56,57). Their angular dependence could be measured and analysed.

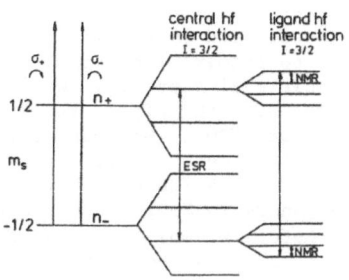

Fig. 36. Level scheme to explain the mechanism of the ODENDOR
experiment (see text).

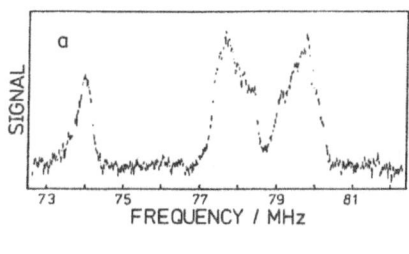

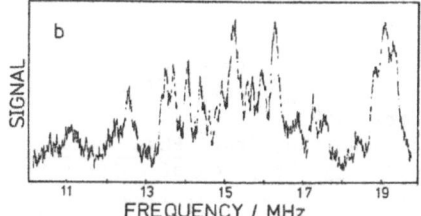

Fig. 37. (a) Part of the ENDOR spectrum of AsAs$_4$ antisite defects
in 'as-grown' s.i. GaAs:Cr resolving the shf interaction
of nearest ^{75}As neighbours. ($B_o \parallel 3^o$ off [110], B_o = 989 mT,
ν_{ESR} = 24.1 GHz, T = 1.6 K, measured at λ = 1320 nm),
(b) ENDOR lines due to second nearest ^{75}As neighbours.
After (56,57).

It turned out that in GaAs:Cr a regular antisite defect with 4 equivalent
next neighbours exists as well as distorted antisite defects in approximately the same concentration (56). Attempts to measure the conventional
ENDOR on this sample failed due to the very low signal to noise ratio of
the ESR (see also section 8).

Unlike in conventional ENDOR it was found that the ENDOR effect is of
the same order as the ESR effect, which is apart from the sensitivity
enhancement through the optical detection a great advantage compared to
conventional ENDOR, where the effect is usually smaller than 1% of the ESR
effect. However, quantitatively, the size of the observed signals is not
yet understood. It should also be mentioned, that a technical difficulty
is to bring enough rf power into the cavity, which is at 1.4 K pumped helium
to saturate the NMR transitions. Low temperature is necessary according
to equ. (7.7) to increase the MCD and one works with less noise if the
temperature is below the λ point of helium. (The MCD can be very small).

Another advantage of the optically detected ENDOR is that similarly
to the 'MCD-tagged by ESR' one can also measure the MCD tagged by 'ENDOR'.
This experiment was successfully done in the case of AsAs$_4$ antisite defects
and shows that indeed the total MCD of Fig. 29 belongs to the antisite
defects and that they have two optical intracenter transitions
(see Fig. 33) (49). This possibility is particularly interesting in that
it enables to establish a direct correlation between optical bands and
details of the defect structure as revealed by the highly resolving ENDOR
method.

Conclusion. It was the aim of this article to show that for the
structure determination of point defects the use of multiple magnetic
resonance methods are very powerful and that the availability of modern
experimental techniques makes it possible to tackle also difficult problems

239

interesting for materials science. The development of these methods is by no means finished. In particular, the optical method can be extended further to time resolved measurements and to experiments, where a spatial resolution of defect distributions can be measured. The increase in sensitivity can be driven further, which is of particular interest for defect studies in thin layers or interfaces.

REFERENCES

1. W.B. Fowler, Ed., Physics of Color Centers, Acad. Press, N.Y. and London (1968)
2. W. Gellermann, K.B. Koch and F. Lüty, Laser Focus (April) 71, 1981
3. G.H. Dieke, Spectra and Energy Levels of Rare-Earth Ions in Crystals, Interscience, New York 1968
4. A.E. Hughes and W.A. Runciman, Proc. Phys. Soc. London, 86, 615 (1965)
5. D.B. Fitchen, Chapter 5 in: "Physics of Color Centers", Ed., W.B. Fowler, Acad. Press, N.Y. and London, 1968
6. A.E. Hughes, "Optical techniques and an introduction to the symmetry properties of point defects' in: "Defects and their Structure in Non-metallic Solids", Eds. B. Henderson and A.E. Hughes, Plenum Press, Nato Advanced Study Inst. Series B, 19 (1976)
7. D.L. Dexter, C.C. Klick and G.A. Russell, Phys. Rev. 100, 603 (1955)
8. R.H. Bartram and A.M. Stoneham, Solid State Comm. 17, 1593 (1975)
9. G.E. Pake and T.L. Estle, The Physical Principles of Electron Paramagnetic Resonance, W.A. Benjamin, Inc., Reading, Mass., 1973
10. C.P. Slichter, Principles of Magnetic Resonance, Harper and Row, N.Y. 1963
11. H.G. Grimmeiss, E. Janzen, H. Ennen, O. Schirmer, J. Schneider, R. Wörner, C. Holm, E. Sirtl and P. Wagner, Phys. Rev. B 24, 4571 (1981)
12. E. Goovaerts, J.A. Andriessen, S.V. Nistor and D. Schoemaker, Phys. Rev. B 24, 29 (1981)
13. J.-M. Spaeth, "Atomic hydrogen as a model defect in alkali halides" in: "Defects in Insulating Crystals", Eds. V.M. Tuchkevich and K.K. Shvarts, Springer, Berlin, Heidelberg, New York 1981, p. 232
14. G. Feher, Phys. Rev. 114, 1219, 1249 (1959)
15. H. Seidel, Z. Physik 165, 218, 239 (1961)
16. G. Heder, J.R. Niklas and J.-M. Spaeth, phys. stat. sol. (b) 100, 567 (1980)
17. Y. Ueda, J.R. Niklas, J.-M. Spaeth, R. Kaufmann and J. Schneider, Solid State Comm. 46, 127 (1983)
18. S. Greulich-Weber, J.R. Niklas, E. Weber and J.-M. Spaeth, Phys. Rev. B 30, 6292 (1984)
19. J.R. Niklas, Habilitationsschrift, Paderborn 1983
20. M.U.A. Bromba and Horst Ziegler, Analytical Chemistry 51, 1760 (1979)
21. H. Ziegler, Applied Spectroscopy 35, 88 (1981)
22. M.U.A. Bromba and H. Ziegler, Analytical Chemistry 55, 648 (1983)
23. U.U.A. Bromba and H. Ziegler, Analytical Chemistry 56, 2052 (1984)
24. Y. Ueda, J.R. Niklas and J.-M. Spaeth, Solid State Comm. 46, 121 (1983)
25. S. Greulich-Weber, J.R. Niklas and J.-M. Spaeth, J. Phys. C: Solid State Phys. 17, L911 (1984)
26. H. Söthe, P. Studzinski and J.-M. Spaeth, phys. stat. sol. (b) 130, 339 (1985)
27. H. Söthe, P. Studzinski and J.-M. Spaeth, to be published
28. Ch. Hoentzsch and J.-M. Spaeth, phys. stat. sol. (b) 94, 479 (1979)
29. P. Studzinski, J.R. Niklas and J.-M. Spaeth, phys. stat. sol. (b) 101, 673 (1980)
30. P. Studzinski and J.-M. Spaeth, Radiation Effects 73, 207 (1983)
31. P. Studzinski, Dissertation, Paderborn 1985
32. J.R. Niklas and J.-M. Spaeth, phys. stat. sol. (b) 101, 221 (1980)
33. J.-M. Spaeth and J.R. Niklas, Condensed Mater Physics 1, 393 (1981)

34. R.C. DuVarney, J.R. Niklas and J.-M. Spaeth, phys. stat. sol. (b) 97, 135 (1980)
35. R. Biehl, M. Plato and K. Möbius, J. Chem. Phys. 63, 3515 (1975)
36. N.S. Dalal and C.A. McDowell, Chem. Phys. Letts. 6, 617 (1970)
37. J.R. Niklas, R.U. Bauer and J.-M. Spaeth, phys. stat. sol. (b) 119, 171 (1983)
38. R.C. DuVarney, J.R. Niklas and J.-M. Spaeth, phys. stat. sol. (b) 128, 673 (1985)
39. P. Edel, C. Hennies, Y. Merle d'Aubigné, R. Romestain and Y. Twarowsky, Phys. Rev. Letts. 28, 1268 (1972)
40. P. Dawson, C.M. McDonagh, B. Henderson and L.S. Welch, J. Phys. C: Solid State Phys. 11, L983 (1978)
41. F.J. Ahlers, F. Lohse and J.-M. Spaeth, Solid State Comm. 43, 321 (1982)
42. J.D. Dunstan and J.J. Davies, J. Phys. C: Solid State Phys. 12, 2927 (1979)
43. K.P. O'Donnell, M.K. Lee and G.D. Watkins, Solid State Comm. 44, 1015 (1982)
44. N. Killoran, B.C. Cavenett, M. Godlewski, A.T. Kennedy and N.D. Wilsey, Physica B 116, 425 (1982)
45. B.K. Meyer, Th. Hangleiter, J.-M. Spaeth, G. Strauch, Th. Zell, A. Winnacker and R.H. Bartram, J. Phys. C: Solid State Phys. 18, 1503 (1985)
46. D. Block, A. Hervé and R.T. Cox, Phys. Rev. B 25, 6049 (1982)
47. L.F. Mollenauer and S. Pan, Phys. Rev. B 6, 772 (1972)
48. F.J. Ahlers, F. Lohse, J.-M. Spaeth and L.F. Mollenauer, Phys. Rev. B 28, 1249 (1983)
49. B.K. Meyer, J.-M. Spaeth and M. Scheffler, Phys. Rev. Letts. 52, 851 (1984)
50. J.-M. Spaeth and B.K. Meyer, Festkörperprobleme XXV, 614 (1985)
51. F.J. Ahlers, Dissertation, Paderborn 1985
52. F.J. Ahlers, F. Lohse and J.-M. Spaeth, J. Phys. C: Solid State Phys. 18, 3881 (1985)
53. F.J. Ahlers, F. Lohse, Th. Hangleiter, J.-M. Spaeth, R.H. Bartram, J. Phys. C: Solid State Phys. 17, 4877 (1984)
54. B.K. Meyer, D. Hofmann and J.-M. Spaeth, to be published
55. U. Kaufmann and J. Schneider, Festkörperprobleme XX, 87 (1980)
56. D.M. Hofmann, B.K. Meyer, F. Lohse and J.-M. Spaeth, Phys. Rev. Letts. 53, 1187 (1984)
57. J.-M. Spaeth, D.M. Hofmann and B.K. Meyer, Proc. of the Mat. Research Society Conf. in San Francisco 1985.

PRINCIPLES OF NMR : ITS USES IN DEFECT STUDIES

J. H. Strange

Physics Laboratory
University of Kent
Canterbury, Kent, U.K.

INTRODUCTION

Nuclear magnetic resonance (NMR) is a well established and powerful technique for the study of molecular motions. In crystalline solids the movement of molecules, particularly the translational motion which results in self-diffusion, frequently occurs by mechanisms involving defects that are present. As discussed elsewhere in this volume, studies of mass transport properties in crystals provide very valuable insight into their defect structure. The various techniques of line-width and relaxation time measurement in NMR can therefore give very useful information which often complements data from other techniques (1). The methods of NMR can tell us about large scale motions, related to radiotracer diffusion and electrical conductivity in ionic crystals. They are also sensitive to localized movement (2,3) such as motion of atoms bound to defects giving information related to that obtained from other techniques such as dielectric relaxation and ionic thermocurrent (ITC). The NMR techniques are also nucleus specific so that we may distinguish which atomic species is mobile and even find out about systems where more than one type of atomic motion exists among the same (chemical) atomic species due, for example, to the existence of inequivalent lattice sites (4). Obvious advantages of NMR are that it can be used to study a very wide range of motional frequencies (see figure 1), so that the one basic

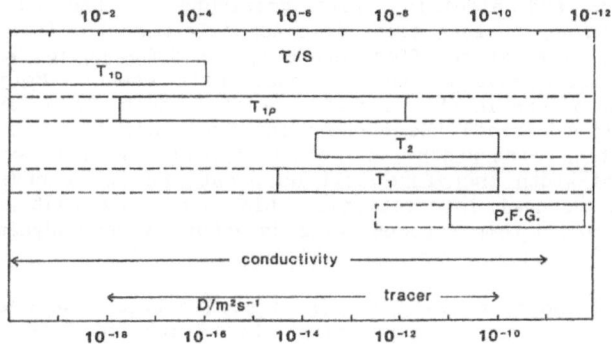

Figure 1. Ranges of applicability of the various techniques for studying diffusion. The upper scale is τ, the mean residence time for ions on a lattice site. The lower scale is D, the diffusion coefficient.

technique can be used to study specific elements in a specific sample non-destructively over a very wide temperature (or pressure) range.

The techniques of NMR are probably more familiar as tools of the analytical chemist for the determination of composition and structure in liquids. Recently attention has been directed towards special techniques for similar studies in the solid state (reviewed by Chadwick (5)) which opens the possibilities of defect structure determination at least in systems of high defect concentration. Booth and McGarvey (6) have already shown that it is possible in ionic solids, using paramagnetic interactions, to detect and characterise defect structures and provide evidence for clustering of rare-earth ions in CaF_2 from line shape analysis. Very elegant and sensitive nuclear double resonance experiments were used long ago by Hartland (7) to investigate Na^+ ions substituted in KF but this method has not been generally exploited, probably because of its complexity. NMR alone is certainly useful but when measurements by various techniques are employed together with NMR on the same samples the combination can become very powerful indeed.

BASIC PRINCIPLES OF NMR

The nuclei of many elements possess nuclear spin $I\hbar$ (quantum number I) and a collinear magnetic moment μ, which are related by $\mu = \gamma\hbar I$ where γ is the magnetogyric ratio characteristic of the nucleus. The observable component of spin (in the z direction, say), $I_z\hbar$, takes values given by the magnetic quantum number m which can have $(2I+1)$ values from $-I$ to $+I$ in integer steps. When placed in a magnetic field B_0 the magnetic dipoles align in the magnetic field, each with energy $\mu.B_0 = -\gamma I_z\hbar B_0 = -\gamma m\hbar B_0$. This then produces energy levels separated by $\gamma\hbar B_0$ which, in a sample containing many nuclei at thermal equilibrium, are populated according to the Boltzmann distribution. The NMR experiment basically consists of detecting transitions between these levels. These can be produced by electromagnetic radiation at the resonant frequency ω_0 (the Larmor frequency) where $\hbar\omega_0 = \hbar\gamma B_0$, i.e. $\omega_0 = \gamma B_0$.

A simple absorption experiment can be performed whereby the frequency of the radiation, ω, (or the main magnetic field B) is swept through resonance ($\omega_0 = \gamma B_0$) which is detected by the absorption of energy. The radiation is normally in the form of an oscillating magnetic field, denoted by B_1. The resonance absorption line will have a finite width, ΔB, due to small variations in the energy levels of the nuclei. Two common causes for this variation are local magnetic fields, B_{loc}, produced by nuclear magnetic dipole-dipole interactions (C.P. Slichter, Chapter 3) (8) and electric field gradients at the nuclear sites interacting with any nuclear electric quadrupole moments, (C.P. Slichter, Chapter 9) (8). The latter effect can only exist for nuclei with $I > 1/2$ as spin 1/2 nuclei have no nuclear quadrupole moment. Motion of the nuclei, such as occurs in atomic self-diffusion, produces an averaging of the local field (provided the correlation frequency for motion exceeds the rigid lattice line-width) and a (motional) line-narrowing occurs. Studies of line-width therefore tell us about the local environment of the nuclei (structural information). Line-narrowing that occurs, for example, at elevated temperatures tells us about the microdynamics of the atoms.

As stated above, nuclei are distributed between the energy levels and if the equilibrium population is disturbed, the Boltzmann distribution is re-established by an exchange of energy between the nuclei and their surroundings ("lattice") in a characteristic **spin-lattice relaxation time**, T_1. If, for example, all magnetization M_z were destroyed M_z would grow back according to

244

$$M_z(t) = M_0(1 - e^{-t/T_1})$$

were M_0 is the equilibrium z magnetization equal to $\Sigma\mu_z$. Since nuclei remain in a particular energy level no longer than T_1 (on average) the minimum line-width is determined by the Uncertainty Principle $\Delta E \Delta t \gtrsim h$ so that $\Delta\nu \gtrsim 1/T_1$. Except for very mobile atoms such as found in liquids the other contributions to $\Delta\nu$ usually dominate. In fact we can define another relaxation time, the transverse or **spin-spin relaxation time** T_2 in terms of the dephasing of the nuclei produced by the local fields, i.e. related to the line-width $\Delta\nu$ so that

$$\Delta\nu \cong 1/T_2.$$

(For a Lorentzian line shape, the transverse relaxation is exponential $M_{xy} = M_{xy}(0)\exp{-t/T_2}$, and $\Delta\nu = 1/\pi T_2$). From the discussion above, it is clear that $T_1 \gtrsim T_2$. Two other relaxation times $T_{1\rho}$ and T_{1D} are less commonly used but are particularly useful for the study of molecular motion in solids. $T_{1\rho}$, known as the spin-lattice relaxation time in the **rotating frame** measures the relaxation of spins among energy levels produced by quantization in a resonant rotating magnetic field B_1 applied at ω_0 in a direction perpendicular to the static field B_0. Spin quantization in solids can also be defined in the direction of the local dipolar fields and reordering of spins in these fields is characterized by the **dipolar relaxation time** T_{1D}.

ATOMIC MOTION

The movements of atoms containing magnetic nuclei relative to one another will affect the NMR parameters and often provide a very convenient method for studying these movements. The motion is usually characterized by a correlation time, τ_c, which may be thought of as the mean time between atomic jumps. The relationship between τ_c and the experimentally determined relaxation parameters will now be discussed.

Motional Narrowing and T_2

The narrowing of the NMR line-width, $\Delta\nu$, mentioned above, occurs when $\tau_c < T_{2r\ell}$. The "rigid-lattice" value of T_2 is typically 10^{-4}s. In this region

$$\Delta\nu \propto 1/T_2 \propto \tau_c.$$

For thermally activated motion $\tau_c = \tau_0 \exp E_a/kT$ and measurement of T_2 (or $\Delta\nu$) as a function of temperature T allows determination of the activation energy E_a.

Spin-Lattice Relaxation

If the Boltzmann distribution of spins between their energy levels has been upset by resonant absorption, relaxation will occur by stimulated emission. The mechanism for this to happen requires a magnetic field fluctuating at ω_0 to provide matching photons of $\hbar\omega_0$. One possible source of such a field can arise from the modulation of the local fields due to the dipole-dipole interaction by relative motion of the atoms. Thermal movement provides a randomly fluctuating field whose frequency components are similar to a 'noise' spectrum (figure 3).

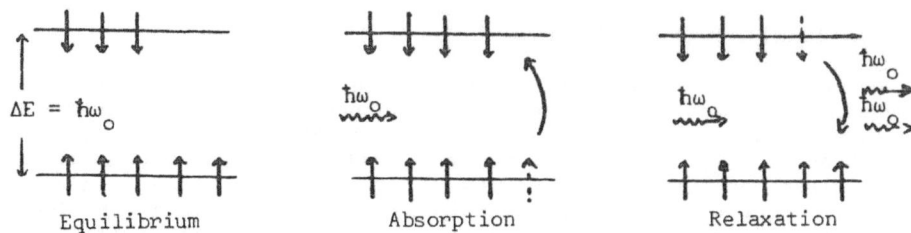

Figure 2. Energy levels (I=1/2) populated by nuclear spins. Upper and lower levels are m = -1/2 and m = +1/2 respectively.

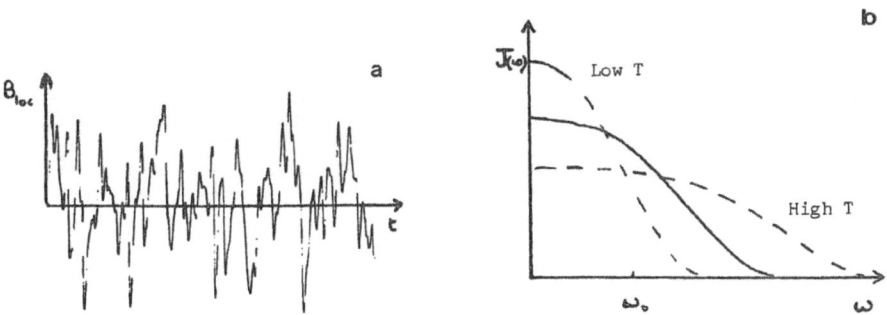

Figure 3(a). Local field fluctuations produce a "noise" spectrum.

(b) Spectral density of local field fluctuations.

If we analyse this spectrum into its Fourier components we obtain a spectral density function, $J(\omega)$, (figure 3) which depends on the atomic motion (i.e. the temperature of the sample **and** its defect structure). Bloembergen, Purcell and Pound (BPP) (9,10) showed that such fluctuations, if characterized by an exponential correlation function, produce a relaxation rate given by an expression of the form *

$$\frac{1}{T_1} \propto J(\omega_O) \propto B_{loc}^2 \frac{\tau_C}{1 + \omega_O^2 \tau_C^2}$$

where B_{loc} here is that part of the local field modulated by the motion. This function has a maximum (T_1 minimum) when $\omega_O \tau_C = 1$. Typically, values for ω_O are about 10^8 to 10^9 s^{-1} so T_1 relaxation is sensing fluctuations (atomic jumps) occurring at this frequency. Similarly $T_{1\rho}$ senses a spectral density $J(\omega_1)$ where $\omega_1 = \gamma B_1$, which is typically 10^5 to 10^6 s^{-1} (10) (Farrar & Becker, Ch.4), and is of the form*

$$\frac{1}{T_{1\rho}} \propto J(\omega_1) \propto B_{loc}^2 \frac{\tau_C}{1 + \omega_1^2 \tau_C^2} \quad .$$

---

*The precise form of these equations depends on the details of the relaxation mechanism (8,10)

This parameter is therefore sensitive to much slower motion and is particularly valuable in the study of solids. $T_{1\rho}$ exhibits a minimum when $\omega_1\tau_c \cong 1$. The dipolar relaxation time, T_{1D}, is usually studied in the long correlation time limit, $\tau_c > T_{2r.\ell}$, complementing T_2 measurements and has the general form

$$T_{1D} = K \tau_c$$

where K is a constant of order unity and depends on the detail of the motion (11). A classic set of data (12) for ^{19}F relaxation in BaF_2 as a function of temperature is shown in figure 4.

Full interpretation of relaxation time data normally requires a model for the atomic motion. The assumption of an exponential correlation function is then unnecessary; the precise function can be calculated from the model and τ_c defined in terms of the mean residence time, τ, of an atom or ion on a lattice site. An example here is the fluorine ion diffusion in fluorites (13) and this theoretical approach can be used to evaluate τ and hence the diffusion coefficient D using

$$D = f\frac{\langle r^2\rangle}{6\tau}$$

where $\langle r^2\rangle$ is the mean atomic jump distance and f is a factor of order unity, depending on lattice geometry and diffusion mechanism. The assumed model can then be checked against macroscopic measurements of D by, for example, radiotracer or conductivity methods and D calculated

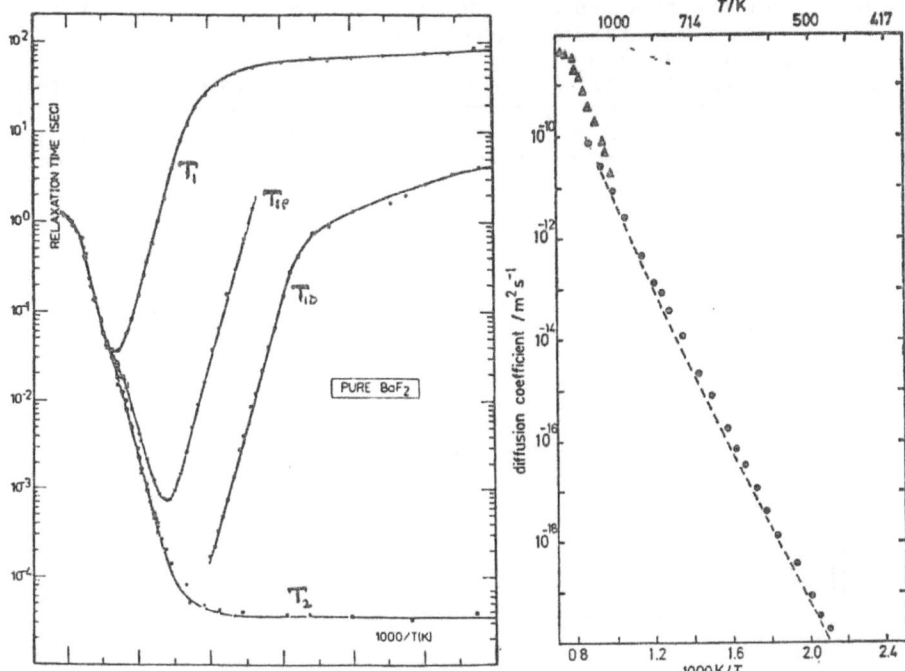

Figure 4
NMR relaxation times T_1, $T_{1\rho}$ T_1 and T_2 as a function of inverse temperature for ^{19}F in BaF_2.

Figure 5
Temperature dependence of F^- diffusion coefficients in BaF_2: crystals calculated from o n.m.r. relaxation; Δ n.m.r. pulsed field gradient; ---- ionic conductivity.

when a vacancy diffusion mechanism is assumed. Such measurements allow a detailed test of the diffusion mechanism, confirmation of diffusion/conduction being a bulk process, identification of the mobile species (F^- in this case) and evaluation of defect parameters such as formation and migration energies. Furthermore, motion of any 'neutral' defects would here have shown up as a major discrepancy between the two types of measurement.

EXPERIMENTAL METHODS

Measurement of T_2

It has already been explained that T_2 is inversely related to the absorption line-width. Now T_2 is the dephasing time of the transverse magnetization and can be mesured directly using pulsed methods. A short pulse of resonant magnetic field, B_1, can be made to tip the macroscopic equilibrium nuclear magnetization from the z-direction (B_0) into the x y plane. This is called a 90° pulse (Farrar & Becker, Chapter 2). The transverse magnetization will now precess about B_0 at the Larmor frequency, ω_0, (which is the same frequency as that required for resonant absorption as described previously). The local fields will produce variations in precessional frequencies and the transverse magnetization will decay as the nuclei dephase. The precessing magnetization can be made to induce a small voltage in a coil surrounding the sample and, when detected, this gives a decaying signal proportional to $M_{xy}(t)$, known as the Free Induction Decay (FID) (figure 6). The signal decays with the characteristic transverse or spin-spin relaxation time T_2 which can be measured from the free induction decay. We have here assumed that no dephasing occurs due to inhomogeneities in B_0. If such affects are present they can be cancelled by using a spin-echo pulse sequence in which a 180° pulse is applied at a time τ following the initial 90° pulse (figure 7). This effectively reverses the coherent dephasing due to B_0 field variation and results in the formation of a spontaneous signed or spin-echo (Farrar and Becker, Chapter 2) at time 2τ whose amplitude is determined by T_2.

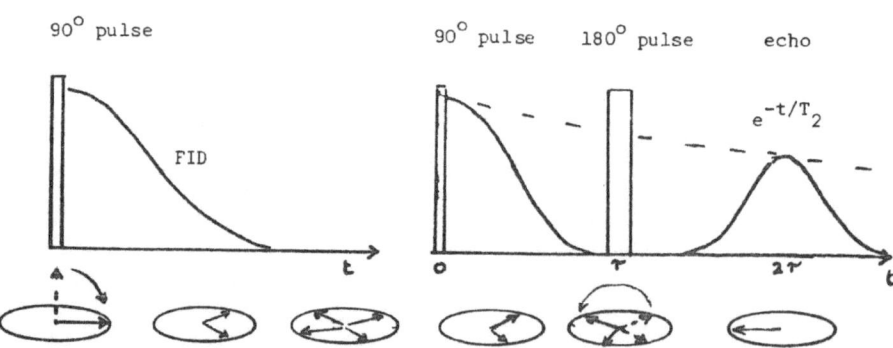

Figure 6. Free Induction Decay schematic shows nuclei (in rotating frame of reference).

Figure 7. Spin Echo produced by 90° τ 180° sequence.

Measurement of T_1

Several pulse sequences can be used to measure spin-lattice relaxation (10) but the simplest is the 90° τ 90° sequence. The first pulse, in tipping the magnetization from the z into the xy direction, destroys the z component of magnetization, M_z. This is equivalent to equalizing the populations of the energy levels. M_z is re-established by spin-lattice relaxation to reach its equilibrium value M_O according to

$$M_z(t) = M_O(1 - \exp{-t/T_1}).$$

The second 90° pulse is used to observe the magnitude of M_z at time t = τ since the amplitude of the free induction decay is proportional to M_z at the time of the pulse. To measure T_1 the signal following the second 90° pulse is recorded for various values of τ and T_1 can they be determined.

Measurement of $T_{1\rho}$ and T_{1D}

The measurement of $T_{1\rho}$ is always made in the presence of a resonant r.f. magnetic field B_1 which rotates in the xy plane at the precession frequency of the nuclei, i.e. at the same rotation frequency of the transverse magnetization. The transverse magnetization is produced with a 90° pulse which is followed immediately by a long B_1 field pulse (of length τ) aligned with M_{xy} which "spin-locks" the magnetization (Farrar & Becker, Chapter 6). The equilibrium magnetization along B_1 is essentially zero and M_{xy} decays with a spin-lattice relaxation mechanism according to

$$M_{xy}(\tau) = M_{xy}(0) \exp - \tau/T_{1\rho}.$$

The value of $M_{xy}(\tau)$ is obtained from the amplitude of the free induction decay following the B_1 pulse and $T_{1\rho}$ obtained from measurements at various τ values.

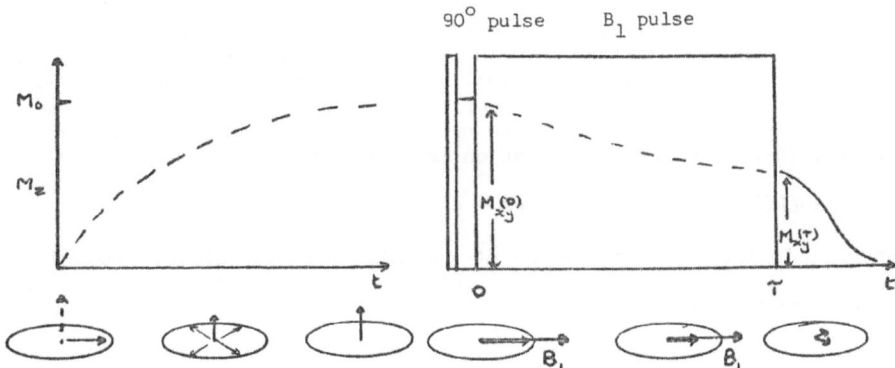

Figure 8. Recovery of magnetization   Figure 9. Pulse sequence for the
M_z following a 90° pulse.   measurement of $T_{1\rho}$.

T_{1D} is equivalent to $T_{1\rho}$ measured in zero field B_1 and may be measured in a similar manner to $T_{1\rho}$ but with an adiabatic demagnetization and remagnetization during the τ period, i.e. B_1 is reduced to zero and then at the end of the τ period it is restored (14). Alternatively a special pulse sequence, 90° τ 45° t 45°, known as the Jeener-Broekaert sequence (15) may be used where the second r.f. pulse is phase shifted by 45° from the others.

Direct NMR measurement of diffusion

If self-diffusion is rapid ($D \geq 10^{-12}m^2s^{-1}$) it is possible to measure the diffusion coefficient directly by an NMR method that is equivalent to a tracer diffusion experiment. This, then, is a measure of the macroscopic atomic migration as opposed to the microscopic jump frequency that is detected in the relaxation experiments described previously. The method relies on identifying the spatial distribution of spins using a known magnetic field gradient, G, superimposed on the main field B_0. If the nuclear spins are prepared with a 90° pulse so that their transverse components of magnetic moment are all in phase they precess about B_0 with the Larmor frequency ω_0. Suppose that a field gradient, G, is now applied across the sample. Spins in different locations will experience different fields. Since $\omega_0 = \gamma B_0$ the precessional frequencies will differ and the spins will dephase in a manner determined by G. This dephasing can be reversed by applying a second pulse (a 180°) at a time τ later which produces a spin-echo at time 2τ as mentioned above. The echo amplitude, A, is determined by the transverse (spin-spin relaxation) decay of the nuclear magnetization so that

$$A(2\tau) \propto \exp - (2\tau/T_2).$$

This assumes all coherent dephasing is refocussed. If, during the time of the experiment the atoms change their positions in the field gradient the dephasing and rephasing processes will not match, refocussing will be incomplete and the echo amplitude will be attenuated. It has been shown (16) that an additional term enters the equation for T_2

$$A(2\tau) \propto \exp[-(2\tau/T_2) - 2\gamma^2G^2D\tau^3/3]$$

and therefore D can be measured directly.

It would appear that D can be measured over a wide range by varying G and τ appropriately. In fact experimental limitations make τ values of less than $100\mu s$ and G values greater than $0.1Tm^{-1}$ very difficult to use. For slower diffusion rates (as for example in mobile solids) a modified experiment using a pulsed gradient has been developed by Stejskal and Tanner (17) which significantly extends the range of the simple experiment. Large field gradients, g, are applied for a short time, d, between the 90° and 180° pulse and between the 180° pulse and echo (figure 10). The echo attentuation is then given by

$$\ln[A(2\tau)/A(0)] = -\gamma^2Dd^2Tg^2$$

where T is the time separation of the two gradient pulses. Much larger field gradients can be used in this case and the technique has been successfully applied to the study of ionic diffusion in solids (18).

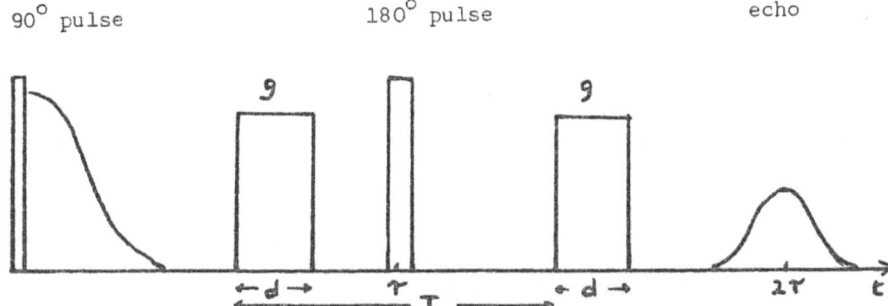

Figure 10. Pulsed field gradient sequence for diffusion measurement.
g indicates the field gradient pulses.

SOME DEFECT STUDIES

Point Defects

The study of point defect concentrations and mobility by NMR
relaxation time and direct diffusion measurement has already been
mentioned. The dipolar interaction, modulated by atomic diffusion was
shown to allow diffusion rates and defect parameters to be obtained and
the example of F^- diffusion in fluorite crystals was given. ^{19}F has a
nuclear spin I = 1/2 and therefore has no nuclear electric quadrupole
moment. An additional relaxation mechanism can enter for quadrupolar
nuclei since electric field gradients at neighbouring atomic sites will
clearly be produced by point defects in an ionic crystal. Fluctuations
in this interaction produced, for example, by defect motion will provide
a relaxation mechanism in a similar way to the modulation of the
dipole-dipole interaction. When present the quadropolar mechanism is
usually very powerful and dominates other mechanisms. Two recent
examples clearly demonstrate its great value. Kanert and Mali [19] have
shown that the localized movement of off-centre Ag^+ ions in RbCl may be
studied from 1.4K to 50K, including the effects of imposing an external
electric field [20]. Barton and Seymour [21] have developed a mean-field
theoretical treatment for quadrupolar relaxation due to self-diffusion
and applied it to the motion of deuterons in transition metal deuterides.

Paramagnetic centres have (electronic) magnetic moments which are
about three orders of magnitude greater than nuclear magnetic moments.
They therefore produce very large local fields. This can be exploited in
two ways to study defects in solids. Firstly, the nuclei surrounding the
paramagnetic impurity suffer a large resonant frequency shift due to the
local field. Measurement of the NMR spectra gives a direct measure of
the defect environment and Booth and McGarvey [6] were able to
demonstrate cluster formation in fluorites doped with a variety of
paramagnetic rare-earth ions. The defect structure could be clearly
explored. Secondly paramagnetic centres have a profound effect on
relaxation rates and this mechanism is often present due to residual
paramagnetic impurities in crystals. It will become evident as a weakly
temperature dependent contribution to T_1 when other mechanisms such as
the diffusion modulated dipolar interaction become weak. This is thought
to be the reason for the low temperature T_1 behaviour in figure 1. This
normally unwanted effect can be turned to great advantage as a sensitive
probe of atomic motion for both long range diffusion [21] and localized
defect-bound movement [4].

In order to measure defect parameters, such as formation and migration energies, temperature dependence studies are undertaken. Another important but less exploited variable is pressure. Variations in relaxation times ($T_{1\rho}$) with pressure have proved particularly valuable for molecular crystals (22) where activation volumes equal to one molar volume were consistently found for many solids, indicating a vacancy mechanism for self-diffusion. The large size of organic molecules (and hence vacancies) results in a large dependence of diffusion rates on pressure (Figure 11) (23). A variant on this is the pressure-quench experiment by which formation and migration volumes and annealing rates can be determined (24). Very little has been reported on pressure dependent NMR studies in ionic solids.

Refinements in the theoretical and experimental techniques of NMR relaxation have led to the determination of diffusion of atoms among inequivalent sites in solids. In crystals of LaF_3, for example, the structure approximates to one in which fluorine ions occupy two sets of equivalent sites. NMR shows that ions on one site are more mobile. A general theoretical formalism has been developed (4) for movement on and between inequivalent sites and has been applied to the case of LaF_3. It was possible to measure motion on and exchange rates between the two sub-lattices. It is also possible to detect anisotropic diffusion (25). In the superionic conductor, lithium nitride, there are two lithium ion sites and a two-dimensional layered structure. Motion within layers involving one site and interlayer motion parallel to the hexagonal axis involving both sites were determined and vacancy parameters for the intra-layer motion obtained. The study used single crystals and observed both ^6Li and ^7Li resonances. A detailed analysis in terms of dipolar and quadrupolar interactions proved very productive.

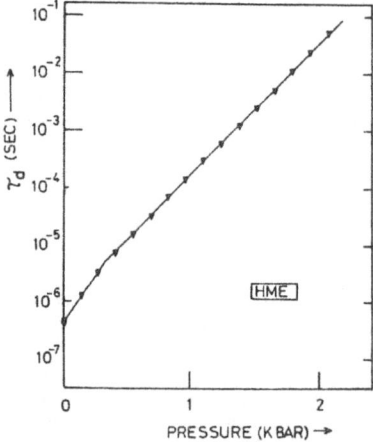

Figure 11. The pressure dependence of the correlation time for molecular self-diffusion, τ_d, in the molecular crystal hexamethylethane at 347K, determined from $T_{1\rho}$.

Dislocations

Motion of dislocations does not normally produce relaxation by modulation of the magnetic dipole-dipole interaction. Dislocations in ionic solids do, however, produce, large distortions in local electric fields and this may be observed in NMR via interaction with the nuclear electric quadrupole moment (for $I > 1/2$). Common methods of observing dislocations are decoration, etch-pit and transmission electron microscopy. These methods require transparent crystals or are surface observations only. They also usually require immobile defects. Pulsed NMR has been developed as a powerful method for studying dislocation density but, more dramatically, Kanert and co-workers have measured mean free paths of dislocation movement and mean times between jumps.

Relaxation can be induced by fluctuation in the electric field gradient tensor surrounding a dislocation when that dislocation moves. This is similar in principle to the relaxation mechanism for self-diffusion via point defects in ionic solids containing quadrupolar nuclei. The additional dynamic study, undertaken for example in polycrystalline aluminium metal and in alkali halide crystals, involves relaxation measurement as a function of strain rate, strain, compression axis, temperature and impurity content. This work is described in an excellent recent view on dislocations and NMR by Hosson, Kanert and Sleeswyk (26). The techniques of NMR examine the bulk of the crystals and are sensitive to the microdynamics of atomic motion. We can expect to learn much more about plastic deformation and the microstructure of metals, alloys and ionic crystals.

Future developments

It is clear that experimental technique and theory are now developing to give reliable and consistent information about both point and line defects. The potential for study and application of NMR in this area are enormous, but considerable investment in time and money is required to benefit properly from exploitation of these techniques. Temperature dependence studies are far more numerous than pressure dependence work and we might expect a greater effort in this direction in the future.

We are only now seeing commercial instruments become available for high-resolution NMR studies in solids using magic-angle - spinning of the sample. This could provide a fertile field for the spectroscopic study of structures in defective solids, although high concentrations of defects will be necessary to overcome the inherent insensitivity of the technique.

Another development, that is even farther in the future, is to employ magnetic resonance imaging to the study of solids. This, if made to work, will provide a non-destructive method for the study of both structural and dynamic properties. We are tackling this problem in our own laboratories. The technical problems are considerable but the potential is enormous and we are optimistic that NMR will continue to grow and become even more important in the study of the solid state.

REFERENCES

1. A.V. Chadwick. This volume : (1985).
2. S.H.N. Wei and D.C. Ailion, Phys.Rev.B. 19 : 4470 (1979).
3. A.F. Aalders, H.G.M. Lochs, A.F.M. Arts and H.W. de Wijn,
    J.Phys.C : (to be published).
4. G.A. Jaroszkiewicz and J.H. Strange, J.Phys.C. 18 : 2331 (1985).
5. A.V. Chadwick, Int.Rev.Phys.Chem. (to be published).
6   R.J. Booth and B.R. McGarvey, Phys.Rev.B. 21 : 1627 (1980).
7   A. Hartland, Proc.Roy.Soc.A. 304 : 361 (1968).
8   C.P. Slichter, Principles of Magnetic Resonance (Springer-Verlag),
    (1978).
9. N. Bloembergen, E.M. Purcell and R.V. Pound, Phys.Rev. 73 : 679
    (1948).
10. T.C. Farrar and E.D. Becker, Pulse and Fourier Transform NMR
    (Academic Press) (1971).
11. D.C. Ailion, Advances Mag.Res. 5 : 177 (1971).
12. D.R. Figueroa, A.V. Chadwick and J.H. Strange, J.Phys.C. 11 : 55
    (1978).
13. D.R. Figueroa, J.H. Strange and D. Wolfe, Phys.Rev.B. 19 : 148
    (1979).
14. C.P. Slichter and D.C. Ailion, Phys.Rev. 135 A : 1099 (1964).
15. J. Jeener and P. Broekaert, Phys.Rev. 157 : 232 (1967).
16. H.Y. Carr and E.M. Purcell, Phys.Rev. 94 : 630 (1954).
17. E.O. Stejskal and J.E. Tanner, J.Chem.Phys 42 : 288 (1965).
18. R.E. Gordon and J.H. Strange, J.Phys.C. 11 : 3213 (1978).
19. O. Kanert and M. Mali, Phys.Lett. 69A : 344 (1979).
20. O. Kanert and R. Kuchler, Radiation Effects, 73 : 37 (1983).
21. P.M. Richards, Phys.Rev.B. 18 : 6358 (1978).
22. J.M. Chezeau and J.H. Strange, Physics Reports 53 : 1 (1979).
23. S.M. Ross and J.H. Strange, J.Chem.Phys. 68 : 3078 (1978).
24. S.M. Ross and J.H. Strange, Molec.Phys. 32 : 775 (1976).
25. D. Brinkman, M. Mali, J. Roos, R. Messer and H. Birli, Phys.Rev.B.
    26 : 4810 (1982).
26. J.Th.M. De Hosson, O. Kanert and A.W. Sleeswyk, Dislocations in
    Solids (North Holland) (Ed. F.R.N. Nabarro), Chapter 32 (1983).

TRANSPORT STUDIES USING NMR

M. Terenzi

Chemistry Department
University of Calabria
Arcavacata di Rende, Rende, Italy

INTRODUCTION

A renewed interest has been devoted in the past several years to the investigation of diffusion in solids, mainly because of the influence transport phenomena have on a variety of solid-state properties of technological relevance, and possibly because of the challenging physics involved. The interest in this field is rapidly growing as evidenced by a number of international conferences being held at frequent intervals and by the large number of articles and reviews for which it is not possible to give a comprehensive and up-to-date bibliography. Possibly, from this point of view, a convenient suggestion is to refer the reader to reference [1], a periodical journal reporting the abstracts of up-to-date published papers on diffusion (in solids as well as in liquids). Considerable progress has been made with the development of new and convenient experimental techniques suitable for the study of diffusion in solids. In the same time the development of theoretical models and analysis afford a deeper insight of the diffusion processes.

At the present the study of diffusion in solids is performed by a variety of experimental techniques, usually classified as macroscopic (or direct) techniques (e.g. ionic conductivity, tracer methods, NMR spin-echoes in the presence of a magnetic field gradient), and microscopic (or indirect) techniques (e.g. NMR relaxation times , Mossbauer linewidths, incoherent quasi elastic neutron scattering). Essentially macroscopic techniques gives the direct measurement of the diffusion coefficient D (or of the self-diffusion coefficient D in "zero" driving force), and microscopic methods allows the determinations of atomic quantities related to the microscopic nature of diffusive motions. All techniques should be considered complementary rather than competitive, since all are limited in their applicability, depending on the material being studied and on different circumstances. As a matter of the fact most of the information concerning diffusion in solids come from experimental studies by the

255

different techniques, since normally their cooperative and complementary use may give comparative information enabling one to relate the atomic microscopic movement to the macroscopic flow of matter.

NMR AND DIFFUSION

Nuclear Magnetic Resonance (NMR), when applicable provides a simple experimental technique for charaterizing and measuring the rate of molecular motion (e.g. rotational and translational motions), as it has been recognized since the early period [2]. However the applications of NMR to study diffusion in solids has grown considerably only recently and is today frequently used in many laboratories interested in materials science. This is possibly due to the progress of experimental techniques, the availability of high-quality commercial apparatus and to the increasing theoretical understanding of the NMR diffusion experiments.[3]

The basic principles and techniques of NMR have been given elsewhere in this volume [4] and are described at length in a number of books [5-9]. The application of NMR to diffusion in solids are discussed in specialized reviews [10-11] that should be particularly useful to the reader. Therefore we will neither describe the NMR methods nor enter in the details of spin relaxation. Our main aim here is to give the individuals interested in diffusion studies in solids some appreciation of the scope and limitation of NMR techniques.

There are few restrictions to be accounted for by a researcher wishing to apply magnetic resonance to solid state diffusion. First, the sample must be transparent to radiofrequency, which restricts it to non metallic solid, or to powder metal and metallic thin film. Second, NMR is a rather insensitive experiment. The real problem is the signal-to-noise ratio (S/N), which, in a pulsed NMR experiment, is related to a number of parameters accounting for the electronics design the kind and the number of nuclear spins, their relaxation times. This ratio (S/N) can be estimated for a given solid sample. Here it is enough to observe that the NMR experiment requires about 10^{21} to 10^{23} spins. It is also difficult to observe nuclei with low gyromagnetic ratio γ, or with low concentration. (e.g. ^{17}O isotope in oxygen ion conductors, due to its extremely low natural abundance). However, in suitable circumstances, it is possible to study these nuclei, by investigating their effect on other nuclei [12]. The limitations due to low γ nuclei could possibly be overcome with the use of a superconducting magnet, which would allow experimenting on a higher magnetic field, therefore increasing the ratio S/N. At the present very few examples of NMR application to the direct study of low nuclei can be found in literature, although it is worthwhile mentioning the direct measurement of Ag self-diffusion constant, in superionic conductors, to be possibly considered the first application of the pulsed field gradient technique to a weak resonance [13]. Finally it must be considered that the linewidth in solid can be quite large, especially for nuclei with quadrupole moment (e.g.

256

I>1/2), sitting in a non-cubic environment, and therefore the signal-to-noise ratio can be very low. Within the above mentioned restrictions, NMR can be applied to the study of diffusion in different solid materials such as, for example, metals [14], ionic solids (see below), superionic solid [15], glasses [16], molecular crystals [17].

An important point to be considered is that NMR is successful in giving informations on the diffusion processes, only when the spin relaxation is controlled by translational motion. It is infact clear that the nucleus inside the materials is subject to a number of different interactions, and therefore a variety of relaxation mechanisms may be competitive over a large temperature range. One of the problems in NMR is to separate the contributions to relaxation due to diffusion-dependent and diffusion-independent mechanisms. This can be done, in suitable circumstances, but the procedure is neither always possible nor always obvious. In such a situation the information on diffusion are not available, a situation prevailing, for example, in solids containing sizeable amount of paramagnetic impurities or nuclei with large quadrupolar interactions.

However, in many solids in a sizeable temperature range, most frequently, the spin relaxation is controlled by translational motion, and therefore is mainly affected by the relative motion of nearby nuclei (due to the short range of interaction) allowing a microscopic picture of diffusion processes. In these situations a typical diffusion study by NMR will involve experimental measurements of relaxation times (T_1, T_2, $T_{1\varrho}$) as a function of temperature, as suggested elsewhere in this volume [4]. Therefore the most practicable region to be in is near a minimum of T_1 and $T_{1\varrho}$. Such minimum occurs whenever $\omega_0 \tau \sim 1$ or $\omega_1 \tau \sim 1$, for T_1 and $T_{1\varrho}$ respectively, where τ is the atomic jump time (e.g. the mean time a nucleus sits on a lattice site between jumps) and ω_0 and ω_1 are the Larmor frequency associated with the external field H_0 and the rotating field H_1, respectively. Provided the interaction controlling the relaxation is dipolar coupling modulated by translational diffusion, usually a symmetric minimum will be observed in the plot of T_1 and for $T_{1\varrho}$ as a function of the temperature (normally expressed as 1/T). From the slope of the relaxation times, near the minimum, one obtains the activation energy associated with the diffusion mechanism, controlling the relaxation, and therefore the values of τ in the same interval of temperature. Although examples in the literature can be found where this analysis does not apply, this is the typical and most commonly encountered situation. Such types of studies are at the present being carried out on different solids in many laboratories interested in material sciences.

Furthermore NMR can be most powerful when it carefully tests detailed theories of diffusion in solids. The theoretical treatment of NMR on this subject has recently received great attention and the possibility exists of testing the microscopic mechanism of diffusion in suitable materials, and therefore of distinguishing the dominant mechanism from alternate possibilities. However, at present, such studies have been performed in few laboratories, using at least partially home made apparatus designed to perform specific functions. Also these experiments may be time-consuming requiring great experimental accuracy and their analysis need a rather considerable NMR background, so that they are mainly for the NMR specialist than for the user.

Since, as previously observed, the literature contains a great number of NMR applications to the study of diffusion in solids, the omission of many significant contributions to the subject is regretable but necessarily unavoidable.

In the following examples are discussed concerning the fluorine diffusion in some fluorides compounds ($KCaF_3$, BaF_2 and PbF_2), as studied by NMR methods, aiming to cover many but by no means all the important aspects. NMR techniques are particularly suitable for investigating these materials since the isotope ^{19}F has a spin of 1/2, and therefore quadrupole effects are not present and only dipolar interaction has to be considered. This simplifies the analysis and interpretation of the experiment. Also the high natural abundance (100%) and the large γ value of ^{19}F provides us with good sensitivity for signal detection. Diffusion in these materials is intimately related to the presence of point defects, e.g. vacancies and interstitials, which affect the NMR relaxations. Being a bulk technique, NMR is sensitive to diffusion occurring through crystal lattice, e.g. volume diffusion, and less through grain boundaries, dislocations or surfaces.

$KCaF_3$

This crystal belongs to the perovskite structure and the open lattice suggest fast ionic conductivity. The predominant point defects are thought to be of the anion Frenkel type, ^-F_v vacancy and ^-F_i interstitial. In order to establish the ionic mobility and defect structure, the combination of a.c. electrical conductivity and NMR relaxation are applied to single crystal of $KCaF_3$, nominally pure and O^{2-} doped[18].The O^{2-} ions irrispective of the site that they occupy in the lattice, will be compensated by fluorine vacancies, and therefore the doping provides a method of controlling their population and of investigating their migration.

The ^{19}F spin-lattice relaxation time T_1, and spin-lattice relaxation time in the rotating frame, $T_{1\varrho}$, are shown in Fig.1, for pure and O^{2-} doped $KCaF_3$. Only the oxygen doped sample provides a reliable set of data points, while measurements in the nominally pure sample are not reproduceable on thermal cycling. This is due to unwanted oxidation of the sample resulting in the uncontrolled production of fluorine vacancies, an effect unsurprising since the fluorine reactivity is known to be rather high. The reduced temperature dependence of the relaxation times, on the low temperature side, can be considered due to residual paramagnetic inpurities controlling the relaxation. At higher temperature,on the other hand, the spin relaxation is controlled by translational motion of fluorine ions and

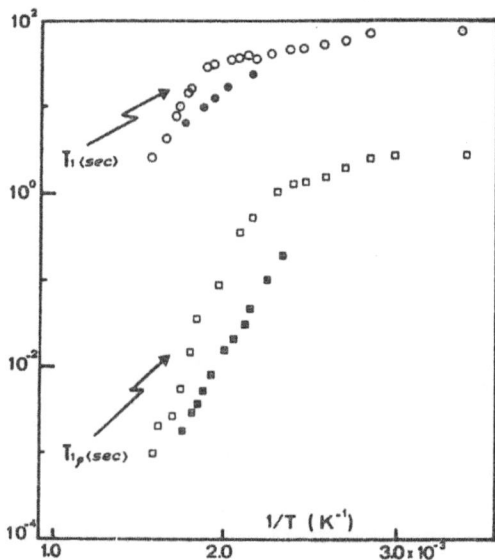

Fig.1. Temperature dependence of ^{19}F relaxation times T_1 and $T_{1\rho}$ in $KCaF_3$ pure (open marks) and O^{2-} doped crystal (black marks).

therefore reflects their diffusion. Through a suitable analysis, appropriate for f.c.c. lattice, which approximate the fluorine sublattice, assuming a powder average, since the crystal orientation is unkown, it is possible to obtain, τ, the fluorine jump time. Then on the assumption of a vacancy diffusion mechanism the fluorine self-diffusion coefficient is calculated:

a) From the NMR data using: $D^{nmr} = 2a_0^2/6\tau$ where the nearest neighbour distance $\sqrt{2}a_0$ is assumed to be the ionic jump distance.

b) From the ionic conductivity data using: $D^{\sigma} = \sigma T k/N_F e^2$, where σ is the ionic conductivity, N_F is the number of fluorine per unit volume, the other symbols having the usual meaning.

The results from the two techniques, considering only the reliable data of the oxygen doped sample, are in fair agreement, the small displacement certainly due to the approximation made in the NMR calculation (Fig.2). The slope gives an activation energy of about 0.66 eV to be considered the enthalpy of migration of fluorine vacancies.

The experiments described above should not be considered a satisfactory example of a rigorous study of diffusion in solids. The NMR analysis is infact approximate since the information needed is lacking (orientation of crystals), or not properly considered (precise lattice structure), however enough to allow the conclusion that the fluorine diffusion is

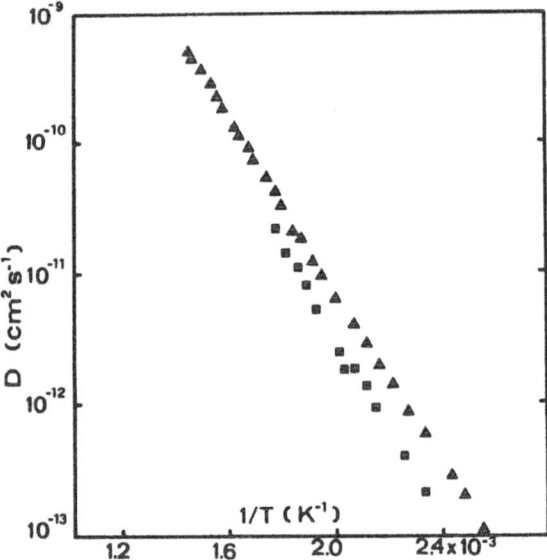

Fig.2. Fluorine self-diffusion coefficient in oxygen doped KCaF$_3$ calculated from conductivity (▲) and T$_{1\varrho}$ (■) data.

mainly assisted by a vacancy mechanism. Nevertheless these experiments are indicative of the strategy usually involved in the study of diffusion in solids, that is the coordinate application of macroscopic techniques, e.g. ionic conductivity and NMR relaxation times respectively in our example. It should be considered that conductivity here is particularly appropriate and convenient, due to its simplicity and reliability, and to the strong ionic character of the materials. On the other hand radiotracer methods are pratically not applicable here due to the too short half-life of the only available fluorine radio-isotope (^{18}F).

BaF$_2$ and PbF$_2$

These materials are representative of a class of crystals with the fluorite structure, where the predominant intrinsic point defects are known to be the anion-Frenkel pairs, that is fluorine anion vacancy and complementary anion interstitial, (see for example ref. 19). It should be observed that in a pure crystal at a given temperature, there is an equal concentration of fluorine vacancies and interstitials, (intrinsic region). However it is possible to introduce in the lattice foreign atoms (doping), provided the electrical neutrality is preserved, and therefore the addition of ions having charges different from those of the host ions will alter the defect concentration from its intrinsic value. For example, doping with trivalent cation (e.g. La^{3+}) increases the concentration of fluorine interstitials, whereas monovalent cation (e.g. K$^+$) as well as divalent cation (e.g. O^{2-}) will produce fluorine vacancies. Selective doping then allows a method of controlling defects population and of investigating their migration properties. Fluorine diffusion in crystals with the fluorite structure proceeds by the migration of point defects essentially through the vacancy and the non-collinear intersticialey mechanisms [19]. Therefore a possible question is to investigate whether these two mechanism are competitive or which one is prevailing over the other. The question has been investigated in BaF$_2$ single crystals, nominally pure and doped with aliovalent La^{3+}, K$^+$ and O^{2-}, using the combination of ionic conductivity and NMR relaxation techniques [20]. Some of the results for pure BaF$_2$ can be found in the article by prof. J.H. Strange in this volume.

Here Fig.3 shows the fluorine self-diffusion coefficients as evaluated from conductivity data and relaxation times (T$_1$, T$_{1\varrho}$, T$_{1D}$), for samples doped with La^{3+} and K$^+$. A knee clearly separates the intrinsic and extrinsic diffusion region, corresponding to the high and low temperature site of the diagram, respectively. The analysis of NMR data assuming a vacancy and a intersticialcy mechanism, gives self-diffusion coefficients which differ too little (less then 4%) to allow distinction between the two

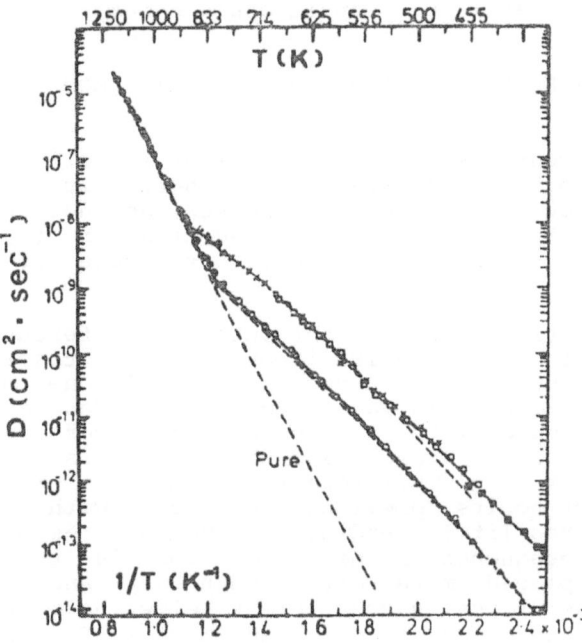

Fig.3. Fluorine self-diffusion coefficient BaF$_2$ single crystal containing cations impurities (La^{3+} and K$^+$). After [20].

mechanisms. The self-diffusion coefficient is evaluated from conductivity data assuming a vacancy diffusion mechanism. They fairly agree with those from NMR data. Larger discrepancies are found assuming an intersticialcy mechanism, which on the other hand should be considered appropriate at least for the La^{3+} doped sample. An important point is that the temperature dependance studies allows the evaluation of other parameters pertinent to diffusion: the formation enthalpy for a Frenkel pair (1.91 eV) and the migration enthalpies for free fluorine vacancies and interstitials (0.57 and 0.76 eV, respectively).

A similar study of the fluorine self-diffusion in PbF_2 is described in ref.21. It follows the same methodology, that is, the combination of NMR and ionic conductivity measurements, a similar analysis of the experimental data, therefore it will not be discussed here. However it is worthwhile mentioning that here we have an example of the measurement of the fluorine self-diffusion coefficient by the spin-echo pulsed field gradient technique. The experiment gives a direct measurement of D, and neither assumptions about the kinetics nor models of the diffusion are needed. Application of this method in solids is limited by the usual strong spin coupling (short T_2). If T_2 is too short the diffusion of nuclei in the field gradient would not be able to produce a discernible echo attenuation, also due to the experimental limitation in the maximum available strength of the gradient. For these reasons the direct measurement of D are limited to the temperature interval 570-1100 K, were the favourable experimental conditions are met.

CORRELATION EFFECTS IN NMR. THE ENCOUNTER MODEL IN FLUORITE LATTICE

Experimental studies of diffusion in solids by NMR, as well as by other techniques, measures quantities related to the migration of the atoms themselves, and not to the point defects. At the usual defect concentration ($\leqslant 10^{-4}$), the jumps of defects are random, but those of atoms are correlated.

Correlation effects enter into the different experiments in different ways. Their importance in tracer diffusion experiments has long been recognized [23]. One may consider the commonly occurring vacancy mechanism. Here an atom which has just jumped into a vacancy is necessarily still a neighbour of this vacancy, and the vacancy is in the position the atom just left. Therefore the atom has a high probability to jump back to its previous position, with greater than random probability, leading to no net displacement. A physical consequence of the correlations is that atoms are left undisplaced more often than at random, therefore reducing the diffusion coefficient below the value that would be expected on the basis of a random walk. For example, in a cubic lattice, where each atomic jump has the some length r, the tracer diffusion coefficient D will be written as: $D = f_T r^2 / 6\tau$, where τ is the mean time between jumps, and f_T the tracer correlation factor. For point-defects mechanism in solids $f_T \leqslant 1$, and only few mechanisms does not entails

A variety of methods have been used in order to calculate the correlation factor, and values of f_T have been worked out for most cases of practical interest. For anion self-diffusion in fluorite lattice these values are: 0.653 and 0.986 for vacancy and intersticialcy non-collinear mechanism, respectively [24]. An important point to be considered is that f_T is characteristic of the crystals structure and of the diffusion mechanism. Therefore experimental measurements of f_T can serve a more important function than just correcting the diffusion coefficients, since it may establish the diffusion mechanism.

The correlation effects so far mentioned are the only relevant in tracer experiment, where the observed effect is the net displacement of atoms resulting from many diffusive jumps occuring over long time-scale. They entail a spatial (or directional) correlation.

In NMR experiments, due to the inherent time scale, we have to consider in addition, temporal aspects of correlated motion. Let us consider, for example, a given nuclei exchanging with a passing nearby vacancy. Following the initial exchange, there is significant probability, that the vacancy will return once or more to the same nuclei before wandering away. The sequence of such jumps of an atom, due to the same vacancy, is called an "encounter" [25]. During an encounter a spin may be displaced by one like-neighbour distance, or more, or it may jump back to the original position, resulting in no net displacement. This is the spatial part of the correlation. The individual jumps of nuclei within the encounter occurs too quickly (at the vacancy jump frequency) to cause appreciable relaxation. This represents the temporal part of correlation. Also the actual duration of an encounter, Δt, is of the order of the jump time of the vacancy, τ_v. Therefore, considering that the vacancy concentration is usually low ($\leqslant 10^{-4}$), it may be assumed that this duration is short compared to the average time interval, τ_{enc}, between two consecutive encounters of the same nuclei with different vacancies. Consequentely different encounters are uncorrelated, that is random. A pictorial draft of the local field, as seen at a given lattice site, due to a neighbouring spin randomly jumping between different encounters is shown in Fig.4. The rapid fluctuation occuring during an encounter do not affect nuclear relaxation, which is instead sensitive to the local field fluctuation between encounters. The mean time between succesive encounters, τ_{enc}, is the correlation time of the fluctuating local field.

So far we have discussed correlation effects, in time and space for individual atoms. However it should be considered that in the NMR experiments, we are observing the effects of the relative motion of pairs of spin. A passing vacancy exchanging with a given spin may exchange, in a very short time, with any other spin in the neighbourhood. As a result some of the surroundings nuclei will jump once or more, others will not, according to a conditional probability that can be evaluated. Therefore a relative displacement of a

263

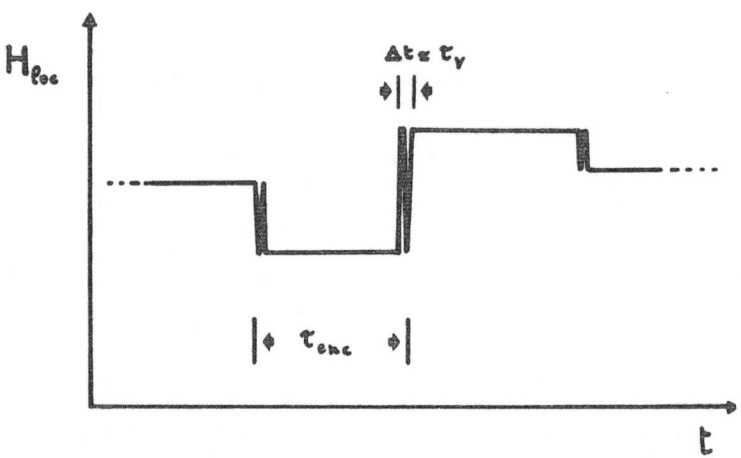

Fig.4. Temporal fluctuation of the local field due to successive encounters of the same spin.

pair of nearby nuclei may occurs and, since they are due to the same vacancy they are correlated (pair correlation). All that matters in the NMR experiments is the relative position of the pairs of spin once that the vacancy has wandered away. The correlation time, τ_{nmr} , characterizing the mean time between succesive changes of the dipolar interactions between a pair of spin, is the mean time between encounters. As discussed in detail in ref.26-27, pair correlation gives an important contribution to the dipolar correlation functions, and, therefore, to the NMR relaxation times. The encounter concept is not limited to a vacancy mechanism but may as well be applied to other point defect mechanisms of self-diffusion (e.g. interstitial, intersticialcy). It also offers an unified point of view to discuss correlation effects and how they enter in different experiments [28].

The encounter model has been applied and tested experimentally for anion diffusion on fluorite lattice as studied by NMR relaxation experiments [29-30]. The significant parameters are evaluated by a computer simulation of the relative and correlated atomic rearrangement due to the random motion of point defects. The theory and calculations are rather involved, however there are two main conclusions:
a) in a single crystal, the orientation dependences (e.g. angular relationship between the external magnetic field and the crystal axes) of the relaxation times are fairly sensitive functions of the diffusion mechanism.
b) in a single crystal (as well as in powder) the shape and the width of T_1 and $T_{1\varrho}$ near the minimun depends on the diffusion mechanism.

These conclusions are not restricted to fluorite lattice since they can apply to other crystal lattice [3]. They are

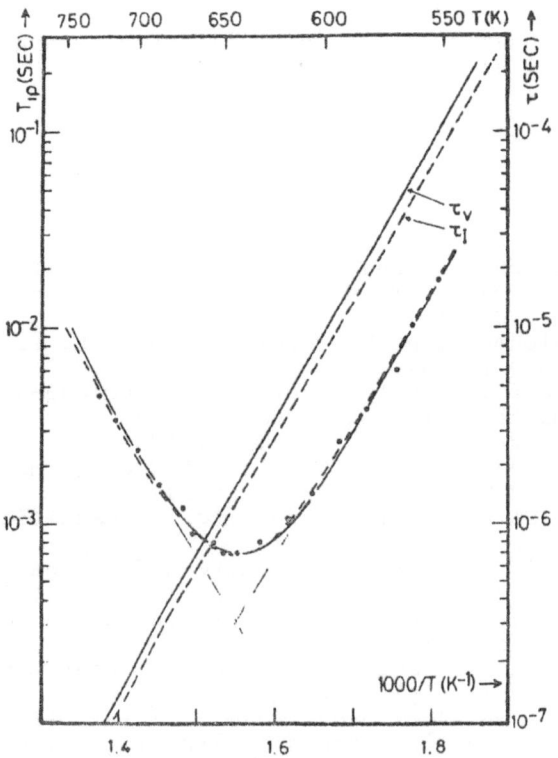

Fig.5. Experimental $T_{1\varrho}$ (●) in BaF_2, oxygen doped. Distinguishing theoretical prediction between vacancy (——) and intersticialcy mechanism (---), demand an unachievable experimental accuracy. Also shown are the jump times for the two mechanisms. After [30].

important in that they possibly offer a method for distinguishing different mechanism of diffusion. In fluorite lattice, as already observed, vacancy and intersticialcy are supposed to be dominant. Detailed measurements of the relaxation times were carried out in BaF_2 single crystal pure and doped (La^{3+}, K^+, O^{2-}), with the purpose of verifying the theoretical predictions, (point (a) and (b) above). The experiments investigate the angular dependence of the relaxation times [29] and the shape and width of $T_{1\varrho}$ near the minimum as a function of the temperature [30].

The analysis of these experiments confirms the prediction of the encounter model in the respect that a correlated self-diffusion mechanism is the origin of the nuclear spin relaxation, clearly ruling out the random walk mechanism. However the experiments are not able to distinguish between the fluorine vacancy and intersticialcy, since the required accuracy is not experimentally achievable. An example of this situation is shown in Fig.5, where the theoretical predictions for the two models are indistinguishable, since their differences are less than the experimental error.

REFERENCES

1. F.H. Wohlbier and D.J. Fisher eds., Diffusion and Defect Data, (Trans. Tech. Publications, Rockport, Mass., U.S.A.).
2. N. Bloembergen, E.M. Purcell and R.V. Pound, Phys. Rev. 73: 679 (1948).
3. D. Wolf., Spin Temperature and Nuclear-Spin Relaxation in Matter (Clarendon Press) (1979).
4. J.H. Strange. This Volume: (1985).
5. A. Abragam, The Principles of Nuclear Magnetism, (Oxford University Press 1961).
6. C.P. Slichter, The Principles of Magnetic Resonance, (Harper and Row) (1973).
7. M. Goldman, Spin Temperature and Nuclear Magnetic Resonance in Solids, (Academic Press) (1971).
8. T.C. Farrar and E.D. Becker, Pulse and Fourier Trasform NMR, (Academic Press) (1971).
9. E. Fukushima and S.B.W. Roeder, Experimental pulse NMR (Addison-Wesly) (1981).
10. D.C.Ailon, Adv. Magn. Reson. 5: 1977 (1971).
11. D.C. Ailon, in Methods of Experimental Physics, Y.N. Mundy, S.Y. Rothman and H.J. Fluss, eds. 21: 439 (1981).
12. S.R. Hartmanm and E.L. Hahn, Phys. Rev. 128: 2042 (1962).
13. H. Looser, M. Mali, Y. Roos and D. Brinkam, Solid State Ionics 9&10: 1237 (1983).
14. D.C. Ailon and C.P. Slichter, Phys. Rev., A137:235 (1961).
15. I. Chung, H.S. Story and W.L. Roth, J. Chem. Phys. 63: 4903 (1975).
16. S.G. Bishop and P.J. Bray, J. Chem. Phys. 48: 1709 (1968).
17. J.M. Chezeau and J.H. Strange, Physics Report 53: 1 (1978).
18. A.V. Chadwick, G.A. Ranieri, J.H. Strange and M. Terenzi, Solid State Ionics 9&10: 555 (1983).
19. A.B. Lidiard in Crystals with the Fluorite Structure, W. Hayes ed. (Clarendon Press) (1974).
20. D.R. Figueroa, A.V. Chadwick and J.H. Strange, J. Phys. C 11: 55 (1978).
21. R.F. Gordon and J.H. Strange, J. Phys C 11: 3213 (1978).
22. E.O. Stejskal and J.E. Tanner J. Chem. Phys. 42: 288 (1965).
23. J. Bardeen and C. Herring in Imperfections in Nearly Perfect Crystals, W. Schokley ed. (Wiley) (1952).
24. K. Compaan and Y. Haven, Trans. Farad. Soc. 54: 1498 (1958).
25. M. Eisenstadt and A.G. Redfield, Phys. Rev. 132: 635 (1963).

26. D. Wolf, Phys. Rev. B 10: 2710 (1974).
27. D. Wolf, Phys. Rev. B 10: 2724 (1974).
28. D. Wolf in Mass Transport in Solids, F. Beniere and
    C.R.A. Catlow eds. NATO ASI Series B: Physics 97: 149
    (1984).
29. D.R. Figueroa, J.H. Strange and D. Wolf, Phys. Rev. B
    15: 2545 (1977).
30. D.R. Figueroa, J.H. Strange and D. Wolf, Phys. Rev. B
    19: 148 (1979).

COMPUTER SIMULATION OF DEFECTS IN SOLIDS

C.R.A. Catlow

Department of Chemistry
University of Keele
Staffordshire ST5 5BG, U.K.

1. Introduction

The study of defect properties of insulating crystals
has been a particularly successful area of theory during the
last few years. This chapter reviews the methodology and
achievements of this exciting and rapidly developing field
which has been characterized by the extent of its interaction
with experiment and the range of theoretical techniques used.

In the account which follows, we first discuss the types
of information which are provided by the calculations. We
then summarise the different techniques that are employed.
The methodologies of each is described, which we follow by an
account of recent applications. We consider only calculations
relating to bulk defect properties. The growing area of
surface simulation is discussed in this book by Colbourn. We
will give emphasis to work on ionic crystals, which has
probably enjoyed the greatest success in recent years,
although we shall refer to studies of metals and semiconductors. The interesting field of point defects in
molecular solids is outside the scope of our discussion, and
we refer the reader to the review of Chadwick (1982) and the
work of Meyer (1982).

2. Aims and Techniques

Calculations have been used in both quantitative and
qualitative studies of defect processes. Of the former
category the principle applications are as follows:

(i) Calculation of the enthalpies and entropies of formation
of defects, i.e. the quantities which, as discussed by Corish
in Chapter 1, control defect concentrations in the intrinsic
defect regime.

(ii) Calculations of the energies and entropies of defect
migration which control the rate of defect transport.

(iii) Calculation of the binding energies of defects and of
defects and impurities.

(iv)   Calculation of electronic energy levels and of excitation energies within defects, and study of the interaction of the wave-functions of electrons within defects and the surrounding lattice (manifested experimentally by hyperfine interactions in spin resonance studies, as discussed in the chapter of Spaeth).

Qualitative applications of theory have, however, been particularly important in defect studies. Some of the main areas of present and future application are

(i)   Evaluation of the critical parameters (e.g. polarisability, ion-size) controlling defect parameters.

(ii) Prediction of the effects of temperature and pressure on defect properties.

(iii)   Development of models of defect aggregation in heavily defective materials.

This list of applications is of course far from exhaustive; but it gives a general indication of the direction and scope of theoretical studies.

The techniques used in defect studies can be subdivided into simulation methods and quantum mechanical calculations. The former which are based on interatomic potential models of the material comprise three main categories: first, static lattice defect simulations which involve calculation of the relaxation of the lattice around the defect and the determination of the energies and entropies of the defect configuration, without, however, any explicit inclusion of the effects of thermal motions.   In contrast, the second technique, molecular dynamics simulations include thermal effects explicitly, by solving the equations of motion of a dynamic ensemble of particles representing the system and to which periodic boundary conditions are applied.   Monte-Carlo methods provide the third simulation technique;   the method which is based on statistical sampling procedures is particularly effective for heavily disordered systems.

Quantum mechanical methods all involve attempts at varying levels of approximation to solve the Schrodinger equation for the defect configuration, which may be treated in two ways.   First, the calculation may be performed on a cluster, i.e. the defect centre and a small number of surrounding atoms; recent work has shown that it is generally desirable to 'embed' the cluster in some simple representation of the surrounding lattice - a point to which we return in section 3.2.   Alternatively, a defect super-cell may be studied, i.e. the calculation is repeated on a periodically repeating array of defects.   These contrasting strategies are illustrated diagrammatically in Figure 1.

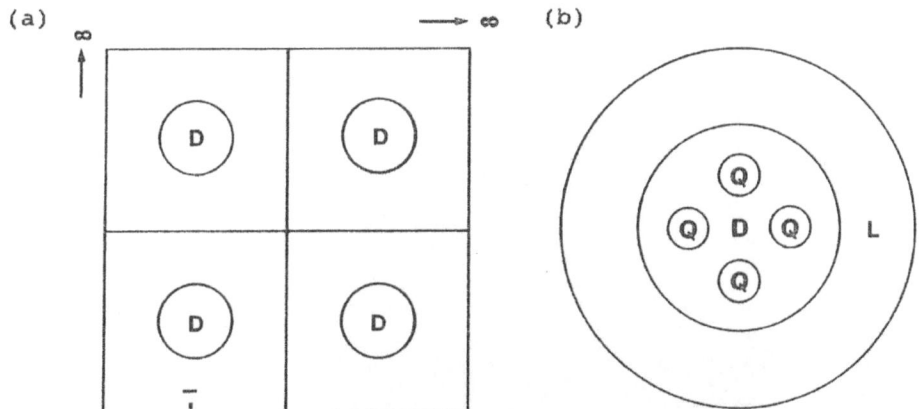

Figure 1: Schematic representation of (a) supercell of defects (D) and (b) cluster calculation in which defects (D) and surrounding atoms (Q) are embedded in a simple representation of the surrounding lattice (L).

Further sub-classification of quantum mechanical approaches is possible on the basis of the approximations used in solving the Schrodinger equation. These will be discussed in the latter part of the next section in which we give more details of the procedures used in the calculations.

3. Methodologies

3.1 Simulations

A. Static Defect Simulations

The central problem in static-lattice defect calculations is the treatment of lattice relaxation around the defect. This is particularly important for ionic and semi-ionic crystals as defects are generally charged species in these solids, which owing to the long-range of the Coulomb forces leads to a long-range relaxation field. It has, however, proved possible to handle lattice relaxation around defects effectively by a procedure which derives ultimately from the work of Mott and Littleton (1938). This approach, known as the 'two-region' method is illustrated diagramatically in Figure 2. It consists of the division of the lattice surrounding the defect into an inner region (I) which is treated atomistically and in which the coordinates of all ions are adjusted until they are at 'zero-force', i.e. there are no net forces acting upon them. This explicit, atomistic method is essential for the region immediately surrounding the defect where the defect forces are strong.

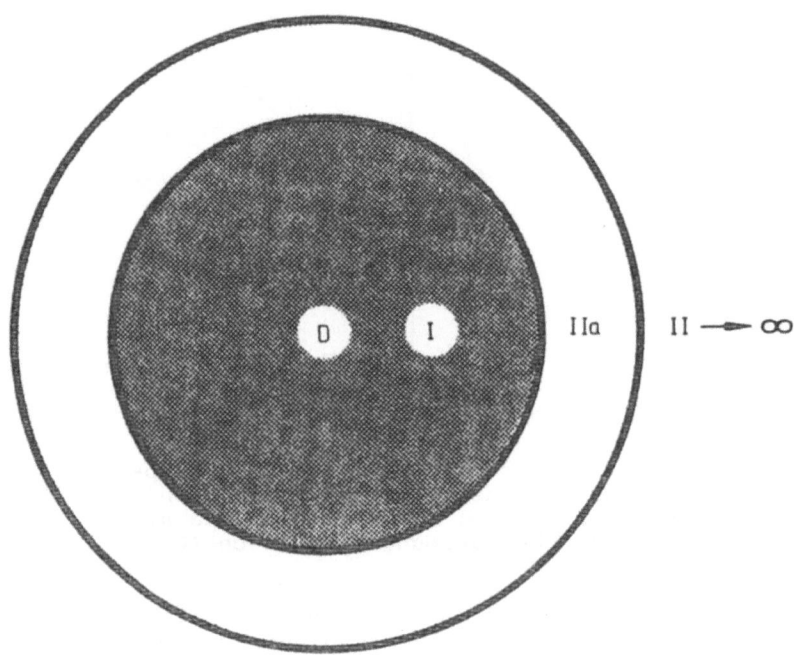

Figure 2: Two region strategy for defect calculations

In contrast, for the more distant, weak-field regions, the
defect forces are relatively weak, and lattice relaxation may
be treated by more approximate methods. The Mott-Littleton
approach, appropriate to ionic materials, calculates the
polarisation, $\underline{P}(r)$, at a point $\underline{r}$ with respect to the defect of
charge, q, according to the expression

$$\underline{P}(r) = \frac{q\underline{r}}{r^3} \left[1 - \epsilon_0^{-1} \right] , \tag{1}$$

where ϵ_0 is the static dielectric constant of the crystal.
Equation (1) is only strictly applicable to cubic materials;
for non-cubic crystals, more complex expression are used as
discussed by Catlow et al (1982) and Catlow and Mackrodt
(1982).

Within the two-region approximation, the energy, E_D, of
defect formation may be written as:

$$E_D = E_I(\underline{x}) + E_{I,II}(\underline{x},\underline{y}) + E_{II}(\underline{y}), \tag{2}$$

where E_I is the energy arising solely from interaction of
atoms within region I, whose coordinates make up the vector $\underline{x}$;
E_{II} is the self energy of region II for which $\underline{y}$ is the vector

of coordinate displacements. $E_{I,II}$ is an interaction energy term.

If the displacements y are sufficiently small, then we may assume the validity of the harmomic approximation and write:

$$E_{II} = \frac{1}{2} \underline{y} \underline{\underline{A}} \underline{y}. \tag{3}$$

Therefore, applying the equilibrium condition to region II we have

$$\left\{ \frac{\partial E_{I,II}(x,y)}{\partial y} \right\}_{y=\underline{y}} = - \underline{\underline{A}} \underline{y}. \tag{4}$$

Thus, substituting equation (4) into (3) and (1), we have:

$$E_D = E_I(x) + E_{I,II}(x,y) - \frac{1}{2} \left\{ \frac{\partial E_{I,II}(x,y)}{\partial y} \right\}_{y=\underline{y}} \underline{y}. \tag{5}$$

We have therefore removed the explicit dependence of E_D on E_{II}, which is convenient.

Defect calculations therefore consist of the relaxation of region I, followed by the evaluation first of $E_I(x)$ by direct summation and secondly of $E_{I,II}(x,y)$ and its derivative. For the latter, a subdivision is made of region II: for the inner part (region IIa) immediately surrounding region I, the displacements are calculated by the Mott-Littleton procedure as the sum of those due to all component defects in region I, and $E_{I,II}$ and its derivative are calculated by direct summation of the interactions between the two regions. For the remainder of region II, the interaction is treated as arising purely from the net effective charge of the defect in region I, and the appropriate summations may be evaluated analytically. The use of the interface, region IIa, is found to be necessary in obtaining accurate results with inner regions of a modest size.

Further details of the mathematical development of the procedure outlined above are given in the report of Norgett (1974) and the articles of Catlow and Mackrodt (1982) and Catlow et al (1982). It is, however clear, that, within the inherent limitation of the static-lattice approximation, the procedures summarised above yield reliable defect formation energies given a sufficiently large size of the inner region (generally 100 atoms or more) and, of course, given reliable interatomic potentials. Moreover, automatic computer codes are now available for performing defect calculations; principally the HADES II code (Norgett, 1974) which is confined to crystals of cubic symmetry, HADES III (Catlow et al, 1982) which may be used for crystals of any symmetry and CASCADE (Leslie, 1981) which has been optimised for use on CRAY computers. In section 4.1 we will illustrate the success that has been achieved by calculations using these codes.

More recently, important progress has been made in the
calculation of the vibrational contribution to defect
entropies, S_{vib}, i.e. the entropic term arising from the
change in the vibrational frequencies of the surrounding
lattice due to the defect; for which the following expression
may be written at temperatures well above the Debye
temperature:

$$S_{vib} = -k \ln \left[\prod_{i=1}^{3\bar{N}'} w_i' \Big/ \prod_{i=1}^{3\bar{N}} w_i \right] + 3k (\bar{N}'-\bar{N}) \{1-\ln h/kT\},$$

$$(6)$$

where the primes indicate the defective crystal; $\bar{N}$ is the number
of atoms in the perfect crystal and w_i the vibrational
frequencies of the normal modes. The methods used in
calculating the ratio of perturbed to unperturbed vibrational
frequencies in the above equation are discussed by Harding in
this volume. His work which builds on that of Gillan and
Jacobs (1983) promises to have the same impact on defect
studies as has been achieved by the development of reliable
defect energy calculations.

Static simulation methods may also be used to study
defect transport. The approach rests on the validity of the
Absolute Rate Theory (ART) approach to defect transport
developed by Vineyard (1957). This gives the rate ν, of
defect jumps as

$$\nu = \nu_0 \exp(-\Delta G_{ACT}/kT),$$

$$(7)$$

where ν_0 is the attempt frequency and ΔG_{ACT} is the free energy
of activation which has, of course, an enthalpy, h_{ACT} and
entropy, s_{ACT}, component. h_{ACT} and s_{ACT} may then be
calculated as the differences between the enthalpies and
entropies of the saddle-point for the defect migration process
and the 'ground state' or equilibrium configuration of the
defect. The procedure which has been widely and successfully
used is acceptable provided the 'hopping' model of defect
transport is a valid approximation. The hopping description
is, in general, acceptable when $\Delta G_{ACT} \gg kT$. For most solids
this condition holds even at temperatures approaching the
melting point. For one important class of materials, i.e.
superionic conductors, i.e. solids with exceptionally high
ionic conductivities, this condition may not hold, and
recourse must be made to other methods, which include thermal
energies explicitly. These are discussed in the next section.

B. Molecular Dynamics Simulations

Molecular dynamics (MD) is a well established technique
in liquid physics, and there have been several applications to
the study of molten salts (Sangster and Dixon, 1976). A
number of applications to solids have been reported in the
last ten years. For reasons that will be explained below,
these applications are largely confined to superionic
materials.

The essence of the MD technique is the specification of an ensemble of particles (commonly referred to as the 'simulation box') to which periodic boundary conditions are applied thus generating an infinite system. For crystalline solids, the simulation box will normally be constructed from several unit cells. At the start of the simulation all particles are assigned velocities and coordinates: the former are chosen in line with a target temperature (although this drifts in the early stages of the simulation); while the latter, for solids, are usually chosen to be at, or close to, the crystallographically determined sites. The simulation proceeds by a succession of time-steps, Δt, which must be shorter than the time-scale of any important process in the system, for example the period of an atomic vibration. After each time step the velocities, v, and coordinates x, are updated, according to the simple, classical equations of motion. In the limit of an infinitesimally small time step, the updating formulae would be:

$$x_i' = x_i + v_i \, \Delta t \tag{7a}$$

$$v_i' = v_i + \frac{f_i}{m_i} \, \Delta t \tag{7b}$$

where m_i and f_i are the mass and the force acting upon the ith particle. For finite Δt, the above expressions become inaccurate, and the updating formulae must include higher powers of Δt. A commonly favoured algorithm is that of Beeman (1976); other algorithms are discussed by Sangster and Dixon (1976).

In the early stages of the simulation, the system equilibrates, i.e. it achieves a Maxwellian distribution of velocities and equilibrium between potential and kinetic energy; 500-1000 time-steps are normally needed. The simulation is then generally run for several thousand time steps and the successive configurations stored on magnetic tape for subsequent analysis.

A rich variety of information is potentially available from analysis of the particle trajectories given by the simulations. These include:

(i) <u>Structural properties</u> from radial distribution functions (rdfs)

(ii) <u>Transport coefficients</u>. The diffusion coefficient, D_α may be obtained from the Einstein expression

$$D_\alpha = \frac{1}{N_\alpha} \sum_\alpha \frac{R_\alpha^2(t)}{6t} \,, \tag{8}$$

where the sum is over all particles, N_α, of type α and $R_\alpha(t)$ is the displacement of a given particle of type α at time t. This approach allows an immediate test of whether diffusion is occurring. If this is the case, a plot of

$\overline{R_\alpha^2(t)} = \frac{1}{N_\alpha} \sum_\alpha R_\alpha^2(t)$ against t will, at longer t, be a straight line (with, of course, small deviation due to statistical fluctuations) with a slope proportional to D_α. If diffusion is not occurring, that is the particles simply vibrate about fixed lattice sites, the plot of

$R_\alpha^2(t)$ vs. t consists, at low t, of a rapid increase to a value equal to the mean square amplitude of thermal motions, after which the value remains constant. A nice illustration of these two types of behaviour is given by Dixon and Gillan's (1980a,b) study of superionic CaF_2, in which there is a mobile F^- sub-lattice but an immobile cation sub-lattice. This is apparent from the plots of $\overline{R_\alpha^2(t)}$ vs. t for the two types of ion that are shown in Figure 3.

Given knowledge of the correlation factor, f, which requires knowledge of the migration mechanism, the conductivity may be obtained from the diffusion coefficient via the Nernst-Einstein relationship. A priori calculations of conductivity are possible via the perturbed M.D. technique discussed by Cicotti et al (1976). These methods have not, however, been extensively applied to defective solids.

(iii) <u>Correlation Factors and Scattering Functions</u> The simplest, positional correlation function, i.e. the r.d.f. was discussed in (i) above. Important dynamical correlation functions include, $Z_\alpha(t)$, the velocity auto-correlation function (v.a.f.) and van Hove self-correlation function $G_\alpha^s(r.t.)$.

The former is given by

$$Z_\alpha(t) = \frac{1}{N_\alpha} \sum_{i=1}^{N_\alpha} \langle v_i(t) \cdot v_i(0) \rangle / \langle v_i(0) \cdot v_i(0) \rangle, \qquad (9)$$

where N_α is the number of particles of type α, and $v_i(t)$ is the velocity of particle i at time t; $v_i(0)$ is the velocity at t=0. The VAF is the ensemble average of the projection of the velocity vector of a particle along its velocity vector at a time t earlier. If a particle is vibrating about a lattice site $Z_\alpha(t)$ will show oscillatory behaviour, which will, however, decay to zero if the oscillations are damped. For a particle exercising random diffusive motion the VAF rapidly decays to zero.

276

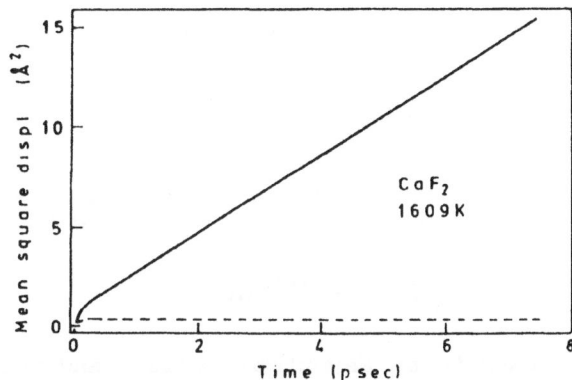

Figure 3: Plot of mean square displacement vs. time for F^- (full-line) and Ca^{2+} (dotted line) in CaF_2.

The function $G_s^\alpha(r,t)$ is the probability of finding a particle of type α at a point r at time t, when that particle was at position 0 at time 0. As seen below, the behaviour of $G_s^\alpha(r,t)$ can yield information on mechanistic aspects of ion transport. In addition, Fourier transformation of $G_s^\alpha(r,t)$ yields the incoherent inelastic neutron scattering cross-section of the system; this is an important relationship as neutron scattering techniques have yielded a great deal of valuable information on particle dynamics in condensed matter; a recent review is available from Lechner (1983).

A further function of importance is the intermediate scattering function $F^\alpha(k,t)$ defined as:

$$F^\alpha(k,t) = \langle \sum_i \sum_j \exp\{i\underline{k}\ r_i^\alpha(t) - r_j^\alpha(0)\}\rangle, \qquad (10)$$

where r_i is the position of the ith particle at time t, and $\underline{k}$ is the scattering vector defined as $\underline{k} = (2\pi/\lambda)\ \hat{\underline{k}}$, in which λ is the scattering length and $\hat{\underline{k}}$ is the unit scattering vector. The self part of $F^\alpha(k,t)$ is written as $F_s^\alpha(k,t)$ and is simply defined by setting $i=j$ in equation (10). G_s^α and

F_α^s are in fact related by the simple transformation:

$$F_s^\alpha(k,t) = \int_0^\infty e^{iQr}\, G_s^\alpha(k,t)\, dr, \tag{11}$$

so that the incoherent neutron scattering cross-section, $S^{INC}(\underline{k},w)$ for scattering from nuclei of type α may be written as

$$S^{INC}(\underline{k},w) = \int_0^{+\infty} F_s^\alpha(k,t)\, e^{-iwt}\, dt \ . \tag{12}$$

$F_s^\alpha(k,t)$ decays to an asymptotic value, and the nature of its decay may yield useful mechanistic information. It will attain the asymptotic value when a typical atom of type α has had sufficient time to sample the time average distribution along direction $\underline{k}$ and over distance λ. Thus if λ is the vibrational amplitude of an atom in a solid, then vibrational motion is mainly responsible for the decay of F_s^α; but if λ is taken to be approximately a nearest neighbour distance in the crystal, and if diffusion is occurring, the initial vibrational decay of F_s^α will be followed by a slower diffusional decay.

(iv) <u>Mechanistic information</u> Analysis of correlation functions may yield valuable information on atomic migration mechanisms. A good example is provided by the behaviour of the van Hove function $G_\alpha^s(r,t)$ Hansen and MacDonald (1976) have shown that G_α will have a Gaussian structure if diffusion proceeds via 'continuous' migration. Moreover, we may test whether G_α is Gaussian like by taking moments of $r_\alpha(t)$ (Rahman, 1976); in particular if

the ratio $P_\alpha(t) = \dfrac{3\langle r_a(t)^4 \rangle}{5\{\langle r_\alpha(t)^2 \rangle\}^2}$ tends to unity with increasing t,

then the Gaussian approximation is acceptable for G_α. Strong deviation of the above ratio from unity indicates the failure of this approximation which in turn implies non-continuous migration mechanisms. Dixon and Gillan (1980a,b) used this criterion as evidence for a hopping migration mechanism for F^- ions in superionic CaF_2. We will see in section 4.2, how calculations of the moment ratio show that there is a change in mechanism with temperature in $Na-\beta''-Al_2O_3$.

A more straightforward way of elucidating mechanism is by simply plotting the particle trajectories – a procedure that is often surprisingly powerful in showing details of migration processes. Good illustrations are given in the study of $L_{13}N$ by Wolf et al (1984a,b).

The above account does not exhaust the range of information that can be obtained from MD simulations; and the technique is evidently one of considerable power. Nevertheless, molecular dynamics has severe limitation when

278

applied to solids. The first of these concerns the <u>time-scale</u> of the simulations. Even with large amounts of computer time on modern super-computers it is only possible to run the simulations for 10-100 pico second. We may therefore only obtain useful information on atomic migration processes if a large number of migration events occur in the simulation box during this time scale. The <u>size of the simulation box</u> raises the second difficulty, with consequences closely related to the first: to date, most simulations have used ~ 100-500 particles; this will only yield useful information on atomic migration if the number of mobile particles (which in many cases will mean the density of defects) is large. The consequence of both these factors is that we can only simulate transport reliably in crystals which contain high densities of mobile particles. In practice this means that full unrestricted applications of MD simulations in the study of solids are confined to superionic materials; restricted applications of MD simulations to non-superionics are, however, reported. Thus Jacobs and Moscinski (1985) obtained useful information on interstitial migration in rock-salt structured halides essentially by injecting excess interstitials (which are mobile but whose equilibrium concentration is low) into the simulation box. The 'rare-events' problem has also recently been studied by Gillan and Harding (1986) who performed MD studies of vacancy motion in rock-salt structured oxides - a process which owing to its low frequency, would not normally be accessible to MD techniques. Their procedure, which involves starting the simulation with an ion at the saddle-point for the migration mechanism, leads to valuable information on the rates and transmission coefficients for migration across the saddle point. Knowledge of the saddle point is, however, a pre-requisite for this kind of application.

A third limitation of the MD technique follows from the use of periodic boundary conditions which generate an infinite system <u>without surfaces</u>. However, the creation of Schottky disorder requires the displacement of ions in stoichiometric ratios to the surface of the crystal. This process cannot therefore be simulated by molecular dynamics. For Frenkel disorder, involving the displacement of ions from regular to interstitial sites there is no problem; but for Schottky disordered crystals transport effected by intrinsic disorder may only be simulated by artificially introducing vacancies at the beginning of the simulation.

A fourth problem arises from the very considerable computational expense of including polarisability in dynamical simulations. Since the induced dipole moments on atoms or ions respond, in effect, instantaneously to changes in nuclear positions, dipole moments must be calculated for all atoms for each time-step. The evaluation of these dipole moments is a complex many-body problem requiring an iterative numerical solution. This addition of a separate iterative procedure together with the doubling of the number of degrees of freedom which follows from allowing all atoms to have dipole moments, will increase the c.p.u. time required for the calculation by a factor of ~ 5-10. Since MD simulations, even with rigid atoms, are expensive in c.p.u. time (typical runs taking several hours on a modern supercomputer e.g. the CRAY-1S), the additional expense incurred on including polarisability is normally prohibitive. However, static calculations suggest that polarisation energies are often a significant component of defect energies. There must therefore be doubt as to the quantitative reliability of simulations of ion transport in solids in which polarisability has been omitted; although we note that the limited number of cases in which simulations of ionic systems have included polarisability encourage the view that its omission may not be too serious. Thus Sangster and Dixon's (1976) study of molten salts found little effect of including polarisability in the simulation, as was the case in a study of superionic CaF_2 by Dixon and Gillan (1980a,b).

To summarise, MD techniques are most appropriate when applied to superionics; and indeed for non-superionics static sumulations generally yield most of the required information. The problems concerning the generation of Schottky disorder and the omission of polarisability remain for superionics; and the technique is probably at its most powerful when applied to the elucidation of qualitative features concerning ion transport mechanisms.

Monte-Carlo Simulations

Monte-Carlo techniques are essentially a procedure for statistical sampling via random number generation. The field has been extensively reviewed recently by Murch (1984), and our account is therefore brief. When applied to the study of atomic diffusion processes the procedure is normally as follows:

(i) A simulation box is defined and periodic boundary conditions are applied, as with M.D. techniques. It is generally, however, possible to include larger numbers of particles.

(ii) Defects are created in the lattice. Thus vacancies may be introduced by abstracting atoms from a certain number of sites.

(iii) A defect is selected at random, as is a corresponding jump direction for the defect. The environment of the defect is examined and the jump frequency, W_i, obtained. The jump frequency appropriate to all possible defect environments must be supplied as input to the program or they must be calculated from e.g. Arrhenius energies appropriate to the given environment. To ensure maximum efficiency of the simulation all jump frequencies are scaled so that the maximum value is unity.

(iv) A random number, R, in the range 0 to 1 is generated and the value of the number compared with W_i. If $R < W_i$ the jump is deemed to be successful; if $R > W_i$ it is considered unsuccessful. This procedure effectively weights the probability of a jump according to its frequency.

(v) The procedure in (iii) and (iv) above is repeated, several thousand times. In the initial stages of the

simulation the system 'equilibrates' as in M.D. simulations.
After equilibration the successive configurations generated by
the simulation can be used to follow the process of defect
migration.

The special value of M.C. simulations of transport are
in complex systems containing several types of jump frequency.
Examples are provided by alloys and non-stoichiometric
compounds. The technique is particularly effective at
calculating correlation coefficients, f_α, and which are
obtained from the simulation using the Einstein expression:

$$f_\alpha = \frac{\langle R_\alpha^2 \rangle}{n\ a^2} \qquad\qquad (13)$$

where $\langle R_\alpha^2 \rangle$ is the mean square displacement of atoms of type α afte
an average of n jumps each of length a. We also consider that
it may become a very powerful technique for studying transport
in heavily defective compounds where the frequencies W_i are
obtained from Arrhenius expressions employing activation
energies calculated by the static simulation techniques. An
example of such an application will be given later in this
chapter.

M.C., in common with other simulations techniques rests
ultimately on the quality of the description of the
interatomic forces; the models in current use are discussed
in the next section.

3.2 Interatomic Potentials

An interatomic potential model is a mathematical
representation of the potential energy of the system as a
function of particle coordinates. This total potential,
$V(r_1, \ldots, r_N)$ is normally decomposed into a sum of functions of
2,3 or more particle coordinates:

$$V(r_1, \ldots, r_N) = \sum_{i>j} \varphi_{ij}(r_i, r_j) + \sum_{i>j>k} \varphi_{ijk}(r_i, r_j, r_k)$$

$$+ \ldots \qquad\qquad (14)$$

Where the φ_{ij} and φ_{ijk} are analytical or numerical functions.
The vast majority of simulations both of solids and liquids
take only the first term on the right hand side of equation
(14), i.e. the properties are purely 'two-body' in nature. A
further simplification assumes 'central-force' models in which
the φ_{ij} depends only on the distance, r_{ij} between particles i
and j, i.e. $\varphi_{ij}(r_i, r_j) = \varphi(r_{ij})$. In ionic materials it is
then convenient to decompose $\varphi(r_{ij})$ into Coulombic and non-
Coulombic components, i.e.

$$\varphi_{ij}(r_{ij}) = \frac{q_i q_j}{r_{ij}} + V_{ij}(r_{ij}), \qquad\qquad (15)$$

where q_i and q_j are the charges on ions i and j, and $V_{ij}(r_{ij})$
is usually referred to as the 'short-range' potential. The
simulation codes then handle the Coulomb term separately,

normally via the Ewald summation, which involves a transformation into reciprocal space (see Tosi (1964) for a good discussion); the short range terms are, in contrast, handled in real space, and are normally cut off (i.e. set to zero) beyond a specified distance. A number of analytical functions have been used for $V_{ij}(r_{ij})$; these include Lennard-Jones and Morse functions, but the most popular, in ionic crystal simulations, has been the Buckingham potential:

$$V(r) = Ae^{-r/\rho} - C\,r^{-6}, \qquad\qquad (16)$$

in which it is tempting to associate the attractive r^{-6} term with genuine dispersion (i.e. induced dipole-induced dipole forces). But in practice, this term will normally include contributions from other attractive forces including small covalent terms. There is, however, no ambiguity about the interpretation of the exponential repulsive term which describes the 'Pauli repulsion' which comes into play when atomic charge clouds overlap. We should note that there is no need other than convenience, to use analytical short-range potentials, and that numerical potentials have been used extensively and successfully by Mackrodt and co-workers (1980a).

In simulating defect properties it has been found essential to include a description of ionic polarisability, as charged defects extensively polarise the surrounding lattice. Earlier studies using point dipole models were found to be unsatisfactory, and the pioneering work of Faux and Lidiard (1970) showed the value of the shell model in defect calculations. The model, originally developed by Dick and Overhauser (1958) describes a polarisable atom or ion in terms of a core into which the mass of the ion is concentrated, which is connected by an harmonic spring to a mass-less electron shell, the latter representing the polarisable valence shell electrons. When an electric field is applied to the atom, the shell will be displaced from the core and hence a dipole moment will develop. Short-range forces are normally taken as acting between the shells, thus giving rise to a coupling between short-range energies and polarisation – an important factor whose omission in simpler point dipole treatments was largely responsible for the inadequacies of the potentials.

Ionic, two-body, shell-model potentials thus form the basis of almost all static and dynamic simulation work on polar crystals in recent years. Such models must, of course, be 'parameterised', i.e. the variable parameters adjusted so as to correspond to the particular crystal under investigation. These parameters include (i) ionic charges (ii) short range parameters (e.g. A, ρ and c in equation (16) (iii) shell charges and spring constants.

As regards ionic charges, in most studies of ionic and semi-ionic systems, these have been fixed at the fully ionic, i.e. integral values. We should stress that, as argued by Catlow and Stoneham (1983) the use of integral ionic charges does <u>not</u> imply an electron distribution corresponding to that of a fully ionic system, and that the validity of a potential model is assessed purely by its ability to reproduce known properties of the crystal. Nevertheless, as the field extends to the study of semi-covalent materials such as silicates there may be increasing use of potential models based on partial charges, and indeed a successful model for Mg_2SiO_4 has already been developed by Parker and Price (1984).

The short-range repulsive terms may be parameterised by two procedures: first, _empirical methods_ in which variable parameters are adjusted via a least squares fitting routine to achieve the best possible match of calculated and experimental crystal properties, the latter including structural properties, elastic and dielectric constants and, where available, phonon dispersion curves. By 'fitting' to the structural properties, we mean the adjustment of potential parameters until the potential gives the observed structure (including cell dimensions and atomic coordinates) as close as possible to equilibrium; and we should note that complex, low symmetry structures may contain several variables, each of which is a datum to be used in the fitting procedure. Empirical potential have been developed for several halide and oxide crystals (see e.g. Catlow et al (1977a), and Sangster et al (1978) for studies of the alkali halides, and Sangster and Stoneham (1981) and Lewis and Catlow (1985) for work on transition metal oxides); and these models have been successfully used in defect calculations. However, the empirical parameterisation procedure is clearly limited in that it can only be applied to crystals for which empirical data are available (although in some cases extrapolation procedures may be used - as in the study of Lewis and Catlow (1985)). A more fundamental problem is that empirical methods only yield information on potentials at internuclear spacings close to those observed in the perfect lattice. But in defect configurations, especially those involving interstitials, the internuclear separations will differ considerably from perfect lattice values. In such cases, empirically derived potentials will be reliable only if the form of the analytical potential is accurate over a wide range of internuclear distances - a point on which there may be considerable uncertainty.

For this reason there have been considerable efforts in recent years in developing _theoretical methods_ for deriving interatomic potentials in ionic materials. The major contribution has been made here by Mackrodt and co-workers who have studied both 'electron gas' and _ab initio_ methods. The former approach derives from the work of Gordon and Kim (1972) and Wedepohl (1967). First, electron densities are obtained for the isolated interacting atoms. This is normally achieved by solving the Hartree-Fock equations, for each atom, and Mackrodt and co-workers have stressed the importance of 'crystal-adapted' wave functions, i.e. of solving the equations in the Madelung potential appropriate to the crystal; this is of particular importance for anions for which the wave functions are more diffuse. Having obtained these electron densities, the Coulomb interactions are calculated and approximate expressions, based simply on the total electron density, are used to calculate the kinetic energy, exchange and correlation contributions to the interaction energy. For details we refer to the paper of

Gordon and Kim (1972). In <u>Ab-initio</u> studies of interatomic potentials, calculations are performed on a 'super molecule', i.e. a molecule comprising those atoms, ions or molecules whose interaction we require; the calculations are performed as a function of the internuclear spacings from which the interatomic potential is extracted. The method has been used recently by e.g. Mackrodt et al (1980b) and Saul et al (1985) in solid state studies; the latter work, which concerned the derivation of a potential suitable for modelling the OH^- ion in ionic solids, also stressed the importance of performing the calculations in the Madelung field appropriate to the solid.

In performing <u>ab-initio</u> studies of interatomic potentials, considerable care must be exercised concerning two further aspects. The first concerns the question of basis sets of the component species in the super-molecule. These must be large and flexible to allow for the electron density redistribution which occurs when the species interact. Moreover, if the wave functions yield energies for the component species that are close to the Hartree-Fock limit, then this will minimise the 'basis-set superposition error', which arises from the lowering of the intra-atomic energies of the interacting atoms due to the greater flexibility of the composite wave functions in the super-molecule compared with the single-atom wave-function. The second technical aspect concerns the 'level' of the calculation: the Hartree-Fock approximation does not include a representation of the effects of electron correlation. If it is desired to include such effects then recourse must be had to multi configurational wave-functions, obtained from either CI or MCSCF procedures (see e.g. Szabo and Ostlund (1984) for a discussion of these methods).

On going beyond the Hartree-Fock approximation, the computer time required for the calculation will increase very considerably. And for heavier atoms, the Hartree-Fock method itself may become prohibitively expensive in computer resources. Nevertheless with the continuing expansion in computer power, we may expect increased use of <u>ab-initio</u>, Hartree-Fock methods in deriving reliable interatomic potentials for solids.

At present only empirical methods may be used to derive shell model parameters (although recent work of Fowler and Pyper (1985) suggests that polarisabilities may be accurately calculated). Measured static and high frequency dielectric constants must be available if these are to be fitted with any reliability; otherwise extrapolation methods must be used. Recent work of Cormack (1986) has demonstrated the sensitivity of calculated defect energies in complex oxide to shell model parameterisation. The derivation of reliable and general procedures for deriving these parameters is therefore one of the most urgent requirements of the field.

As noted earlier, the great majority of interatomic potential models for ionic solids are of the two-body form. Our recent work has shown that this is clearly inadequate for one class of materials, namely framework structures silicates.

Thus for e.g. α-SiO$_2$, two-body central-force models cannot reproduce the observed crystal properties. However, Sanders et al (1984) found that a simple extension of the model, i.e. the inclusion of 'bond-bending forces around the O-Si-O bond-angles dramatically improved the potential which was then capable of reproducing the observed elastic, dielectric and lattice dynamical properties with reasonable accuracy. The form of the additional terms used was particularly simple: the bond-bending energy, $E(\theta)$ is written as:

$$E = K_B(\theta - \theta_0)^2, \tag{17}$$

where θ_0 is the tetrahedral angle and k_B is the bond-bending force-constant. This type of function may be relatively simply incorporated into the simulation codes, and seems unquestionably to improve the performance of potentials in several silicate crystals. It is at present an open question as to how useful such bond-bending terms will be in the study of other types of compound.

The development of interatomic potentials for metals is a far less straightforward procedure than for ionic crystals. For 'simple' metals such as Na,Ca, the pair potential approximation appears to be useful, although for transition metals it is clearly less satisfactory (Taylor 1982,1985). 'Two-region' methods may be used for point defect calculations (although region II is treated as an elastic rather than a dielectric continuum). Such calculations have enjoyed limited success as discussed in section 4.1.

3.3 Quantum Mechanical Methods

Calculations on colour centres in ionic crystals and the majority of defects in covalent semi conductors require solution of the Schrodinger equation for the defect and at least part of the surrounding lattice. As outlined in the Introduction there are two general strategies for such calculations; i.e. defect super-cell and embedded cluster methods. A detailed account of the techniques used in these calculations is beyond the scope of the present chapter; an excellent general account is given by Stoneham (1975).

In attempting to solve the Schrodinger equation by either of the two strategies the Hartree-Fock approximation is generally used; few calculations are reported in which configurational interaction is included (the study by Colbourn and Mackrodt, 1982 of chemisorption at surface defects on MgO being a notable exception). For large clusters or larger super-cells, a full solution of the Hartree-Fock equations may become computationally prohibitive, and approximate methods must be used; indeed for super-cell calculations this has to date generally been the case. The various approximations involve the omission and/or parameterisation of certain of the integrals required in the secular determinant. The principal classes of approximate methods are Extended Huckel Theory (EHT). Complete Neglect of Differential Overlap (CNDO)) and Intermediate Neglect of Differential Overlap (INDO), detailed discussions of which are given by Pople and Beveridge (1970).

All three of the above approximations have been used on defect studies. A good account of earlier work on semi-conductors is given by Lidiard (1973). Extensive use of the CNDO method has been made in recent years by Stoneham and co-workers who have studied problems as diverse as the behaviour of hydrogen in solids (Mainwood and Stoneham, 1982), and the behaviour of water in SiO_2 (Hagon, 1986). However, with growth in computer power, we anticipate that much greater use will be made of <u>ab-initio</u> Hartree Fock methods.

Recent work using the cluster strategy has stressed the importance of embedding and of an accurate interfacing of the cluster with the surrounding lattice which is normally represented by point charges.[1] A detailed study of this problem has been reported by Vail et al (1984) who considered the case of the F^+ centre in MgO. They paid special attention to the question of the consistency between the quantum mechanical cluster and the surrounding lattice, the relaxation of which was handled by classical defect simulations available in the HADES and CASCADE codes. The procedure adopted was as follows;

(i) The quantum mechanical (q.m.) cluster is 'simulated' by a set of point charges and the lattice is relaxed to equilibrium around the point-charge simulators.

(ii) The Hartree-Fock equations are solved for the q.m. cluster in the field of the relaxed surrounding lattice which is represented by point charges.

(iii) From the electron density distribution obtained for the q.m. cluster a new set of point charges simulators is defined. These are chosen so as to reproduce the electrostatic multipole moments of the cluster; but in practice moments higher than the quadrupole moment have not been considered.

(iv) The relaxation calculation is repeated with the new simulators.

(v) The q.m. calculation is repeated with the new relaxed point charge distribution.

(vi) The iterative process is continued until we have full consistency between the q.m. cluster and the relaxed, embedding point charges. In practice it is found that this is achieved after not more than three q.m. calculations have been performed.

This procedure is currently being automated, and it is hoped that a general code (ICECAP) will soon be available. The approach offers the possibility of reliable quantum mechanical cluster calculations on a wide-range of electronic defects in solids.

---

[1] Note that if a shell model treatment is used, two point charges (i.e. the core and the shell) will be needed for each ion.

This completes our account of the various methodologies used in defect calculations. We should stress that it has not been possible to cover many matters of detail, for which the reader should consult the monograph edited by Catlow and Mackrodt (1982), the reports of Norgett (1974) and Catlow and Norgett (1976), and Stoneham's (1975) general account of the theory of defects in solids.

4.   Applications

Our account will focus on recent examples of defect simulations which illustrate both the 'state-of-the-art' of the field, and the problems that may be encountered in performing reliable studies.

4.1 Static Defect Simulations

An extensive range of defect-energy calculations are now available in the literature for several classes of ionic and semi-ionic material, including both oxides and halides.   Good reviews of recent results are available from Mackrodt (1982,1984).   Table 1 gives a small selection of results on simpler materials, for which we compare calculated and experimental defect energies.   The comparison between theory and experiment is seen to be very satisfactory;   and, in general, it is found that for those compounds for which good interatomic potentials are available, accurate defect energies may be calculated.

The current direction of the field is increasingly toward complex and semi-ionic materials.   Examples are provided first by recent work of Lewis (1984) (see also Lewis and Catlow, 1986) on $BaTiO_3$ - an important electronic ceramic material - some of the results of which are summarised in Table 2. The calculations show that the basic disorder in the material is of the Schottky type;   the defect energies are, however, high: $\sim$ 2-3 eV per defect suggesting that intrinsic disorder will play a very minor role except possibly at the highest temperatures.   Lewis (1984) and Lewis and Catlow (1986) also undertook a detailed study of the behaviour of impurities in the material which play a vital role in determining its electronic properties.

A second example is provided by Doherty's recent work on Mg_2SiO_4 - a material of great importance as it forms the major component of the upper part of the earth's mantle - a summary of which is given in the recent article of Catlow et al (1985).   Doherty examined several models for the intrinsic defect structure of the material.   His work suggests that the lowest energy defects are Mg^{2+} Frenkel pairs (with a formation energy of $\sim$ 5 eV);   although Schottky pairs comprising Mg^{2+} and 0^{2-} vacancies have only slightly higher formation energies Doherty found a low value of 0.3 eV for the Mg^{2+} interstitial migration activation energy;   a larger value of $\sim$ 2 eV was found for the Mg^{2+} vacancy migration energy.   There is at present very little definitive data with which to compare the results of the calculations which have therefore a clear predictive value.   There does, however, seem to be some evidence for Arrhenius energies for Mg^{2+} diffusion in the range of 3-4 eV (see Catlow et al 1985 for details), which would be compatible with the results of the calculations.

Table 1 Defect energy calculations

Crystal Energy* (eV)	Process	Calculated
NaCl (Catlow et al., 1979)	Schottky pair formation	2.4-2.7 (2.3-2.7)
	Cation vacancy migration	0.66 (0.7-0.8)
CaF_2 (Catlow et al., 1977b)	Anion Frenkel pair formation	2.6-2.7 (2.6-2.7)
	Anion vacancy activation	0.35 (0.5-0.6)
	Anion interstitial activation	0.91 (0.9-1.0)
MgO (Mackrodt and Stewart, 1977)	Scottky pair formation	7.5-7.7 (5-7)
	Cation vacancy activation	1.8-2.2 (2.0-2.3)
NiO (Catlow et al., 1977c)	Schottky pair formation	6-7
	Cation vacancy activation	1.86 (1.5)

*Experimental values are given in brackets and discussed in the literature cited.

Table 2   Defect energies for undoped $BaTiO_3$

(a) Schottky (eV per defect)		(b) Frenkel (eV per defect)	
$BaTiO_3$	2.29	Oxygen	4.49
TiO_2	2.90	Barium	5.94
BaO	2.58	Titanium	7.56

As noted in the Introduction, the calculations have been applied in a more 'qualitative' sense to the study of complex modes of defect aggregation. Ealier applications to non-stoichiometric oxides, e.g. $Fe_{1-x}O$, UO_{2+x} and TiO_{2-x} are reviewed by Catlow (1981). A particularly relevant recent study concerned the widely investigated rare-earth doped alkaline earth fluorides. Controversy has surrounded the nature of the aggregates formed by the substitutional rare-earth dopants and their charge compensating interstitials in heavily doped crystals (i.e. crystals containing > 5 mole% rare-earth). Calculations of Bendall et al (1983) were of value in suggesting that there is a change in cluster structure on going from larger rare earths (e.g. La^{3+} and Nd^{3+}) to smaller cations, e.g. Er^{3+}. For the former, relatively small clusters comprising two dopant ions and two or three interstitials (see Figure 4 a,b) were calculated to have the greatest stability; we note that these clusters are stabilised by a coupled lattice-interstitial relaxation mode that was identified in an early theoretical study of Catlow (1973). In contrast, for the smaller dopant ions, the beautiful, symmetrical cubo-octahedral cluster shown in Figure 4b has the greatest stability. The cluster consists of 6 rare-earth ions grouped around a central interstitial site. The eight F^- lattice sites of the cube are vacant, and $12F^-$ ions are situated each above one of the cube edges. Greatest stability is achieved when the central cube contains a pair of F^- interstitials orientated along the $\langle 111 \rangle$ direction.

The predictions of the calculations have recently been strongly supported by an EXAFS study of the local environment of the rare-earth cation in CaF_2. The EXAFS technique (see e.g. Hayes and Boyce, 1976, for a good review) allows us to probe the local structure of particular atomic species. EXAFS spectra were collected for a range of dopants in 10 mole% doped CaF_2, the data being collected using the synchrotron radiation source (SRS) at the SERC Daresbury Laboratory, U.K. The spectra for the larger rare-earth ions (e.g. Nd^{3+}) could be fitted accurately assuming the formation of the type of clusters shown in Figure 4a; whereas the spectra of the smaller ions (e.g. Er^{3+}) indicated the presence of the cubo-octahedral clusters. The work (details of which are available in Catlow et al, 1984) is a nice illustration of the way in which simulations and experiments may be used in a concerted way to investigate complex problems in defect physics and chemistry.

Simulation techniques have been less extensively applied to point defects in metals, which, as discussed in section 3.2, is largely attributable to difficulties in deriving suitable interatomic potentials. A selection of results are collected in Table 3, and are compared with experiment where data are available. For further discussion we refer again the the excellent review of Taylor (1982).

Table 3   Calculated and experimental vacancy formation energies in metals[*]

Metal	Li	Na	K	Al
Calculated energy (eV)	0.48	0.29	0.31	0.19
Experimental energy (eV)	0.34	0.36	0.39	0.66

[*]For details of calculations and references to theoretical and experimental work we refer to Taylor (1982).

As noted in section 3.1, reliable methods are now available for studying defect entropies as well as energies. A number of recent applications to ionic materials have been reported by Harding, Jacobs, Gillan and co-workers; these are discussed separately by Harding in this book. Their work opens up an exciting future for this area of defect physics, as it will enable us to calculate both absolute concentrations of defects and absolute rates of defect transport.

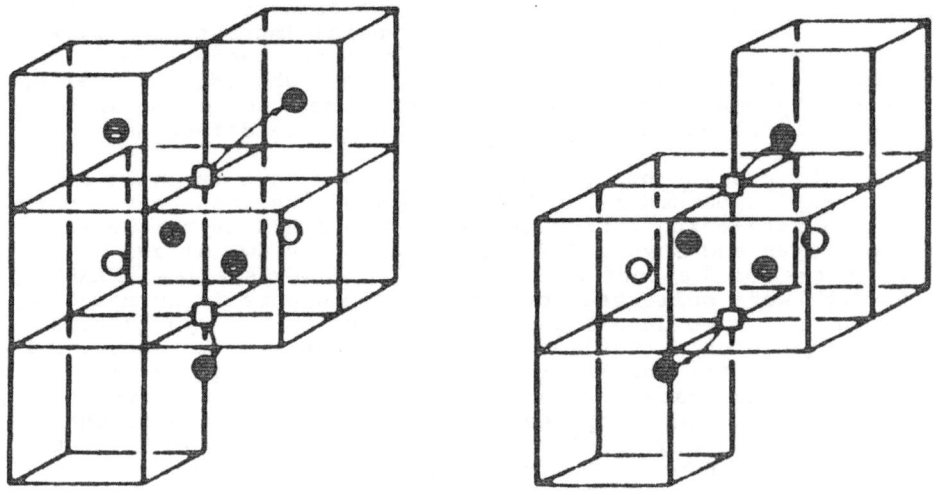

Figure 4a: Clusters of pairs of dopants with charge compensating interstitials.

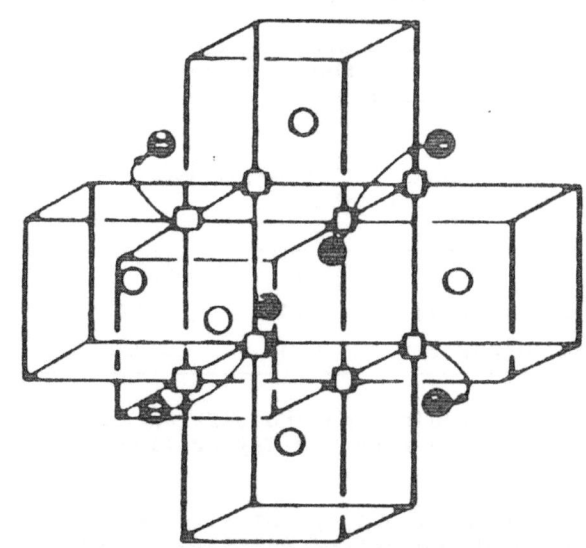

Figure 4b: Cubo-octahedral dopant cluster.

4.2 Molecular dynamics Simulations

As argued in section 3.2, the main applications of MD
simulations to solids have concerned superionics. Simulations
have been reported for superionic materials including high
temperature CaF_2 (Jacucci et al, 1978; Dixon and Gillan,
1980a,b) AgI (Vashishta1 et al, 1979), Ag_2S (Vashishta, 1986),
Li_3N (Wolf et al, 1984a,b), $\beta"Al_2O_3$ (Wolf et al, 1984c; Wolf,
1984). In this section we concentrate on the latter material;
a number of reviews (e.g. Gillan, 1985; Vashishta, 1986) are
available discussing work on the fluorite structured materials
and on the superionic silver compounds.

The study of $\beta"Al_2O_3$ (Wolf 1984; Wolf et al 1984c was
particularly illuminating as it showed how MD techniques can
be used to obtain subtle mechanistic information.

$\beta"Al_2O_3$ is an important layer structured superionic. As
illustrated in Figure 5, conduction planes containing the
mobile Na^+ ions are sandwiched between spinel structured
alumina layers which also contain Mg^{2+} (the formula of the
stoichiometric compound being $Na_2OMgO5Al_2O_3$). The compound is
closely related to the celebrated $\beta-Al_2O_3$, in which, however,
the spinel blocks are thicker and in which there is no Mg^{2+}.

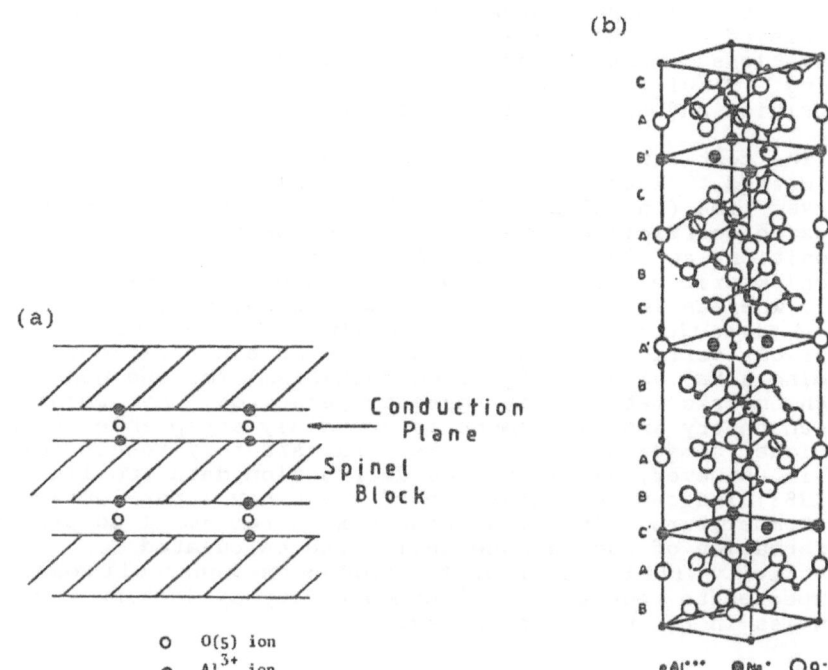

Figure 5: Structure of Na $\beta"-Al_2O_3$: a) schematic
illustration of layer structure with conduction planes
b) actual structure.

As normally prepared, both β and $\beta''Al_2O_3$ are non-stoichiometric. In the latter case the non-stoichiometry arises from the replacement of Mg^{2+} by Al^{3+} with the consequent creation of Na^+ ion vacancies in the conduction plane. And amongst the many unanswered questions concerning this material one of great importance concerns the role of non-stoichiometry in controlling the conductivity of the material. Other questions concern the variation of the conductivity with temperature, as there is good evidence from experimental studies that there is a change in the ion migration mechanism at about 500 K (Farrington and Briant, 1979). And in general the nature of the dynamics of the mobile Na^+ ions is poorly characterised.

To investigate these problems, Wolf (1984) and Wolf et al (1984c) performed an illuminating dynamical simulation study. They used a simulation box in which the triply primitive hexagonal unit cell of the crystal (containing 90 ions) had been quadrupled normal to the C-axis. The interatomic potentials were taken from those of the appropriate binary oxides; the type steps used in the simulation were 5×10^{-14} sec. or 4×10^{-14} sec. depending on the temperature which was varied from 300K to 700K. The simulations were performed on the CRAY-1S computer at the University of London Computer Centre, U.K.

The first problem to be investigated concerned the effect of non-stoichiometry on the conductivity. In non-stoichiometric β-Al_2O_3, one in six of the Na^+ sites are vacant; the closest approach to this composition which could be achieved with the box size which could be used in the calculations was one in which one in eight of the Na^+ ions are missing. Simulations were therefore performed on the latter composition and on the stoichiometric material.

The contrast between the behaviour of the two systems is very marked. Plots of the mean square displacement of Na^+ ions vs. time (the slopes of which are proportional to the diffusion coefficient) are illustrated for the two compositions at 550K in Figure 6. They show that in the stoichiometric material, the Na^+ diffusion coefficient is small, whereas in the non-stoichiometric solid it is high. Table 4 gives the calculated conductivities (obtained from the diffusion coefficient via the Nernst-Einstein relationship assuming a Na^+ vacancy migration mechanism) for the non-stoichiometric material at three temperatures. Agreement between theory and experiment is certainly acceptable at the two higher temperatures. The less satisfactory result for 300K is, however, explicable as diffraction data (Collin et al, 1983) indicate the presence of a vacancy super-lattice at this temperature. The simulation box is too small to permit the formation of such a supercell. The calculated conductivity will therefore be too high as the supercell would be expected to reduce the Na^+ conductivity by locking the Na^+ vacancies in to an ordered array.

Table 4   Calculated and experimental conductivities of
non-stoichiometric Na^+-β''-aluminia

T/K	Calculated $D/cm^2 \, s^{-1}$	Calculated $\sigma/\Omega^{-1} \, cm^{-1}$	Observed $\sigma/\Omega^{-1} \, cm^{-1}$
300	1.41×10^{-5}	1.53	0.014–0.160
550	2.24×10^{-5}	1.23	0.80
700	5.78×10^{-5}	2.49	1.24

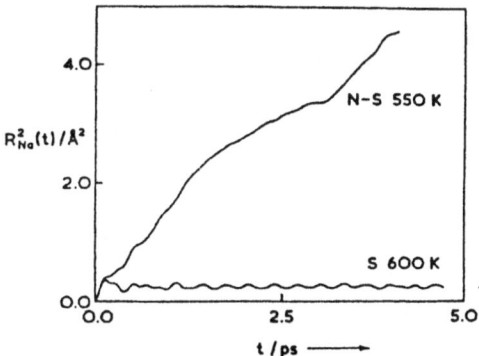

Figure 6:   Mean square displacement vs. time for Na^+ ions:
stoichiometric system(s) at 600K; non-stoichiometric (N-S) at
550K.

Information on the structural consequences of the non-stoichiometry is provided by the $Na^+...Na^+$ radial distribution functions, which are illustrated in Figure 7 for both stoichiometric and non-stoichiometric solids at 300K. The greater diffuseness of the r.d.f. for the latter is indicative of the greater disorder in the conduction plane of the non-stoichiometric compound.

We also note that the first peak in the r.d.f. of the non-stoichiometric material has a shoulder at high r, which is indicated by the arrow shown in the diagram. As argued by Wolf (1984) and Wolf et al (1984c) this feature can be explained in terms of relaxation of the n.n. Na^+ ions towards a Na^+ vacancy, and its occurrence indicates that there are well defined vacancies in the conduction plane. Migration would therefore be expected to take place via a conventional vacancy hopping mechanism. In contrast, at 700K, the shoulder has disappeared suggesting that vacancies are present at lattice sites for too short a period for the occurrence of appreciate relaxation of the surrounding lattice, indicating a much more 'continuous' type of migration mechanism.

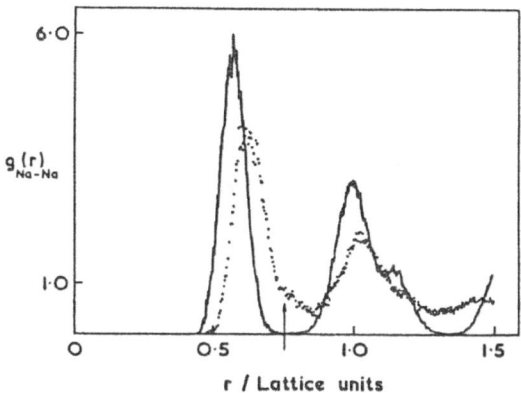

Figure 7: Radial distribution function for $Na^+...Na^+$ in non-stoichiometric Na-β"Al_2O_3 at 300K.

Further evidence for this fascinating change in the nature of the Na$^+$ ion dynamics is provided by study of the moment ratio, $P_\alpha(t)$, defined as

$$P_\alpha(t) = \frac{3\langle r_\alpha(t)^4\rangle}{5\{\langle r_\alpha(t)^2\rangle\}^2} \quad .$$

As noted, it has been shown by Hansen and MacDonald (1976) and Rahman (1976) that this ratio will tend to unity at large t when a continuous diffusion mechanism is operative. In contrast, strong deviations from unity indicate a hopping mechanism. Figure 8 shows P(t) vs. t plots for three temperatures. At higher temperatures P(t) approaches unity from which, however, it deviates strongly at the lower temperature.

The work of Wolf (1984) and Wolf et al (1984c) provides far greater detail of the nature of the Na$^+$ ion migration revealed by the simulations. It is clear that subtle effects may be predicted which it would be of great interest to investigate experimentally.

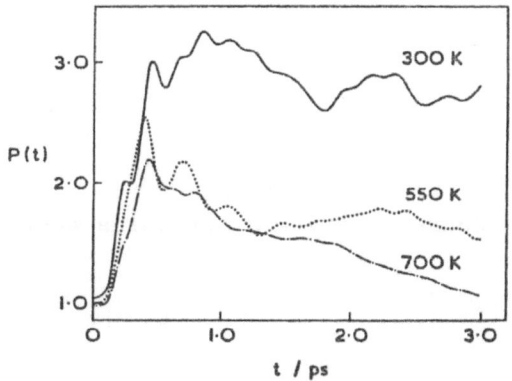

Figure 8: P(t) vs. t for Na$^+$ in Na$^+\beta''$-Al$_2$O$_3$.

4.3 Monte Carlo simulation

As noted, the MC technique has been extensively applied to complex systems; and the reviews and papers of Murch and co-workers (Murch 1975, 1982a,b; 1984; de Bruin and Murch, 1973) give a good indication of the scope and range of the technique. To illustrate the potential of the method we have chosen a recent study of Murch et al (1986) which also combines MC techniques with static simulation procedures. They studied the intriguing problems posed by Y^{3+} doped CeO_2 which in common with other doped fluorite oxides (e.g. Ca^{2+} doped ZrO_2) shows a maximum in the plot of conductivity vs. dopant concentration. Data obtained by Wang et al (1981) are shown in Figure 9. The effect is difficult to understand as the conductivity would normally be expected to increase with dopant concentration until the concentration of the mobile

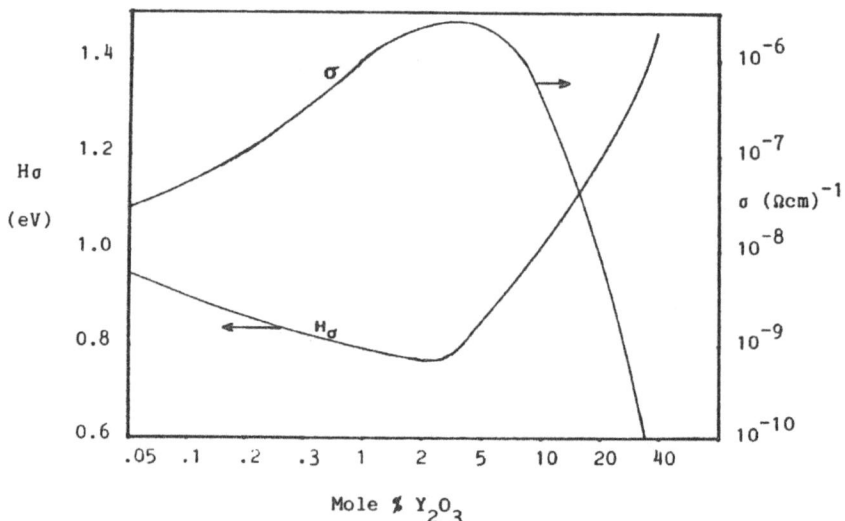

Figure 9: Conductivity and Arrhenius energies in Y_2O_3 doped CeO_2.

oxygen vacancies which are created as charge compensators for
the low valence impurity ions is 50 mole%. However, as noted
in Figure 9, the maximum occurs at a relatively low impurity
concentration (16 mole% equivalent to 8 mole% vacancies) and
the decrease in the conductivity after the maximum is
dramatic. We also note that the maximum in the conductivity
corresponds to a minimum in the Arrhenius energy of the
conductivity.

To investigate this problem Murch et al (1986) undertook
an MC simulation study of Y/CeO_2 for a wide range of dopant
concentrations, assuming a random distribution of dopant ions.
The procedure used was essentially that described in section
3.3; but the jump frequencies were taken as proportional to
the Arrhenius factor, $exp(-\Delta E_i/kT)$, with ΔE_i being calculated
using the static simulation techniques, for all possible
dopant environments for oxygen ions jumping by a vacancy
mechanism. In defining the dopant environment only first
neighbour cation sites with respect to the two oxygen ions
involved in the jump were considered. A weak d.c. electric
field was applied enabling us to calculate an important and
interesting quantity, i.e. the conductivity correlation
factor, f_I, defined as

$$f_I = \frac{2 kT \langle X \rangle}{nqa^2 E}$$ (18)

where $\langle X \rangle$ is the drift distance of ions of charge q after an
average of n jumps of length a in an applied field E. f_I
represents the efficacy of ion jumps in effecting
conductivity.

The results of the calculation are summarised in
Figures 10-12. We note that the maximum in the conductivity
is reproduced qualitatively. As regards the enthalpy, the
increase at higher dopant concentrations is well reproduced
but the calculated behaviour at low concentrations is less
satisfactory. Of particular interest is the variation of f_I
with dopant concentration where a pronounced decrease is
observed. Thus at higher concentrations vacancy jumps are
becoming decreasingly effective in leading to bulk ionic
conductivity - an observation that yields qualitative insight
into the origins of the observed maximum.

Clearly, further work remains to be done on these
fascinating systems. But the results obtained to date
indicate the power of the MC technique in examining diffusion
in complex solids, and the way in which that power is enhanced
by combination of MC with static simulation techniques.

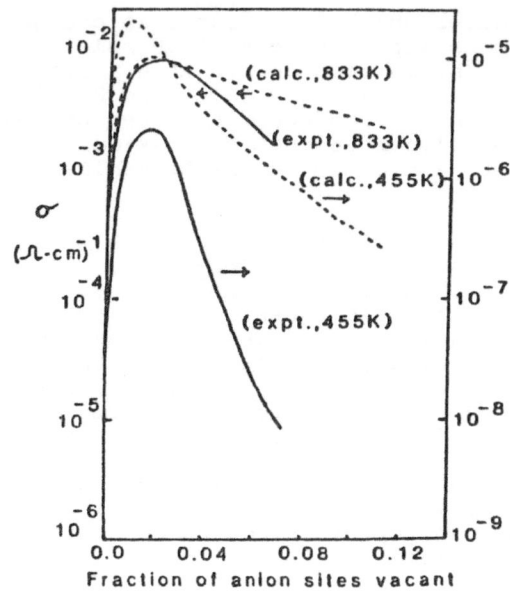

Figure 10: Calculated and experimental conductivities.

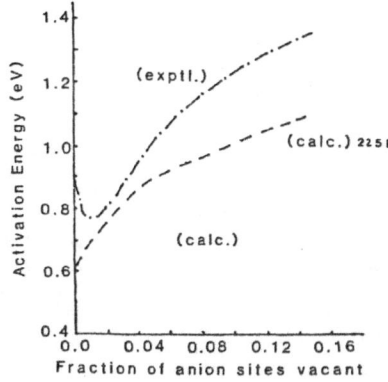

Figure 11: Calculated and experimental Arrhenius energies.

298

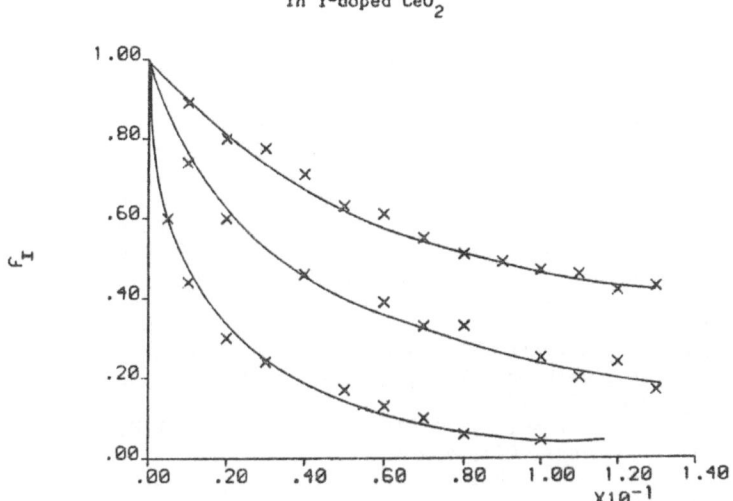

Figure 12: Plot of f_I vs. dopant concentration for
3 temperatures: 1200K, 833K and 455K (lowest curve).

5.   Summary and Conclusions

Computer simulation of defect properties of solids is
now a well established technique. The field is currently
moving in the direction of materials of greater complexity and
materials for which simple Born model potentials are
inadequate. A particularly important development is the
combination of classical simulation with quantum mechanical
calculations, where considerable progress can be expected in
the near future.

References

Beeman, D. (1976), J. Comput. Phys. 20, 130.
Bendall, P.J., Catlow, C.R.A., Corish, J. and Jacobs, P.W.M. (1984),
    J. Solid State Chem. 51, 159.
Catlow, C.R.A. (1973), J. Phys. C6, L64 (1973).
Catlow, C.R.A. (1981), in Non-Stoichiometric Oxides (ed. O.T. Sorensen),
    Academic Press.
Catlow, C.R.A. and Norgett (1976), UKAEA Report AERE-M. 2936.
Catlow, C.R.A. and Mackrodt, W.C. (Eds.), (1982), Computer Simulation of
    Solids, Lecture Notes in Physics, Vol. 166 (Springer-Verlag, Berlin).
Catlow, C.R.A. and Stoneham, A.M. (1983), J. Phys. C. 16, 4321.
Catlow, C.R.A., Diller, K.M. and Norgett, M.J. (1977a), J. Phys. C10,
    1395.
Catlow, C.R.A., Norgett, M.J. and Ross, T.A. (1977b), J. Phys. C10, 1627.
Catlow, C.R.A., Mackrodt, W.C., Norgett, M.J. and Stoneham, A.M. (1977c),
    Phil. Mag. 35, 177.
Catlow, C.R.A., Corish, J., Diller, K.M., Jacobs, P.W.M. and Norgett, M.J.
    (1979), J. Phys. C12, 451.
Catlow, C.R.A., James, R., Mackrodt, W.C. and Stewart (1982), Phys. Rev.
    B25, 1006.
Catlow, C.R.A., Chadwick, A.V., Greaves, G.N. and Moroney, L.M. (1984),
    Nature 312, 601.
Catlow, C.R.A., Doherty, M., Price, G.D., Sanders, M.J. and Parker, S.C.
    (1986), Mat. Sci.Forum 7, 163.
Chadwick, A.V. (1982), in 'Mass Transport in Solids' (eds. F. Bénière
    and C.R.A. Catlow), Plenum Press.
Cicotti, G., Jacucci, G. and MacDonald, I.R. (1976), Phys. Rev. A13, 428.
Colbourn, E.A. and Mackrodt, W.C. (1982), Surf. Sci. 117, 571.
Collin, G. et al (1983), in 'Solid State Ionics 83' (eds. M. Kleitz,
    B. Saporal and D. Ravaire) (North Holland), p. 311 (1983).
Cormack, A.N. (1986) - to be published.
de Bruin, H.J. and Murch, G.E. (1973), Phil. Mag. 27, 1475.
Dick, B.G. and Overhauser, A.W. 1958, Phys. Rev. 112, 90.
Dixon, M. and Gillan, M.J. (1980a), J. Phys. C13, 1901.
Dixon, M. and Gillan, M.J. (1980b), J. Phys. C13, 1919.
Farrington, G.C. and Briant, J. (1979), in 'Fast Ion Transport Solids',
    (eds. P. Vashishta, J.N. Mundy and G.K. Shenoy), North Holland.
Faux, I.D. and Lidiard, A.B. (1971), Z. Naturforsh. A26, 62.
Fowler, P.W. and Pyper, N.C. (1985), Proc. Roy. Soc. A398, 377.
Gillan, M.J. (1985), Physica, 131B, 157.
Gillan, M.J. and Jacobs, P.W.M. (1983), Phys. Rev. B28, 759.
Gillan, M.J. and Harding, J.H. (1986) - in press.
Gordon, R.G. and Kim, Y.S. (1972), J. Chem. Phys. C12, 431.
Hagon, J. (1986), Ph.D. Thesis, University of Newcastle (also available
    as UKAEA Report AERE TP 1177).
Hansen, J.P. and MacDonald, I.R. (1976), in 'The Theory of Simple
    Liquids' (Academic Press).
Hayes, T.L. and Boyce, R. (1983), Solid State Physics, 37, 173.
Jacobs, P.W.M. and Moscinski, J. (1985), Physics, B131, 175.
Jacucci, G. and Rahman, A. 1978, J. Chem. Phys. 69, 4117.
Lechner, R.E. (1983), in 'Mass Transport in Solids' (eds. F. Bénière
    and C.R.A. Catlow), Plenum Press.
Leslie, M. (1982), SERC Daresbury Laboratory Report, DL-SCI-TM31T.
Lewis, G.V. (1984), Ph.D. Thesis, University of London.
Lewis, G.V. and Catlow, C.R.A. (1985), J. Phys. C18, 1149.
Lewis, G.V. and Catlow, C.R.A. (1986), J. Phys. Chem. Solids, 47, 89.
Lidiard, A.B. (1973), Proc. International Conference on Defects in
    Semiconductors, Reading, UK.

Mackrodt, W.C. (1982), in 'Computer Simulation of Solids' (Eds. C.R.A. Catlow and W.C. Mackrodt), Lecture notes in Physics, Vol. 166 (Springer-Berlin).

Mackrodt, W.C. (1984), in 'Transport in Non-Stoichiometric Compounds' (Eds. G. Petot-Ervas, Hj Matzke and C. Monty),(North Holland).

Mackrodt, W.C. and Stewart, R.F. (1979), J. Phys. C12. 431.

Mackrodt, W.C., Stewart, R.F., Campbell, J.C. and Hillier, I.M. (1980), J. de Phys. C7, 64.

Mackrodt, W.C., Colbourn, E.A. and Kendrick, J. (1980a), Corporate Laboratory Report CL-R/81/1637/A.

Mainwood, A. and Stoneham, A.M. (1982), Physica B116, 101.

Meyer, M. (1982), in 'Computer Simulations of Solids' (Eds. C.R.A.Catlow and W.C. Mackrodt), Lecture Notes in Physics, Vol. 166.

Mott, N.F. and Littleton, M.J. (1938), Trans. Farad. Soc. 34, 485.

Murch, G.E. (1984), in 'Diffusion in Crystalline Solids' (Ed. G.E. Murch and A.S. Nowick), Academic Press.

Murch, G.E. (1975), Phil. Mag. 32, 1129.

Murch, G.E. (1982a), Phil. Mag. A46, 151.

Murch, G.E. (1982b), Phil. Mag. A46, 575.

Murch, G.E., Murray, A.D. and Catlow, C.R.A. (1986), Solid State Ionics 18, 196.

Norgett, M.J. (1974), UKAEA Report, AERE-R.7650.

Parker, S.C. and Price, G.D. (1984), Phys. Chem. Mineral 10, 209.

Pople, J.A. and Beveridge, D.L. (1970), 'Approximate Molecular Orbital Theory' (McGraw-Hill, New York).

Rahman, A. (1976), J. Chem. Phys. 65, 4845.

Sanders, M.J., Leslie, M. and Catlow, C.R.A. (1984), J. Chem. Soc. Chem. Comm. 1271.

Sangster, M.J. and Dixon, M. (1976), Adv. Phys. 25, 247.

Sangster, M.J. and Stoneham, A.M. (1980), Phil. Mag. 43, 597.

Sangster, M.J., Schroder, U. and Atwood, R.M. (1978), J. Phys. C11, 1523.

Saul, P., Catlow, C.R.A. and Kendrick, J. (1985), Phil. Mag. B51. 107.

Szabo, A. and Ostlund, N.S. (1984), 'Modern Quantum Chemistry' (MacMillan).

Stoneham, A.M. (1975) in 'The Theory of Defects in Solids', Oxford University Press.

Taylor, R. (1982), in 'Computer Simulation of Solids' (Eds. C.R.A.Catlow and W.C. Mackrodt), Lecture Notes in Physics, Vol. 166 (Springer-Berlin).

Taylor, R. (1985), Physica B151, 103.

Tosi, M. (1964), Solid State Physics (Eds. F. Seitz and S. Turnbull), 161.

Vail, J.M., Harker, A.H., Harding, J.H. and Saul, P. (1984), J. Phys. C17, 3401.

Vashishta, P. and Rahman, A. (1979), in 'Fast Ion Transport in Solids' (Eds. P. Vashishta et al), North-Holland.

Vashishta, P. (1986), Solid State Ionics - in press.

Vineyard, G. (1957), J. Phys. Chem. Solids, 3. 157.

Wang, D.Y., Park, D.S., Griffiths, J. and Nowick, A.S. (1981), Solid State Ionics 2, 95.

Wedepohl, P.T. (1967), Proc. Phys. Soc. 92, 79.

Wolf, M.L. (1984), Ph.D. Thesis, University of London.

Wolf, M.L., Walker, J.R. and Catlow, C.R.A. (1984a), J. Phys. C17, 6623.

Wolf, M.L. and Catlow, C.R.A. (1984b), J. Phys. C17, 6635.

Wolf, M.L., Walker, J.R. and Catlow, C.R.A. (1984c), Solid State Ionics 13, 33.

CALCULATION OF DEFECT PROCESSES AT HIGH TEMPERATURE

J. H. Harding

Theoretical Physics Division
AERE Harwell
Didcot, Oxon OX11 0RA, England

1. INTRODUCTION

In the past most calculations on defects have been of defect energies. The problem of defect entropies (except for the configurational terms; see the lecture of Corish in this volume) has been ignored. Further, most calculations of these energies have been performed using three basic assumptions:

a)   the defect energy calculated at constant volume (usually taken to mean constant lattice parameter) using the zero temperature lattice parameter may be compared with enthalpies derived from experiment,

b)   any effect of the defect on the lattice vibrations may be ignored for the purpose of calculating defect energies,

c)   in calculating defect migration and activation energies we may use Vineyard's[1] expression for the hopping rate

$$\Gamma = \tilde{\nu} \, \exp(- \frac{\Delta E}{kT}) \qquad\qquad (1)$$

where ΔE is the activation energy and $\tilde{\nu}$ is an effective frequency given by the ratio of the normal mode frequencies of the defect in its stable site $\{\nu_i\}$ and at the saddle point $\{\nu_i'\}$ i.e. $\tilde{\nu} = (\prod_i \nu_i)/(\prod_i \nu_i')$. The unstable mode corresponding to motion of the defect across the saddle plane is excluded from the denominator.

To investigate assumptions (a) and (b) in detail we must calculate energies and entropies of defects as a function of temperature. Throughout this paper we shall use the quasi-harmonic approximation; that is we assume that the temperature dependence of internal energies and lattice frequencies is due only to the lattice expansion. To investigate (c) we must compare the results of such calculations with a molecular dynamics simulation. We shall first consider the calculation of defect parameters (sections 2 and 3) before turning to consider defect motion in section 4.

2. THERMODYNAMICS OF POINT DEFECTS

Most experiments measure the Gibbs free energy of defects at constant pressure, g_p. It is, however, simpler to calculate the Helmholtz energy at constant volume, f_v. These quantities may be shown to be equal by the following argument[2]. The relation between the Gibbs free energy at constant pressure and at constant volume for the defect process is

$$g_p = g_v - \left(\frac{\partial G}{\partial P}\right)_T \Delta P = g_v - V\Delta P \tag{2}$$

where ΔP is the pressure change caused by the defect process at constant volume. Since by definition $g_v = f_v + V\Delta P$ the required relation $g_p = f_v$ follows.

To proceed further we require the following relation,

$$\left(\frac{\partial f_v}{\partial T}\right)_P = \left(\frac{\partial f_v}{\partial T}\right)_V + \left(\frac{\partial V}{\partial T}\right)_P \left(\frac{\partial f_v}{\partial V}\right)_T \tag{3}$$

We can use this and the relation proved above to obtain the following two relations which connect the experimentally measured enthalpy and entropy (h_p, s_p) to the constant volume quantities that can be calculated.

$$h_p = u_v - T\left(\frac{\partial V}{\partial T}\right)_P \left(\frac{\partial f_v}{\partial V}\right)_T \tag{4a}$$

$$s_p = s_v - \left(\frac{\partial V}{\partial T}\right)_P \left(\frac{\partial f_v}{\partial V}\right)_T \tag{4b}$$

These equations may easily be shown to be equivalent to those used elsewhere in this volume. Using equation (4a) we can show that the first assumption made in defect calculations may often be reasonable[2,3]. If we ignore the volume variation of the entropy, s_v, and assume that the internal energy u_v may be written as a Taylor expansion about the internal energy at zero temperature

$$u_v = u_v^o + \left(\frac{\partial u_v}{\partial V}\right)^o \delta V \approx u_v^o + \left(\frac{\partial u_v}{\partial V}\right)^o T\left(\frac{\partial V}{\partial T}\right)_P \tag{5}$$

then, assuming that all quantities vary linearly with temperature and/or volume we arrive at the result $h_p \approx u_v^o$. This is not only our assumption (a) but also shows why the experimental assumption that h_p is constant is frequently correct. Applying similar arguments to (4b) we obtain

$$s_p \approx s_v - \left(\frac{\partial V}{\partial T}\right)_P \left(\frac{\partial u_v}{\partial V}\right)^o . \tag{6}$$

This suggests that an important (frequently in fact dominant) contribution to the measured entropy has nothing to do with lattice vibrations but comes from the variation of the internal energy with volume. This is one origin of a correlation, often observed in diffusion measurements, between the logarithm of the pre-exponential factor and the activation energy; the so-called Meyer-Neldel rule. This rule has been discussed by Sangster and Stoneham (to be published).

3. CALCULATING DEFECT ENTROPIES

To proceed further, we must calculate the effect of point defects on the lattice vibrations. Within the quasi-harmonic approximation we may write the vibrational entropy as

$$S = k \sum_{i=1}^{3N} \left\{ \frac{\hbar\omega_i}{kT} \left(\exp\left(\frac{\hbar\omega_i}{kT}\right) - 1 \right)^{-1} - \ln\left(1 - \exp\left(-\frac{\hbar\omega_i}{kT}\right)\right) \right\} \qquad (7)$$

Using the obvious definition of the defect entropy of $\Delta S_d = S$(defect crystal) − S(perfect crystal) where the defect crystal contains a single defect and assuming the high temperature limit ($\hbar\omega_i \ll kT$) we obtain

$$\Delta S_d = -k\ln \frac{\prod_{i=1}^{3N'} \omega_i'}{\prod_{i=1}^{3N} \omega_i} + 3k(N'-N)(1 - \ln\frac{\hbar}{kT}) . \qquad (8)$$

where the primes denote the defect crystal. The second term corrects the dimensions of the first term for $N' \neq N$. Equation (8) defines defect entropies with respect to ions brought in from rest at infinity or taken to there and so corresponds to the definition used in static lattice codes such as HADES.

We calculate entropies from equation (8) using the large crystallite model. The crystal is divided into two regions; an inner region containing the defect and surrounding ions where the ions may vibrate and an outer region where the ions are held fixed in their static relaxed positions. In calculating the entropies of charged defects there is a correction which arises because the effect of the long-range distortion field does not vanish no matter how large a crystallite is chosen. The problem of calculating this correction has been considered by Gillan and Jacobs[4] and we apply it here.

The SHEOL code[5] thus calculates entropies in two stages. Firstly the relaxed positions of the ions around the defect are obtained by a calculation of the HADES type (see the lecture of Catlow). Then, using these positions the frequencies of the defect and perfect lattice crystallite are calculated. The shell model is used to obtain a good representation of the phonon spectrum. Such a calculation gives the internal energy, u_v, and entropy at constant volume, s_v, of the defect.

Since we require these quantities as functions of temperature we must investigate whether the interionic potentials available, which are usually obtained either by fitting to low temperature data or by calculations ignoring thermal effects, are suitable for calculations at high temperature. We can do this by calculating high temperature properties with these potentials and comparing with experiment. A useful test is to calculate the thermal expansion within the quasi-harmonic approximation. Using standard codes we may readily calculate the internal energy and entropy of the crystal as a function of volume. We can then obtain the thermal expansion by requiring

$$P = -\left(\frac{\partial F}{\partial V}\right)_T = 0 . \qquad (9)$$

since atmosphere pressure is negligible. This gives a volume-temperature relation. Examples of the calculation are shown in Fig. 1 for KCl and CaF$_2$. The agreement can be very good, suggesting both that the potential and that the quasi-harmonic approximation are valid. It is possible, obviously, to calculate other properties. Fig. 2 shows a calculation of the isothermal (κ_T) and adiabatic (κ_s) compressibilities for AgBr. However, care should be taken to ensure that the correct quantity is calculated. The definition of the quantities shown in Fig. 2 are

$$ S = k \sum_{i=1}^{3N} \left\{ \frac{\hbar\omega_i}{kT} \left(\exp(\frac{\hbar\omega_i}{kT}) - 1 \right)^{-1} - \ln(1 - \exp(-\frac{\hbar\omega_i}{kT})) \right\} \tag{7} $$

Using the obvious definition of the defect entropy of ΔS_d = S(defect crystal) - S(perfect crystal) where the defect crystal contains a single defect and assuming the high temperature limit ($\hbar\omega_i$ <<kT) we obtain

$$ \Delta S_d = -k\ln \frac{\prod\limits_{i=1}^{3N'} \omega_i}{\prod\limits_{i=1}^{3N} \omega_i} + 3k(N'-N)(1-\ln\frac{\hbar}{kT}) . \tag{8} $$

where the primes denote the defect crystal. The second term corrects the dimensions of the first term for $N' \neq N$. Equation (8) defines defect entropies with respect to ions brought in from rest at infinity or taken to there and so corresponds to the definition used in static lattice codes such as HADES.

We calculate entropies from equation (8) using the large crystallite model. The crystal is divided into two regions; an inner region containing the defect and surrounding ions where the ions may vibrate and an outer region where the ions are held fixed in their static relaxed positions. In calculating the entropies of charged defects there is a correction which arises because the effect of the long-range distortion field does not vanish no matter how large a crystallite is chosen. The problem of calculating this correction has been considered by Gillan and Jacobs[4] and we apply it here.

The SHEOL code[5] thus calculates entropies in two stages. Firstly the relaxed positions of the ions around the defect are obtained by a calculation of the HADES type (see the lecture of Catlow). Then, using these positions the frequencies of the defect and perfect lattice crystallite are calculated. The shell model is used to obtain a good representation of the phonon spectrum. Such a calculation gives the internal energy, u_v, and entropy at constant volume, s_v, of the defect.

Since we require these quantities as functions of temperature we must investigate whether the interionic potentials available, which are usually obtained either by fitting to low temperature data or by calculations ignoring thermal effects, are suitable for calculations at high temperature. We can do this by calculating high temperature properties with these potentials and comparing with experiment. A useful test is to calculate the thermal expansion within the quasi-harmonic approximation. Using standard codes we may readily calculate the internal energy and entropy of the crystal as a function of volume. We can then obtain the thermal expansion by requiring

$$ P = -(\frac{\partial F}{\partial V})_T = 0 . \tag{9} $$

since atmosphere pressure is negligible. This gives a volume-temperature relation. Examples of the calculation are shown in Fig. 1 for KCl and CaF$_2$. The agreement can be very good, suggesting both that the potential and that the quasi-harmonic approximation are valid. It is possible, obviously, to calculate other properties. Fig. 2 shows a calculation of the isothermal (κ_T) and adiabatic (κ_s) compressibilities for AgBr. However, care should be taken to ensure that the correct quantity is calculated. The definition of the quantities shown in Fig. 2 are

306

$$\frac{1}{\kappa_T} = V \left(\frac{\partial^2 F}{\partial V^2}\right)_T \ ; \ \frac{1}{\kappa_s} = V \left(\frac{\partial^2 U}{\partial V^2}\right)_s \ ; \ \frac{1}{\kappa_{stat}} = V \left(\frac{\partial^2 \Phi}{\partial V^2}\right)^o$$

where Φ is the crystal potential energy alone. At zero temperature all these definitions amount to the same thing. However in general they do not. In particular, the compressibility calculated simply by expanding the static lattice corresponds to no measurable quantity at all.

When we are satisfied that the potential is usable over the temperature range we may calculate u_v and s_v as a function of temperature. Table I shows the result of such a calculation for the Schottky defect in KCℓ.

Table 1 Free energy of Schottky defect in KCℓ

Temperature (K)	u_v(eV)	s_v(k)	g_p(eV)
856.2	2.233	-0.97	2.305
903.3	2.190	-1.23	2.286
944.2	2.146	-1.49	2.267
979.1	2.102	-1.78	2.252
1007.6	2.056	-2.07	2.236

$$h_p = 2.70eV, \ u_v^o = 2.64eV, \ s_p = 5.35k$$

As can be seen both from Table 1 and from Fig. 3; although both u_v and s_v vary strongly with temperature g_p is a linear function of temperature and so h_p and s_p are constant. Also the approximate equality $h_p \simeq u_v^o$ works well. However s_p and s_v are quite different. The positive sign of s_p is due to the temperature variation of u_v. Agreement with the latest experiments is fair, the results of Acuna and Jacobs[6] give $h_p = 2.50eV$, $s_p = 7.5$-$7.9k$.

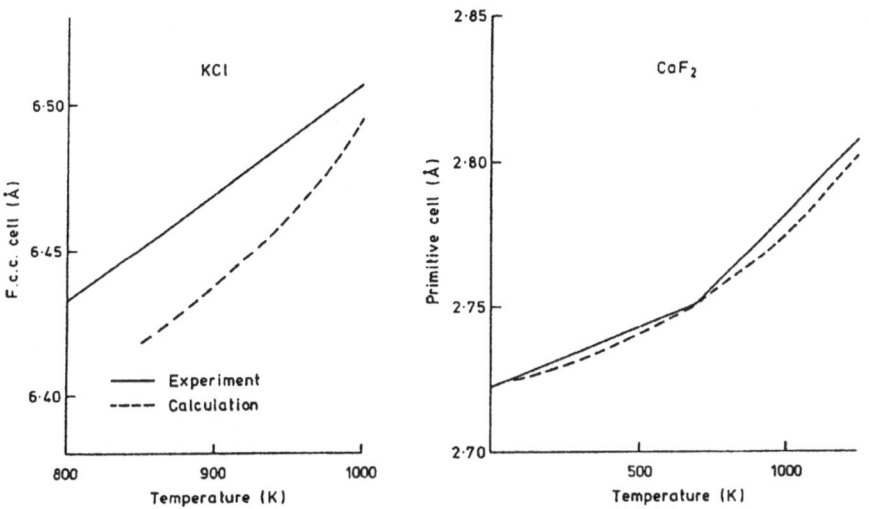

Fig. 1. Linear expansion of ionic crystals.

Figure 3 also shows the free energy of formation of the Frenkel defect in CaF_2. Here the assumption (a) breaks down. The free energy curve fits to two straight lines. Below 800K we have h_p = 2.81eV, s_p = 5.4k. Above this temperature the values are quite different; h_p = 3.17eV, s_p = 10.8k. This allows us to resolve an ancient experimental controversy. The latest experimental data[7] suggests h_p = 2.7eV, s_p = 5.5k apparently contradicting the older work of Ure[8] which gave h_p = 2.82eV, s_p = 13.1k. The calculations show that the experiments, done at different temperatures, do not contradict each other. The error lies in assuming that h_p and s_p are constant over the whole temperature range (for a fuller discussion see ref. 3).

Since we can calculate u_v and s_v as functions of volume, it is clearly possible to calculate volumes of formation $v_p = - \kappa_T V (\frac{\partial f_v}{\partial V})_T$. Examples of such calculations are shown in Table 2.

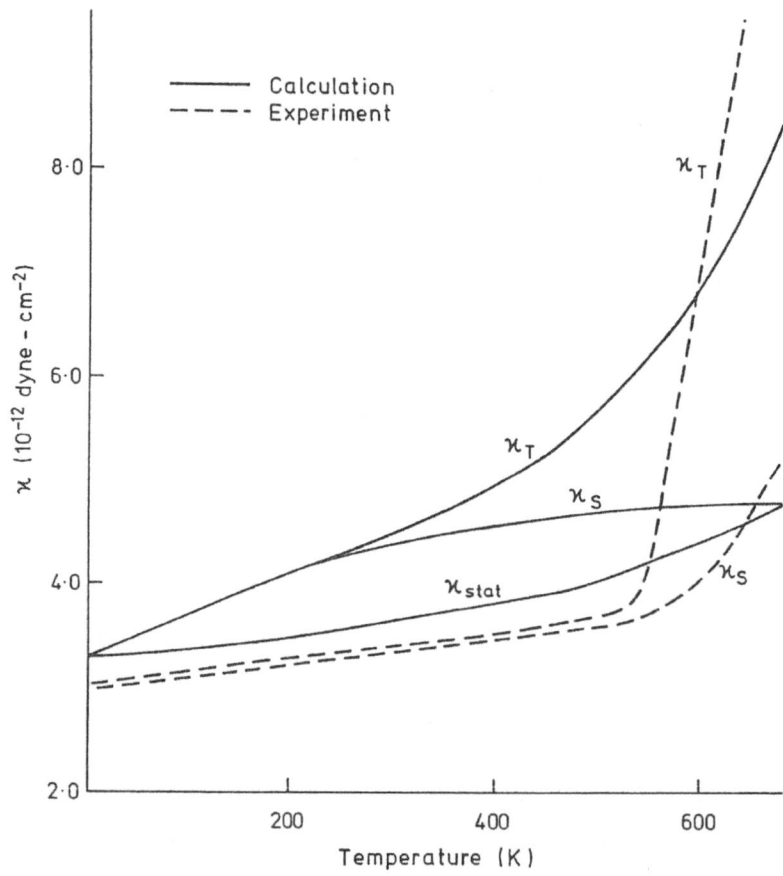

Fig. 2. Compressibilities for Ag Br.

Table 2   Volumes of defect processes (cm^3 $mole^{-1}$)

System	Calculation	Expt.	Ref.
Schottky defect formation KCℓ	56.9	63.8±5	9
cation vacancy migration KCℓ	7.5	8±1	9
bound vacancy migration CaF_2	0.79	1.74	10
anion Frenkel formation CaF_2	10.5	9.8±1.6	11

It is clear from Table 2 that atomistic methods are capable of calculating these quantities and that the various semi-empirical schemes that have been proposed[12] are not necessary.

4.   THE VINEYARD RATE THEORY

The third assumption made in defect calculations is that we may use equation (1) for the hopping rate. Vineyard derived this by considering a constant volume ensemble and making two further assumptions:

i)   it is sufficient to expand the potential energy about the saddle point to quadratic order (and so obtain the frequencies v_i)

ii) every crossing of the saddle plane represents a successful defect jump.

The first assumption may be tested by static lattice calculations, plotting the potential energy surface in the saddle plane. Two recent calculations[5,16] have shown that in some simple but important cases the assumption may break down entirely. The saddle point can on occasion even be a maximum in the saddle plane rather than a minimum. It is however possible[5] to overcome this problem while still remaining within the spirit of the Vineyard method.

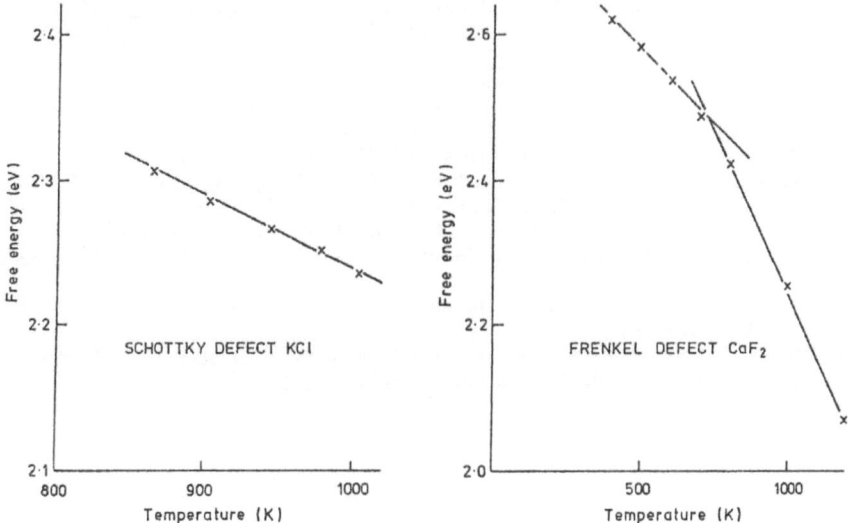

Fig. 3.   Defect free energies.

Of more interest is assumption (ii). This has been investigated for rare gas solids by Bennett[13] and Jacucci et al.[14] and more recently for ionic solids by Gillan et al.[15] All these investigations use molecular dynamics, but the simulation is constrained so that the system spends most of the time near the saddle point rather than executing (to us uninteresting) vibrations about the equilibrium position. This allows the generation of a set of trajectories of defect hopping at the saddle plane. Details of the methods are given in the references.

The calculations of Gillan et al. on a rigid ion model of CoO suggest that the Vineyard result is qualitatively correct, but it appears to overestimate the migration rate by about a factor of 2. The migration energies are very similar however. It is clear that more work needs to be done in this area, possibly by the consideration of suitable model systems[17].

5. CONCLUSIONS

We have considered ways of calculating defect processes at high temperature, and investigated the assumptions made in conventional static lattice calculations. Clearly the calculations we have described here are more time-consuming than simple static lattice methods and it will not always be necessary to go to the level of sophistication considered here. However the value of the more detailed calculations lies in pointing out that the conventional calculations have their limitations and in indicating where these limitations are likely to be found.

REFERENCES

1. G. H. Vineyard (1957) J.Phys.Chem.Solids 3 121.
2. M. J. Gillan (1981) Phil.Mag. 43 301.
3. J. H. Harding (1985) Phys.Rev.B. in press.
4. M. J. Gillan and P. W. M. Jacobs (1983) Phys.Rev. 28 759.
5. J. H. Harding (1985) Physica 131B 13.
6. L. A. Acuna and P. W. M. Jacobs (1980) J.Phys.Chem.Sol. 41 595.
7. P. W. M. Jacobs and S. H. Ong (1976) J.de Physique (Colloque) C7 331.
8. R. W. Ure (1957) J.Chem.Phys. 26 1363.
9. D. Lazarus, D. N. Yoon, and R. N. Jeffrey (1971) Z. Naturf.(a) 26 56.
10. J. J. Fontanella, M. A. Wintersgill, A. V. Chadwick, R. Saghefian and C. G. Andeen (1981) J.Phys. C14 2451.
11. J. Oberschmidt and D. Lazarus (1980) Phys.Rev. B21 5823.
12. P. Varotsos and K. Alexopoulos (1977) J.Phys.Chem.Sol. 38 997.
13. C. H. Bennett (1975) in 'Diffusion in Solids: Recent Developments' eds. J.J. Burton and A.S. Nowick (New York, Academic).
14. G. Jacucci, M. Toller, G. Delorenzi and C. P. Flynn (1984) Mat.Sci. Forum 1 187.
15. M. J. Gillan, J. H. Harding and R-J Tarento (1985) Harwell Report No. AERE-M.3494.
16. M. J. L. Sangster and A. M. Stoneham (1984) J.Phys. C17 6093.
17. C. W. McCombie and M. Sachder (1975) J.Phys. C8 L413.

THE NATURE OF DEFECTS ON SOLID SURFACES AS STUDIED BY ELECTRON

SPECTROSCOPY

Victor E. Henrich

Applied Physics, Yale University

New Haven, Connecticut, 06520

INTRODUCTION

The study of bulk defects in solids has been an active area of both experimental and theoretical research for many years, and, as the other articles in this volume attest, a great deal is known about them and their properties. Our understanding of surface defects on solids is, on the contrary, in its infancy. The whole field of surface science has only blossomed during the last 15 - 20 years, largely because the experimental techniques necessary for the controlled study of surfaces, including ultrahigh vacuum (UHV), were not previously widely available. With the advent of commercial UHV systems and surface-sensitive electron spectrometers of various types, a growing segment of the scientific community began trying to unravel the mysteries of the physics and chemistry of surfaces. As with any new field, the logical starting point was the study of simple systems, i.e., nearly perfect single-crystal surfaces. Work on perfect surfaces served as a proving ground for both the new experimental techniques and theoretical approaches. An extremely important aspect of surface science is the interaction of surfaces with adsorbed molecules, and studies in that area also began by using perfect surfaces as substrates.

Surface science is still a relatively young field, and even to date only a small fraction of the surface science literature deals directly with defects. Everyone knows that they are there, even on the best surfaces, but they are complicated to treat theoretically and difficult to handle experimentally. The net result is that little attention has been paid to surface defects.

This paper will give a brief overview of what is known about surface defects. The types of defects that can exist on surfaces are considered in Section II. Some of the experimental techniques that can be used to study surface defects are described in Section III, and theoretical approaches to surface defects are considered in Section IV. Section V presents some of the major conclusions that have been drawn about defects on solid surfaces.

TYPES OF SURFACE DEFECTS

There are three major classes of defects that can occur on single crystal surfaces: defects that are extended in the plane of the surface, such as steps and kinks; point defects such as adatoms, vacancies or substitutional or interstitial atoms; and the termination at the surface of defects that are extended in the bulk, such as line dislocations. To get a feeling for the differences between these types of defects we will first consider idealized models of their geometry.

Extended Defects

No real single crystal surface can be atomically flat over macroscopic distances. Even the best cleaved surfaces contain a large number of atomic height steps that separate one geometrically perfect terrace from another. An example of a surface step is shown in Fig. 1, which is an idealized model of {111} steps separating (111) terraces for a face-centered-cubic (FCC) lattice. (The notation used here for steps indicates the crystallographic plane that intersects the surface and on which the step-edge atoms lie.) This surface is used for illustration since it is the most densely packed, and hence generally most stable, surface for FCC metals. The primary characteristic of a step is the change in ligand coordination for atoms at the step edge. In the FCC structure bulk atoms have twelve nearest neighbors. On a (111) surface plane the coordination is reduced to nine, while atoms along the upper edge of the step shown have only seven nearest neighbors. Atoms along the bottom of the step have ten nearest neighbor ligands, more than on terraces. The bottom of a step also provides sites for adsorption in which an atom or molecule can coordinate to more substrate atoms than on terraces. The step in Fig. 1 contains a kink to show the further reduction in ligand coordination when two surface steps intersect.

The atomic geometry at steps becomes more complex in compounds. Figure 2 shows {100} steps separating (100) terraces for the perovskite

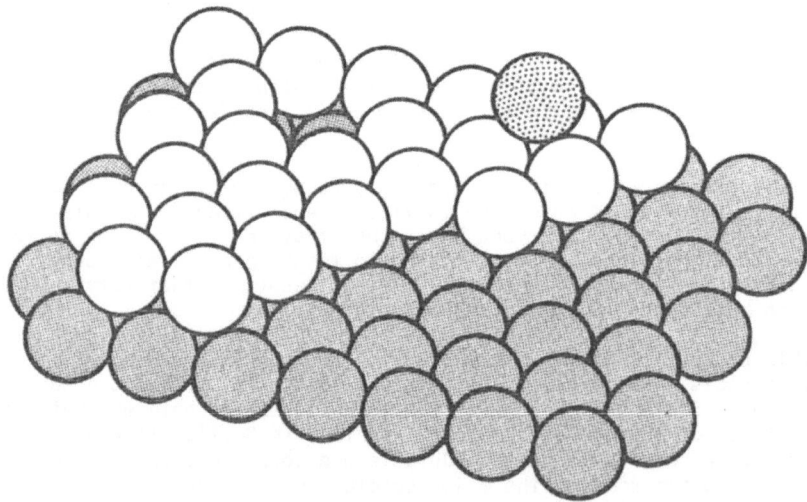

Fig. 1. Model of a face-centered-cubic (111) surface containing a step, kink, vacancy and adatom.

structure, ABO_3. There are two types of (100) surfaces in perovskite, one having an AO composition and the other BO_2. The reduction in ligand coordination for all three types of atoms at step edges and kinks can be seen. A similar step geometry is seen in Fig. 3 for a {100} step on a (100) surface of the rocksalt structure.

The above models of steps are idealized in that they assume that there is no distortion (relaxation) of the bulk lattice at the surface or at steps. The former is largely true for most real surfaces, but the latter may not be. Experimental studies of geometric distortion at steps is not available, but calculations for ionic compounds indicate significant relaxation may occur at steps and kinks.[1,2]

Point Defects

Point defects on surfaces are localized on an atomic scale. Figure 1 shows idealized models for a single vacancy on the FCC (111) surface and for a single adatom. A surface O vacancy is shown on the BO_2 plane of the (100) perovskite surface in Fig. 2. Somewhat more complicated surface anion vacancies are shown for the rutile (110) surface in Fig. 4, and both anion and cation vacancies are shown for the rocksalt (100) surface in Fig. 3.

The effect of point vacancies or adatoms on the electronic and chemical properties of a surface range from relatively small for some metals to extremely large for ionic compounds; these will be discussed in Section V. Substitutional or interstitial atoms can also have a

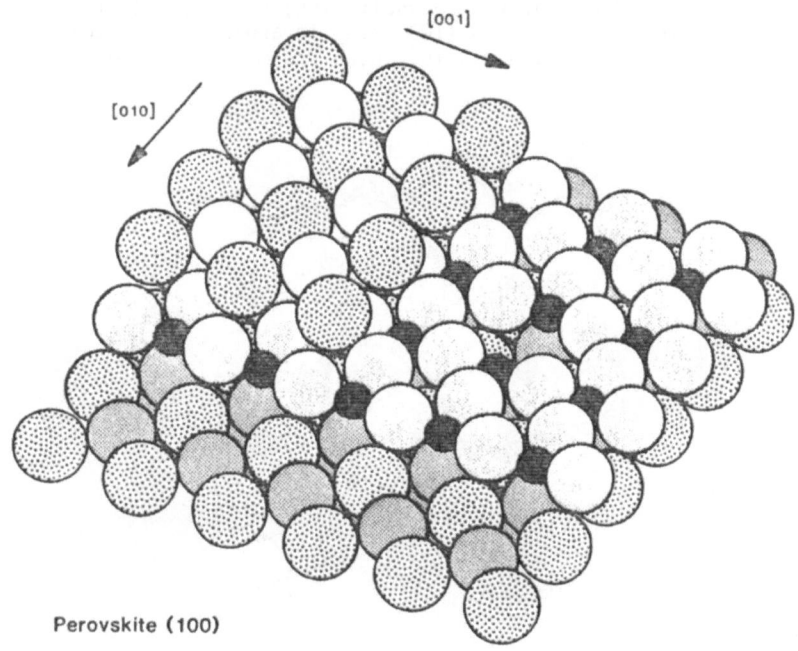

Perovskite (100)

Fig. 2. Model of the perovskite (100) surface. Small solid circles are B cations, large speckled circles are A cations, remainder are O anions. Two {100} steps to (100) terraces are shown, as is an O-vacancy defect.

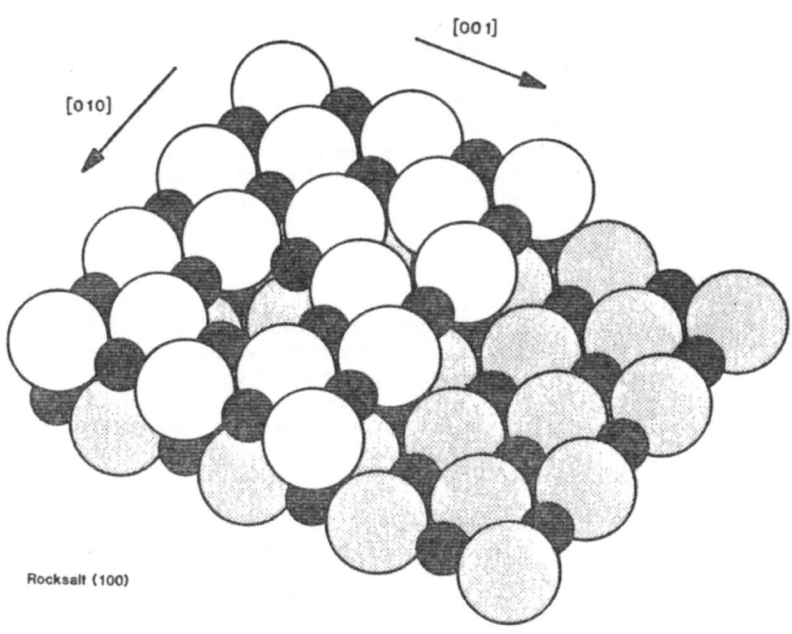

Fig. 3. Model of the rocksalt (100) surface. Small circles
are metal cations. A {100} step to another (100) terrace is
shown, as are both missing anion and missing cation defects.

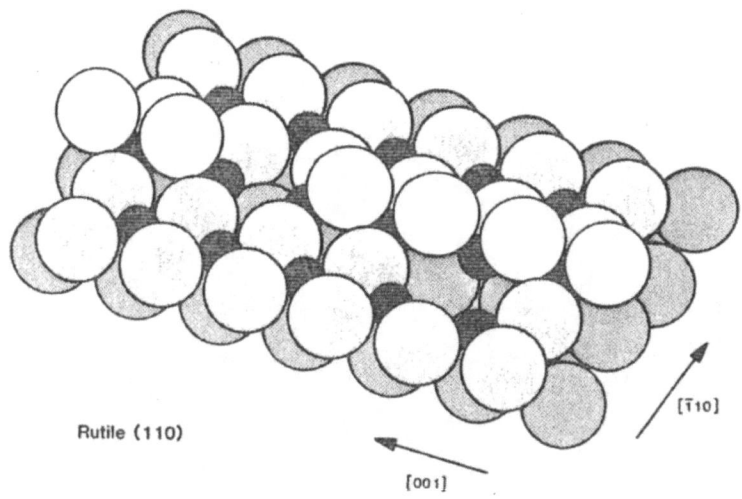

Fig. 4. Model of the rutile (110) surface. Two types of anion
vacancy defects are shown.

major effect on the properties of many surfaces, particularly if their valence state upsets local charge neutrality.[2,3]

Termination of Bulk Defects

Extended bulk defects such as edge and screw dislocations cannot simply end inside a crystal. They must either close on themselves, forming loops, or they must terminate at surfaces. Although essentially nothing is known either theoretically or experimentally about this type of defect on surfaces, they must be present, particularly on metal surfaces, and may exist in numbers large enough to affect surface chemical properties. They are, after all, the sites at which pits form when single-crystal surfaces are chemically etched.

Calculations of the positions of atoms around both edge and screw dislocations in the bulk FCC lattice have been performed by Cotterill and Doyama,[4] and models for the surface geometry that would result when those dislocations intersect the (111) surface are shown in Figs. 5 and 6, respectively. Both dislocations have <110> Burgers vectors and {111} slip planes and intersect the (111) surfaces shown slightly off from normal incidence. The models in Figs. 5 and 6 are again idealized in that, although account was taken of atomic relaxation about the dislocations in the bulk, no further relaxation at the surface is included; in fact, it hasn't even been calculated.

That a great deal of strain is present in the vicinity of the termination of an edge dislocation can be seen clearly in Fig. 5. For a few atoms at the core of the dislocation, the close packing of the (111) surface is destroyed, leading to the formation of different types of surface sites. While the perfect (111) surface offers on-top (onefold), bridge (twofold) and threefold hollow sites for adsorption, distorted fourfold sites exist at the dislocation line. The combination of strain and unique adsorption sites gives rise to the relative chemical activity of dislocations during surface etching and undoubtedly has a major effect on the adsorption and catalytic properties of the surface.

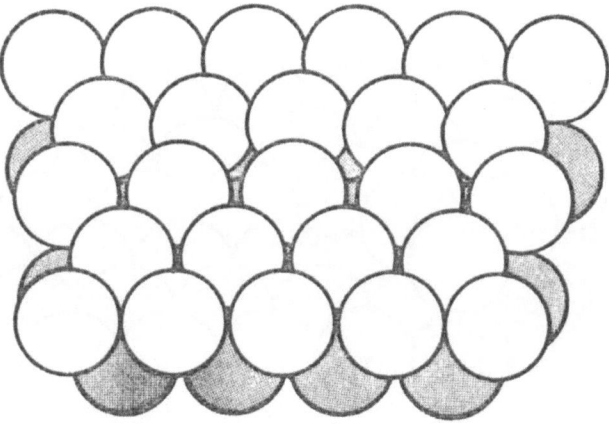

Fig. 5. Model of an edge dislocation intersecting a (111) surface in a face-centered cubic lattice.

A different type of distortion occurs when screw dislocations terminate at surfaces (Fig. 6). Screw dislocations generate essentially an atomic spiral staircase within the crystal, which results in the formation of steps at the surface. In addition there is some distortion of surface adsorption sites at the dislocation line, but it is less dramatic than that at edge dislocations.

Since nothing beyond such idealized models is known about the electronic or geometric structure of dislocations at surfaces, we will unfortunately have nothing further to say about them. Calculations of the changes in electronic structure at these types of surface defects would be extremely valuable, particularly for understanding the chemistry of real surfaces.

EXPERIMENTAL TECHNIQUES FOR THE STUDY OF SURFACE DEFECTS

Since the types of surface defects that we are considering only exist within the first one or two atomic planes of a solid, the probes used to study their properties must be sensitive primarily to that region. The probes that are most commonly used are low energy (less than a few thousand eV) electrons, ions or atoms.[5] Ions and atoms are the most surface-sensitive in that they are not able to penetrate the surface at all. For example, He atoms having thermal energies are scattered by the potential a fraction of a lattice constant above the surface plane.[6,7] Electrons are able to penetrate into the solid but are strongly Coulomb scattered by the cloud of electrons on the atoms, resulting in very short mean free paths. For electrons having kinetic energies between about 50 and 200 eV, the mean free path is about 3 - 5 A. All of the surface sensitive techniques discussed here are performed in vacuum, generally at pressures of 10^{-10} - 10^{-11} Torr.

Most electron, ion and atom spectroscopies are local probes; i.e., they are sensitive to one atom and its immediate neighbors. Low energy electron diffraction (LEED), however, is sensitive to the long range order of the surface and is thus useful for studying ordered arrays of defects. Some information can also be obtained about extended defects (steps) even if they are not ordered.

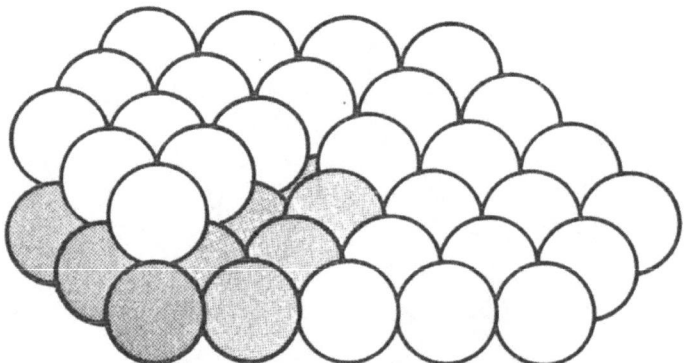

Fig. 6. Model of a screw dislocation intersecting a (111) surface in a face-centered cubic lattice.

Low Energy Electron Diffraction (LEED)

In LEED a collimated monoenergetic beam of electrons having kinetic energies between about 20 and 300 eV is incident on a single-crystal surface. If the surface layer is well ordered, the electrons, which behave as waves, are constructively scattered in specific directions from the regular array of surface atoms. In the simplest LEED spectrometers, the diffracted beams are displayed as bright spots on a fluorescent screen.

If anything such as adsorbed atoms or molecules, point defects or steps forms a commensurate ordered structure on the surface, the LEED patterns remain sharp. Fractional order spots will appear between those due to the perfect surface if the periodic repeat distance on the surface is larger than the substrate unit cell.[8-10] Some useful information on the geometry of regularly stepped metal and semiconductor surfaces has been obtained in this way,[11] but the interpretation of the data is not unequivocal. The orientation and height of steps can be determined in favorable cases, but no information can be obtained about atomic distortions at step edges.

Generally surface defects do not occur in ordered arrays, and LEED is of limited use. The orientation of random steps on surfaces can be determined by the direction in which LEED spots broaden.[8-10] Since the size of a LEED spot (i.e., its angular divergence in real space) increases as the size of the perfectly ordered terraces on the surface decreases, an average terrace size, and hence step density, can be estimated from spot shape and size. Figure 7 shows a LEED pattern from an Ar ion bombarded MgO (110) surface.[12] The spot shape indicates that steps are formed preferrentially parallel to the (001) direction (vertical in the photograph); the resulting terraces are roughly 30 A wide in the (001) direction but only 15 A wide in the (110) direction. When random point defects occur on a surface, the LEED spots become

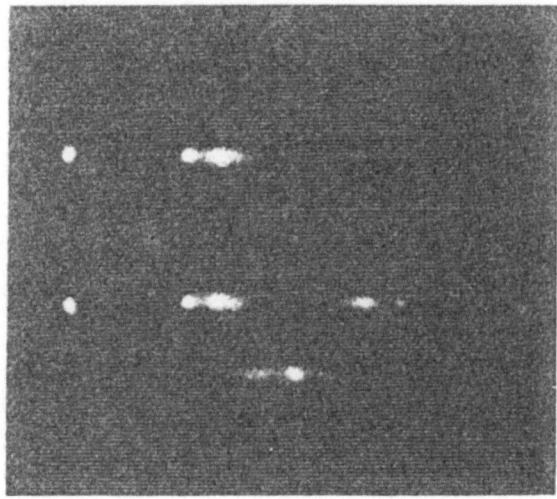

Fig. 7. LEED pattern from MgO (110) surface after Ar ion bombardment. Incident electron energy = 142 eV. (from Ref. 12).

broader but remain circular, and the background intensity increases due to the presence of random scattering centers. Only crude estimates of point defect density can be made from spot size.

Photoemission Spectroscopy

One of the most useful tools for the study of surface defects is photoemission spectroscopy, which is shown schematically in Fig. 8(a). Monoenergetic photons having energies in the range of 20 - 1500 eV are incident on a sample and excite electrons from their ground state to excited states $h\nu$ higher in energy. Some of the photoexcited electrons that originate within a few mean free paths of the surface escape from the sample and are analyzed by an electron spectrometer. The resulting photoelectron spectra mirror the joint density of filled and empty electronic states on the surface. By suitable choices of photon energy and by using tunable photon sources such as synchrotron radiation, it is often possible to separate the contributions of the initial and final states to the spectra.

Valence orbitals can be studied by using photons of any energy, although generally vacuum ultraviolet or soft x-ray energies are used. The process is then usually referred to as ultraviolet photoelectron spectroscopy (UPS). Atomic core levels can be studied by using higher photon energies, and one then refers to it as x-ray photoelectron spectroscopy (XPS). Since UPS and XPS are local probes, the spectra are linear combinations of electrons emitted from defects and those from perfect regions of the surface. It is thus generally necessary to vary the defect density in order to identify those spectral features originating from defects.

Inverse Photoelectron Spectroscopy

A relatively new technique as applied to surfaces is inverse

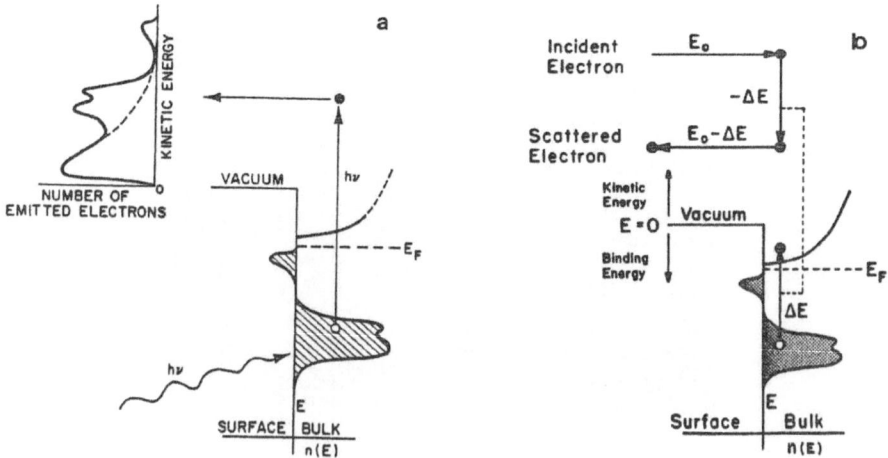

Fig. 8. Schematic diagrams of (a) photoelectron spectroscopy (UPS and XPS) and (b) electron energy loss spectroscopy (EELS).

318

photoelectron spectroscopy, or bremsstrahlung isochromat spectroscopy[13-15]. A monoenergetic beam of electrons is incident on a
surface, and the photons emitted as these electrons are scattered
inelastically into empty states in the band structure above the Fermi
level, E_F, are detected with a spectrometer or Geiger counter. Inverse
photoemission is a relatively direct way of determining the spectrum of
surface electronic states above E_F, particularly those lying below the
vacuum level that are inaccessible by other techniques. To date inverse
photoemission has been applied primarily to perfect surfaces.

Electron Energy Loss Spectroscopy

A technique that is well suited to study the joint density of
surface states, electron energy loss spectroscopy (EELS), is
illustrated schematically in Fig. 8(b). The amount of energy lost by
an electron incident on the sample as a result of generating some type
of excitation in the solid is measured, making EELS the electron analog
of Raman scattering (but with different selection rules, of course).
As with photoemission, EELS is a local probe, and the spectra are thus
the sum of signals from defects and from the perfect surface. The
types of excitations that can be observed in EELS include interband
electronic transitions, bulk and surface plasmons, bulk and surface
phonons, and both electronic and vibrational transitions in adsorbed
molecules. The spectrometers used to study molecular vibrations have
resolutions of 5 meV or less, making EELS a powerful technique for the
study of adsorbed atoms and molecules.

Auger Electron Spectroscopy

In Auger spectroscopy, an incident electron or photon having an
energy generally greater than 1000 eV removes core level electrons from
surface atoms by impact ionization. In the subsequent rearrangement of
the remaining electrons on the atom, Auger electrons are emitted that
have characteristic energies determined by the binding energies of the
core and valence electrons. Auger electron spectra thus contain
information about the electronic structure of surface and near-surface
atoms, and in compounds the relative number of Auger electrons from the
different types of atoms gives a measure of surface stoichiometry. The
latter information is generally more straightforward to obtain, making
Auger a valuable tool for determining surface composition and thus the
atomic nature of point vacancy defects on surfaces. It is possible in
some cases, however, to study the electronic structure of surface
defects by changes in the energy and lineshape of features in Auger
spectra.[16]

Scanning Tunneling Microscopy

The scanning tunneling microscope is a new and potentially
extremely powerful technique for determining the geometry of surface
defects.[17,18] An extremely sharp metal tip, comparable to field
emission tips, is brought close enough to a surface (several Angstroms)
in vacuum that electrons can tunnel from the tip to the substrate when
a voltage difference is applied between them. The distance of the tip
from the surface (in the z-direction) is controlled mechanically
(usually piezoelectrically) and kept constant by monitoring the
tunneling current, which is an extremely strong function of distance.
The z position of the tip can thus be controlled to within a few tenths
of an Angstrom. The tip is then rastered across the surface in the x-
and y-directions, and the z position necessary to maintain constant
current is recorded. This gives a picture of the topography of a
surface with a resolution in the z-direction of a fraction of an

Angstrom and in the x- and y-directions of about one Angstrom. There
are some complications, such as a lack of detailed knowledge of the tip
geometry and the fact that a change in sample work function changes the
tunneling current for constant distance, but the technique still
promises to be an extremely powerful one for the study of the geometry
of surface defects.

Other Techniques

In addition to the surface analysis methods described above,
other techniques can be useful for the study of surface defects.
Ion scattering spectroscopy, in which a beam of monoenergetic ions
having energies of a few keV are scattered from a surface, yields
information on short range order of the surface and on surfaces
composition.[19,20] Ion scattering spectroscopy has been applied to the
study of steps on surfaces.[21]

As mentioned above, very low energy atoms and molecules (typically
He and H_2) yield information about the potential above the surface when
they are either elastically or inelastically scattered from surfaces.[6,7]
Atom scattering has been used to study regular arrays of steps on
surfaces.[22-24]

A novel technique for the study of surfaces is positron
scattering.[25] Positrons incident on a surface can either directly
annihilate with electrons, producing two 511 keV gamma rays, or they
can combine with electrons to form positronium. In the latter case,
the positronium has been found to bond selectively to defect sites,
making the technique very sensitive to the presence of defects.
Positron scattering has to date found rather limited application to
surface studies.

THEORETICAL APPROACHES

The symmetry possessed by the infinite periodic arrangement of
atoms in perfect crystalline solids makes the calculation of bulk
electronic structure relatively straightforward. The destruction of
some of that symmetry at defect sites in the bulk significantly
complicates electronic and geometric structure calculations, but some
of the other papers in this volume show that powerful methods have been
developed to handle bulk defects. The existence of even a perfect
surface also destroys some of the crystal's symmetry, and theoretical
methods for calculating the electronic and geometric properties of
perfect surfaces are less mature than their bulk counterparts. Surface
defects have the lowest symmetry of all, and relatively few
calculations have been performed for specific surface defects.

The atomic geometry of steps and kinks on metal surfaces should be
rather close to a termination of the bulk lattice, and most
calculations involving steps on metals have assumed the step geometry
and then calculated interaction energies between steps or between
adatoms and steps.[26,27] Such calculations are particularly valuable in
modelling crystal growth. The geometry of steps on ionic compounds is
more complex, however, due to changes in the Madelung potential for
ions at surfaces and defects. One of the most fruitful methods of
determining the equilibrium configuration of defects on ionic compounds
has been the defect lattice method,[1-3,28] which is discussed in the
paper by Dr. Cattow in this volume. Basically, the method uses
realistic interatomic potentials and minimizes the force on each ion
surrounding a defect by allowing the ionic positions to relax from

320

their bulk values. Electronic polarization is accounted for by using shell models for the ions. This method has been applied to both point and extended defects on surfaces.

The electronic energy structure in the vicinity of surface defects has been calculated by several methods. For metals, the jellium model is often used, in which the solid is modeled as a background of fixed, uniform positive charge in the presence of which the electronic charge density is calculated.[27,29] While specific atomic orbitals do not exist in such theories, changes in electron density at point defects, steps, kinks etc. can be calculated. While jellium methods work best for free electron metals such as Al, Mg and Be, the calculations have been modified to take account of more localized d states in transition metals.[29]

Discrete atomic models including the full crystal structure have also been used to study surface defects on some metals.[30,31] Such models can determine changes in charge density on specific atoms at defects, information which is important in order to understand most heterogeneous catalytic reactions.

One of the most useful calculational techniques for determining the electronic structure of atoms and ions at defects on metals, semiconductors and insulators is the cluster method.[3,32,33] In a cluster calculation, the atoms or ions in some finite region enclosing the defect are treated as a single molecule, and a full molecular orbital calculation is performed. The result of the calculation is a discrete set of electronic energy levels for the "molecule," which can be correlated with the band structure of the solid. In order to accurately represent a defect, the cluster must contain at least all of the atoms significantly perturbed by the defect, and the atoms on the periphery of the cluster must be terminated in some way that simulates the present of the remainder of the solid.

Cluster methods have also been combined with other types of models to more realistically represent the semi-infinite solid. For metals, for example, the "embedded cluster" method treats the atoms in the vicinity of a surface or defect as a cluster, but embeds the cluster in a jellium model for the remainder of the solid.[34]

PROPERTIES OF SURFACE DEFECTS

In this section we will summarize some of the results, both experimental and theoretical, that have been obtained for surface defects in specific materials. Only a few cases can be discussed in the space available, and these have been chosen to illustrate general principles and trends. Many other excellent studies have been performed, and their omission is in no way a reflection on their quality or importance.

Extended Defects

The geometry of steps on surfaces has been studied for a number of materials. Some of the first work was performed by Somorjai et al. on stepped Pt surfaces.[35-38] Single-crystal Pt (111) surfaces and surfaces cut slightly off of (111) were prepared and characterized by LEED. The predominant step height was found to be a single atom (see Fig. 1) and various average terrace widths were prepared. Using molecular beam scattering techniques, differences in chemical reactivity of the various surfaces were found, with the stepped surfaces being more reactive. It was also found that certain terrace widths were more active for certain catalytic reactions than others, and these trends were correlated with geometrical effects in molecular chemisorption on the terraces.

Theoretical calculations of the electronic density of states on individual atoms in the vicinity of a step on the Cu (100) surface have been performed by Sohn et al.[31] In addition to finding significant differences in the local density of states for various atoms, there were also differences in the total electronic charge on the atoms. For example, Cu atoms at the top of a step had a charge 0.037 electron _less_ than bulk atoms, while atoms at the bottom of the step had 0.031 electron _more_. These differences, together with those in the density of states, could play a role in the increased catalytic activity often exhibited by steps.

The relation between the electronic structure of steps and terraces on Ni (100) surfaces has been addressed in a series of LEED and UPS studies by Cinti et al.[39] They prepared nearly perfect Ni (100) surfaces and two types of stepped surfaces, having three- and five-atom-wide (100) terraces, respectively. Angle-resolved UPS spectra taken for the three surfaces did not show any remarkable differences, from which Cinti et al. concluded that the local surface densities of states were similar for flat and stepped surfaces. Since such experiments average over the electronic structure of all surface and some bulk atoms, however, they are probably not sensitive enough to detect the type of effects calculated by Sohn et al.[31]

Steps on semiconductor surfaces have been prepared by cleavage of single-crystal samples. In one of the early studies by Henzler,[8] Ge was cleaved nominally along the (111) plane. On regions of the surface where the sample did not cleave well, arrays of steps were produced, as indicated in Fig. 9(a). Based upon the changes in splitting of spots in LEED patterns with incident electron energy, the height of the steps was found to be one double layer, as shown in Fig. 9(b). The step spacing, also determined by LEED, was constant in any one area, but varied between five and fifteen atomic distances for different areas.

Similar step arrays have been produced on Si (111) surfaces, where again the step height was found to be one double layer.[40] UPS measurements on cleaved surfaces having high step densities and ones having low step densities permitted identification of an occupied

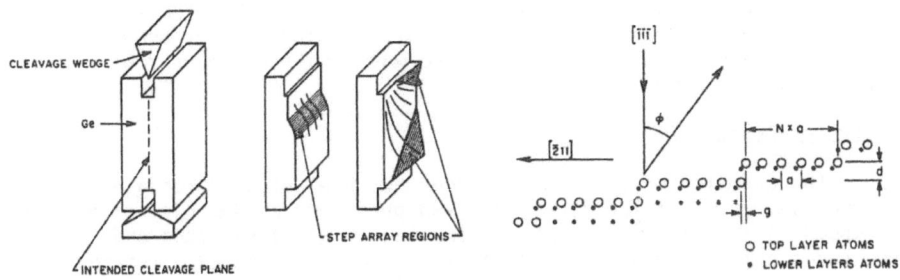

Fig. 9. (a) Cleavage of Ge crystals that result in step arrays on the cleavage face. (b) Model of the atomic positions on stepped Ge surfaces (from Ref. 8).

electronic surface state associated with the steps which had an energy 0.4 eV higher than the dangling bond surface state for the flat surface. Subsequent cluster calculations of the electronic structure of steps on Si (111) showed that localized electronic states, distinctly different than the perfect surface dangling bond state, were present on the doubly bonded Si atoms along the top edge of the step.[33]

Regular step arrays have also been observed on insulator surfaces. When ZnO, which has the hexagonal wurzite crystal structure, is cleaved parallel to the c-axis, yielding nonpolar surfaces (i.e., having no net dipole moment normal to the surface), only steps having a height of one double layer are observed.[41] However, cleavage normal to the c-axis, which produces polar surfaces, resulted predominately in steps whose height was two double layers. Subsequent experiments on nonpolar ion bombarded and annealed ZnO surfaces also showed the existance predominately of steps having one double layer height.[42]

Calculations of the atomic geometry at steps on insulator surfaces have been performed for rocksalt oxides.[1-3] Figure 10 shows, to scale, the results for {100} steps on an MgO (100) surface. Significant relaxation can be seen for step-edge ions.

Less is known about the electronic structure of ions at steps on the surfaces of insulators and ionic compounds. UPS measurements on vacuum-fractured surfaces of TiO_2 and $SrTiO_3$, which contain a fairly large density of steps, do not exhibit evidence for surface electronic structure measurably different than that of the bulk.[16,43-46] Owing to the high step density, it is inferred that there is not a significant difference between the electronic structure of steps and terraces, and that both have nearly the bulk structure. Theoretical studies of CO chemisorption on perfect and defect MgO surfaces are consistent with a similar electronic structure for steps and terraces, since little difference was found between the strength of CO bonding to fourfold coordinated step-edge cations and to fivefold cations on perfect terraces.[3,47] The above evidence is indirect, however, and controlled studies on stepped insulator surfaces are necessary before any definitive conclusions can be drawn.

The influence of surface steps on electron diffraction patterns has recently become of technological importance in the growth of extremely high quality semiconductor single crystals by molecular beam epitaxy (MBE). In MBE, a nearly perfect single crystal surface is used as a substrate upon which additional layers of either the same material or closely related materials are grown epitaxially, one monolayer at a

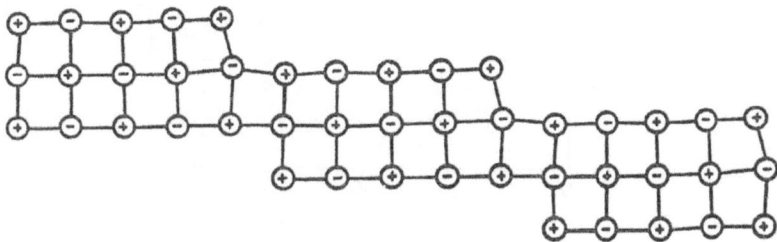

Fig. 10. Atomic structure of {100} steps on an MgO (100) surface (from Ref. 1).

time, by vapor deposition in ultrahigh vacuum.[48] The topography of the sample surface is monitored by reflection high energy electron diffraction (RHEED), in which a beam of roughly 10 keV electrons impinges at almost grazing incidence onto the sample.[49] Reflected beams are detected on a phosphor coated fluorescent screen. Electrons that are incident at a Bragg angle for scattering from the sample are relatively insensitive to the presence of surface steps, but when the electron beam is incident between two Bragg angles, the shape and intensity of the reflected beam varies significantly with surface step density. When the intensity of the specularily scattered RHEED beam is monitored during atom deposition, oscillations are observed, as shown in Fig. 11. They are interpreted as a periodic variation in the density of steps on the surface, where a partially completed monolayer has a larger step density than does a completed monolayer. The oscillations are thus monitoring the layer-by-layer growth, with the period of the oscillations corresponding to the monolayer formation time. Such exquisite control of crystal growth parameters opens the door to the fabrication of extremely sophisticated semiconductor structures.

Also germain to crystal growth is the interaction of adatoms with extended defects on surfaces. The forces experienced by a metal atom adsorbed on a terrace on a metal surface as it approaches a step have been considered theoretically within the jellium model.[27] It was found that an atom approaching a step from the lower terrace experiences an attractive force, with the adatom's binding energy increased by about 40% at the step. Conversely, an adatom approaching a step from the upper terrace is repelled from the step, since its binding energy at the step edge is about 40% smaller than on the flat terrace. This effect, which is consistent with an increase in binding energy with an increase in the number of ligands surrounding the atom (i.e., approaching the bulk ligand coordination) has been observed experimentally in field ion

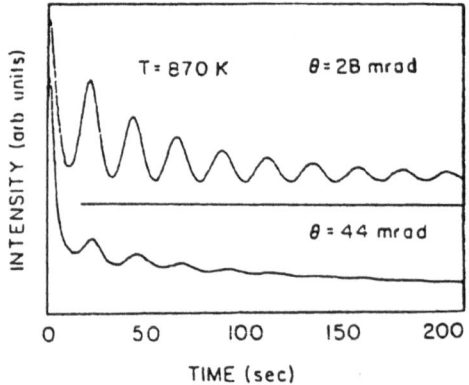

Fig. 11. Intensity of the specularily reflected RHEED beam as a function of time during GaAs growth by MBE for two angles of incidence (from Ref. 49).

microscopy for adatoms of the substrate metal and for different types of metal atoms.[50]

A quite different type of adatom has been employed to study steps, kinks and other types of surface defects. Xe atoms have been adsorbed at low temperatures (~ 100 K) on metal surfaces containing defects.[51,52] Xe is found to have different binding energies for different types of surface sites, and UPS measurements of the Xe core level binding energies and thermal desorption measurements, in which the flux of desorbing Xe atoms is monitored as the sample temperature is raised, have been used to titrate and characterize defect sites.

Point Defects

While extended defects such as step arrays may be present on cleaved or other surfaces, they do not have to be. Point defects, on the other hand, must be present at any finite temperature in order for a system to be in thermodynamic equilibrium.[53] The free energy, F, of a system is given by

$$F = U - TS,$$

where U is the internal energy of the system, S is its entropy and T is the absolute temperature. Although the internal energy of a surface vacancy or adatom is higher than that for a perfect surface, the entropy increases with increasing disorder as surface defects are created. The $-TS$ term then reduces the free energy at finite temperatures when some defects are present. This is shown schematically in Fig. 12, which presents the results of Monte Carlo calculations of equilibrium surface roughness for a stepped surface for various values of kT/E, where E is the interatomic bond energy.[54] This is an extreme example, since kT/E > 0.5 is above the melting point of most materials, but the general idea is correct.

Direct evidence is lacking for the details of point defects on metal surfaces. Significant changes occur in UPS or EELS spectra when nearly perfect single-crystal metal surfaces become disordered, but part of that arises from the highly directional properties of electron emission from single-crystal surfaces. Creation of point defects disrupts the lattice structure and hence destroys some of the angular dependence, making comparison of spectra from perfect and defect surfaces difficult. That there are important differences, however, is attested to by the increased activity of defect surfaces for adsorption and catalysis. But detailed models of these changes have not been reported.

One aspect of point defects on metal surfaces that has received quite a bit of attention is surface segregation in alloys.[53] Most work has been done on substitutional metal alloys, but some calculations and a few experiments have even been done on doped ionic insulators. We will not, however, consider surface segregation here.

The status of our understanding of point defects on semiconductors is similar to that of metals. There are probably larger geometric changes associated with defects such as vacancies on semiconductors due to the covalent, and sometimes partially ionic, bonding, but nothing is known in detail about the geometry of such defects. Some calculations have been made of the electronic structure at single atom vacancies on semiconductor surfaces, and localized surface states, shifted in energy from the ideal surface states by a fraction of an eV, have been predicted.[55]

An area that has received a great deal of attention is the reconstruction of semiconductor surfaces. Many semiconductor surfaces exhibit either a spontaneous or thermally assisted lowering of the symmetry of the surface either through bond rearrangement or actual atom diffusion. This reduction in symmetry is referred to as reconstruction. It does not really involve defects in the sense that we are considering them, however, since reconstructed surfaces still have long range periodicity in two dimensions. Reconstructed surfaces are usually in registry with the bulk lattice, although occasionally incommensurate overlayer structures are seen. The reader is referred to the LEED literature for more information on reconstruction.[5]

Ionic compounds exhibit the largest electronic effects in the presence of point defects, and they are thus easier to study.[46] To understand the reason for this, consider the formation of an O-vacancy defect on the rutile (110) surface, Fig. 4, by removing one of the bridging O^{2-} ions. Since the perfect surface is neutral, removal of an O^{2-} ion would leave the surface with a net charge of +2. But Coulomb forces are extremely strong, and local charge neutrality must thus be maintained. So two additional electrons (usually obtained by removing the O^{2-} ion as a neutral O atom) must be localized in the vicinity of the defect to balance the excess positive charge. Since the adjacent O^{2-} ions have closed shell electronic configurations, the electrons reside on the two cations. Cluster calculations have been performed for defects of this type on TiO_2 and $SrTiO_3$, and the excess electronic charge is found to be shared in a covalent manner between the cations.[32]

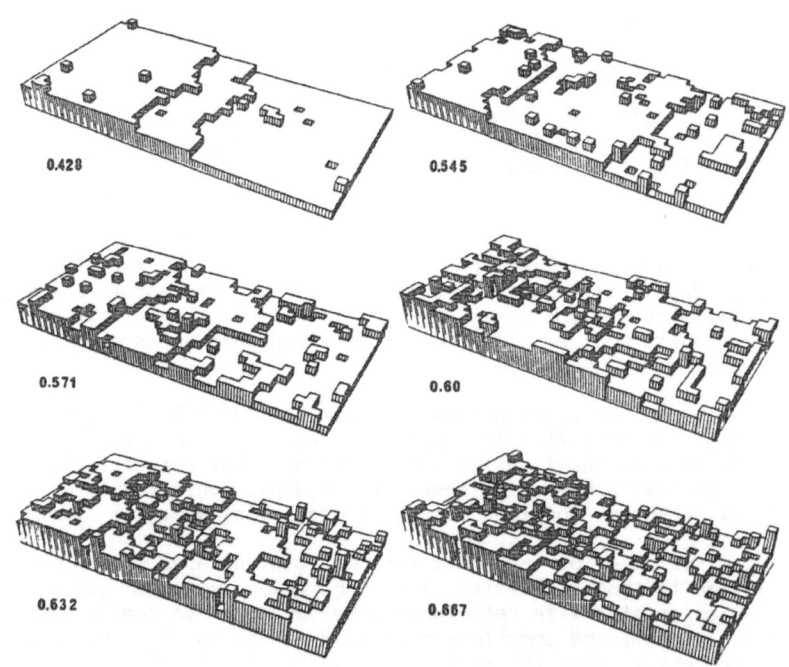

Fig. 12. Computed equilibrium surface configurations using the Monte Carlo method for various values of kT/E. See text for details (from Ref. 54).

Such redistribution of electronic charge can have large effects on atomic geometry at defect sites. Pope et al.[56] have used defect lattice techniques to calculate the displacement of O ions that would occur at a surface cation vacancy (V center) on rocksalt MgO (100) (see Fig. 3). Figure 13 shows the way in which the five O ions adjacent to the vacancy repel each other compared to their positions in the bulk lattice. (The actual ionic displacements are one-third of the length of the vectors shown in Fig. 13.) In a more conjectural vein, we can ask what might happen around a surface anion vacancy (F center) on a rocksalt transition metal oxide such as NiO.[57] Figure 14(a) shows a model of the unrelaxed geometry of such a center. In bulk NiO the Ni ions are essentially Ni^{2+}, and Fig. 14(a) is drawn assuming that cationic radius. But we know that two additional electrons must reside at the defect, and that they will be shared by the four surrounding Ni ions (we will assume equally). But when transition metal ions acquire additional electrons they also increase in radius, so Fig. 14(b) has been drawn with each of the four Ni ions 25% larger (retaining the concept of ionic radii and space filling models). Such a picture is pure speculation, but the covalent bonding that arises when the O^{2-} ion separating such cations is removed, coupled with the additional electron density, might lead to similar changes in surface structure. No calculations have yet been reported on such a defect.

Most of what is known about point defects on ionic surfaces concerns electronic structure, and most of that has been obtained from electron spectroscopies such as UPS and EELS.[43,45,46] An example of the type of information that can be obtained is shown in Fig. 15 for point defects created on the TiO_2 (110) surface by Ar ion bombardment.[58] Perfect TiO_2 is an insulator having a 3.1 eV bandgap, and the UPS spectrum in (a) exhibits only emission from the O 2p valence band. The creation of O-vacancy defects on the surface gives rise to an emission peak in the bulk bandgap, near 3 eV in Fig. 15(b), arising from the localized electrons on the pairs of Ti ions adjacent to the vacancy that were discussed above. Defect creation also induces a new feature, labelled A, in the EELS spectrum that corresponds to the

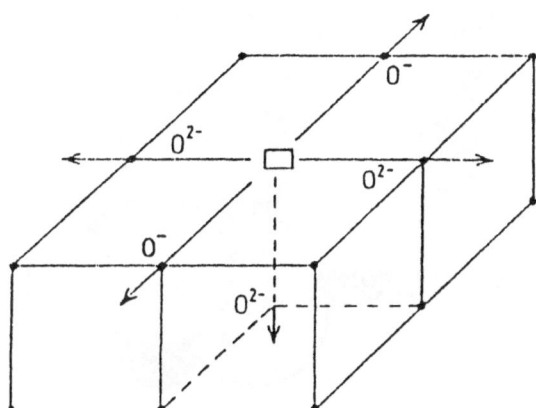

Fig. 13. Nearest neighbor anion relaxation around the Mg^{2+} surface V center. Length of arrows is three times the atomic displacement (from Ref. 56).

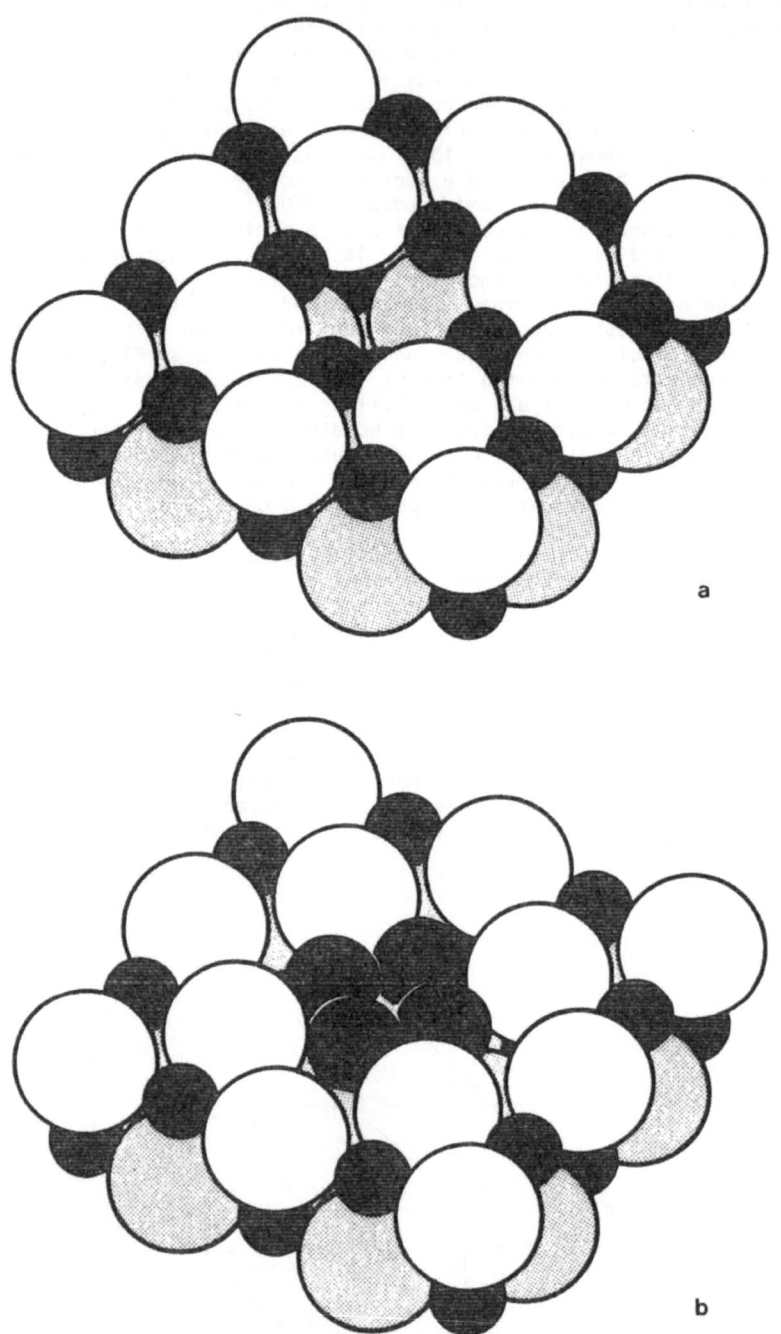

Fig. 14. Schematic model of an anion vacancy on the rocksalt (100) surface assuming that the cations adjacent to the defect have (a) their bulk ionic radii or (b) radii 25% larger.

328

bonding-to-antibonding transition of the excess charge on the Ti pairs. We will return later to consideration of the chemical activity of this type of surface defect.

Surface O-vacancy defects are common to most oxides, and the electronic structure changes associated with them are qualitatively similar. For example, stoichiometric NiO is an insulator whose Ni[2+] ions contain eight 3d electrons. Figure 16 shows the UPS spectrum from both a nearly perfect NiO (100) surface (solid curve) and for the (100) surface containing about 15% O-vacancies (dashed curve).[57] The valence band is composed of the eight Ni 3d electrons and the six O 2p electrons, as indicated. The primary change that surface defects produce in the electronic structure is the creation of a shoulder about 2 eV above the highest Ni 3d level. This corresponds to the extra electronic charge on the surface cations (see Fig. 14). Note also that

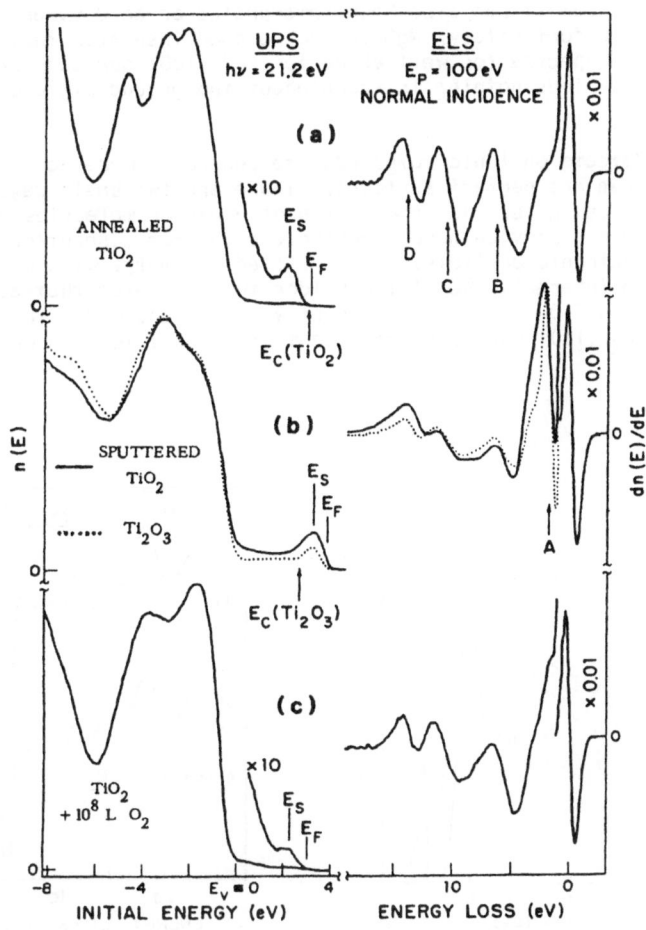

Fig. 15. UPS and ELS spectra for (a) annealed TiO_2 (110), (b) Ar ion bombarded TiO_2 (110) (solid curves) and vacuum fractured Ti_2O_3 (dotted curves), and (c) surface in (b) after exposure to 10^8 Torr-sec of O_2 (from Ref. 58).

there are now occupied electronic levels at the Fermi energy, E_F, indicating that the presence of surface defects has made the surfaces metallic; the creation of a conducting surface layer on defect TiO_2 surfaces can also be seen in Fig. 15(b).

Surface defect states on very wide bandgap insulators such as MgO, CaO and SrO have been identified by their ground-to-excited state transitions in EELS.[59-63] For MgO, these defect transitions have even been seen on vacuum-cleaved (100) surfaces,[63] as shown in Fig. 17. The bulk bandgap in MgO is 7.8 eV, so any EELS transitions having energies less than that must correspond to surface states. The feature at about 6.5 eV in Fig. 17 is an intrinsic surface state transition for the perfect surface. The feature at about 2 eV in Fig. 17(a), however, is due to some type of surface defect. It is present with varying amplitude on cleaved surfaces and can be created by electron or ion bombardment.[62] It can be removed by annealing or exposure to O_2, the latter shown in Fig. 17(b). The exact nature of the surface defect responsible for the 2 eV peak is not completely understood. It most probably corresponds to a surface Mg vacancy (surface V center, see Fig. 13),[59-61] but it has also been interpreted as an O vacancy, or F center.[62,63] Unfortunately, MgO is such a good insulator that meaningful UPS spectra for well characterized (100) surfaces have not been reported. Thus nothing is known about the ground state of the defect.

Point defects on ionic compounds are generally more active chemically than are perfect surfaces. There are two basic ways in which defects can alter the interaction of atoms or molecules with a surface: they can provide novel geometric sites for adsorption and/or different electronic orbitals, either filled or empty, with which the molecules can interact. We will consider two molecules that adsorb at defect sites on TiO_2 (110) in different ways: O_2 and H_2O. Vacuum fractured TiO_2 (110) surfaces are essentially inert to both molecules

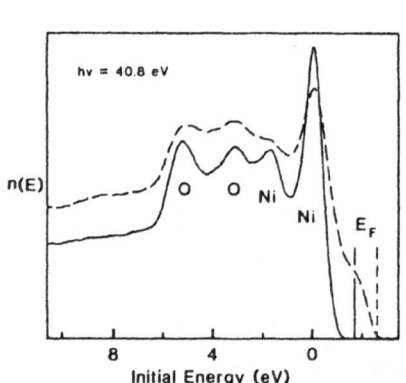

Fig. 16. UPS spectra of cleaved NiO (100) (solid curve) and the equilibrium 500 eV Ar ion bombarded surface (dashed curve).

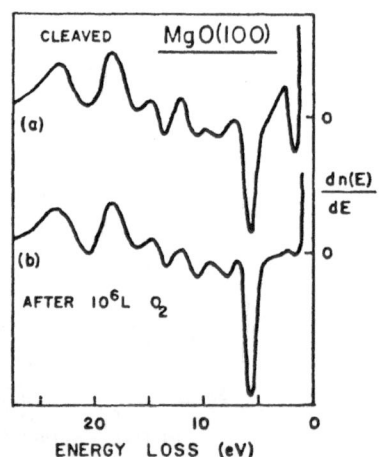

Fig. 17. EELS spectra for vacuum cleaved MgO (100) (a) before and (b) after exposure to 10^6 Torr-sec of O_2 (from Ref. 63).

at room temperature, even though, as discussed above, they contain a significant density of steps and kinks.[45] Both molecules interact strongly with O-vacancy defects on TiO_2, however.[58,64] Figure 15(c) shows the UPS and EELS spectra for defect TiO_2 (110) after exposure to O_2.[58] The electrons that were localized on the cations adjacent to the defect are removed by the oxygen, resulting in negatively charged adsorbed species. For small O_2 exposures, the O_2 molecules dissociate, adsorbing as negative atoms (i.e. O^-, O^{2-} or some intermediate valence state). For larger exposures, it is believed that a negative molecular specie adsorbs, probably O_2^-. The depopulation of the surface defect state shows that the excess charge at the site plays an important role in adsorption.

Quite different things happen when the same surface is exposed to H_2O, as shown in Fig. 18.[65] There is strong adsorption at defect sites, and detailed analysis of the UPS spectra indicates that H_2O initially dissociates, adsorbing as OH^- radicals. For larger exposures, molecular adsorption is observed. But the excess electrons on the defect site seem to play no part in the process; no reduction in the amplitudes of the surface state emission is seen.

Somewhat more complex adsorption of O_2 and H_2O has been observed for O-vacancy defects on NiO (100).[57] The perfect surface is inert with respect to both molecules at room temperature. O_2 adsorbs readily at defect sites, depopulating the defect state shown in Fig. 16. H_2O, on the other hand, will not adsorb at defect sites on the clean surface. If the defect surface is first exposed to a small amount of O_2, however, H_2O readily adsorbs, as shown in Fig. 19. It dissociates upon adsorption, bonding to the surface as OH^- radicals.

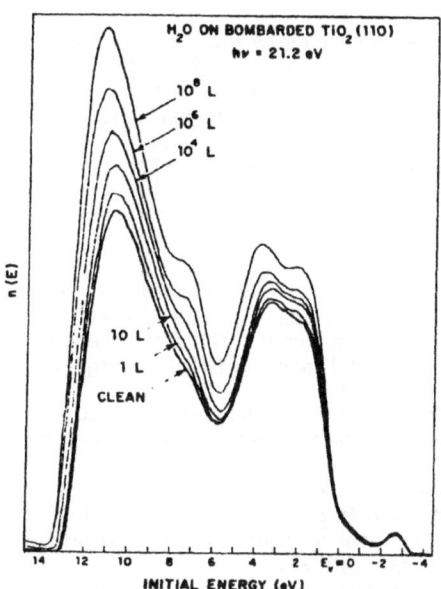

Fig. 18. UPS spectra of clean, Ar ion bombarded TiO_2 (110) surface after successive exposures to H_2O (from Ref. 65).

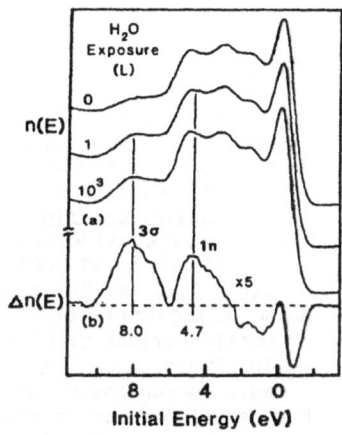

Fig. 19. (a) UPS spectra of Ar ion bombarded NiO (100) pre-exposed to 10 Torr-sec O_2 as a function of H_2O exposure. (b) Difference between spectra for 10^3 and 0 Torr-sec H_2O.

An interesting effect has been found for both O_2 and H_2O adsorption on the corundum oxides Ti_2O_3 and V_2O_3.[66-68] Both materials cleave along a (1012) plane. The cleaved surfaces interact strongly with O_2 and H_2O for small exposures, exhibiting sticking coefficients close to unity. When the surfaces are damaged by Ar ion bombardment, however, the sticking coefficients are smaller by as much as an order of magnitude. This is contrary to one's usual view of the role of defects in adsorption, and it has not yet been explained.

CONCLUSION

Our knowledge of the nature and properties of defects on solid surfaces is less mature than that for bulk defects, but surface sensitive electron spectroscopic techniques have begun to elucidate some of their properties. This paper has attempted to give an overview of what currently is known about the geometric and electronic structure of both point and extended defects on surfaces. Techniques are available to probe both the long range order of surface defects such as regular step arrays, as well as the properties of localized defects. Some controlled experimental studies have addressed the chemical activity of surface defects. Calculational techniques that have proved useful in studying bulk defects and/or perfect surfaces are now being applied to defects on surfaces. In the near future new experimental and theoretical methods promise to greatly expand our understanding of all types of surface defects.

ACKNOWLEDGEMENTS

The author would like to thank the many people who have helped to broaden his horizons in surface science, among them P.A. Cox, G. Dresselhaus, R.R. Gay, J.B. Goodenough, G.L. Haller, R.L. Kurtz, J.M. McKay, M.H. Mohamed, H.R. Sadeghi, K.E. Smith, E.I. Solomon, T. Wolfram and H.J. Zeiger. Some of this work was supported by National Science Foundation Grant DMR-82-02727.

REFERENCES

1. P. W. Tasker and D. M. Duffy, The Structure and Properties of the Stepped Surfaces of MgO and NiO <u>Surf. Sci.</u> 137:91 (1984).
2. J. Kendrick, E. A. Colbourn and W. C. Mackrodt, The Calculated Defect Structure of Planar and Non-Planar Surfaces of MgO and FeO, <u>Radiation Effects</u> 73:259 (1983).
3. E. A. Colbourn and W. C. Mackrodt, A Theoretical Study of CO Chemisorption at (001) Surfaces of Non-Defective and Doped MgO, <u>Surf. Sci.</u> 143:391 (1984).
4. R. M. J. Cotterill and M. Doyama, Energies and Atomic Configurations of Line Defects and Plane Defects in FCC Metals, <u>In</u>: "Lattice Defects and Their Interactions," R. R Hasiguti, ed., Gordon and Breach, New York, (1967).
5. For detailed descriptions of the various techniques and their nuances, see "Electron Spectroscopy for Surface Analysis," H. Iback, ed., Springer-Verlag, Berlin (1977).
6. R. B. Doak and J. P. Toennies, Inelastic Molecular Beam Scattering From Solid Surfaces, <u>Surf. Sci.</u> 117:1 (1982).
7. G. Blatter and T. M. Rice, Scattering of Atomic Beams Off Stepped Surfaces, <u>Phys. Rev. B.</u> 27:7050 (1983).

8. M. Henzler, LEED Investigation of Step Arrays on Cleaved Germanium (111) Surfaces, Surf. Sci. 19:159 (1970).

9. D. G. Welkie and M. G. Legally, Analysis of Surface Structural Defects by Low Energy Electron Diffraction, Thin Solid Films 93:219 (1982).

10. M. Henzler, Measurement of Surface Defects by Low-Energy Electron Diffraction, Appl. Phys. A 34:205 (1984).

11. L. K. Verheij, J. Lux and B. Poelsema, A New Approach for the Analysis of Diffraction from Randomly Stepped Surfaces, Surf. Sci. 144:385 (1984).

12. V. E. Henrich, Thermal Faceting of (110) and (111) Surfaces of MgO, Surf. Sci. 57:385 (1976).

13. V. Dose, Ultraviolet Bremsstrahlung Spectroscopy, Prog. Surf. Sci. 13:225 (1983).

14. J. B. Pendry, New Probe for Unoccupied Bands at Surfaces, Phys. Rev. Lett. 45:1356 (1980).

15. D. P. Woodruff, P. D. Johnson and N.V. Smith, Inverse Photoemission, J. Vac. Sci. Technol. A 1:1104 (1983).

16. V. E. Henrich, G. Dresselhaus and H. J. Zeiger, Surface Defects and the Electronic Structure of $SrTiO_3$ Surfaces, Phys. Rev. B 17:4908 (1978).

17. G. Binnig, H. Rohrer, Ch. Gerber and E. Weibel, Surface Studies by Scanning Tunneling Microscopy, Phys. Rev. Lett. 49:57 (1982).

18. A. M. Baro, G. Binnig, H. Roher, Ch. Gerber, E. Stoll, A. Baratoff and F. Salvan, Real-Space Observation of the 2 x 1 Structure of Chemisorbed Oxygen on Ni (110) by Scanning Tunneling Microscopy, Phys. Rev. Lett. 52:1304 (1984).

19. H. H. Brongersma, Surface Structure Analysis by Ion Scattering, J. Vac. Sci. Technol. 11:231 (1974).

20. D. G. Armour, Applications of Ion Scattering in Surface Analysis, Vacuum 31:417 (1981).

21. S. B. Luitjens, A. J. Algra, E. P. Th. M. Suurmeijer and A. L. Boers, Argon (10 keV) Scattered from Structures, Induced by Bombarding a Cu (100) Surface: Ionization and Neutralization, Surf. Sci. 100:315 (1980).

22. J. Lapujoulade, Y. Lejay and N. Papanicolaou, Diffraction of Helium from a Stepped Surface: Cu (117) - An Experimental Study, Surf. Sci. 90:133 (1979).

23. G. Comsa, G. Mechtersheimer, B. Poelsema and S. Tomoda, Direct Evidence for Terrace Bending From He Beam Scattering on Pt (997), Surf. Sci. 89:123 (1979).

24. J. Harris, A. Liebsch, G. Comsa, G. Mechtersheimer, B. Poelsema and S. Tomoda, Refraction Effects in Atom Scattering from Stepped Surfaces, Surf. Sci. 118:279 (1982).

25. K. G. Lynn, Slow Positrons in the Study of Surface and Near-Surface Defects, Proc. Int. Sch. Phys. "Enrico Fermi," 609 (1981).

26. G. H. Gilmer and J. D. Weeks, Statistical Properties of Steps on Crystal Surfaces, J. Chem. Phys. 68:950 (1978).

27. M. D. Thompson and H. B. Huntington, Adatom Binding at the Surface Ledges of a Jellium Metal, Surf. Sci. 116:522 (1982).

28. C. R. A. Catlow and W. C. Mackrodt, In: "Computer Simulation of Solids," C. R. A. Catlow and W. C. Mackrodt, ed., Springer-Verlag, Berlin (1982).

29. L. L. Kesmodel and L. M. Falicov, The Electronic Potential in a Metal Close to a Surface Edge, Solid State Commun. 16:1201 (1975).

30. M. C. Desjonqueres and F. Cyrot-Lackmann, On the Local Densities of States on Flat and Stepped Pt Surfaces, Solid State Commun. 18:1127 (1976).

31. K. S. Sohn, D. G. Dempsey, L. Kleinman and G. P. Allredge, Electronic Structure of Steps on the (001) Surface of Copper, Phys. Rev. B 16:5367 (1977).

32. M. Tsukada, H. Adachi and C. Satoko, Theory of Electronic Structure of Oxide Surfaces, Prog. Surf. Sci. 14:113 (1983).

33. A. Redondo, W. A. Goddard III and T. C. McGill, Electronic Structure of Steps on Silicon (111) Surfaces from Theoretical Studies of Finite Clusters, Phys. Rev. B 24:6135 (1981).

34. J. P. Muscat, Embedded Cluster Model Studies of Impurities at Metal Surfaces, Prog. Surf. Sci. 18:59 (1985).

35. G. A. Somorjai, R. W. Joyner and B. Lang, The Reactivity of Low Index [(111) and (100)] and Stepped Platinum Single Crystal Surfaces, Proc. Roy. Soc. Lond. A331:335 (1972).

36. R. W. Joyner, B. Lang and G. A. Somorjai, Low Pressure Studies of Dehydrocyclization of N-Heptane on Platinum Crystal Surfaces Using Mass Spectrometry, Auger Electron Spectroscopy and Low Energy Electron Diffraction, J. Catal. 27:405 (1972).

37. G. A. Somorjai, Catalysis on the Atomic Scale, Catal. Rev. Sci. Eng. 18:173 (1978).

38. G. A. Somorjai and F. Zaera, Heterogeneous Catalysis on the Molecular Scale, J. Phys. Chem. 86:3070 (1982).

39. R. C. Cinti, T. T. A. Nguyen, Y. Capiomont and S. Kennov, LEED and UPS Study of Ni $\sim$ (001) Vicinal Surfaces, Surf. Sci. 134:755 (1983).

40. J. E. Rowe, S. B. Christman and H. Ibach, Photoemission Measurements of Step-Dependent Surface Sites on Cleaved Silicon (111), Phys. Rev. Lett. 34:874 (1975).

41. M. Henzler, The Roughness of Cleaved Semiconductor Surfaces, Surf. Sci. 36:109 (1973).

42. W. Göpel and G. Neuenfeldt, Debye Temperatures and Step Arrays of ZnO (1010) Surface Determined by LEED, Surf. Sci. 55:362 (1976).

43. V. E. Henrich, Ultraviolet Photoemission Studies of Molecular Adsorption on Oxide Surfaces, Prog. Surf. Sci. 9:143 (1979).

44. V. E. Henrich and R. L. Kurtz, Surface Electronic Structure of TiO_2: Atomic Geometry, Ligand Coordination, and the Effect of Adsorbed Hydrogen, Phys. Rev. B 23:6280 (1981).

45. V. E. Henrich, The Nature of Transition-Metal-Oxide Surfaces, Prog. Surf. Sci. 14:175 (1983).

46. V. E. Henrich, The Surfaces of Metal Oxides, Rep. Prog. Phys. (In press).

47. E. A. Colbourn and W. C. Mackrodt, Theoretical Aspects of H_2 and CO Chemisorption on MgO Surfaces, Surf. Sci. 117:571 (1982).

48. W. S. Knodle and P. E. Luscher, Recent Developments in Device Fabrication by MBE, Semiconductor International, (Nov., 1980).

49. J. M. Van Hove, C. S. Lent, P. R. Pukite an P. I. Cohen, Damped Oscillations in Reflection High Energy Electron Diffraction During GaAs MBE, J. Vac. Sci. Technol. B1:741 (1983).

50. H. W. Fink and G. Ehrlich, Interaction of Individual Adatoms with Surface Steps, Forty-third Annual Conference on Physical Electronics, Albuquerque, NM (June, 1983) (unpublished).

51. J. Küppers, K. Wandelt and G. Ertl, Influence of the Local Surface Structure on the 5p Photoemission of Adsorbed Xenon, Phys. Rev. Lett. 43:928 (1979).

52. K. Wandelt, J. Hulse and J. Küppers, Site-Selective Adsorption of Xenon on a Stepped Ru (0001) Surface, Surf. Sci. 104:212 (1981).

53. J. M. Blakely and M. Eizenberg, Morphology and Composition of Crystal Surfaces, In: "The Chemical Physics of Solid Surfaces and Heterogeneous Catalysis," D. A. King, ed., Elsevier, Amsterdam (1981).

54. H. J. Leamy and G. H. Gilmer, The Equilibrium Properties of Crystal Surface Steps, J. Cryst. Growth 24/25:499 (1974).

55. S. Erkoc, M. Tomak and S. Ciraci, Vacancies in a Si (111) Thin Film, Solid State Commun. 40:919 (1981).

56. S. A . Pope, M. F. Guest, I. H. Hillier, E. A. Colbourn, W. C. Mackrodt and J. Kendrick, Ab Initio Study of the Symmetric Reaction Path of H_2 with a Surface V Center in Magnesium Oxide, Phys. Rev. B28:2191 (1983).

57. J. M. McKay and V. E. Henrich, Surface Electronic Structure of NiO: Defect States, O_2 and H_2O Interactions, Phys. Rev. B (in press).

58. V. E. Henrich, G. Dresselhaus and H. J. Zeiger, Observation of Two-dimensional Phases Associated with Defect States on the Surface of TiO_2, Phys. Rev. Lett. 36:1335 (1976).

59. P. R. Underhill and T. E. Gallon, The Surface Defect Peak in the Electron Energy Loss Spectrum of MgO (100), Solid State Commun. 43:9 (1982).

60. A. R. Protheroe, A. Steinbrunn and T. E. Gallon, The Electron Energy Loss Spectrum of CaO, J. Phys. C 15:4951 (1982).

61. A. R. Protheroe, A. Steinbrunn, and T. E. Gallon, The Electron Energy Loss Spectra of Some Alkaline Earth Oxides, Surf. Sci. 126:534 (1983).

62. V. E. Henrich, G. Dresselhaus and H. J. Zeiger, Energy-Dependent Electron-Energy-Loss Spectroscopy: Application to the Surface and Bulk Electronic Structure of MgO, Phys. Rev. B22:4764 (1980).

63. V. E. Henrich and R. L. Kurtz, Intrinsic and Defect Surface States on Single-Crystal Metal Oxides, J. Vac. Sci. Technol. 18:416 (1981).

64. V. E. Henrich, G. Dresselhaus and H. J. Zeiger, Chemisorbed Phases of O_2 on TiO_2 and $SrTiO_3$, J. Vac. Sci. Technol. 15:534 (1978).

65. V. E. Henrich, G. Dresselhaus and H. J. Zeiger, Chemisorbed Phases of H_2O on TiO_2 and $SrTiO_3$, Solid State Commun. 24:623 (1977).

66. R. L. Kurtz and V. E. Henrich, Surface Electronic Structure of Corundum Transition-Metal Oxides: Ti_2O_3, Phys. Rev. B25:3563 (1982).

67. R. L. Kurtz and V. E. Henrich, Chemisorption of H_2O on the Surface of Ti_2O_3: Role of d Electrons and Ligand Geometry, Phys. Rev. B26:6682 (1982).

68. R. L. Kurtz and V. E. Henrich, Surface Electronic Structure and Chemisorption on Corundum Transition-Metal Oxides: V_2O_3, Phys. Rev. B28:6699 (1983).

COMPUTER SIMULATION OF SURFACE DEFECTS

E.A. Colbourn

Imperial Chemical Industries plc
New Science Group
The Heath, Runcorn, Cheshire, WA7 4QE, U.K.

INTRODUCTION

Since the surface of a material interacts directly with the environment, surface processes are important in such areas as corrosion, catalysis, electrode behaviour, sintering and particle growth. Often it is difficult to study surface phenomena experimentally, since the techniques used may also sample the bulk material or may cause extensive surface damage or surface charging, and even when experiments can be carried out it may prove difficult to interpret the results unambiguously. It is then that computer simulation techniques may be useful to provide further information to augment the experimental data or to elucidate the atomic mechanisms of surface processes. With the advent of faster computers and more sophisticated programmes, usage of such modelling techniques to complement experiment is likely to increase.

Over the past fifteen years, methods for studying ionic and quasi-ionic material have been developed and extensively used. They have been shown to be helpful not only for clarifying experimental results[1], but also to have predictive capability. More recently, they have been extended for application to surfaces[2] and surface defects[3]. The work described here pertains only to materials which are largely ionic in nature and concentrates on some oxides which are of relevance to corrosion, ceramics and catalysis. The main emphasis will be on the differences between surface and bulk calculations both in terms of computational procedures and of the types of defects observed.

The surface itself may be considered to be a large-scale defect in relation to the bulk structures, since the termination of the crystal can result in relaxation of the perfect lattice positions. In addition, though, there may be local defects, in particular point defects (vacancies, interstitials and substitutions), defect aggregates (including surface segregation to monolayer or greater coverages) and extended defects (for example ledges, corners and kinks). Grain boundaries, which can be viewed as internal surfaces, can also be important especially in explaining corrosion processes. These will not be dealt with further in this paper, but more information on the simulation of grain boundaries and their defect structure is described by Duffy and Tasker[4][5].

COMPUTATIONAL PROCEDURES

The HADES formalism, discussed elsewhere[6], was originally developed to study defects in the bulk, where there is three-dimensional symmetry about the defect. The solid can be partitioned into two regions I and II, and in region I the ions are treated explicitly. Region II is further partitioned into IIa, which is treated in a quasi-elastic approximation, and IIb, which can be viewed from the defect as a continuum. For bulk defect calculations, it is convenient to choose Regions I and II to be spherical[6]. For the surface the three-dimensional symmetry no longer exists, so the long-range summation of the Coulomb interactions cannot be treated by a straightforward application of the Ewald summation used in the bulk. To date two different approaches have been used to tackle this. One, adopted in early studies by Mackrodt and Stewart[3], was based on the Ewald summation but used truncated Gaussians in the integration to facilitate rapid convergence. Their method has proved very satisfactory for studying low index faces of f.c.c. materials but has not been extended to other than cubic materials. A more generally useful approach is to calculate the Coulomb summation with the procedure outlined by Parry[7] and subsequently used by Tasker in his MIDAS code[8]. This involves a plane by plane summation, so non-cubic materials and high index planes can more easily be investigated, although it is not as efficient as the Ewald summation used for bulk calculations. To date all the surface codes in use assume that only two-body potentials are required, and the bond-bending terms which have been found to be important for silicates and zeolites[9] have not been incorporated.

A further problem which relates to surface calculations concerns the applicability of potentials which are derived for the bulk in studying surface properties. For some materials, such as NiO and MgO, the experimental LEED spectra show that the surface is very like a simple bulk termination[10][11]. In cases like this, it seems reasonable that the bulk potentials can reliably be used for surface calculations. Molecular orbital studies of small clusters representing the surface and bulk of MgO[12] also show that the surface is essentially as ionic as the bulk (over 99% in both cases) which suggests that special potentials need not be used for the surface in this case. Finally, the good agreement between experiment (where available) and calculated surface properties indicate that for the cases studied to date, no special adjustment need be made to the bulk potentials in order to use them reliably for calculations at the surface. For materials like ZnO, the case may be otherwise. Here, the LEED spectra show that the $(10\bar{1}0)$ surface undergoes substantial relaxation[13], with the top layer oxygen ions moving inwards by about 0.5Å.

In the present work, the potentials are modified electron-gas potentials obtained using techniques outlined by Gordon and Kim[14], with the correction factor of Rae[15] included. In general, these potentials predict a bulk lattice parameter which is slightly larger than that observed experimentally. For MgO, the discrepancy is small, on the order of 3%. For NiO, however, the unadjusted GKR potentials predict a lattice parameter which is in error by over 20%. For this system, the anion-anion and cation-anion potentials were shifted along the R (internuclear separation) axis in order that the lattice parameter be reproduced exactly, according to a prescription suggested by Mackrodt[16].

It is possible to include polarisability of the lattice ions using the shell model of Dick and Overhauser[17], parameterised by fitting to experimental dielectric constants. Ion polarisability can play an important role in the calculation of surface defect energies. In the bulk of an f.c.c. material like MgO or NiO, a substituent ion experiences an octahedral environment so the defect energy is the same whether the dopant ion is treated as polarisable or not. At the surface, the environment is not symmetrical with regard to the surface plane. Therefore, calculated defect energies at the surface can be lowered by allowing polarisability of the dopant ions, while bulk defect energies are unaffected. This can be important in studies of surface segregation phenomena.

Because of the increased freedom of ion movement at the surface, these calculations usually do not converge as quickly as those for bulk defects, even when the same minimisation procedures are used. There can also be problems, particularly for heavily-doped surfaces, in seeking a global rather than a local minimum in the energy surface. Normally a global minimum can be found by breaking the symmetry of the starting configuration, but this is not assured. It can be useful, therefore, to do calculations assuming several different starting configurations to see whether they relax to the same equilibrium geometry.

CALCULATIONS OF DEFECT STRUCTURE

Defects can be considered to be of several distinct types. There are local defects which are intrinsic to the pure material, such as cation and anion vacancies, as well as those which depend on the impurities present, such as substitutions. These can in general be modelled satisfactorily for ionic f.c.c. oxides. There are also electronic defects, such as F^+ centres, which may be induced by radiation. Electronic defects will not be treated further in this paper since they are not easily modelled with static lattice simulations. However, Sharma and Stoneham[18] have calculated the electronic structure of the F^+ centre in MgO using ab initio molecular orbital methods, treating a cluster of surface ions, and find good agreement with experiment.

All of the above-mentioned defects can exist both in the bulk and at the surface. In addition, there are defects which can occur only at the surface, and in this category both adsorbed ions and extended defects like ledges and corners spring to mind. These are important for such phenomena as catalysis and crystal growth.

VACANCIES AND SUBSTITUTIONS

The energies needed to form vacancies in the bulk and at the (001) surface of MgO and NiO are listed in Table 1. These are slightly different than those previously reported[19], for two reasons. For the charged defects, i.e. the anion and cation vacancies, the Mott-Littleton

Table 1
Fundamental Defect Energies in MgO and NiO (eV)

| | MgO | | NiO | |
	bulk	surface	bulk	surface
cation vacancy	25.38	24.98	24.22	24.34
anion vacancy	22.82	22.57	24.58	24.62
vacancy pair	45.59	44.50	46.65	45.89

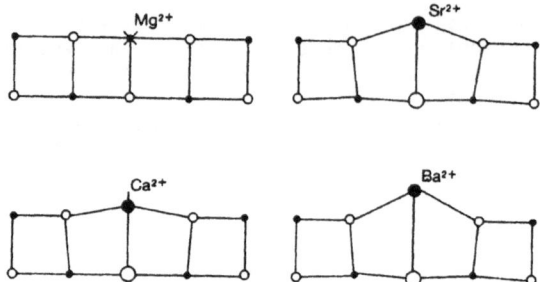

Figure 1.   Lattice deformation about isovalent dopants in NiO: side
view.

method as previously applied did not correctly account for the
polarisation due to the image charge[20]. For all the defects, the change
to a cylindrical region I has also led to a difference in the computed
defect energies. For the results reported previously, it was necessary
to calculate the formation energies for a series of smaller box-shaped
regions I and to extrapolate these to a large region size. For the anion
and cation vacancy, this gives results very similar to those now obtained
with the cylindrical region I. For the vacancy pair, with its associated
lower symmetry, the fit to the formation energies calculated for the
box-like region I was not as good. Additionally, in the new procedure
the cylinder has to be smaller for the vacancy pair so may not give a
completely reliable formation energy. For the bulk calculations, about
100 ions have been found necessary in region I to ensure that the
energies are accurate. Because of the lowered symmetry at the surface,
it is difficult to perform calculations with such a large number of ions
when the defect itself is not symmetrical.

The substitution energies of dopant ions can differ markedly between
the surface and the bulk. Here, the energy difference is a function of
both dopant ion size and charge. The dopant size effect is easily
rationalised since large dopants may be accommodated more easily at the
surface where there are fewer geometrical constraints. This size effect
is especially clear for the isovalent series of ions Mg^{2+}, Ca^{2+}, Sr^{2+} and
Ba^{2+} in NiO. Mg^{2+} is almost the same size as Ni^{2+} so does not perturb
the lattice significantly, but Ba^{2+} distorts the lattice considerably, as
Figure 1 shows.

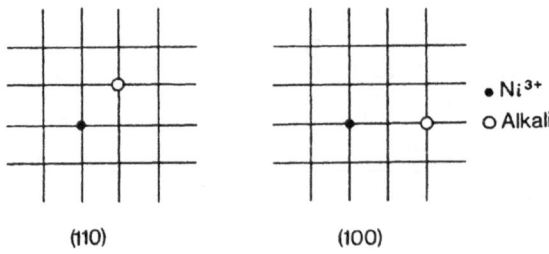

Figure 2. Ni^{3+}-alkali ion pairs in (001) NiO: (110) and (100)
configurations.

Table 2
Substitution Energies of Dopants in NiO

Dopant	Ion Radius(A)*	Bulk (eV)	Surface (eV) (001)	(011)
Li^+	0.76	15.816	15.827	14.497
Na^+	1.02	19.025	17.860	16.049
K^+	1.38	23.584	19.506	-
Mg^{2+}	0.720	0.226	0.200	0.264
Ca^{2+}	1.001	5.938	4.529	3.791
Sr^{2+}	1.18	9.797	6.889	5.345
Ba^{2+}	1.35	14.295	9.087	6.752
Mn^{2+}	0.83(hs) 0.67(ls)	5.731	4.370	3.666
Fe^{2+}	0.78(hs) 0.61(ls)	5.685	4.342	3.654
Co^{2+}	0.745(hs) 0.65(ls)	5.559	4.249	3.582
Al^{3+}	0.535	-30.138	-29.408	-29.025
Y^{3+}	0.900	-16.173	-16.720	-15.850
Ce^{4+}	0.87	-44.880	-44.162	-41.449

* From R.D. Shannon, Acta Cryst. A32, 751 (1976)

For aliovalent impurities, charge and size effects both play an important role. Cations whose charge is less than that of the host lattice cation are not so strongly attracted to the neighbouring ions and tend to sit above the surface, while highly-charged cations will lie below the surface plane. Substitution energies in the bulk and at the (001) and (011) surfaces of NiO are given in Table 2, along with the ionic radii of the impurities, although not all of these ions may exist in NiO as isolated defects. In these calculations, non-polarised dopant ions were used.

In NiO, singly charged cations can be compensated by trapped holes. Following Catlow et al[21], the holes are modelled as small polarons, and calculations on the charge-neutral defect pairs Li^+-Ni^{3+}, Na^+-Ni^{3+} and K^+-Ni^{3+} were carried out. These can exist in several possible geometries; the two considered here are designated as the (110) and (100) dimers and are illustrated in Figure 2. For the (100) configuration, comparison with the bulk was not possible since the bulk calculation did not converge before the Ni^{3+} core-shell separation became excessive. However, the calculations presented in Table 3 show that the (110) configuration of Li^+-Ni^{3+} is more stable in the bulk and at the surface, while the sodium- and potassium-containing pairs are more stable at the

Table 3
Ni^{3+} - alkali pairs in NiO

	(110) pair bulk	surface	ΔE	(100 pair) surface	ΔE
Li^+-Ni^{3+}	-16.303	-15.725	1.224	-16.105	1.624
Na^+-Ni^{3+}	-12.994	-13.657	1.209	-14.097	1.649
K^+-Ni^{3+}	-8.385	-12.096	1.294	-12.580	1.778

surface. The calculations also indicate that at the surface the (100) pair is slightly more stable than the (110) pair, although the difference is small. For both orientations, however, there is a binding energy ΔE for the pair of well over 1 eV.

SURFACE SEGREGATION

As shown above, dopant ions can have considerably different substitution energies at the surface and in the bulk. Provided that a mechanism for ion migration exists, this energy difference will provide a driving force for segregation to the lowest energy sites. Although the majority of existing experimental studies concern segregation in metals and alloys, there is a growing body of literature on segregation in ceramics, including a detailed study of Ca^{2+} in MgO by McCune and Wynblatt[22]. In a series of theoretical papers[23][27], Tasker, Mackrodt and Colbourn have calculated the segregation of the isovalent dopants Ca^{2+}, Sr^{2+} and Ba^{2+} in MgO, considering both the (001) and (011) surfaces.

Except at extremely small impurity concentrations, the point ion defect energies cannot give a measure of the enthalpy of segregation because there will be interactions between impurity ions which significantly affect the calculated segregation energy. As a consequence, calculations at 0.25, 0.5, 0.75 and a full monolayer of dopant were performed[24]. In the most recent calculations, four surface molecules per unit cell were used, and some account was taken of dopant ion polarisability. Tasker and coworkers[25] have derived an expression for the free energy of segregation which accounts for the configurational, but not the vibrational, entropy contribution:

$$\Delta g = y\left\{\Delta h - kT\left[\ln\left(\frac{x}{y}\right) - \frac{1-y}{y}\ln\left(\frac{1-x}{1-y}\right)\right]\right\}$$

where $\Delta h = h(\text{surface}) - h(\text{bulk})$ is the difference in the enthalpies of substitution per ion at the surface and in the bulk, x is the bulk concentration of the impurity and y is the surface concentration. Although the energies produced by the calculations are internal energies, these are a good approximation to the enthalpies of substitution[28]. This allows some estimate of the effect of temperature, although this will only be approximate since the two-body potentials used are independent of temperature.

A number of interesting features arose from their studies. First, it was noted that substitution would occur preferentially at any (011) surface which was present[24]. The undoped (011) surface of MgO, although stable, is higher in energy than the (001) surface by about 1.7 Jm^{-2}. However, since the segregation lowers the energy of the (011) surface more than that of the (001) surface, for Ba^{2+}, the largest dopant, calculations predict that the doped (011) surface will be more stable than the (001) surface at coverages greater than 0.7. Therefore it is possible that at high coverages, facetting might result from impurity segregation. Neither Ca^{2+} nor Sr^{2+} is predicted to have this effect[24].

Another unexpected result was that when Ca^{2+} (or other dopant) completely substituted the second layer, rather than the topmost layer, the calculated energy per defect was as great as for a full monolayer or dopant in the bulk[24]. Provided the top layer was already completely doped, segregation to the second layer would occur.

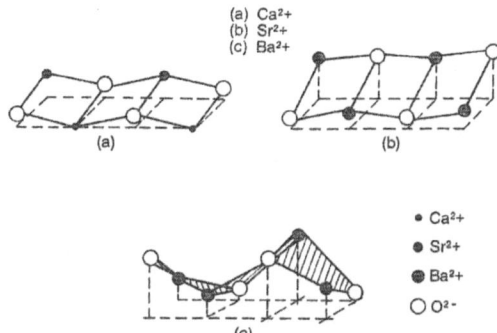

(a) Ca²⁺
(b) Sr²⁺
(c) Ba²⁺

• Ca²⁺
● Sr²⁺
⬤ Ba²⁺
○ O²⁻

Figure 3.   Reconstructed surface of MgO for full coverage of dopant
(a) Ca²⁺   (b) Sr²⁺   (c) Ba²⁺.

     The relief of elastic strain, which is the primary driving force for
segregation, also causes reconstruction of the doped surface. Since all
of the impurity ions are significantly larger than the Mg^{2+} they replace,
their interaction with each other at full monolayer coverage affects the
surface structure, so that a reconstruction of the doped (001) surface is
predicted[24]. For Ca^{2+} and Sr^{2+}, the reconstruction is 2 x 2, even when
four molecules per unit cell are used in the calculation. This is
illustrated in Fig. 3(a) and (b). For Ba^{2+}-doped MgO, as shown in Figure
3(c), all four barium ions in the unit cell are unique, although there
are only two different anion positions[27]. Since the calculations were
not done with more than four molecules in the unit cell, the actual
relaxed surface may involve more extensive reconstruction than this
indicates. Not surprisingly, this large reconstruction affects the
reactivity of the barium-doped MgO surface. The barium-doped surface is
active for N_2O decomposition, although no O_2 evolution is observed[26].
Since the decomposition mechanism of N_2O is believed to involve formation
of $N_2 + O_s^-$, it is expected to occur at sites of reduced Madelung
potential, probably the Ba^{2+} which lie highest out of the surface.
Surface luminescence experiments show one peak for pure MgO, but two for
barium-doped MgO[26], which is consistent with the existence of two
different anion sites predicted by the calculations.

     One of the most surprising implications of the calculations is that
for certain levels of Sr^{2+} or Ba^{2+} there is a barrier to sintering which
persists to high temperatures. The calculations (assuming polarisable
dopant ions) suggest that even for coverages as low as 10% of barium or
35% of strontium, particles will not sinter[27]. This is illustrated in
Figure 4, which shows the change in surface energy as a function of
coverage. When the drop in surface energy arising from the segregation
is identical in magnitude to the surface energy of the pure material
(represented by the dotted line) a barrier to sintering is expected.
Calcium doping, however, will not prevent sintering at any coverage.

     The Ca^{2+}-doped MgO case is of special interest to illustrate how
well the calculations agree with the experimental work of McCune and
Wynblatt. Their experiments were carried out on a sample with about 200
ppm calcium, at temperatures between 1200K and 1700K, and concluded that
the equilibrium value for the calcium segregation occurred between 950°C

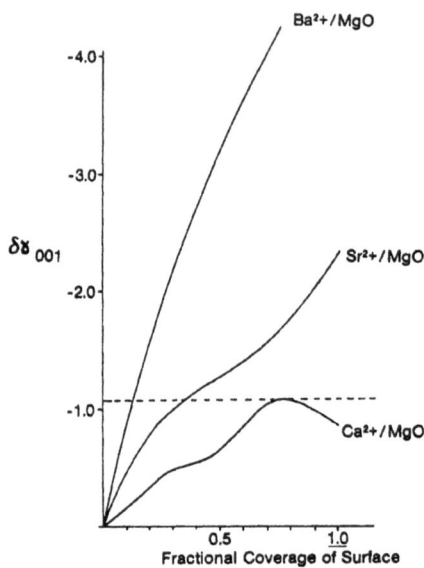

Figure 4. Change in surface energy (Y_{001}) as a function of fractional
coverage for doped MgO (001) surfaces.

(1223K) and 1000°C (1273K) and corresponded to 20% occupation of the
surface cation sites. As Figure 5 shows, the computer simulation also
predicts a minimum in the free energy, Δg, at a coverage of about 20% at
these temperatures[25][27]. The calculated heat of segregation is also in
good agreement with the results of McCune and Wynblatt[23][27].

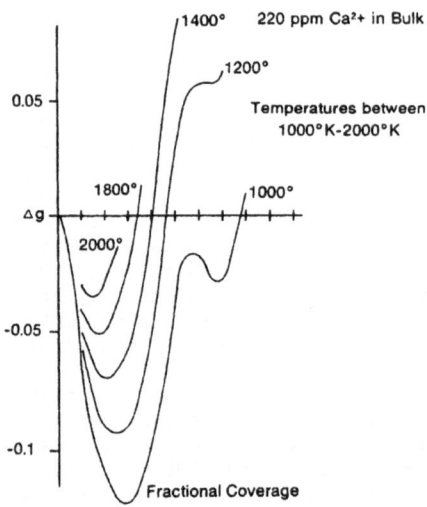

Figure 5. Free energy of segregation as a function of coverage for
220 ppm Ca^{2+}/MgO at temperatures between 1000K and 2000K

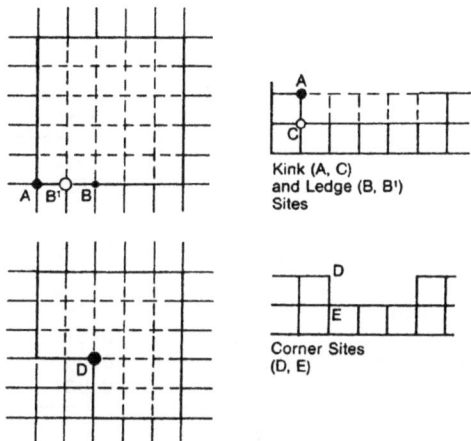

Figure 6. Topological defect sites in (001) MgO surface.

SURFACE TOPOGRAPHY

Real surfaces are not perfect flat planes, but exhibit features such
as kinks, ledges and corners. An enhanced catalytic activity is
frequently attributed to such sites because of their expected reduction
in the Madelung potential, so examination of such topographical defects
is of special interest. The topography of ledge sites has been discussed
previously by Tasker and Duffy[29] for NaCl, MgO and NiO and for Tsang and
Falicov[30] for NaCl and Ar. In both these studies the steps in the (001)
surface were modelled by treating a high Miller index face like (11 0 1),
which resembles a stepped (001) surface with the steps far enough apart
to be non-interacting. Colbourn and Mackrodt[19] have taken a different
approach to modelling kinks, ledges and corners in MgO, CaO, NiO and
NaCl, removing square arrays of ions from the surface (to represent
ledges and kinks) and building up arrays of ions above the surface (for
ledges and corners). Their calculations gave results in good agreement
with those of Tasker and Duffy so both approaches to the modelling of
topographical defects seem adequate.

In the present work, all the defects were modelled by removing
arrays of ions from the surface. This was dictated by our computer
programme, which gives the long-range Coulomb interstitial energy as a
sum of the contributions from all interstitials. Hence to find the
Madelung potential at each site, only one interstitial (but any number of
vacancies) could be used in each calculation; this was done by replacing
an ion at each of the sites by an identical ion. Figure 6 illustrates
the different sites considered, and Table 4 gives the Madelung potentials
at these sites before and after the lattice was allowed to relax.

The extent of relaxation, particularly at kinks and corners, has
been noted previously[19]. As Table 4 shows, in all cases the lattice
relaxation serves to try to make the potential at the kink, corner or
ledge as much like that in the bulk as possible. Therefore, there is a
definite contraction inwards at corners, and some movement inwards at
ledges, while the ions at kinks move away from their nearest neighbours.

345

Table 4
Madelung potentials* at topological defects in MgO

Site	Unrelaxed	Relaxed anion	cation
bulk	0.852	0.852	0.852
(001) surface	0.820	0.820	0.826
(011) surface	0.750	0.786	0.796
A (kink)	0.926	0.791	0.791
B' (ledge)	0.706	0.857	0.868
B (ledge)	0.846	0.875	0.869
C (beneath kink)	0.784	0.852	0.841
d (corner)	0.666	0.818	0.828
E (beneath corner)	0.895	0.856	0.846
adatom	0.040	0.521	0.512

* Absolute values are given

Colbourn and Mackrodt[19] have studied the doping of the different sites of
these topological defects. Perhaps surprisingly, even quite large
dopants do not preferentially substitute corner and ledge sites. This
can be rationalised by noting that the relaxation at these sites causes
the corner and ledge ions to move closer to their nearest and
next-nearest neighbours. This is not so easily accomplished if the
dopant ion is large. However, at kink sites the relaxation tends to be
away from the nearest and next-nearest neighbours, and large dopant ions
are more easily accommodated. Colbourn, Kendrick and Mackrodt [31] have
investigated the reactivity of O^- ions at corner and ledge sites towards
H atom chemisorption and have found the bond strengths to differ by about
10 kcal/mole from the value on the planar surface. With the exception of
the isolated adatom (which is probably unrealisticlly modelled) there is
a good correlation between decreasing Madelung potential at the active
site and increasing OH bond strength. This suggests that the reactivity
of O^- at a kink site, a case not examined by Colbourn, Kendrick and
Mackrodt, might be of special interest.

CONCLUSIONS

Static lattice simulation procedures can be used at the surface not
only to give point defect energies and geometries, but also to look at
the structure of topological defects. Calculations of surface
segregation are in good agreement with available experimental results
[22-25,27] and have also been valuable in interpreting the reactive and
spectroscopic properties of doped materials.

REFERENCES

1. W.C. Mackrodt, in "Computer Simulation of Solids", p.175, ed.
    C.R.A Catlow and W.C. Mackrodt (Springer-Verlag, Berlin, 1982).
2. P.W. Tasker, in "Computer Simulation of Solids", p.288, ed.
    C.R.A. Catlow and W.C. Mackrodt (Springer-Verlag, Berlin, 1982).
3. W.C. Mackrodt and R.F. Stewart, J. Phys. C. Solid State Phys. 10,
    431 (1977).
4. D.M. Duffy and P.W. Tasker, Phil. Mag. A 47, 817 (1983); ibid 48,
    155 (1983); ibid, 50, 1543 (1984); ibid, 50, 155 (1984).
5. D.M. Duffy and P.W. Tasker, Advances in Ceramics 10, 275 (1984).

6.   C.R.A. Catlow and W.C. Mackrodt, in "Computer Simulation of Solids",
     p.3, ed. C.R.A. Catlow and W.C. Mackrodt (Springer-Verlag, Berlin,
     1982); M.J. Norgett, AERE Harwell, Report AERE-R.7015 (1972).

7.   D.E. Parry, Surface Science 49, 433 (1975); Surface Science 54, 195
     (1976).

8.   P.W. Tasker, AERE Harwell, Report AERE-R.9130 (1978).

9.   M.J. Sanders, C.R.A. Catlow and J.V. Smith, J. Phys. Chem. 88, 2796
     (1984).

10. C.G. Kinniburgh and J.A. Walker, Surface Science 63, 274 (1977).

11. M.R. Welton-Cook and W. Berndt, J. Phys. C. Solid State Phys. 15,
     5691 (1982).

12. E.A. Colbourn and J. Kendrick in "Computer Simulation of Solids",
     p.67, ed. C.R.A. Catlow and W.C. Mackrodt, (Springer-Verlag, Berlin,
     1982).

13. C.B. Duke, R.J. Meyer, A. Paton and P. Mark, Phys. Rev. B18, 4225
     (1978).

14. R.G. Gordon and Y.S. Kim, J. Chem. Phys. 56, 3122 (1972); Y.S. Kim
     and R.G. Gordon, J. Chem. Phys. 60, 1842 (1973).

15. A.I.M. Rae, Chem. Phys. Lett. 18, 574 (1973).

16. W.C. Mackrodt, E.A. Colbourn and J. Kendrick, unpublished results.

17. B.G. Dick and A.W. Overhauser, Phys. Rev. 112, 90 (1958).

18. R.R. Sharma and A.M. Stoneham, J. Chem. Soc. 72, 913 (1976).

19. E.A. Colbourn and W.C. Mackrodt, Solid State Ionics 8, 221 (1983).

20. D.M. Duffy, J.P. Hoare and P.W. Tasker, J. Phys. C. Solid State
     Physics 17, L195 (1984).

21. C.R.A. Catlow, W.C. Mackrodt, M.J. Norgett and A.M. Stoneham, Phil.
     Mag. 35, 177 (1977).

22. R.C. McCune and P. Wynblatt, J. Amer. Ceram. Soc. 66, 111 (1983).

23. E.A. Colbourn, W.C. Mackrodt and P.W. Tasker, J. Mater. Sci. 18,
     1917 (1983).

24. P.W. Tasker, E.A. Colbourn and W.C. Mackrodt, J. Amer. Ceram. Soc.
     68, 74 (1985)

25. E.A. Colbourn, W.C. Mackrodt and P.W. Tasker in "Proceedings of the
     3rd International Conference on Transport in Non-Stoichiometric
     Compounds", ed. V.S. Stubican and G. Simkovitch.

26. J. Nunan, J. Cunningham, M.A. Deane, E.A. Colbourn and W.C.
     Mackrodt, Proceedings of an International Conference in memory of
     A.J. Tench (Brunel University, 1984).

27. E.A. Colbourn, W.C. Mackrodt and P.W. Tasker, Physica (in press).

28. J.H. Harding, Physica (in press).

29. P.W. Tasker and D.M. Duffy, Surface Science 137, 91 (1984).

30. Y.W. Tsang and L.M. Falicov, Phys. Rev. B 12, 2441 (1975).

31. E.A. Colbourn, J. Kendrick and W.C. Mackrodt, Surface Science 126,
     550 (1983).

ATOMIC DIFFUSION IN METALS

Jean Philibert

Laboratoire de Métallurgie Structurale
Université de Paris-Sud, 91405 Orsay, France

INTRODUCTION

The purpose of this paper is to show the kind of information that diffusion and mass transport studies are able to bring on defects in metals. It is well known that classical observations like the Kirkendall effect prove that point defects are intervening in the process. More generally every time mass transport is observed (Kirkendall shift in mutual diffusion, electrotransport, diffusion creep, sintering, cavity growth or cavity annihilation,...), diffusion processes are controlled by point defect creation and/or migration. From this simple evidence it is difficult to infer the nature of the defect: vacancy, interstitial, or some more complex entity...?

In this lecture we shall mainly address the identification of the defect and the determination of its main characteristics. The first three parts are devoted to pure metals and dilute alloys, a very well known field, we shall then discuss the case of concentrated alloys, because of their importance and the many questions which still remain open. Finally, a last part will briefly discuss the contribution of diffusion studies to our knowledge of extended defects.

1. PURE METALS, SELF-DIFFUSION

1.1.- Random walk and correlation

Self-diffusivity – i. e. the coefficient which characterizes the mobility of atoms of species A in metal A – is often mistakingly identified with the tracer A^* diffusivity. Actually,

$$D_{A*}^{A} = f \ D_{A}^{A} \qquad (1)$$

where f is the correlation factor; $f \neq 1$ everytime diffusion proceeds by a mechanism involving at least three different "particles", i. e. A atoms, A^* atoms and vacancies (or interstitials).

Let us recall the main results of the random walk model for particles diffusing in a lattice. For diffusion along the x-axis, the diffusion coefficient is given by:

$$D_x = \langle X^2 \rangle \ / \ 2t \qquad (2)$$

where $\langle X^2 \rangle$ is the mean square (projected) displacement of particles during time t. X is the sum of individual jump lengths:

$$X(t) = x_1 + x_2 + \ldots\ldots + x_n,$$

(n jumps during time t, n is a very large number). In the absence of external force $\langle X \rangle = 0$ and

$$\langle X^2 \rangle = \sum_{i=1}^{n} \langle x_i^2 \rangle + 2 \sum_i \sum_j \langle x_i x_j \rangle \qquad (3)$$

If successive jumps are independent of each other, the double sum is zero and $\langle X^2 \rangle = n \langle x^2 \rangle$: whence

$$D_x = (1/2) \, \Gamma \, \langle x^2 \rangle \qquad (4)$$

where $\Gamma = n/t$ is the average number of jumps per unit time, i.e. the jump frequency. Generally, only jumps to the Z near neighbours are possible, and:

$$\langle x^2 \rangle = \sum_{s=1}^{Z} \frac{\Gamma_s}{\Gamma} x_s^2 \; .$$

In this formula s holds for a given direction. Γ is not to be confused with Γ_s, as $\Gamma = Z \, \Gamma_s$.

In a cubic structure let us drop the subscript of D (as $D_x = D_y = D_z$); one easily derives for BBC and FCC structures:

$$D = \Gamma_s \, a^2, \qquad (5)$$

where a is the lattice parameter. But in diamond-like structures:

$$D = (1/8) \, \Gamma_s a^2, \qquad (6)$$

Let us now consider diffusion of A atoms via vacancies or interstitials. Clearly the total number of defect jumps per unit volume and unit time is equal to the total number of atom jumps:

$$n_d \, Z \, w_d = n_a \, Z \, \Gamma_A \qquad (7)$$

whence

$$\Gamma_A = N_d \, W_d$$

Here n holds for concentrations (number per unit volume) and N for mole fractions. W_d is the jump frequency of the defect to a neighbour site.

We have now to introduce the correlation factor. Roughly speaking, for a defect based mechanism a fraction of jumps are lost, because of some "memory" of the past configuration. Once a jump has been made, the tagged atom sees again a defect on a neighbour site (fig. 1). So it has a probability higher than the average of coming back to its initial position. The defect "sees" equivalent neighbour sites and makes jumps in any direction: defect jumps are not correlated[*]. In order to calculate $\langle X^2 \rangle$, all movements of the defect which bring it later on as a neighbour of the tagged atom have to be considered. From eq. (3), one obtains:

$$\langle X^2 \rangle = f \, \Sigma \, \langle x_i^2 \rangle = \frac{1}{2} f \sum_{s=1}^{Z} \Gamma_s x_s^2 \qquad (8)$$

where f ($\leqslant$ 1) is a purely geometrical factor which depends on the crystal structure and the mechanism.

---

[*] This is not exactly true: the defect sees for instance (Z-1) atoms A and one atom A[*] whose mass is different – whence the isotope effect (see below).

a)

b)

Fig.1: Correlation effect for a
vacancy mechanism. First
exchange in b, second
exchange in c̲.

c)

Table 1
Correlation factor for self diffusion in cubic structures

Mechanism	Crystal structure	f
Vacancy	honey comb	0.333
	square	0.467
	triangle	0.56
	simple cubic	0.653
	BCC	0.727
	FCC	0.7815
	Diamond cubic	0.5
divacancy	FCC	0.458
interstitial (direct)	any structure	1
split interstitial <100>	FCC	0.88

1.2.- Diffusion Mechanisms
1.2.1.- Vacancy mechanism

Let us assume vacancies are in thermal equilibrium, then:
$$N_V = \exp(-\Delta G_{1V}^f / kT)$$
where ΔG holds for the free enthalpy of formation of single vacancies (1V). For vacancy jumps:
$$w_V = \bar{\nu} \exp(-\Delta G_{V}^m / kT)$$
with $\bar{\nu} \sim$ Debye frequency, and ΔG_{V}^m the free enthalpy of migration.

Finally tracer diffusivity is given by:
$$D^* = \beta \nu\, a^2\, f_o \cdot \exp\left(\frac{\Delta S_{1V}^f + \Delta S_{1V}^m}{kT}\right) \exp\left(-\frac{\Delta H_{1V}^f + \Delta M_{1V}^m}{kT}\right) \quad (9)$$

with $\beta = 1$ (BCC or FCC) or $1/8$ (diamond cubic). D^* follows an Arrhenius law, with an activation enthalpy:
$$\Delta H = -k \frac{d \ln D}{d(1/T)} = \Delta H_{1V}^f + \Delta H_{1V}^m \quad (10)$$

and a pre exponential factor:
$$D_o = \beta \nu\, a^2\, f_o \cdot \exp\left(\frac{\Delta S_{1V}^f + \Delta S_{1V}^m}{k}\right) \quad (11)$$

The D expression may still be written:
$$D^* = f_o N_{1V} D_{1V}$$
with the vacancy diffusivity being given by:
$$D_{1V} = \beta a^2 \nu \exp(-\Delta G_V^m / kT)$$
D^* and D_{1V} are not to be confused. Their activation enthalpies differ by the formation enthalpy of vacancies. Sometimes D^* is written as D_V^* to show it

means self diffusion by a vacancy mechanism: $D^*_{1V} \neq D_{1V}$.

1.2.2. – Divacancy mechanism
In a FFC lattice, at thermal equilibrium:
$$N_{2V} = 6 \exp(-\Delta G^f_{2V}/kT)$$
and the following formulas easily obtained:

$$D^*_{2V} = 4a^2 \, f_{2V} \, \bar{\nu}_z \, \exp\left(-\frac{\Delta G^f_{2V} + \Delta G^m_{2V}}{kT}\right) \qquad (12)$$

1.2.3. – Interstitial mechanism
Generally the migration proceeds according to the so-called "interstitialcy" mechanism, where the interstitial pushes out one atom, which itself becomes an interstitial, and so on... . Real situations are still more complicated as the interstitial structure is not isotropic, but forms a dumbbell configuration with a neighbour atom. In these conditions, jump length of the "defects" and the tagged atom are different. This, together with the correlation factor, is taken into account in the so-called Haven ratio H_i; and tracer diffusivity becomes:
$$D^*_i = H_i \, N_i \, D_i \qquad (13)$$
where N_i and D_i are the mole fraction and the diffusivity of interstitials.

1.2.4. – When several mechanisms are simultaneously active, tracer diffusivity is a sum extended to all types of defects:
$$D^* = \sum_d D^*_d = \Sigma_{Hd} \, H_d \, N_d \, D_d \qquad (14)$$

with $H_d = f$ for vacancies.

1.3. – Diffusion and Point Defect Characteristics

What kind of information will diffusion coefficient measurements reveal about point defects?

1.3.1. – Temperature dependence of self-diffusivity

A straight Arrhenius plot allows eq. (10) to be checked. The migration and formation enthalpies are determined by independent quenching and annealing experiments. Self- diffusion activation energy only gives their sum. But the consistency of both kinds of result is very important. Actually, the situation is a little more involved, as quenching could introduce both single and di-vacancies. The comparison with self-diffusion results allows the choice of the convenient range of temperatures for the quenching procedure in order to avoid di vacancy presence. A coupled set of experimental results is necessary for an unambiguous determination of ΔH^m and ΔH^f for single vacancies. Such a procedure has been successfully applied to FCC metals: Au, Cu, Al, Ni (Table II).

However in several FCC metals, when D measurements are performed over a very large range of temperatures, they disclose near the melting temperature T_M a slight upwards deviation from an Arrhenius straight line (fig. 2).

Two kind of explanations have been proposed for this deviation:
- a two-defect model, with an increasing contribution from di-vacancies at increasing temperature,
- a one-defect model, where vacancy properties are temperature dependent. According to Varotsos et al. (1978) this behaviour is generally to be expected because of anharmonicity effects. Gilder and Lazarus (1975) suggested a detailed model where the thermal expansion and compressibility of the defects appear explicitly.

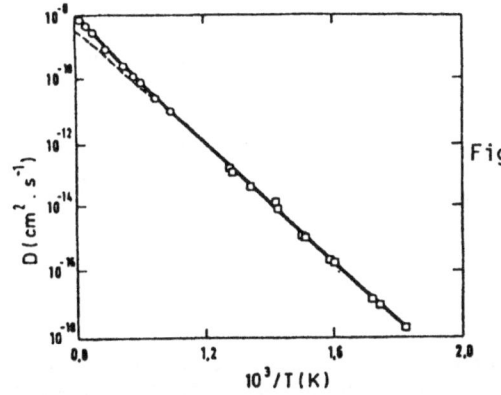

Fig.2: Self-diffusion in silver. Experimental results compiled by Le Claire (1976).

Table II
Self-Diffusion In FCC Metals
(From G Brebec, 1983, and N. Peterson, 1978)

	$D_o(cm^2 s^{-1})$	Q(kcal/mole)	$\Delta H^f_{1V} + \Delta H^m_{1V}$	ΔV	ΔK (with f = 0.78)
Al*	0.035	28.75	31.8	1.3-0.71	
Ag*	0.44	44.3	44.4	0.7-0.9	0.92 (low T)
Cu*	0.62	49.6	49.2	0.9	0.87
Ni*	1.9	68.0	66.8		
Au	0.091	41.7	40.7-42.1	0.72	
Pb	1.37	26.0	22.8	0.73	
Pd	0.21	63.6			1
Fe γ	0.49	67.9			0.68

star (*) means a slight curvature of the Arrhenius graph.

The Gilder-Lazarus model seems to be inadequate, as recent very accurate results on β-Thallium show a slight "normal" upwards curvature instead of the expected downwards curvature (Chiron and Faivre 1985).

The two-defect model, with single and di-vacancies, is presently the more popular. By careful measurements down to low temperature (T < 0.4 T_M di-vacancy contribution is negligible), properties of the single vacancy can be observed. For di-vacancies, the results depend more sensitively on the adjustment technique and the formula used. At least three kinds of "theoretical" formula have been used:
        - the simplest one $D = D_{10} \exp(- Q_1/kT) + D_{20} \exp(- Q_2/kT)$ with a variant where the second term of the R.H.S. is split in two parts (for nearest and next nearest vacancy pairs),
        - a more complex one, where activation entropies and enthalpies are temperature dependent, in agreement with thermodynamic rules:
$$h(T) = h(T_0) + \alpha k (T - T_0)$$
$$s(T) = s(T_0) + \alpha k \ln (T/T_0)$$
If the same α is used for both defects, a minimum of five adjustable parameters are required.

In FCC metals the di-vacancy contribution may be as high as 50% near T_M in Cu or Ag – but quite lower in Au. A few results are given in table III.

Table III

Metal	D_{10} $cm^2 s^{-1}$	Q_{10} (eV)	D_{20} $cm^2 s^{-1}$	Q_{20} (eV)	ΔV_{1V} (Ω)	ΔV_{2V} (Ω)	Reference
Ag	0.046	1.76	3.3	2.28	0.65	1.25	Rein Mehrer (1982)
Cu	0.13	2.06	4.6	2.48			Bartdorff et al. 1978
Au	$2.7.10^{-2}$	1.71			0.70		Mehrer (1983)

Similar results and interpretations apply to most of BCC metals which show a weak Arrhenius anomaly. The consistency of the model sometimes relies on other data, such as the isotope effect or the activation volume (see below and fig. 3 and 4) or to non-tracer data, for instance NMR experiments or quasi-elastic neutron scattering, techniques which permit the observation of the diffusive motion on an atomic scale and provides information on both temporal and spatial correlations of the atomic jump process.

Among BCC metals, Na is one of the more documented: tracer diffusivity has been measured over 5 orders of magnitude along with pressure dependence (ΔV) and isotope effect. The Arrhenius graph shows a slightly upwards curvature. The two-defect model fits the data well, with strongly relaxed defects: $\Delta v_{1V} = 0.32\Omega$, $\Delta V_{2V} = 0.59\Omega$. The kinetic factor Δk of the isotope effect $E = f\Delta K$ varies accordingly with temperature. But the situation is probably more complicated, as the divacancy can move by different mechanisms, as any atomic jump "dissociates" the vacancy pair. This interpretation is confirmed by NMR (Göltz et al 1980) and quasi elastic neutron scattering (Brünger et al 1980). Finally the diffusivity consists of 3 terms: monovacancies, divacancies at distance a $\sqrt{3}/2$ and a, and divacancies at a and a $\sqrt{11}/2$, the last divacancy configurations prevailing and with a slight dependence of the activation enthalpies (because of the correlation factor ?).

Some BCC metals (group IV: Tiβ, Zrβ, Hfβ and lanthanides) show a highly anomalous behaviour. The Arrhenius graph presents a strong curvature and the low temperature activation energy and D_0 look very weak ($Q \sim 10T_m$, $D_0 \sim 10^{-4}$ $cm^2.s^{-1}$ in β-Zr). Many assumptions have been proposed to explain this anomalous behaviour. Quenching experiments cannot be carried out in these metals because of the $\beta \to \alpha$ phase transition. In the absence of complementary data, it is difficult to draw conclusions on the basis of actual data limited to tracer diffusion results and there does not yet exist a general agreement about the nature of the defects and jump mechanism.

Apparently in metals an extrinsic regime, as classically observed in ionic crystals, does not exist. Let us recall that for this regime the defect concentration is fixed by the impurity concentration, so that the diffusion activation energy only consists in the migration part. Comparison of intrinsic and extrinsic regimes then allow a complete determination of ΔH^m and ΔH^f. Such a situation is not found in metals.

The possible role of self-interstitials, at least in BCC metals would deserve more attention, which to date has been neglected.

1.3.2. - Isotope effect

Actually the tracer atom behaves like an impurity, because of the mass difference with host atoms. This means that, as for impurity diffusion, the correlation factor is not a pure number but depends on the various exchange frequencies of the vacancy with A or A* atoms.

Let us consider two tracers α and β of the same element A, with w_α and w_β the corresponding jump frequencies, D_α and D_β the self-diffusivities.

Under certain restrictive conditions (which are true for a vacancy mechanism in cubic crystals),

$$\frac{D_\alpha - D_\beta}{D_\beta} = f. \left[\frac{w_\alpha}{w_\beta} - 1 \right] \qquad (15)$$

In a first approximation, if atoms are looked at as harmonic oscillators:

$$\frac{w_\alpha}{w_\beta} = \left(\frac{m_\beta}{m_\alpha} \right)^{1/2} \qquad (16)$$

where m_α and m_β are the isotope masses. Actually a jump has to be considered as a many-body problem, and theoretical analysis leads to:

$$\frac{w_\alpha}{w_\beta} - 1 = \Delta K \left(\frac{m_\beta}{m_\alpha} - 1 \right)^{1/2} \qquad (16 \text{ bis})$$

where $\Delta K < 1$ is, roughly speaking, a measure of the fraction of kinetic energy localized on the jumping atom. Because of anharmocity, $\Delta K < 1$. The isotope effect is defined as:

$$E = \frac{D_\alpha/D_\beta - 1}{(m_\beta/m_\alpha)^{1/2} - 1} = f \Delta K \qquad (17)$$

The isotope effect is a measure of the correlation factor multiplied by ΔK value. For a double mechanism:

$$E = \frac{D_1}{D_1 + D_2} f_1 \Delta K_1 + \frac{D_2}{D_1 + D_2} f_2 \Delta K_2 \qquad (18)$$

For a single mechanism E does not depend on temperature. In the case of a single + di-vacancy mechanism , E is expected to decrease with increasing temperature (fig. 3).

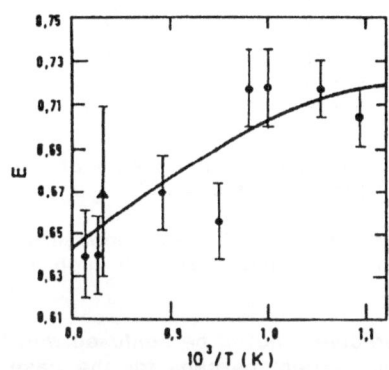

Fig.3: Isotope effect versus temperature for Ag self-diffusion (Peterson 1975).

The isotope effect is a rather small. For ^{55}Fe and ^{69}Fe for instance $\Delta D/D \sim 4\%$. Much care has to be exercised in order to get reliable data.

A slightly different expression holds for E when several atomic jumps are involved as the mass difference is damped (the interstitialcy mechanism is an example).

The isotope effect is a good criterion to assess the nature of point defects responsible for diffusion.

1.3.3. – Activation volume

Let us come back to the D expression:
$$D = \beta a^2 \, \nu \, \exp(-\Delta G/kT).$$
Because of a free enthalpy in the exponential term, it is possible to calculate the pressure derivative. Since:
$$\Delta V = (\partial \Delta G/\partial p)_T,$$
differentiation gives:

$$\left(\frac{\partial \ln D}{\partial p}\right)_T = -\frac{\Delta V}{kT} + \left(\frac{\partial \ln a^2}{\partial p}\right)_T + \left(\frac{d \ln \tilde{\nu}}{\partial p}\right)_T$$

The second term is zero as measurements are performed at ambient pressure; the third term has been showed to be negligible. This yields:
$$\Delta V \approx - kT \, (\partial \ln D/\partial p)_T, \qquad (19)$$
with:
$$\Delta V = \Delta V^f + \Delta V^m, \qquad (20)$$

Pressure dependence of self-diffusivity will allow the determination of the activation volume ΔV. Physical meaning of ΔV^f is simple. In order to create a vacancy (or an interstitial) an atom has to be moved from the bulk to the surface (or the converse): the volume change is thus $+ \Omega$ (vacancy) or $- \Omega$ (interstitial) where Ω is the atomic volume. To this quantity must be added the relaxation around the defect ΔV_{rel}, corrected for image forces (Eshelby factor):
$$\Delta V^f = \pm \, \Omega + f_{Esh} \, \Delta V_{rel}, \qquad (21)$$

Estimates of ΔV^f are given in Table IV.

Table IV
Formation volume of defects
(from Lazarus 1983) (in units of Ω)

Crystal Structure	Defect	v^f	ΔV_{rel}	f_{Esh}	ΔV^f
FCC	vacancy	+ 1	− 1/5	1.5	0.7
	divacancy	+ 2	− 2/3		~ 1
	interstitial	− 1	+ 2/3		~ 0
BCC	vacancy	+ 1	− 1/3	1.6	0.4
	divacancy	+ 2	− 3/4		0.8
	interstitial	− 1	+ 1/2		− 0.2 ?

The meaning of ΔV^m is less clear, as it will depend on the theory of atomic jump process (many-body problem). It is believed that $0 \leqslant \Delta V^m \leqslant 0.2\Omega$.

They are only a very few cases where ΔV^f or ΔV^m were measured by quenching experiments, because of experimental difficulties with high pressures.

According to table IV, the activation volume – not to be confused with the relaxation volume – is always positive or nil, except perhaps for the case of interstitials in BCC metals. Predictions for single and di-vacancies in noble metals are well verified (tables II and III). In silver, relation (20) seems well obeyed ($\Delta V^f = 0.52-0.65\Omega$ and $\Delta V^m = 0.15\Omega$). In the two defect models the measured ΔV increases with temperature (fig. 4). A slight variation of ΔV_{1V} with temperature is questionable.

In aluminium: $\Delta V \simeq 1.3.\Omega$ which means a negative relaxation around the vacancy in this metal - a result perhaps in agreement with the shape of the screened electrical potential round a defect (Friedel's oscillations).

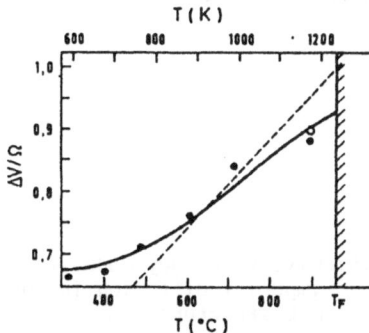

Fig.4: Activation volume for Ag self-diffusion (Rein and Mehrer (1982)

Negative activation volumes were found in Pu-ϵ and some BCC rare earths, perhaps a not unexpected result since in these metals the derivative dT_M/dp of the melting temperature is negative.

2. IMPURITY DIFFUSION

2.1.- Diffusion model and results

Impurity diffusion - also called heterodiffusion - can be analysed along the same lines as self-diffusion. There are two main differences:
- the probability of a defect to be present on a site neighbour to the impurity atom is different from the equilibrium concentration because of their elastic and electronic interactions (cf. eq. (7),
- the correlation factor depends on the frequencies of all kinds of jumps allowed to the defect.

Let us consider diffusion of impurity B in metal A. For a cubic crystal and a vacancy mechanism:

$$D_B^A = \beta a^2 . N_V . f(w_i) . w_2 \qquad (22)$$

where w_2 represents the vacancy-atom B exchange frequency. N_V is just the equilibrium concentration in the pure metal $\bar{N}_V$ times a binding factor:

$$N_V = \bar{N}_V \exp - (\delta g^f/kT).$$

The number of w_i frequencies to be taken into account depends on the crystal structure (fig. 5). In FCC metals the 5-frequency model is well known, where exchange jumps with host atoms are distinguished depending on whether they break the vacancy-impurity pair (dissociative jumps w_3) or they do not (w_1) or they create such a pair (w_4).

It is interesting to compare impurity and self-diffusion in the same metal, which yields:

$$\frac{D_B^A}{D_A^A} = \frac{N_V}{\bar{N}_V} . \frac{f(w_i)}{f_o} . \frac{w_2}{w_o} . \qquad (23)$$

where w_2 is the vacancy-impurity exchange frequency.

It follows that the difference of the activation energies is given by:

$$\Delta Q = Q_{imp} - Q_{self},$$
$$\Delta Q = \delta h^f + \delta h^m - C \qquad (24)$$

where δh^f comes from $N_V/\bar{N}_V$ and is the binding impurity-vacancy enthalpy, δh^m measures the difference in migration enthalpy for a "free" vacancy (w_o)

and a vacancy exchanging with an impurity atom (w_2), and C results from the temperature dependence of the correlation factor. The presence of this last term is rather unfortunate, as it makes any comparison with theoretical calculations not straightforward.

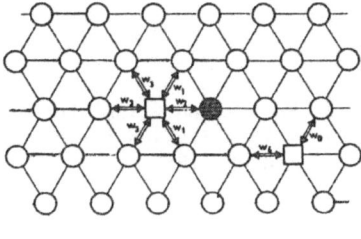

a   □ vacancy
    ● solute atom

Fig.5: Five-frequency model for impurity diffusion in the FCC structure.

a) close packed plane
b) unit cell

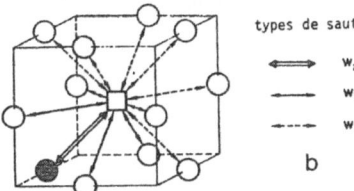

types de saut

⟺ w_2
⟶ w_1
⟶ w_3

b

Comparison of activation energies for impurity and self-diffusion does not allow the determination of the binding energy between vacancies and the impurity atom. Nevertheless methods were developed for a comparison of theoretical computations with measurements. Electronic theory makes it possible to compute the binding enthalpy δh^f. The model used to calculate the δh^m for the various kinds of jumps is not so reliable. With a screened Coulomb potential to describe the impurity atom, computed ΔQ values are in good agreement with experimental ones for diffusion of column-B solutes in noble metals. In other systems where the vacancy mechanism is known (alkaline metals, Mg, Al) the values disagree, probably for several reasons: shape of the electronic screened potential around the impurity atom, size effect, saddle point configuration,...

It would be interesting to obtain more data about the various frequencies involved. This is partly possible by studying A self-diffusion in a series of dilute AB alloys. For low concentration the self-diffusivity follows a linear relation with the molar fraction N_B:

$$D_{A\star}^{AB} = D_{A\star}^{A} [1 + b_1 N_B] \qquad (25)$$

where b_1 is a calculated function of the various jump frequencies w_i. Also f_2 can be determined for impurity diffusion by isotope effect measurements (assuming ΔK has the same value for self and impurity jumps). From all these sets of data three frequency ratios can be determined. Several systems based on FCC metals have been analysed (see e.g. Bocquet et al. 1984). Table V gives two examples.

Table V
Vacancy jump frequency ratios for Zn-vacancy pairs in Ag and Cu alloys

System	T(K)	$D_{B\star}^{A}/D_{A\star}^{A}$	f_B	b_1	w_2/w_1	w_3/w_1	w_4/w_0
AgZn	1010	4.1	0.52	12.6	1.53	0.27	1.15
	1153	3.9	0.57	12.7	1.54	0.39	1.30
CuZn	1168	3.56	0.47	7.3	2.5	0.5	1.2

As Zn diffuses faster than the solvent Ag or Cu, one accordingly finds $w_2 >$ w_1 and since it attracts the vacancy, $w_4 > w_0$.

This is the maximum information which can be extracted from diffusion data. It must be realized how large is the number of data required for such as very sophisticated analysis. Unfortunately the ratio w_4/w_3 which would give the binding enthalpy as $w_4/w_3 = \exp(-\delta g^f/kT)$, cannot be extracted. According to Le Claire (1983) the reason of this inadequacy is that this ratio relates to vacancy jumps in opposite directions between a pair of sites (next nearest and nearest neighbour of an impurity atom (fig. 5), whilst diffusion data can only give information on frequency ratios for jumps from the same lattice site.

2.2. - Fast Diffusion

Diffusion studies have revealed a quite unexpected behaviour in some metals, which show the vacancy model is sometimes irrelevant.

Generally for impurity and self diffusion, the ratio of the D_0's lies between 0.75 and 1.25 and the ratio of the activation energies between 0.75 and 1.25, a result which suggests the same mechanism is operative for both processes.

But in some cases $D_B^A/D_A^A \sim 10^2 - 10^5$ near Tm. A classical example is diffusion of gold in lead ($D_{Au}^{Pb}/D_{Pb}^{Pb} \sim 10^5$ at 175°C).

As it has been clearly demonstrated, these data do not result from short-circuits (grain boundaries or dislocations) but are truly typical of the bulk.

This very fast diffusion is observed in group III-B (In, Tl) and IV (Sn, Pb) metals. Fig. 6 compares the Arrhenius lines for several elements in lead. Activation energy for lead self-diffusion is 1.1eV, but it is as low as 0.35 and 0.43eV for Cu and Au diffusion in this metal.

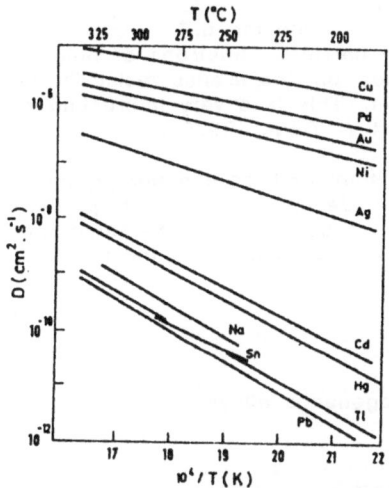

Fig.6: Arrhenius graph for Pb self-diffusion and impurity diffusion in Pb. (Warburton, Turnbull, 1975).

These important results suggest that certain solute atoms dissolve not only as substitutionals but also as interstitials, although the atomic radii ratio lies in between 0.62-0.93, values higher than the Hägg ratio of 0.59.

Several reasons rule out a vacancy mechanism. At least three interstitial mechanisms are possible:
- simple interstitial
- mixed dumbbell (sometimes called "diplons", see Warburton 1975),
- vacancy-interstitial pairs.

Measurements of isotopic effect ($f\Delta K \sim 0.25$-0.30) and impurity effect on self-diffusion (b_1 factor, cf. eq. 25) allow a detailed discussion. As a conclusion impurity is found either as a mixed dumbbell or as an interstitial bound to a vacancy – or both.

Other experimental evidence confirms the presence of interstitials in these alloys: centrifugation (Au in K, Na, Pb), chanelling in AuPb, Mössbauer effect for in Co,...

The same type of mechanism seems also active for transition metals dissolved in the "anomalous" BCC metals (Tiβ, Zrβ, or V, Nb) and for noble metals dissolved in alkaline metals.

3. ELECTROMIGRATION AND THERMOMIGRATION

The application of an electric field or a thermal gradient gives rise to a net transport force in a pure metals (as observed by the shift of inert markers) or in dilute alloys (as observed from the migration and redistribution of solute species).

These phenomena deserve mention, as they bring some information on the defects involved in the transport. Electro-transport force and corresponding flux in pure metals are proportional to the effective charge of the atoms: $Z_A^* e$. From the sign of the observed flux, this charge is found to be positive or negative:

Metal	Ag	Zrβb	Uγ
Z_A^*/f	-20.5	$+0.3$	-1.6

The transport force results from two effects:
- a direct electrostatic force acting on the ion $q_i E$,
- a friction force from the charge carriers, electrons or holes. The corresponding effective charge $-q_i^*$ depends on the excess resistivity due to the jumping ion at the saddle point position. This is a very interesting result which has not been sufficiently exploited (N.V. Doan, 1970).

Thermomigration seems to involve too many and too complex processes to yield understandable information on point defect structure or properties, although the phenomenological analysis of the process follows the same lines as for the electromigration. (For references on this topics, see the proceedings "Atomic Transport in Solids and Liquids", 1970).

4. CONCENTRATED ALLOYS

We shall focus our attention on homogeneous alloys.

4.1. – Classical diffusion studies

The same experimental methods used for self-diffusion and impurity diffusion in pure metals and dilute alloys apply to homogeneous concentrated alloys. In a binary AB alloy, these methods allow the determination of D_{A^*} and D_{B^*}, the self diffusion coefficients of tracers A and B versus the alloy composition. The description of diffusion processes is an extension of that in pure metals. Point defect mechanisms are still valid, but theoretical analysis is very far from a simple one, due to the large number of local configurations and jump frequencies which must be considered around a vacancy or an interstitial. However tracer diffusivities can always be written in the same way as in pure metals:

$$D_I = f_I \; p_I \; \widetilde{w}a^2 \qquad\qquad I = A \text{ or } B \qquad\qquad (26)$$

where $\widetilde{w}_I$ is an average jump frequency, p_I is the so-called vacancy availability (also an average quantity) and f_I a correlation factor. At thermodynamic equilibrium, the degree of order of the solid solution determines directly p_I and indirectly $\widetilde{w}_I$.

A very crude but successful model was suggested by Manning in 1968, the random alloy model. It allows a description of diffusion in primary phases without long range order, as far as the short range order degree is low. Manning just considered two vacancy jump frequencies, according to whether it exchanges with an A or B atom:

$$w_V = N_A \; w_A + N_B \; w_B \qquad\qquad (27)$$

This model allows correlation factors f_A and f_B to be computed from measured D_{A^*} and D_{B^*}. This conclusion was checked with success in disordered $\beta CuZn$ solution and in α–CuZn (30%) primary solution (Peterson, Rothman 1967, 1978). Moreover it appears that the contribution to the activation energy of the temperature dependence of the correlation factor is generally negligible.

These results demonstrate the consistency of the Manning model at temperatures high enough for SRO to be negligible, with a vacancy mechanism. Unfortunately they do not allow any derivation of the characteristic properties of these defects. Any "classical" diffusion measurement will not overcome this difficulty. Fortunately there exists a way out, namely the study of short range diffusion, although it applies to a lower range of temperatures.

4.2. – Short range diffusion

Contrary to classical diffusion where migration over large distances and a high number of jumps are involved, it is possible to study processes related to only a very few atomic jumps, through changes in short range order due to vacancy (or interstitial) migration. This technique was pionneered by Zener in 1953-47 and successfully developed by Nowick (1962), Berry (1968), and others... then extended more recently by a Grenoble group (see e.g. J. Hillairet et al. 1981).

4.2.1. – Principles of the method

The method rests on the study of the kinetics of local ordering, either classical short range ordering (SRO), or not very different, directional ordering, induced either by an applied stress (Stress Induced Order) or by a magnetic field. Clearly the kinetics depend on atomic mobilities via point defects.

Let us consider an alloy in thermodynamic equilibrium, and let us follow the rate at which local order evolves when the temperature is suddenly changed, or where a mechanical stress or a magnetic field is applied – or removed. Let us assume this kinetics be characterized by some relaxation time τ (Actually the situation is not so simple, see infra). Clearly:

$$\tau^{-1} = \alpha \; N_d \; w_d \qquad\qquad (28)$$

where N_d is the mole fraction and w_d the jump frequency of the defect d responsible for atomic migration. α is a constant specific of the method (≈ 1). The relaxation time can be measured in three experimental conditions:

1) at thermodynamical equilibrium:

$$\tau^{-1} = \alpha_V \, \bar{N}_V \, w_V \qquad\qquad (28.1)$$

with the subscript V for vacancies, and upper bar ($\bar{N}$) for thermal equilibrium.

2) in the quenched condition:

$$\tau^{-1} = \alpha_V \, (\bar{N}_V + N_V^q) \, w_V \qquad\qquad (28.2)$$

where N_V^q holds for vacancies retained by quenching.

3) under permanent irradiation:

$$\tau^{-1} = \alpha_V \, (\bar{N} + N_V^{irr}) \, w_V + \alpha_i \, N_i \, w_i \qquad\qquad (28.3)$$

with the subscript i for the self-interstitials.

Through a series of measurements at thermodynamical equilibrium, and a series of annealings after quenching, or under irradiation, it is possible to derive the activation energies for the formation and migration of vacancies and self-interstitials, and for the "atomic diffusion". The latter corresponds to some kind of average of the diffusivities of A and B species. But due to the lack of precise atomic models, self-diffusivities D_A^* and D_B^* cannot be derived from such data. However, contrary to tracer experiments, direct access to point defects is simple.

Briefly, according to equations (28), the following quantities can be measured:
 - $\Delta H_V^f + \Delta H_V^m$ in thermodynamical equilibrium conditions.

 - ΔH_V^m through a series of isochronous annealing treatments after a given quenching treatment. If N_V^q is the molar fraction of retained vacancies and $\tau_a(o)$ the relaxation time measured immediately after quenching at the anneal temperature T_a:

$$\tau_a(o) \propto N_V^q \, . \, \exp \, (- \Delta H_V^m / k \, T_a) \qquad\qquad (29)$$

 - ΔH_V^f through isothermal anneal after quenching from a serie of T_q temperatures:

$$\tau_a(o) \propto N_V^q \propto \exp \, (- \Delta H_V^f / k T_q) \qquad\qquad (30)$$

as T_a remains fixed.

4.2.2. - Experimental methods

4.2.2.1. - Short range order

Several experimental methods can be used to follow SRO kinetics: heat capacity, elastic modulus, electric thermopower, electrical resistivity. The latter two are probably the more popular. Resistivity has to be measured at very low temperature, generally at 4. K. As the concentration of defects is very low, their effect on the resistivity is negligible. Depending on system, any increase in SRO leads either to an increase or a decrease of resistivity. It seems that ρ is a linear function of SRO parameter, whatever its sign (i.e. local segregation or demixing).

4.2.2.2. - Stress induced order (Directional order D.O.).

Stress induce order gives rise to a typical mechanical behaviour, namely

anelasticity, which appears as:
   – elastic after-effect when a static stress or strain is suddenly applied to the specimen (fig. 7),
   – internal friction under cyclic conditions.

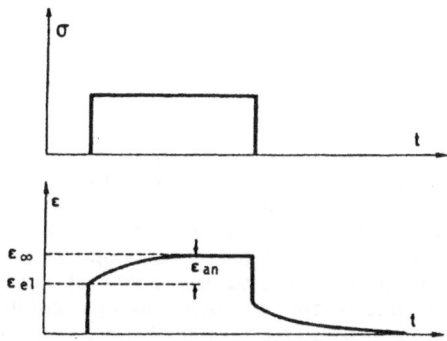

.7: Scheme of anelasticity measurements: variation of and strain ε versus time for loading and unloading.

An early model for anelasticity in substitutional solutions was the Zener pair model. Solute atoms form pairs which act as elastic dipoles. In a free material, these dipoles are randomly oriented. When a stress is applied along some direction, pairs will try to choose orientations which minimize the elastic energy. This reorientation is controlled by the atomic mobility. The original Zener model (1947) was extended by Le Claire and Lomer (1954), Welch and Le Claire (1967) on the basis of short range order (rather than "solute" pairs) induced by an applied stress.

Application is of course limited to those alloys where the "relaxation strength" is sufficiently high.

4.2.2.3. – Induced magnetic anisotropy (Directional order)

A directional order can also be induced by a magnetic field during a "thermomagnetic" treatment. In a polycrystalline specimen, it is characterized by the density ($E(J.m^{-3})$) of magnetic anisotropy energy. E is determined by measuring on a torsion balance the mechanical couple exercised on a disc shaped specimen placed in an external magnetic field H_0.

The method is specific to ferromagnetic alloys at temperatures lower than the Curie temperature. It is the magnetic analog of the Zener anelasticity.

4.2.3. – Relaxation kinetics

Usually SRO kinetics is analyzed in terms of a "chemical" kinetics model. But the phenomenological models developed for anelasticity and D.O. seem to be more appropriate. According to Nowick and Berry (1962), there is no unique relaxation time, because of the local order fluctuations, but rather a relaxation time distribution, centered around the mean $\bar{\tau}$. Several distribution functions are possible. The more popular is the log-normal distribution, i.e. a Gaussian distribution of the logarithms of the relaxation times:

$$\psi (\ln \tau) = \frac{1}{\beta \sqrt{\pi}} \exp \left[- \left(\frac{\ln (\tau/\bar{\tau})}{\beta} \right)^2 \right] \qquad (31)$$

where β is the gaussian half-width.

The evolution of SRO or D.O. follows a kinetics law which is a sum of simple exponential relaxations according to a ψ distribution:

$$\Phi(t) = \int \psi (\ln \tau) [1 - \exp (- t/\tau) d \ln \tau \qquad (32)$$

The initial slope:

$$\left[\frac{d\Phi(t)}{dt} \right]_{t=0} = e^{\beta^{2/4}} \cdot \bar{\tau}^{-1} \qquad (33)$$

allows the determination of $\bar{\tau}$, provided that β has been determined (by a preliminary analysis).

Complications arise when the concentration of defects varies during the relaxation experiment, since $\bar{\tau}$ according to eq. (28) would not be a constant. The procedure of delayed relaxation was initiated by Berry and Orehotzky (1968) to overcome this difficulty. After quenching, the specimen is maintained at the annealing temperature T_a, during various times t_a, then the relaxation experiment is started. One then measures $\tau(t_a)$ through the initial slope of the relaxation curves (eq. 33). Directional order thus allows the study of non-stationnary conditions – and defect concentrations can be varied independently of the degree of order, which makes possible a proper analysis of the decay of a supersaturated vacancy population – whence a determination of the life time of vacancies and information on the sink density and efficiency.

Other measurements can also be performed: for instance in impure alloys, vacancy trapping by impurities.

4.2.4. - Applications

We shall describe the pioneering work of Berry and Orehotsky, before discussing some more recent experiments.

4.2.4.1. - First example: α AgZn

Berry and Orehotsky (1968) reported anelasticity experiments on Ag-Zn alloys (24-33% Zn). They observed, under equilibrium conditions, by internal friction a simple Zener peak whose temperature varies from about 250°C to 340°C when the pendulum frequency changes from 0.618 to 74.6 Hz. For lower temperatures they employed elastic after-effect experiments (173 to 116°C). From these experiments they derived the mean relaxation time τ, the relaxation strength Δ and the distribution width β. Interestingly, Δ varies according to a Curie-Weiss law $\Delta = T_0/(T - T_C)$, where $T_C = 150$ K for 24%Zn. T_C is the "self-ordering" temperature. The temperature dependence of τ followed an Arrhenius law with an "effective diffusion activation energy" Q_{eff} which varies slightly with the alloy composition.

Migration and formation enthalpies were measured on quenched specimens according to the method described above. Since after quenching, the vacancy concentration varies during the anneal, the kinetics $\tau(T_a, t_a)$ was followed by the delayed procedure. Results are given in Table VI.

Table VI
α Ag-Zn solid solutions (Berry et al 1968)

Composition (Zn %)	Q_{eff} (eV)	$\log \tilde{\tau}_o^{-1}$	ΔH_V^f	ΔH_V^m
24	1.46 ± 0.02	14.6 ± 0.02	0.88 ± 0.02	0.56 ± 0.02
27	1.41	14.4	0.89	0.55
30	1.38	14.4	0.84	0.54
33	1.35	14.5	0.83	0.54

364

The ratio $\Delta H^m / \Delta H^f$ is 0.65, which is close to the value 0.75 for this ratio in pure silver.

Berry and Orehotsky were also able to follow the kinetics of vacancy annealing after down quench from 200°C or below to an ageing temperature of 90°C, producing initial quenched-in vacancy mole fraction in the range $5.19^{-9} - 10^{-11}$. By measuring the initial slopes of a family of delayed mechanical relaxation curves, the kinetics of vacancy annealing could be obtained without any restrictive assumptions being made concerning the homogeneity of the annealing process, i.e. the uniqueness of the distribution function of relaxation times: in general this function changes with time, due to the development and eventual smoothing-out of appreciable vacancy concentration gradients so that $\overline{\tau}$ changes accordingly, but also $\psi(\tau)$. The quantity $[\overline{\tau}^{-1}(t_a) - \overline{\tau}^{-1}(\infty)]/\overline{\tau}^{-1}(\infty)$ depicts the relative vacancy supersaturation versus time. It well follows a first order kinetics (exponential decay).

The decay curves appeared to depend on the prior recristallisation temperature (i.e. before quenching procedure). From this it can be inferred that the sink concentration depends on this prior thermal treatment.

In well annealed specimens, the relaxation curves followed a simple exponential decay (first order kinetics). Whatever the initial supersaturation, at a given anneal temperature, the decay time Θ was found constant: Θ measures the vacancy life time $(\sim 2.10^{10}$ jumps). By varying the anneal temperature, Θ was found to follow an Arrhenius behaviour with an activation energy $\Delta H^m_V = 0.51 eV$ a slightly different to the 0.56eV value formerly found. In strained specimens, the life time Θ was reduced. All these results are consistent with a model of simple migration of vacancies to fixed sinks (dislocations).

Differences in measured migration enthalpies reflect differences in the physical process. In directional ordering experiments only a very few vacancy jumps are involved, instead of about 10^{10} jumps in life-time measurements. This difference may be explained on the basis of a very simple model (Caplain and Chambron (1977) as follows:

Let be w^V the jump frequency of a vacancy to a neighbour site; a distinction is made between A and B atoms exchanges:

$$w^V = w^V_A + w^V_B, \qquad (34)$$

The jump frequency of A atoms is accordingly given by (*):

$$n_A \, \Gamma_A = n_V w^V_A,$$

whence,

$$D_A^* = \Gamma_A \, a^z = \frac{n_V}{n_A} = w^V_A \, a^z \qquad (35.1)$$

and similarly

$$D_B^* = \frac{n_V}{n_B} w^V_B \, a^z \qquad (35.2)$$

(*) n holds for concentration in number per unit volume and N for mole-fraction.

But all vacancy jumps do not necessarily contribute to SRO or DO evolution. If the vacancy makes $Z.W^V$ jumps per unit time (Z = number of nearest neighbours), the only effective jumps are jumps with an i-atom (i = A or B) which follow a exchange jump with a j-atom (i ≠ j). Probability of a B jump is ZW_B^V/W^V. The number of B jumps following a A jump is $W_A^V \times ZW_B^V/W^V$. Multiplying this value by 2, in order to take in account A jumps following a B jump, we arrive at:

$$W_{eff} = 2 \, Z \, W_A^V \, W_B^V \, / \, (W_A^V + W_B^V). \qquad (36)$$

The number of AB pairs per unit volume in a disordered alloy is $Z(n_A n_B/n_A + n_B)$. The average life time t_0 for an A-B permutation is given by the ratio of the number of pairs to the number of vacancy effective jumps. With $n = n_A + n_B$, total number of atoms per unit volume, on obtains:

$$t_0 = \frac{Z \, n_A \, n_B}{n.n_V \, W_{eff}} = \frac{n_A \, n_B}{2 n_V . n} \left[\frac{1}{W_A^V} + \frac{1}{W_B^V} \right]$$

$$\qquad (37)$$

$$= \frac{a^2}{2} \left[\frac{N_B}{D_{A*}} + \frac{N_A}{D_{B*}} \right]$$

The relaxation time for SRO is clearly of the order of t_0. Incidently this expression gives, to within an unknown constant factor, the effective diffusion coefficient as measured by anelastic or magnetic anisotropy relaxation.

Let us now consider an experiment where a supersaturation of vacancies is annealed out by migration to fixed sinks, with an average life time Θ.

As far as vacancies are concerned, the nature of exchange (with A or B atoms) is irrelevant. The average number $\langle \nu \rangle$ of vacancy jumps before annihilation to a sink is given by:

$$\Theta = \langle \nu \rangle \, / \, Z \, W^V = \langle \nu \rangle / Z \, (W_A^V + W_B^V)$$

or

$$\Theta = \frac{N_V \, a^2 \, \langle \nu \rangle}{Z \, (N_A \, D_A^* + N_B \, D_B^*)} \qquad (38)$$

$\langle \nu \rangle$ is inversely proportional to sink density.

The activation energies for t_0 and Θ are clearly different according to eqs 37 and 38. Moreover the mean relaxation time τ ($\propto t_0$) is controlled by the slower species diffusing, whilst the vacancy anneal-out is controlled by the faster species.

4.2.4.2. - <u>Second example: α-CuZn alloys and AuNi solutions</u>
(Balanzat 1983)

Fast quenching of α-CuZn alloys results in the freezing of both vacancy concentration in thermodynamical equilibrium at the quench temperature Tq and the SRO state characteristic of this temperature. Fig. 8 shows the resistivity (i.e. the SRO parameter) associated with the quench from various Tq. In region I(Tq < 200°c) atomic mobility is too small for equilibrium to be reached during the 10 min thermal treatment. In region II (200-350°C), resistivity variation is totally reversible and follows the equilibrium SRO (At these very low quench temperatures, $Tq/T_M \leqslant 0.5$, the vacancy concentration is very low ($\sim 10^{-9}$) and does not contribute to the resistivity). In region III (Tq > 350°C) some ordering takes place during the quench.

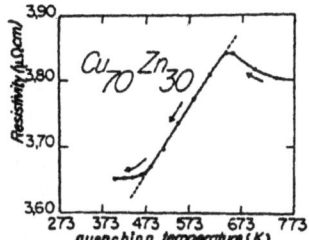

Fig.8: α-CuZn alloy, variation of the quenched-in resistivity with quench temperature. 10 min. isochronous treatments. (Balanzat 1983).

Fig. 9 shows the isochronal evolution of resistivity excess induced by a quench from Tq = 300°C versus the anneal temperature Ta. Behaviour is different according to sink density (i.e. as-rolled, or well-annealed (e.g. a week at 750°C). In the as-rolled specimen, no SRO evolution occurs until about 150°C, because excess vacancies are rapidly annealed out, through a very limited number of jumps. Then the frozen-in disorder is progressively eliminated due to thermal vacancies, until ~ 240°C, where equilibrium SRO is reached within the annealing time. This temperature range is favourable to the study of kinetics of SRO at constant vacancy concentration. By using slightly cold-worked specimens, excess vacancies from Ta are rapidly annealed-out, and SRO is established due to thermal vacancies at the annealing temperature. Thence kinetics law and effective diffusion activation energy are determined: ΔH_{eff} = 1.59eV.

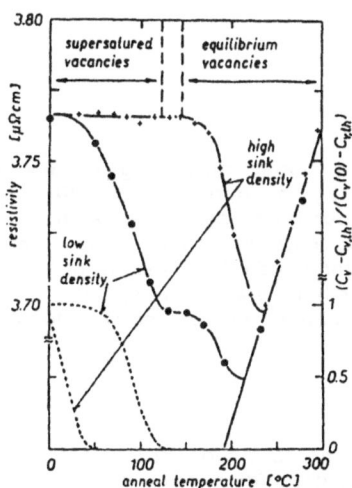

Fig.9: α-CuZn alloy. Isochronal evolution of resistivity induced by a quench from 300°C versus annealing temperature. (Balanzat, Hilairet 1983).

In contrast the well pre-annealed specimens show a very different behaviour. As the excess vacancies have to migrate a long way before annihilation at sinks, they are able to produce at low temperatures a large amount of ordering, until 130°C where they are annealed-out. A second stage of ordering is observed at a higher T_a where equilibrium vacancies reach high enough concentration. The first stage analysis will allow the determination of vacancy formation enthalpy by changing the quench temperature, and the migration enthalpy by changing the anneal temperature. Whence the following results: ΔH_f^V = 1.03eV and ΔH_m^V = 0.64eV. Of course these values refer of course to single vacancies. As the quench temperature is very low, no divacancies are present. The relation $\Delta H_{eff} + \Delta H_f = \Delta H_m$ is well obeyed, as the same physical process (i.e. SRO evolution) is involved in the three series of measurements.

Similar procedures were also applied by Balanzat to α-AuNi (70-30) solid solution where SRO corresponds to some demixing of the solution (Balanzat et al. 1983). Unfortunately, in this alloy the formation energy could not be determined because of its rather low value and the very limited range of available quenching temperatures. Diffusion activation enthalpy was determined by anelastic after-effect at low temperatures (280-310°C), after preliminary equilibration of vacancies at the measurement temperature. This activation enthalpy turns out to be 1.85eV. The migration enthalpy was measured through a series of isothermal anneals carried-out after repetitive quenchs from 325°C (fig. 10): during these experiments, the vacancy concentration was practically a constant (equal to the equilibrium value at Tq = 325°C, as the ordering was completed within a few thousand seconds at 218°C for instance, whilst the decay time (i.e. vacancy life time) measured from a series of delayed anelastic relaxation curves at the same temperature is ~ 10^5 sec.

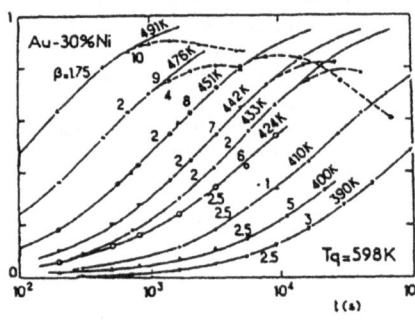

Fig.10: AuNi alloy. Isothermal evolution of resistivity at various annealing temperatures after quenching from 598 K. (Balanzat 1983).

The results for the AuNi solution are rather surprising as the ratio $\Delta H_{1V}^m/\Delta H_{1V}^f = 1.47$ is very different of that ratio for pure gold (0.90). In α-CuZn solutions a "normal" ratio of 0.62 was obtained (as compared to ~1.0 in pure Cu) and 0.65 in AgZn solutions. This high ratio may have important implications, especially for the behaviour under irradiation. Values of this ratio for several alloys and corresponding pure metals are compared in Table VII which shows that the values for pure metals cannot be extrapolated to alloys.

Table VII

Comparison of activation enthalpies for the formation and migration of vacancies in binary alloys and corresponding pure metals
(pure metal values from Brebec (1983))

System	ΔH(eV)	ΔH_{1V}^f(eV)	ΔH_{1V}^m(eV)	$\Delta H_{1V}^m/\Delta H_{1V}^f$	
α-AgZn 76-24	1.46	0.88	0.56	0.65	Berry Halwachs
Ag	1.02	1.10	0.83	0.75	
α-CuZn 70-30	1.59	1.03	0.64	0.62	Balanzat
Cu	2.15	1.05	1.08	1.02	
AuNi 70-30	1.85	–	1.1	1.47	Balanzat
Au	1.81	0.94	0.83-0.89	0.88-0.95	
FeNi (Ni 50-94%)	2.90	1.80	1.10	0.61	Caplain
Ni	2.94	1.40	1.50	1.07	

4.2.4.3. – Third example: Fe–Ni alloys

We shall now discuss similar studies in Fe–Ni alloys (50–94% Ni) where directional order was followed by induced magnetic anisotropy (Caplain and Chambron, 1977). Alloys were always quenched from a temperature higher than the long range order critical temperature (for $L1_0$ and $L1_2$ ordered structures).

Firstly the thermodynamical equilibrium was attained. Specimens were isothermally annealed under a magnetic field (400–500°C) and the evolution of magnetic anisotropy energy, i.e. D.O., was recorded. The vacancy concentration remains practically constant during each experiment. From the initial slope or from the total relaxation curve the relaxation time $\tilde{\tau}$ was derived – and its activation enthalpy determined: it does not in practice depend on alloy composition and is equal to 2.90 ± 0.05eV. This is an "effective" diffusion activation enthalpy (fig. 11).

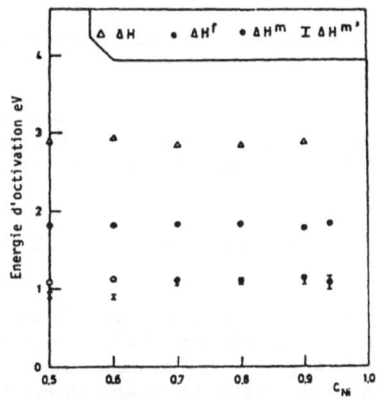

Fig.11: FeNi alloys. Activation energies versus alloy composition. (Caplain 1978).

Other experiments were performed in the as–quenched condition (excess vacancies created in 10 min at Tq). Anneal treatments are then performed in the torsion balance under a magnetic field. Annealing is periodically interrupted to measure the anisotropy $E(t_a, T_a)$.

At a given anneal temperature, the initial slope $(dE/dt)_{t = 0}$ is proportional to $\tilde{\tau}^{-1}$, i.e. to the quenched vacancy concentration $\bar{C}_V$ (Tq). Whence for various Tq's, it yielded a value of $\Delta H_V^f = 1.80$eV (whatever the alloy composition (fig. 11)).

Two methods were used for the study of vacancy migration: 1) For a constant vacancy concentration, the characteristic time of ordering $\tilde{\tau}^{-1}$ varies as $\exp(-\Delta H_V^m/kT)$: $\Delta H_V^m = 1.10 \pm 0.05$eV (fig. 11). The relationship $\Delta H = \Delta H^f + \Delta H^m$ is well obeyed. 2) The second method is based on the kinetics of vacancy anneal–out. The activation enthalpy $\Delta H_V^{m'}$ for the life time Θ varies from 0.90 to 1.10eV according to the alloy composition (fig. 11).

The difference between ΔH^m and $\Delta H^{m'}$ was already commented on earlier in this paper. Another interesting result from these experiments is the absence of a difference in formation enthalpy between the para– and the ferro–magnetic states (in several alloys, the Curie temperature lied in the range of Tq temperatures).

The vacancy–impurity interactions were also studied. Such interactions were inferred from a shift of the temperature stage where vacancies anneal out during a series of isothermal treatments (fig. 12): the higher the carbon content, the higher the vacancy annealing peak. The ratio of vacancy life times in pure and impure alloys was determined versus the impurity concentration at a given annealing temperature.

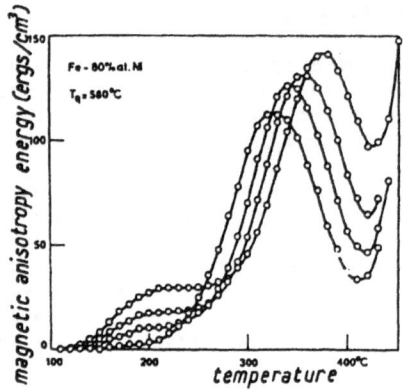

Fig.12: Magnetic anisotropy energy induced by isochronous anneals (10 min) after quenching from 580°C, for various carbon contents (Caplain 1978). The low temperature peak is due to carbon reorientation.

For carbon (up to 600ppm), results do not agree with the simple Lomer model given by:

$$N_V / \bar{N}_V = 1 + Z \, N_C \, exp \, (E^B/kT)$$

where Z is the number of vacancy sites near neighbours of a carbon atom, and N_C the mole fraction of carbon atoms. Trapping of vacancies by several carbon atoms has to be taken in account. Results were discussed with a binding energy of the order of 0.35eV.

4.2.4.5. - Fourth example: Irradiation defects in α-AgZn solution

Our last example will be devoted to a thorough investigation of anelasticity in α-AgZn alloys under irradiation by electrons. (Halbwachs, 1977). For this purpose an inverted torsional pendulum with negligible inertia was adapted to work in-line with a Van de Graaff accelerator in order to follow directional ordering by anelastic relaxation.

Let us first recall some classical results about defect population kinetics under irradiation at temperatures where these defects are mobile. Frenkel pairs are produce at a rate G (per unit time and unit volume) and point defects disappear either by mutual recombination (vacancy + interstitial) or by annihilation to fixed sinks (dislocations jogs).

Corresponding balance equations for concentrations of vacancies and insterstitials are classically written:

$$\frac{\partial C_V}{\partial t} = G - Z \, C_i \, C_V \, (W_i + W_V) - \rho_V \, C_V \, W_V$$

$$\frac{\partial C_i}{\partial t} = G - Z \, C_i \, C_V \, (W_i + W_V) - \rho_i \, C_i \, W_i$$

where the second term on the right hand side corresponds to the mutual recombination: Z is the recombination efficiency factor. The third term expresses the annihilation at sinks of density ρ, assumed to be homogeneously distributed in the material. Numerical solutions of these coupled equations lead to the following pattern with a stationary stage being only attained after a long time. The concentrations of defects are first increasing at an equal rate, then a pseudo-stationary stage is observed, due to prevalent recombinations ($C_i = C_V$). After a new transient the true stationary stage is obtained as defects have enough time to migrate to sinks: the corresponding concentrations are given by:

$$\rho_V \, C_V \, W_V = \rho_i \, C_i \, W_i$$

Generally $\rho_V \neq \rho_i$ as dislocation sinks have not the same efficiency for interstitials and vacancies. But roughly speaking, $C_i/C_V \approx W_V/W_i$. Generally in

370

pure metals C_I <C_V, as W_I> W_V. The relative importance of these stages depends on temperature and dislocation density.

From these results, the relaxation rate $\overline{\tau}^{-1}$ is easily computed. At high temperature, or without irradiation, its activation energy is that for self-diffusion ΔH. When recombinations are prevalent, it is $1/2\,\Delta H_s^M$, the migration enthalpy of the slowest species, whilst for prevalent sink annihilation, the activation energy is zero (fig. 13).

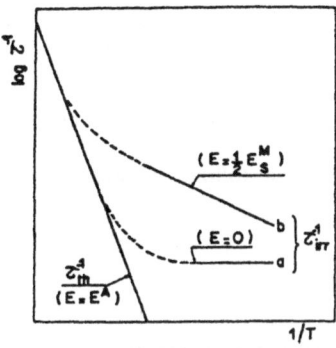

Fig.13: Theoretical variation of the atomic mobility $\overline{\tau}^{-1}$ versus $1/T$. th. is for thermal equilibrium and irr. for irradiation conditions (Halbwachs 1977).

For the quasi-stationary stage:
$$\tau^{-1}_{qs} = (GW_f/Z)^{1/2},$$

whilst for the stationary stage:
$$\tau^{-1}_{st} = (GW_s/Z)^{1/2},$$

where s and f hold respectively for the slowest and fastest species (i.e. W_s < W_f).

The method was applied to the $\alpha-AgZn$ 30% solution. The relaxation process was found to be accelerated under irradiation by electrons. The evolution of $\overline{\tau}^{-1}$, versus irradiation time is shown on fig. 14 for 3 temperatures. It follows the predicted pattern. The stationary stage corresponds to a defect mole fraction of about 10^{-8}. The very long time observed to reach the stationary stage reveals a very low sink density in these alloys – probably related to a strong dissociation of dislocations ($\rho_V = 5.10^{-11}$).

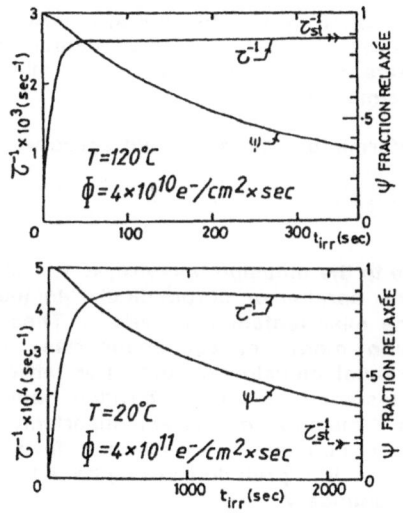

Fig.14: Relaxation rate $\overline{\tau}^{-1}$ versus electron irradiation time in AgZn alloy at three different temperatures. ψ is the relaxation curve, and $\overline{\tau}$ the relaxation rate derived from the ψ-curve (Halbwachs 1977).

Relaxation experiments following the stopping of the electron flux during the quasi-stationary or the stationary stages brought complementary proof of the consistancy of the model. From the kinetics of the quasi-stationary stage (-70 to + 70°C) and the stationary stage (40 to 100°C) the migration enthalpy for the fastest (0.56eV) and the slowest (0.94eV) defects were determined. These figures are to be compared with the migration enthalpy determined after quenching experiments: 0.52 eV, in good agreement with old results of Berry and Orehotsky we described above. (§.4.2.4.1.).

Does this mean that vacancies are more mobile than interstitials? The ratio of the quasi-stationary and stationary rates is unusually small (~2.5) which shows that the respective mobilities only differ by a factor of 6. To answer the question, combined quench + irradiation experiments were performed. They definitely showed vacancies are the fastest defects, and interstitials the slowest.

The results are quantitatively interpreted with the following values:

$$W_i = 3.10^{19} \exp (- 0.94/kT)$$
$$W_V = 2.5.10^{14} \exp (- 0.56/kT)$$
$$Z = 18 \text{ at high temperature}$$
$$\rho_i/\rho_V = 10$$

Dislocations appear to be more efficient sinks for interstitials than for vacancies (dislocation-bias).

The slow mobility of interstitials - an unexpected result and contrary to all experimental evidences in pure metals - could be ascribed to a strong elastic interaction with the solute element (size effect) in dilute alloys, but a more sophisticated explanation is required for concentrated alloys. Moreover the high values of the preexponential factor remains unexplained.

Another interesting result arises from this study: the interstitial defect appears to be as active as the vacancy to induce short range order in the solution.

5. EXTENDED DEFECTS

Diffusion only reveals a very limited information on the detailed structure of extended defects: dislocations, grain boundaries, surfaces. As the number of sites and jump types is very large, any accurate description of atomic migration becomes impossible - except by numerical simulation.

However diffusion data brings a lot of information which, should agree with the predictions from structural models.

5.1. - Grain boundaries

The main result in this respect is relative to the diffusion anisotropy. For tilt sub-boundaries, $D_{//} > D_{\perp}$ (diffusion direction parallel or perpendicular to the tilt axis), this difference decreasing when the misorientation increases. These results agree quite well with the dislocation model of sub-boundaries, if dislocations are seen as pipes where preferential diffusion occurs. For larger misorientation, the diffusivity-measured for instance as the penetration depth along boundaries (fig.15) - shows minima ('cusps') for certain misorientations, whilst the activation energy remains fairly constant. These results favour an organised structure of grain boundaries - an interpretation strengthened by the anisotropy observed in the plane of the boundary ($D_{//} > D_{\perp}$).

372

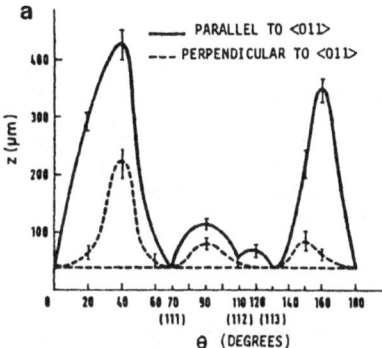 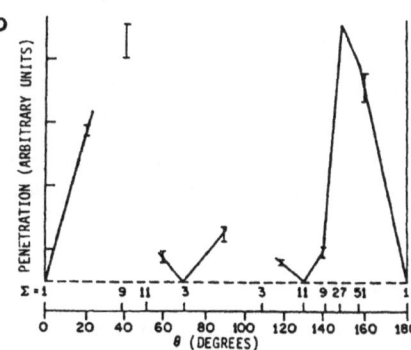

Fig.15: (a) Penetration depth of Zn in Al bicrystals versus the misorientation angle θ ;tilt axis <001> .Annealings 720h at 250°C (Herbeuval 1973).
(b) Same data, with line segments drawn in accordance with structural unit model (Balluffi 1984).

All diffusion results are in agreement with a model of relatively narrow core of grain boundaries consisting in a series of structural units as developed by Vitek and Sutton and by Balluffi (see Balluffi 1984). Diffusion is slower for certain special boundaries: (low coincidence ratios, described as a serie of identical small structural units).

The diffusion mechanism is probably of the vacancy type. Hydrostatic pressure effect gives an activation volume of (1.1 ± 0.2) Ω in silver (as compared to 0.8 for lattice diffusion) (Martin et al., 1967). Isotopic effect in the same metal gives $f\Delta K = 0.46$ for polycrystals and twist bicrystals (Robinson 1972). All these results favour the existence and mobility of localized vacancies ("bona fide defects", following Balluffi's expression), not very much relaxed.

A few more general results confirm this view: 1) net mass transport along grain boundaries is observed in creep, sintering, cavity growth, electromigration, which must occur by a defect mechanism, 2) Some experiments indicate that the two species in a binary system diffuse at different rates in boundaries (as observed as the "Diffusion Induced Grain Boundary Migration" or DIGM).

5.2. – Dislocations

The problem is similar to the above case. But experiments on well defined dislocation structure are still more difficult. One generally confines to a purely phenomenological description, where dislocations are described as pipes of a small radius. But this approach, as useful as it can be, does not give any insight in structural models. The more interesting results are relative to the activity of dislocations as sinks for points defects. Some results were quoted in chapter 4.

5.3. – Surfaces

The crystal surface would apparently constitute a favoured 2-dimensional medium for atomic mobility studies. Actually our degree of knowledge is rather poor, for several reasons:

- a too poor theoretical knowledge of structure defects "in" surfaces,
- the possibility of several, if not many, competing diffusion processes.
- the difficulty to make a safe correspondence between experimental data and elementary jump processes.

From the point of view adopted in this paper, the only relevant experiments were performed with the field ion microscope: this instrument shows directly the migration of adatoms on the surface at low temperature ($< 0.2T_M$). This method is limited to a very few systems: W, Mo, Ta, Re, Ir and Pt on W, or Rh on Rh. Through the mean square path, mobility is measured on surface of various orientations. For instance on Rh {111} (FCC) $\Delta H^m = 0.15eV$, but $0.88eV$ on {100}. On W (BCC), ΔH^m varies from $0.8eV$ on {211} (a channelled surface) to 1.8 on a {111} surface. Adatoms pairs migrate faster than isolated adatoms. Interaction of adatoms with ledges is also observed.

References

- Readers will find useful references in the following text-books or conference proceedings.

C. P. Flynn
Point defect and diffusion, Clarendon Press Oxford 1972.
J. R. Manning
Diffusion kinetics for atoms in crystals, Van Nostrand, Princeton 1968.
J. Philibert
Diffusion et transport de matière dans les solides. Les Editions de Physique, 1985.
Atomic Transport in Solids and Liquids
A. Lodding, T. Lagerwall, ed., Z. Natürforshung, Tübingen (1970).
Diffusion in Solids. Recent Developments
A. S. Nowick, J. J. Burton, ed., Academic Press, New York (1975).
Diffusion in crystalline solids
A. S. Nowick, G. E. Murch, ed., Academic Press, New York (1984).
Dimeta-82, Diffusion in Metals and Alloys.
ed., F. J. Kedves, D. L. Beke., Trans. Tech. Publications (1983).
Mass Transport in Solids
F. Bénière, CRA Catlow, ed., Nato series B, 97 Plenum Press, 1983.
Physical Metallurgy, R. W. Cahn, P. Haasen, ed., North Holland (1984) chap. 8, by J. L. Bocquet, G. Brébec, Y. Limoge.

Balanzat, E., Hillairet, J., (1981), J. Phys. F. Metal Physics,11,1977.
Balanzat, E., (1983), Thèse, Université de Grenoble.
Balanzat E., Halbwachs, M., Hillairet, J., Mairy, C., Guyot, P., Simon, J.P., (1983), Acta Metall.,31, 883.
Balluffi, R.W., (1984), Chap 6 in "Diffusion in crystalline solids" op.cit.
Bartdorff, D., Neumann, G., Reimers, P., (1978), Philos. Mag. 38, 157.
Berry, B.S., Orehotsky, J.L., (1968), Acta Metall. 16, 683 and 697.
Bocquet, J.L., Brebec, G., Limoge, Y., (1984), chap 8 in "Physical Metallurgy" op. cit.
Brebec, G., (1983), Mass Transport in Solids, op. cit., 251.
Brünger, G., Kamert O., Wolf D., (1980), Sol. State Comm. 33, 569.
Caplain, A., Chambron, W., (1977) Acta Metall.25, 1001.
Chiron, R., Faire, G., (1985), Philos. Mag.A, 51, 865

Gilder, H.M., Lazarus, D. (1975). Phys. Rev. B 11, 723

Göltz, G., Heidemann, A., Mehrer, H., Seeger, A., Wolf, D., (1980), Philos. Mag., A, 41, 723.

Halbwachs, M., (1977), Thèse, Université de Grenoble.

Halbwachs, M., Stanley, J.T., Hillairet, J., (1978), Phys. Rev. B, 18, 4938.

Herbeuval, I., Biscondi, M., Goux, C., (1973), Mém. Sc. Rev. Metall., 70, 39.

Hillairet, J., Halbwachs, M., Beretz, D., Balanzat, E., (1981), Czech J. Phys. B31, 1087.

Lazarus, D., (1983), Dimeta-82. op. Cit. 133

Le Claire, A.D., Lomer, W.M. (1954), Acta Metall. 2, 731.

Le Claire, A.D. (1978), Colloque Saclay, La Diffusion dans les Milieux Condensés I, 447.

Le Claire, A.D., (1983) Dimeta-82, op. cit., 82.

Martin, G., Blackburn, D., Adda, Y., (1967), Phys. Stat. Solidi, 23, 223.

Mehrer, H., (1983), Dimeta-82, op. cit., 393.

Nowick, A.S., Berry, B.S. (1962), Acta Metall. 10, 312.

Peterson, N. (1975), in Diffusion in Solids, Recent developments, op. cit., 115.

Peterson, N., (1978), J. Nucl. Materials, 69-70, 3

Peterson, N., Rothman, J. (1967), Phys. Rev. 154, 558.

Rein, G., Mehrer, H., (1982), Philos. Mag. A45, 467.

Robinson, J.T., Peterson, N.L. (1972), Surface Sc. 31, 586

Varotsos, P., Alexopoulos, K., (1977), J. Phys. Chem. Sol. 38, 977.

Warburton, W., Turnbull, D., (1975) in Diffusion in Solids, Recent developments, op. cit., 171.

Welch, D.O., Le Claire, A., (1967), Philos. Mag. 16, 981.

DIFFUSION IN OXIDES

Claude Monty

C.N.R.S. Bellevue, 92195 Meudon, France

1. INTRODUCTION

Oxides are compounds whose interest has been increasing during the twenty last years in solving many application problems. The large spectrum of their properties related to different types of bondings is responsible for such a situation. Oxides are generally semi-ionic but some of them are taken as reference systems for ionic crystals, others are treated as covalent compounds and several oxides exhibit a metal/-insulator transition. The property which identifies most easily the dominant character of a given oxide is the electrical conductivity which has been considered in a previous chapter : it is a way to characterize the electronic structure and the point defect population far from the equilibrium or at a given thermodynamic state /1/, /2/.

In the case of pure binary oxides, the thermodynamic states are defined using two parameters : the temperature, T, and the oxygen partial pressure, P_{O_2}. It is one of the attractions of these materials that they can be easily studied at the thermodynamical equilibrium because T and P_{O_2} are relatively simple to control. The composition of an oxide can change with P_{O_2} and T. Departures from stoichiometry can appear. For example an MO oxides can have the real formula $M_{1-x}O$ where x is well defined for given values of P_{O_2} and T. x can be zero or so small that it is impossible to measure it (MgO, Al_2O_3,...) but in some cases it reaches several per cents ($Fe_{1-x}O$, UO_{2+x},...). Old techniques such as thermogravimetry are nowadays sensitive to density changes of the order of 10^{-5} and x can be determined directly as a function of T and P_{O_2} in that range /3/, /17/,

The departures from stoichiometry are generally directly related to point defect concentrations, one species being in excess in one sub-lattice. In some cases nevertheless the nonstoichiometry is mainly associated to extended defects (WO_3, Nb_2O_5,...) and the point defect concentrations cannot be related easily to x.

In ionic and covalent compounds the point defects are generally charged compared to the normally occupied lattice sites. The main difference between ionic and covalent crystals is, from the point defect point of view, the possibility that electrons and holes can play a role as their concentration can be large in covalents. The knowledge of the charged species which dominates ("majority point defects") is essential; their concentration obey the neutrality equation which is the "identity card" of the compound in given ranges of P_{O_2} and T. There is a balance between positively and negatively charged species, where only the species of highest concentration are important.

Diffusion is an interesting property which can increase our knowledge about point defects at thermodynamic equilibrium. In this chapter we shall be dealing with the main information which is revealed by atomic transport studies in oxides. As the main field of this conference is to look at modern techniques, we shall select recent experiments which have been performed using techniques extensively developed in the last ten years such as : Secondary Ion Mass Spectrometry, Nuclear Reactions Analysis, Micro-Sectioning using ion beams,... Particularly in the case of oxides, they have provided results on single crystals or on polycrystalline material of good purity. The most significant results have been obtained on non stoichiometric oxides in which the impurity effects are masked by large concentrations of point defects. The series NiO, CoO, MnO, FeO whose largest departures from stoichiometry are increasing from NiO ($\simeq 10^{-4}$) to FeO (several per cent) is at the moment a reference series and we shall show several results concerning these oxides. Several other examples can be found in the literature /1/, /2/, /4/.

2. THERMODYNAMICS AND KINETICS OF POINT DEFECTS

2.1. Point defects formation

The formation of defects is a thermodynamic process. In the case of oxides, defects can be considered as the result of an oxidation or a reduction reaction : for example, a V_M'' vacancy in a MO oxide can be considered as created during the reaction :

$$\frac{1}{2} O_2(g) \rightleftharpoons V_M'' + 2 h^\cdot + O_O \tag{1}$$

This quasi-chemical reaction obeys several rules such as the conservation of mass, of charge and of the structure. The mass action law can be applied and we can write :

$$K_{V_M''}^f = \frac{[V_M''] [h^\cdot]^2}{P_{O_2}^{1/2}} = \exp - \frac{\Delta G_{V_M''}^f}{kT} \tag{2}$$

We have considered that the activities are equal to the concentrations which is a first approximation assuming that the concentration of point defects is not too high, and they are no interactions, between defects.

This type of equation describes the formation of one atomic point defect and of two electronic point defects (holes h·).

$\Delta G_{V_M''}^f = \Delta H_{V_M''}^f - T \Delta S_{V_M''}^f$ defines the "enthalpy and the entropy of formation" of a vacancy created during the process described by (1). Other processes have been considered, these parameters in such cases are different. For example, we can create both a cationic and an anionic vacancy by a purely thermal process (no dependence on P_{O_2}) :

$$o \rightleftharpoons V_M'' + V_O^{\cdot\cdot} \tag{3}$$

which defines the "Schottky energy" ΔG_S by :

$$K_S = [V_M''] \, [V_O^{\cdot\cdot}] = \exp - \frac{\Delta G_S}{kT} \qquad (4)$$

Other purely thermal processes can be defined (Frenkel or Anti-Frenkel processes for example) they are often used in the literature and can be related to exchange processes producing only one atomic point defect, see, for example references /1/ and /4/.

The concentration of point defect can be now defined in given situations. For example, in the case were the compound behaves as a p-type semiconductor the following neutrality equation holds for some oxides :

$$2 \, [V_M''] \simeq [h^\cdot] \qquad (5)$$

we can solve both (5) and (2) and write :

$$[V_M''] = A \, (P_{O_2})^l \, \exp \, (\, -\Delta \widetilde{H}^f_{V_M''}/kT) \qquad (6)$$

with

$$\begin{cases} A = (\tfrac{1}{4})^{1/3} \, \exp \, (\, \Delta S^f_{V_M''}/3 \, k \,) \\ l = 1/6 \\ \Delta \widetilde{H}^f_{V_M''} = \Delta H^f_{V_M''}/3 \end{cases}$$

We have related in a particular case the apparent formation enthalpy ($\Delta \widetilde{H}^f_{V_M''}$) characterizing the dependence on T of the concentration of the defect in given T and P_{O_2} ranges to the formation enthalpy $\Delta H^f_{V_M''}$ characterizing the creation of this defect by an "exchange process" (or "oxidation process"). The relation (6) has a general form and shows also a dependence on P_{O_2}, the l exponent being characteristic of the point defect type and its charge, and the neutrality equation involved (here relation 5).

2.2. Point defects interactions

The simple defects we have considered until now are generally charged species. Interactions between them can give new, more complex, species or thermodynamical effects such as non ideal behaviour. Strong interactions between charged atomic point defects and electronic carriers can produce species with a smaller effective charge. For example, considering a MO oxide containing mainly V_M'' and $h^\cdot$, the following reaction can occur :

$$V_M'' + h^\cdot \rightleftarrows V_M' \qquad (7)$$

Depending on the oxide, the holes can be more or less localised. Their existence corresponds to the fact that ions can change their valency ($M^{2+} \rightarrow M^{3+}$). Relation (7) can be written also as :

$$V_M'' + M_M^\cdot \rightleftarrows (V_M'', M_M^\cdot)' \qquad (8)$$

where the complex defect ($V_M'', M_M^\cdot$)', made by the association of a vacancy and a three plus ion, is seen as a new entity that we call the singly charged vacancy V_M'. A neutral vacancy V_M^x can be defined in the same way. More complex associations involving several atomic point defects or impurities can be considered (see Chapter 1). In $Fe_{1-x}O$ for example aggregates of 4 vacancies and one iron interstitial seem to be the elementary units of the "point defect" populations. The standard free energies associated to these reactions are binding energies (but a reaction such as 7 can be considered also as an ionisation process).

Charged species can interact at large distance. A tendency for a species of a given sign to be surrounded by species of the other sign, creates a non randomly distributed population of defects. The interactions between species are weaker than in the case of associations, they modify the thermodynamical properties of the crystal : considering that we have a solution of defects in the solid, the thermodynamical activity, of each species is not equal to their concentration. Using the Debye-Huckel theory proposed for electrolytes, activity coefficients have been calculated and found close to the experimental values /5/. In NiO or CoO for example properties related to point defects (electrical conductivity, thermogravimetry, diffusion...) can be explained as well by assuming that two main defects $V_M^{\prime\prime}$ and $h^{\cdot}$ are interacting, or considering that the solution of the defects $V_M^{\prime\prime}$, $V_M^{\prime}$, V_M^{X} and $h^{\cdot}$ is ideal. The reality is perhaps between them, and it could be necessary to consider complex defects with non-ideality effects.

2.3. Point defect migration

At equilibrium the point defects are moving in the crystal from one site to another. Considering a vacancy for example, the jump frequency ω_V is proportional to the attempt frequency $\overline{\gamma}$ and to the probability that the defect jumps over the energy barrier existing between two sites : ΔG_V^m. This probability is generally expressed as a Boltzmann factor and ω_V is written on the form :

$$\omega_V = \overline{\gamma} \, \exp - \frac{\Delta G_V^m}{kT} \qquad (9)$$

Such an expression comes from the reaction rate theory and is currently used for all types of materials /6/. ΔG_V^M is the Gibbs energy difference between the "saddle point configuration" considered as an equilibrium state and the stable position for a jumping atom. $\overline{\gamma}$ is generally taken as the Debye frequency.

In oxides, which are compounds, each type of vacancy migrates on its own sub-lattice. Nevertheless, the influence of the other sub-lattice cannot be completely neglected and the jumps between two sites can be more complex than expected in geometrical models. (See later "diffusion mechanisms") /7/.

Another problem is to know how large complex defects migrate : without dissociating or partially destroying to be built in an other place ? Can they contribute to feed the population of fast moving species which are involved alone in diffusion processes ? A related question is : how is the "charge" transfered of $V_M^{\prime}$ and V_M^{X} vacancies during the jumps and how does the migration energy change with the state of charge ? The answer to such questions is essential to the understanding of the diffusion processes, computational techniques are able to support models but we need new experiments to improve them.

3. SELF DIFFUSION

3.1. Experimental evidence

A very simple experiment shows the reality of the atom movements in an oxide : A slice of a single crystal of MO oxide is annealed at a given temperature T under a given oxygen partial pressure P_{O_2} using natural ^{16}O. When the thermodynamic equilibrium is reached, ^{16}O is pumped off and replaced by ^{18}O. After a given time t the sample is cooled. The concentration of ^{18}O can be measured from the surface into the bulk using Secondary Ion Mass Spectrometry (SIMS) or Nuclear Reaction Analysis (NRA). Figure 1 shows an example of such a profile obtained in NiO by SIMS.

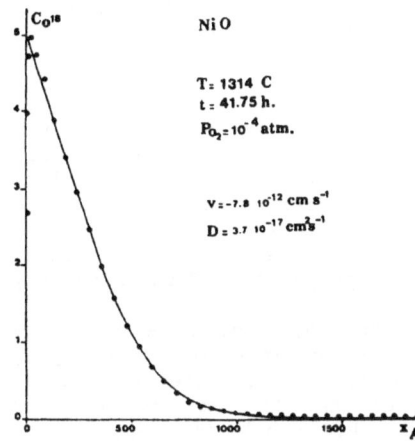

Fig. 1 : ^{18}O diffusion profile in a NiO single crystal /26/

The concentration/penetration curve obtained is the result of an "isotopic exchange" between ^{16}O and ^{18}O atoms. The thermodynamic force responsible for that is the difference in mixing entropy between the initial state and a final state in which the two isotopes would be at the same concentration everywhere. This difference can be reduced, in principle, to a perturbation as small as wanted, using mixtures of dilute ^{18}O in ^{16}O in the gas phase.

Such an experiment is not original in principle. The evidence of atomic movements has been classically done, in a similar way, using radioactive tracers deposited as thin films on the surface of the samples. The originality comes from the techniques used to establish the diffusion profiles of oxygen : SIMS is really a modern technique which is beginning to be extensively used in diffusion studies.

The principle of SIMS consists of sputtering the sample by a primary ion beam and analysing by mass spectrometry the secondary ions which are produced. A flat bottomed hole can be obtained on the surface of the samples. An area of 20 µm in diameter or more can be selected. Recent devices (CAMECA IMS 3F) can work with a resolution power in mass of about 10000 or more and give images of the distribution of elements in the sample. It is possible to look at all profiles, particularly at short ones (several tenth of Angströms), and to separate stable isotopes /8/, /9/.

The method of using ion beams for the microsectioning of the diffusion samples has been also applied to measurements using radio-tracers. Such technics provided recently results on short distances diffusion processes /10/, /11/.

Nuclear Reactions (NRA) or Rutherford Back-scattering Analysis (RBA) are also powerful techniques, but more limited and less easy to use. The method consists in irradiating the sample by a particle beam of energies ranging around 1 MeV and to analyse the secondary beam produced by the nuclear reactions induced inside the sample with the elements of interest /12/. In the case of ^{18}O, for example, the reaction ^{18}O (p,α)^{15}N is commonly used with protons of 1,76 MeV. In some cases there is no reaction and the energy lost by the incident and back-scattered particles is only used to determine the distribution of scattering centers inside the sample. Such a process can be efficiently used only in the case of atoms heavier than the solvent atoms as this effect is essentially mechanical.

3.2. Diffusion mechanisms

How do the atoms move in an oxide ?

As in other cristalline materials, the mechanisms involving point defects seem to be the energetically most favourable. Vacancy, interstitial or interstitialcy mechanisms have been mainly considered /13/.

As an oxide is a compound, the diffusion of a given atom is generally considered as occuring on its own sub-lattice and concepts developed for pure metals have been used. Nevertheless, if the possibility of a given atom to be exchanged with a vacancy of the other sublattice has not been seriously proposed (as anti-structure defects are highly improbable), the possibility of assisting the jumps of interstitial atoms in one sub-lattice by putting a vacancy of the other sub-lattice near the jumping atom has been proposed. ("Counter-vacancy mechanism").

More generally, the second sub-lattice may not have the same symmetry as the first sub-lattice in many cases. In Cu_2O, for example, all the cationic jumps to the first neighbouring sites are not equivalent as it could be expected from the fact that Cu atoms are on a FCC sub-lattice, because the oxygens are on a BCC sub-lattice and do not influence all the possible trajectories on the same way /15/. Such effects, depending on the structure, have not been considered in detail at the moment. They should lead to the introduction of different jump frequencies for the atoms, leading to curvature in the Arrhenius plots and modifying the correlation factors (see later).

3.3. Atomic jump frequencies

For a vacancy mechanism, the jump frequency of a marked atom between two sites, Γ_S, is proportional to the probability that a vacancy jumps, i.e. the jump frequency of vacancies, ω_V, and to the probability of finding a vacancy on a neighbouring site, i.e. the atomic fraction of vacancies in the site, X_V.

Using expressions which have been established previously we can write in an oxide :

$$\Gamma_S = \omega_v \cdot X_v = B \, \tilde{\gamma} \, (P_{O_2})^1 \, \exp \, (- \frac{\widetilde{\Delta} H_v^f + \Delta H_v^m}{kT}) \qquad (10)$$

where B contains entropic terms and a numerical constant. Such an expression can be generalised to other mechanisms.

3.4. Self diffusion coefficients

In a lattice where there is only one atomic jump frequency Γ_S. The diffusion coefficients can be written :

$$D = \beta \, \Gamma_S \, a^2 \qquad (11)$$

where β is a geometrical constant, a is related to the jump distance between two sites, it is a characteristic distance of the structure (the lattice parameter in a cubic lattice). Using (11) and (10) one can write in the case of a vacancy mechanism :

$$D = \beta \, a^2 \, \tilde{\gamma} \, B \, (P_{O_2})^1 \, \exp \, - \frac{\widetilde{\Delta} H_v^f + \Delta H_v^m}{kT} \qquad (12)$$

which is of the general form :

382

$$D = D'_o \; (P_{O_2})^{\frac{1}{1}} \; \exp - \frac{\Delta H^D}{kT} \tag{13}$$

Such an expression holds for other mechanisms when only one of them appears to be dominant.

The self diffusion coefficients in a MO oxide give access to the type and the charge of the defects responsible for the migration of the considered atoms (M or O) (given by the $\frac{1}{1}$ exponent) and to the formation and migration energies of these defects.

3.5. Self-diffusion studies, some examples :

There is generally good information about the majority point defects in simple non stoichiometric oxides. However oxides such as MgO and Al_2O_3 and more generally ionic oxides, are not fully understood because of the high sensitivity of the defect populations to the impurity content in such materials /16/.

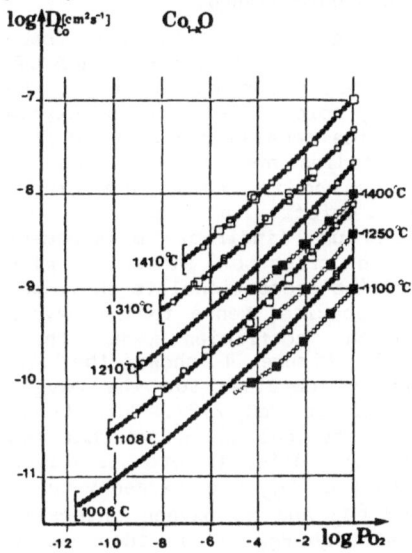

Fig. 2 : Co self-diffusion coefficient in CoO. Dependence on P_{O_2} at several temperatures /19/.

The series of transition metal oxides : NiO, CoO, MoO, FeO is a reference series for non stoichiometric oxides. They have a NaCl structure and the departures from stoichiometry are increasing from NiO to FeO. The defect structure of these oxides is relatively well-understood, $Fe_{1-x}O$ except where aggregates have been identified, the dominant defects are vacancies $V^{\alpha'}_M$ and electronic holes $h^{\cdot}$. The effective charge depends on P_{O_2} and T. Figure 2 shows measurements of the cationic self-diffusion coefficients in CoO. It is striking to see the continuous curvature of the curves $\log D_{Co} / \log P_{O_2}$. The positive curvature, associated to a continuously varying P_{O_2} exponent, means that the effective charge α of the vacancies decreases with $\alpha \leqslant 1$, which means that the vacancies involved in diffusion are V^X_M and V'_M in variable proportion or that we are faced on a problem of Debye-Huckel interactions between V''_M and $h^{\cdot}$.

The activation energy deduced from these results is around 1.8 eV which means that $\Delta H^D = \Delta H^f + \Delta H^M = 1.8$ eV.

The temperature dependence of the departure from stoichiometry x of $Co_{1-x}O$ is known. The ratio D^*_{Co}/x gives the diffusion coefficient of the vacancies /17/ :

$$D_{V_M} = \frac{D^*_{Co}}{x} = 0.113 \; \exp - \frac{1.41(eV)}{kT}$$

which means that the migration enthalpy of the $V^{\alpha'}_M$ vacancies is $\Delta H^m_{V_M} \simeq 1.41$ eV, their apparent formation enthalpy is $\Delta H^f_{V_M} = \Delta H^D - \Delta H^m_{V_M} \simeq 0.4$ eV. (Other self-diffusion measurements lead to 0.25 eV /18/). These values correspond to $\alpha \simeq 1$. That means that the formation enthalpy of V'_{Co}

vacancies created by an oxidation process is around 0.5 to 0.8 eV, ($\widetilde{\Delta} \tilde{H}{}^f_{V_M} \simeq$ $2 \Delta H^f_{V_M}$ in the case where $[V'_M] \simeq [\overset{.}{h}]$).

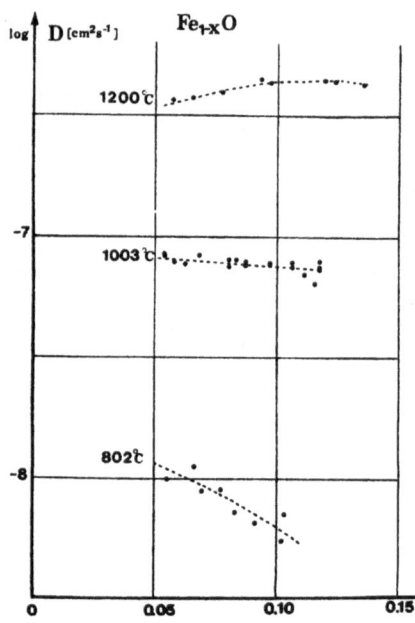

Fig. 3 : Fe self-diffusion coefficient in $Fe_{1-x}O$ as a function of x /40/, /18/.

Figure 3 shows the self-diffusion coefficient of iron atoms as a function of the departure from stoichiometry x in $Fe_{1-x}O$, x is the concentration balance of dominant point defects in the lattice. We would expect an increase of D_{Fe} with x. As seen, it is not the case and the explanation proposed is the importance of interactions between point defects known to be large aggregates. The smallest ones may be the (4:1) complexes made of 4 vacancies and one Fe interstitial. At the moment any detailed model has not been established.

Oxygen diffusion has been investigated too in these oxides but the state of our knowledge of the minority point defects of the oxygen sublattice remains poor. The diffusion studies are the only way to identify and to characterize these point defects at thermodynamic equilibrium (indirect measurement such as creep or sintering give quantities proportional to the lower moving species, hence to the oxygen diffusion coefficient in these oxides). Figure 4 shows the P_{O_2} dependence obtained for the oxygen diffusion in CoO /19/. This behaviour is typical and shows that two types of defects are involved depending on P_{O_2} : An oxygen vacancy at low P_{O_2} and an oxygen interstitial at high P_{O_2}. It is interesting to compare the diffusion coefficients of the oxygen and cobalt measured in the same conditions. At T = 1200°C, under P_{O_2} = 10^{-4} atm, $D_{Co}/D_O \simeq 10^5$ to 10^6. This justifies considering oxygen as immobile compared to the metal.

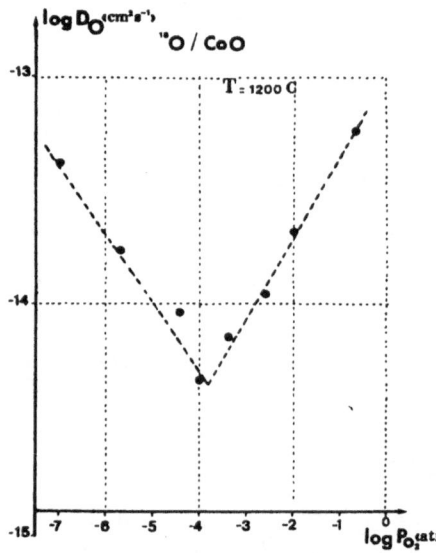

Fig. 4 : P_{O_2} dependence of oxygen diffusion in CoO. /19/

3.6. Correlation factor

Equation (12) has been established assuming a random walk of the tracer. As it is more easy for the tracer, after a first jump, to exchange again with the same neighbouring vacancy, than to wait for a fresh one, there are series of in efficient jumps which have to be taken into account in the expression of atomic jump frequency. There is only a part ,f, of useful jumps. f is the "correlation factor" which appears only when a given "marked" atom is followed. In this case we define a diffusion coefficient D* related to D by :

$$D^* = f\ D \tag{14}$$

f depends on the diffusion mechanism and on the structure of the crystal. It can be measured using two isotopes α and β of mass M_α and M_β. The diffusion coefficients D_α and D_β are related to f by :

$$f\ \Delta K = \frac{D_\alpha / D_\beta - 1}{(M_\beta / M_\alpha)^{\frac{1}{2}} - 1} \tag{15}$$

where ΔK is a kinetic factor. It is related to the part of the kinetic enrgy of the jumping atom which is transferred to the crystal during the jump. (ΔK = 1 means no transfer). This factor is not well known, particularly in oxides, but it is assumed to be close to 1.

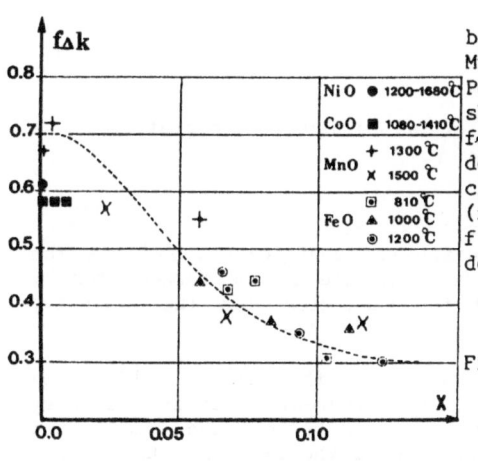

Measurements of f. ΔK have been done in the series NiO, CoO, MnO, FeO by the staff of Norman Peterson in Argonne. The figure 5 shows the results obtained /18/ : fΔK is a decreasing function of the departure from stoichiometry and is close to 0.6-0.7 at low departures. (f = 0.78 for a vacancy mechanism). f ΔK is strongly dependent of the defects interactions.

Fig. 5 : Isotopic effect for cationic self-diffusion in the transition metal oxide series as a function of departure from stoichiometry /18/

3.7. "Chemical Self-diffusion"

A particular transport phenomenon appears in non stoichiometric oxides when one of the thermodynamic parameters, P_{O_2} or T, is suddenly changed : a transitory state during which the concentration of point defects vary, is induced and can be studied by following a property sensitive to the non stoichiometry.

As defects are created or eliminated at the surfaces of the sample to reach the new equilibrium, it is mainly their diffusion which controls the concentration changes. Solving the Fick's equation for the boundary conditions of this problem, a chemical diffusion $\tilde{D}$ can be determined (we prefer "chemical self-diffusion" to avoid the ambiguity with an inter-diffusion process). As only the defect movements are involved, $\tilde{D}$ characterizes their mobility. Several values of the migration enthalpies ΔH_v^m

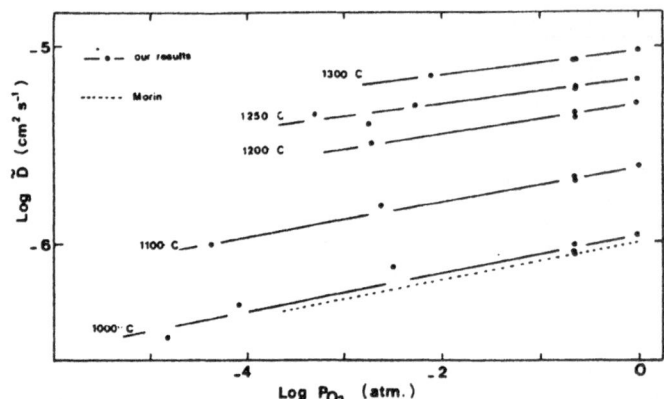

Fig. 6 : P$_{O_2}$ dependence of the "chemical self diffusion coefficient" in CoO /20/, /21/.

have been obtained in this way. Measurements of $\tilde{D}$ following the electrical conductivity in CoO have given $\Delta H_v^M \simeq 1.2$ eV and have shown that D$_v$ depends on P$_{O_2}$ (Figure 6) /20/. Nevertheless, the migration enthalpy itself seems to be insensitive to the P$_{O_2}$ parameter which seems to show that ΔH^m depends only slightly on the apparent charge state /21/.

4. IMPURITY DIFFUSION

4.1. Impurity diffusion coefficient

The diffusion coefficient D_F^* of an impurity F migrating by a vacancy mechanism can be written :

$$D_F^* = f_{V,F} \quad D_{V,F} \quad [V]_F \tag{16}$$

D_F^* has a similar form to the self-diffusion coefficient. $[V]_F$ is the concentration of vacancies in a neighbouring site to the marked atom F, $D_{V,F}$ is the diffusion coefficient of the vacancies near the impurity, $f_{V,F}$ is the correlation factor.

Similar expressions can be written as in the case of self-diffusion but the characteristic parameters are different as it is necessary to take into account the possible interactions between the vacancy and the impurity in the stable position and during the exchange jumps. That changes the migration and the formation enthalpy, and also the correlation factor which now is temperature dependent. To a first approximation, we can write :

$$f_{V,F} = f_{V,F}^o \ \exp - \frac{\varphi}{kT} \quad \text{with} \quad \varphi = \frac{\partial \ln f_{V,F}}{\partial (-1/kT)} \tag{17}$$

but $f_{V,F}^o$ generally depends on temperature, φ can be negative. As $D_{V,F}$ and $[V]_F$ can be written :

$$[V]_F = A_F (P_{O_2})^{1_F} \ \exp \ (- \frac{\Delta \tilde{H}_{V,F}^f}{kT}) \tag{18}$$

$$D_{V,F} = \beta \omega_{V,F} \ a^2 = \beta \ a^2 \bar{\mathfrak{I}} \ \exp \ (\ - \ \frac{\Delta G_{V,F}^m}{kT}) \tag{19}$$

where the parameters have been defined in the case of self-diffusion, the subscript indicates that they are different when the vacancy is close to the F impurity than when it is "free", exchanging with solvent atoms. Applying the relation (16), the general expression is obtained :

$$D_F^* = D'_{o,F} \, (P_{O_2})^{\frac{1}{F}} \quad \exp - \frac{\Delta H_F^D}{kT} \tag{20}$$

with
$$D'_{o,F} = N \cdot a^2 \cdot f_{V,F}^o \, \tilde{\nu} \quad \exp \left(\frac{\Delta \tilde{S}_{V,F}^f + \Delta S_{V,F}^m}{k} \right) \tag{21}$$

$$\Delta H_F^D = \Delta \tilde{H}_{V,F}^f + \Delta H_{V,F}^m + \varphi \tag{22}$$

where N is a numerical constant, $\Delta \tilde{S}_{V,F}^f$ and $\Delta S_{V,F}^m$ are the formation and the migration entropy of the vacancies close to the impurity F.

In oxides the correlation factor for impurity diffusion is generally treated, as in metals, using the five frequencies model (Lidiard and Le Claire /22/).

$$f_{V,F} = \frac{\omega_1 + \frac{7}{2} S \omega_3}{\omega_1 + \omega_2 + \frac{7}{2} S \omega_3} \tag{23}$$

where $\omega_1, \omega_2, \omega_3$ are vacancy exchange frequencies respectively, with a solvent atom remaining close to the impurity, with the impurity itself, with a solvent atom whose jump dissociates the complexes (V_M, F_M). S is a known function of the ratio ω_4 / ω_o of the exchange frequencies of solvent atoms, the jumps letting the vacancy free (ω_o) or creating a complex vacancy/impurity (ω_4).

4.2. Comparison between self- and impurity diffusion

The diffusion measurements give diffusion coefficients as a function of the thermodynamic parameters defining the equilibrium states P_{O_2} and T in pure binary oxides. From the function $D (P_{O_2}, T)$ two quantities can be deduced :

$$P = \frac{\partial \, \text{Log} \, D}{\partial \, \text{Log} \, P_{O_2}} \quad \text{and} \quad Q = \frac{\partial \, \text{Log} \, D}{\partial \, (-1/kT)} \tag{24}$$

and one writes the phenomenological relation :

$$D = B \, (P_{O_2})^P \exp \left(- \frac{Q}{kT} \right) \tag{25}$$

In the case of self-diffusion

$$\begin{cases} P = 1 \\ Q = \Delta \tilde{H}_v^f + \Delta H_v^m \end{cases} \tag{26}$$

in the case of impurity diffusion

$$\begin{cases} P_F = 1_F \\ Q_F = \Delta \tilde{H}_{V,F}^f + \Delta H_{V,F}^m + \varphi \end{cases} \tag{27}$$

The P parameter gives an information on the charge state of the vacancies involved in the diffusion mechanism : it can be different in self-diffusion and in impurity diffusion since the interaction between an F impurity and neighbouring vacancies can be different for different states of charge of the vacancies. It is usual to compare the Q quantities by looking at their difference :

$$\delta Q = Q_F - Q = (\widetilde{\Delta H}_{V,F}^f - \widetilde{\Delta H}_V^f) + (\Delta H_{V,F}^m - \Delta H_V^m) + \Psi \qquad (28)$$

$$= \delta H^f + \delta H^m + \Psi$$

The term δH^f is in fact the binding energy of the vacancies with the diffusing impurity, δH^m corresponds to the change in the barrier height for the exchange jumps (a binding energy between vacancies and impurities but in the saddle point configuration), Ψ is the apparent activation energy of the correlation factor ($\Psi = \partial \text{Log } f_{V,F}/ \partial (-1/kT)$).

4.3. Impurity diffusion in NiO and CoO

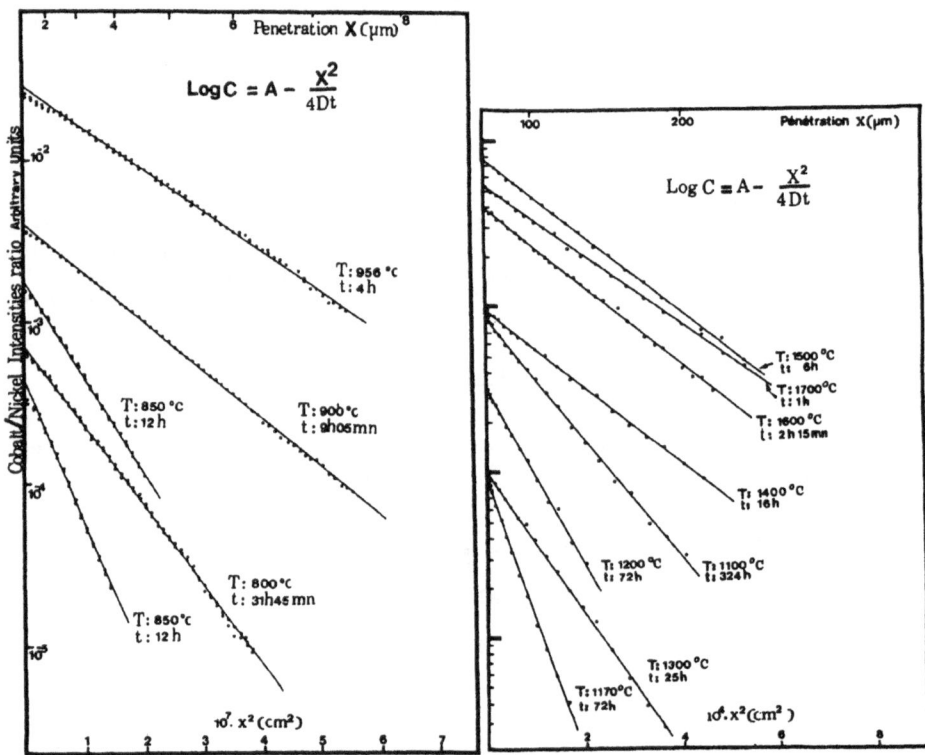

Fig. 7 : Concentration /penetration curves ($\text{log}c/x^2$) for the diffusion of Cobalt in NiO /23/. (SIMS and Electron Microprobe Analysis)

The impurity diffusion of several impurities has been measured in these two reference oxides. Figure 7 shows (concentration c/penetration x) curves in $\text{log}c/x^2$ coordinates which have been obtained in the case of Co diffusion in NiO using the Electron Microprobe Analysis (EMA) or the SIMS technique, particularly when the penetrations were small. The diffusion coefficients deduced from the slope of these curves obey to an Arrhenius law /23/ (figure 8) and are in good agreement with measurements made by radiotracers /23/, this is a good test of the techniques used.

The diffusion of several impurities has been compared to try to establish a general behaviour as a function of the characteristics of the impurities. Figure 9 shows the activation energies measured, Q_i, as a function of the ionic radius r_i, of the diffusing elements. It is surprising to see that Q_i decreases when r_i increases /24, 25, 28/ for a

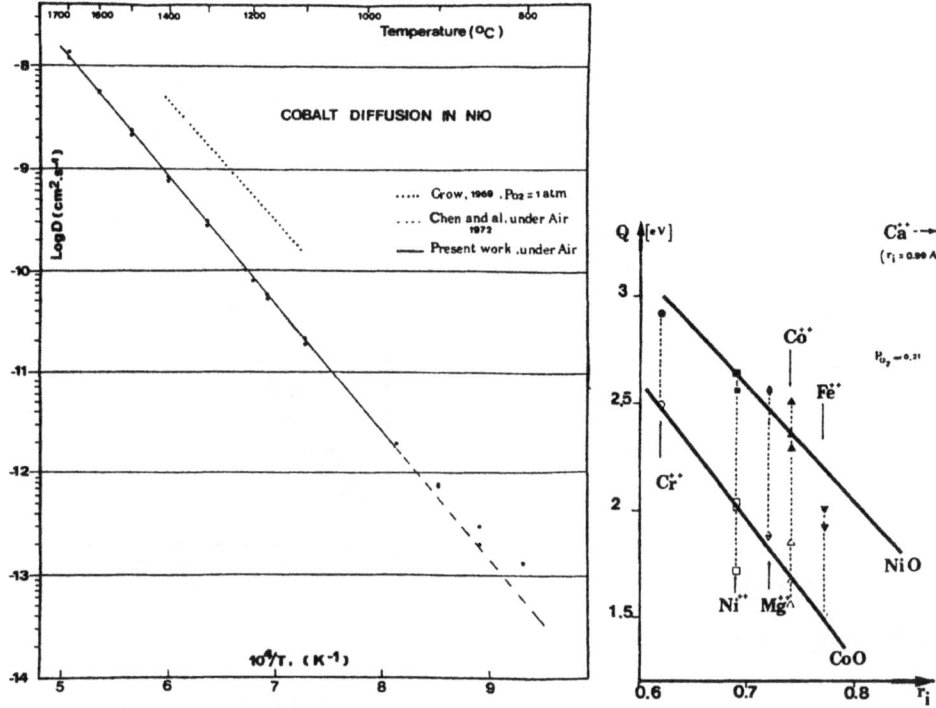

Fig. 8 : Arrhenius diagram for Co
diffusion in NiO /23/.

Fig. 9 : Activation for
diffusion : dependence
on the ionic radius of
diffusing species /28/.

series of ions in which r_i is close from the solvent atom radius (this
phenomenological correlation is not obeyed for Ca^{++} whose radius is
close to 0.99 A and the activation energy close to 3.4 eV /27/). To
understand such a behaviour it is necessary to analyse the activation
energies in more detail. This has been done in the case of cobalt in NiO
whose correlation factor has been measured as a function of the temperature /29/ ($f \simeq 0.2$ eV) and also the binding energy between cobalt ions
and nickel vacancies ($\delta H^f \simeq 0$) /30/. The δQ difference (equation 28) is
experimentaly equal to -0.1 eV. This means that $\delta H^m \simeq -0.3$ eV /23/.
This typical and rare example shows that there is a strong kinetic
effect in replacing Ni by Co in NiO while interactions between Co and Ni
vacancies are negligible in the stable positions.

Oxygen partial pressure dependencies of impurity diffusion for Mg
and Ni in CoO have been studied and compared to self-diffusion results.
Figure 10 shows the comparative behaviour observed /28, 31/. Impurity
diffusion in this case is characterized by a P_{O_2} exponent l_F which is
different from the l exponent characterizing the self-diffusion. This
seems to show that the charge of vacancies involved in the diffusion
mechanism is higher for impurity diffusion than for self-diffusion ($l_F <$
1). In other words, there is a stronger attractive interaction between
Mg or Ni and doubly charged vacancies than with singly charged vacancies.

In conclusion, impurity diffusion is an interesting tool to obtain
information about impurity/point defect interactions and we need more
experiments of this type, particularly in oxides where apart from the
temperature dependence, P_{O_2} dependence and also the effect of doping by
aliovalent impurities can be studied.

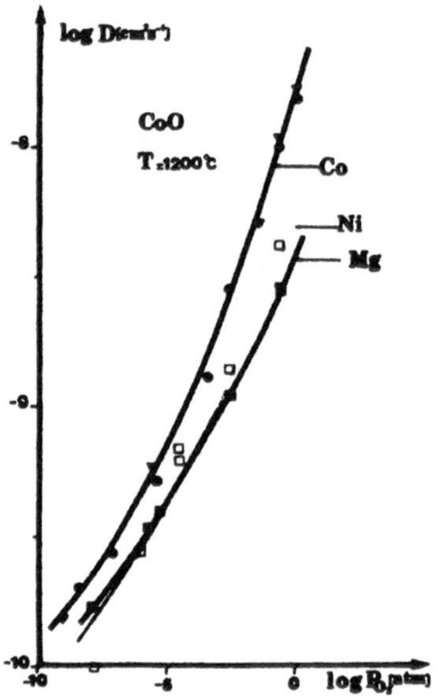

Fig. 10 : Oxygen partial pressure dependence of the diffusion coefficient for Co, Mg and Ni in $Co_{1-x}O$ /31/.

[figure: graph with axes log D(cm²s⁻¹) vertical, log P_{O_2}(atm) horizontal, labeled CoO, T 1200°C, with curves Co, Ni, Mg]

5. DIFFUSION IN EXTENDED DEFECTS

5.1. Evidences of "short circuits"

At relatively low temperatures (T < 0.5 x melting temperature), dislocations, isolated or hogomeneously distributed or forming walls, are preferential paths for diffusion. Grain boundaries are also short circuits which are easy to detect. Figure 11 shows a series of images obtained on mass 18 on a polycrystalline NiO annealed under ^{18}O gas pressure with a SIMS apparatus. It is clear on this figure, which is equivalent to a series of autoradiographies, that near the surface the ^{18}O has diffused homogeneously in the bulk, but that after a given penetration, it entered into the sample following mainly the grain boundaries /32/. When profiles are established, a long tail appears after a "normal" part which corresponds to the bulk diffusion. Figure 12 shows an example of such profile obtained using nuclear reactions (^{18}O in polycrystalline NiO) /32/. Dislocations give profiles which roughly look similar. Figure 13 shows an example obtained after diffusion of radioactive Ni in NiO at T = 700°C after 1 hour of annealing. The dislocation influence is quite clear (the sample was in this case a single crystal and the nickel has been deposited as a thin film) /33/.

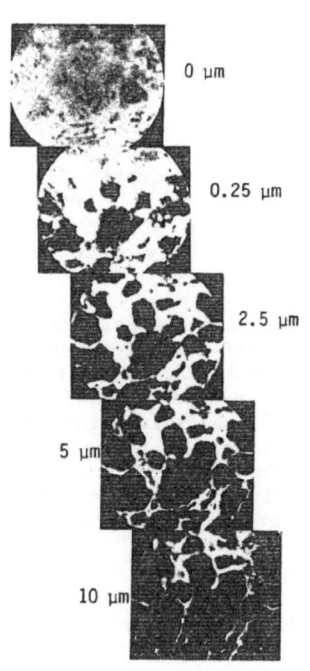

0 μm

0.25 μm

2.5 μm

5 μm

10 μm

Fig. 11 : SIMS images obtained on polycrisline NiO after ^{18}O self diffusion (T = 1300C, 71 hours) /32/.

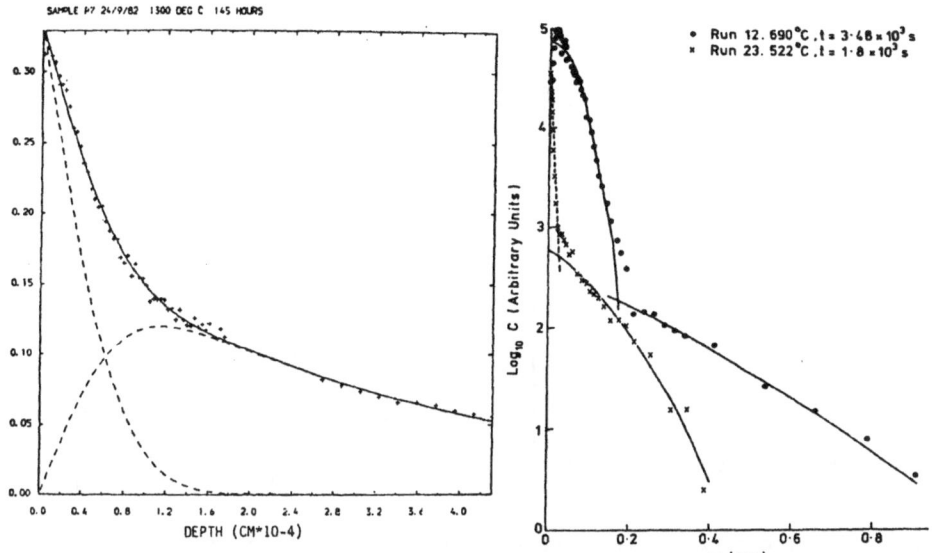

SAMPLE P7 24/9/82 1300 DEG C 145 HOURS

• Run 12. 690°C , t = 3·48 × 10³ s
× Run 23. 522°C , t = 1·8 × 10³ s

Fig. 12 : ^{18}O diffusion profile in
polycristalline NiO /32/

Fig. 13 : Ni diffusion profile
in NiO single crystals
/33/

5.2. Analysis of the phenomenon and application

Recent progress have been done in analysing the diffusion in
dislocations /34/ considering two different boundary conditions : (I) a
constant surface concentration of the tracer, (II) a tracer deposited as
a thin film. The mathematical analysis corresponding to similar boundary
conditions for grain boundary diffusion has been previously performed by
Whipple (condition I) /35/ and by Suzuoka (conditions II) /36/. In both
cases the average concentration $<c>$ measured in a tracer experiment,
sectioning the sample perpendicularly to the direction of diffusion,
obey to the general form :

$$<c> (y,t) = c_B (y,t) + q (y,t)$$

there c_B (y,t) is the solution in a bulk without defects which depends
on the penetration y and the time t and is characterized by the bulk
diffusivity D. q is a contribution from the defect which depends on y,
t, 2R the distance between the defects and on the core dimensions
(dislocations of radius a or grain boundary thickness δ). The diffusion in the extended defects is characterized by D^d or D^j. Numerical
computations of the profiles have been performed in the case of Ni
diffusion in NiO grain boundaries and dislocations /37/ (Figure 14).
Three different regimes can be distinguished as previously proposed by
Harrison /38/ :

A : when there is a recovery of the diffusion zones between
defects ($\sqrt{Dt} > R$)

B : corresponding to independent defects ($\sqrt{Dt} \ll R$)

C : when the diffusion is essentially in the defect cores
($R \gg \sqrt{Dt} < a$ or δ) it corresponds to short annealing times.

391

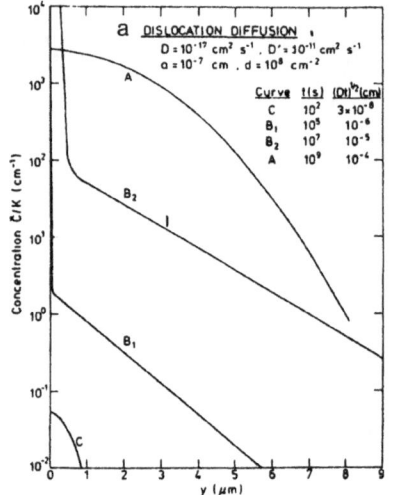
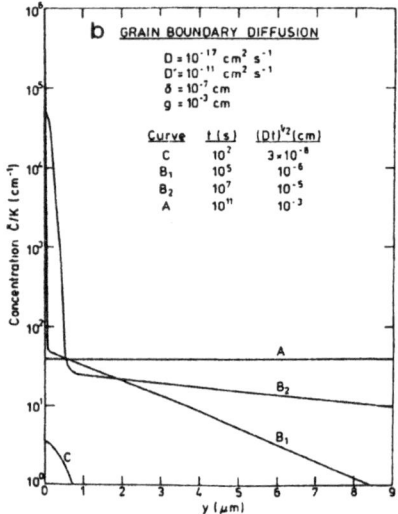

Fig. 14 : Theoretical diffusion profiles for diffusion in dislocations (a), or in grain boundaries (b) /37/.

From the tails in the B type regime it is possible to deduce the product (D^d-D) a^2/D or $D^j \int \sqrt{Dt}/2D$. A good knowledge of D, which can be obtained in the same experiment or independently, gives access to $D^d a^2$ or $D^j \int$. From the C type regime, D^d or D^j can be obtained alone. Using both results obtained in B and C regimes, the characteristic parameters a or δ can be deduced from diffusion experiments. a or δ are "adhoc" parameters necessary for a phenomenological description and difficult to relate to the actual structure, nevertheless they characterize the dimensions of the perturbed zone introduced by the extended defect on the diffusion processes. In the case of impurity diffusion or in less pure materials, segregation effects lead sometimes to large values of the defect core and diffusion studies could be used to investigate those phenomena /39/.

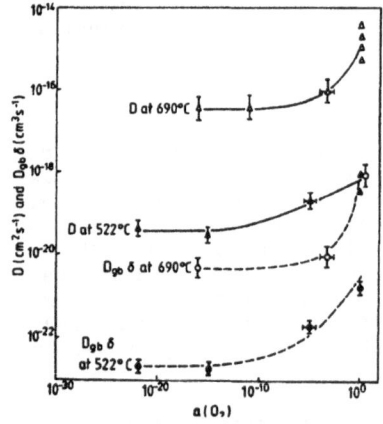

Fig. 15 : Dependence on P_{O_2} of the Ni self diffusion in bulk and in dislocation arrays in NiO /33/.

The parameter P_{O_2} can also be used in oxides to increase our knowledge of the diffusion mechanisms in the extended defects. For example, oxygen partial pressure dependence measurements have been done in the case of Ni-diffusion in bulk and in dislocation arrays (fig. 15). They tend to prove that the same defect mechanisms are involved in the two phenomenon and that measurements have been done in pure NiO showing an intrinsic behaviour /33/.

Systematic studies of the diffusion processes in bulk and in extended defects, for both metal and oxygen atoms, are required to understand important phenomena such as sintering, creep, ... NiO is probably at the moment the best studied oxide (see figure 16). Such investigations should be extended to other oxides in such a way to obtain more general ideas on the mass transport phenomena in these compounds.

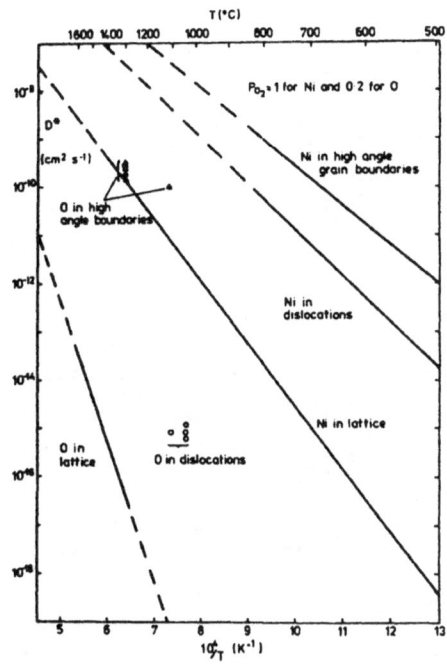

Fig. 16 : Set of values obtained for Ni and O self diffusion in bulk, in dislocations and in grain boundaries of NiO /32/.

REFERENCES

/1/   P. Kofstadt, "Nonstoichiometry, Diffusion and Electrical Conducti-
      vity in binary oxides", Wiley-Interscience (1972)
/2/   F.A. Kroger, "The chemistry of imperfect crystals"
      North-Holland (1973)
/3/   R. Dieckmann, Solid State Ionics 12: 1-22 (1984)
/4/   C. Monty, Ch. XII in "Defauts Ponctuels dans les Solides"
      Ed. de Physique, Orsay (1977)
/5/   R. Farhi, G. Petot-Ervas, J. Phys. Chem. Sol. 39 :1169 (1978)
/6/   A.B. Lidiard, chap. 3 in "Mass Transport in Solids" Ed. F. Benière
      and C.R.A. Catlow. NATO ASI Series B : Physics, Vol. 97
      Plenum Press New York and London (1981)
/7/   P. Varotsos, K. Alexopoulos, Phys. Rev. B, 15: 5994 (1977)
/8/   G. Slodzian, An. Phys. Paris, 9: 591-648 (1964)
/9/   P. Contamin, G. Slodzian, C.R. Acad. Sciences Paris, Série C,
      267: 805-808 (1968)
/10/ D. Gupta., Thin Solid Films 25: 231 (1975)
/11/ A. Atkinson, R.I. Taylor, Phil. Mag A, 39: 581-595 (1979)
/12/ W.K. Chu, J.W. Mayer, M.A. Nicolet, R.M. Buck, G. Amsel, P. Eisen
      Thin Solid Films 17: 1 (1973)
/13/ J. Philibert, "Diffusion et Transport de Matière dans les Solides"
      Ed. de Physique, Orsay (1985)
/14/ W.J. Moore, Y. Ebisuzaki, J.A.Sluss, J. Phys. Chem. 62:
      1438-1441 (1958)
/15/ D. Mekki, Thèse 3è cycle, Paris VI (1984)

/16/ B.J. Wuensch, Chap. 14, in "Mass Transport in Solids"
Ed. F. Benière, C.R.A. Catlow,NATO ASI Séries B. : Physics, Vol.
97 Plenum Press New York and London (1981)
/17/ R. Dieckmann, Zeit. Physik. Chem. Neue Folge 107:189 (1977)
/18/ C. Monty, "Radiation Effects" 74:29 (1983)
/19/ C. Clauss., R.J. Tarento, C. Monty, A. Dominguez-Rodriguez,
J. Castaing, J. Philibert, in "Transport in nonstoichiometric
coumpounds" 3è conf. Penn State Univ. (1984), Ed. G. Simkovich
V. Stubican, Plenum Press (1985)
/20/ F. Morin., R. Dieckmann, Z. Phys. Chem. 2 (1982)
/21/ G. Petot-Ervas, O. Radji, P. Ochin, B. Sossa
Radiation Effects 69:301 (1983)
/22/ A.D. Leclaire, A.B. Lidiard, Phil. Mag. 1:518 (1976)
/23/ N. Tabet, C. Monty, in "Reactivity of Solids - Cracovie 1980"
Ed. Dyrek, Novotny, Haber, Elsevier (1982)
/24/ W. Crow, Ph. D. Thesis, Cornell University (1969)
/25/ N. Peterson, Solid State Ionics 12:201 (1984)
/26/ C. Dubois, C. Monty, J. Philibert
- Phil Mag.A46:419(1982)
- Solid state Ionics 12:75 (1985)
/27/ N. Tabet, C. Dolin, C. Monty, Rev. Int. Hautes Temp. Refract.
19:413-416 (1982)
/28/ C. Monty, in "Physical Chemistry of the Solid State : application
to Metals and their compounds" Ed. P. Lacombe, Elsevier (1984)
/29/ W.K. Chen, N.L. Peterson, J. Phys. Chem. Solids 33:881 (1972)
/30/ W.K. Chen, N.L. Peterson, J. Phys. Chem. Solids 34:1093 (1973)
/31/ R. Gomri, H. Boussetta, C. Bahezre, C. Monty
Solid State Ionics 12:227-233 (1984)
/32/ A. Atkinson, F.E.W. Pummery, C. Monty, in "Transport in non-
stoichiometric compounds" Penn State Univ. 1984, Ed. G. Simkovich,
V. Stubican, Plenum Press (1985)
/33/ A. Atkinson, R.I. Taylor, Phil. Mag. A395:581-595 (1979)
/34/ A.D. Leclaire, A. Rabinovitch, in "Diffusion in crystalline Solids"
Ed. G.E. Murch, A.S. Nowick,Academic Press, Orlando , 257-318 (1984)
/35/ R.T. Whipple., Phil Mag 45:1225 (1954)
/36/ R. Suzuoka, Trans. Jap. Inst. Mat. 2:25 (1961)
/37/ A. Atkinson, Solid State Ionics 12:309-320 (1984)
/38/ L.G. Harrison, Trans. Faraday Soc 57:1191 (1961)
/39/ A. Atkinson, R.I. Taylor, AERE R 11763 (1985)
/40/ W.K. Chen, N.L. Peterson, J. Phys. Chem. Solids 41:647 (1980)

MEASUREMENT AND ANALYSIS OF IONIC CONDUCTANCE IN SOLIDS

A.V. Chadwick and J. Corish*

University Chemical Laboratory, University of Kent, Canterbury
Kent CT2 7NH, England
*Chemistry Department, Trinity College Dublin 2, Ireland

INTRODUCTION

If full use is to be made of the properties which materials possess because of their defect structures and, more importantly, if these properties are to be optomized in the design of materials for specific applications then the nature of the defects and of their distributions, interactions and movements must be fully understood.   The occurrence of ionic conductance and of self- and impurity diffusion in classical ionic conductors, such as the alkali and silver halides, first suggested the existence of point defects in these crystals (Schottky; 1935, Frenkel; 1926).   Careful measurements of ionic conductance and of diffusion coefficients, when accompanied by thorough analyses of these data, have remained the principal source of accurate information on the thermodynamic parameters which govern point defect behaviour and thus provide the basis of our understanding of ionic migration in these crystals.   In this chapter the modern experimental techniques for the measurement and analysis of these kind of ionic conductance data, which, ideally should be measured on single crystal specimens, will be described.   In some very well understood cases these analyses have recently been improved by incorporating temperature-dependent defect parameters derived from theoretical calculations into the data fitting procedures.

Many of the technologically important materials, such as the oxide ceramics, cannot be prepared as single crystals: in addition there are now widespread applications for composite solids.   Because of their nature, ionic motion is more complex in such materials than it is in classical ionic conductors and its interpretation in terms of the movement of isolated point defect species through a crystalline lattice is inadequate.   The presence of different regions in the conductance specimens or of large numbers of grain boundaries, often with differing intergranular phases, will influence the conducting properties.   These are most readily investigated by constructing complex impedance diagrams.   The basis of this technique, which will be described in the latter part of this chapter, is the determination of the impedance of the specimen as a function of the frequency of the alternating current (ac) used.   This frequency dependence is characteristic of the structure of the material under observation: the technique can also be used to study interfacial effects and in electrochemical systems it may assist in the detection of adsorption, diffusion and reaction processes (Heyne; 1983, Rickert; 1982).   The complex impedance plot is interpreted in terms of equivalent circuits composed of resistive and capacitive elements and based on a physical model of the system on which the measurements have been made.

IONIC CONDUCTANCE IN CLASSICAL IONIC CONDUCTORS

The determination of a complete set of defect parameters for a classical ionic conductor from ionic conductance data requires that extensive measurements be made on a number of pure and aliovalently doped specimens. It is also necessary to develop a model to which these experimental data can be fitted. This procedure will be illustrated here with $AgCl:Cd^{2+}$ as an example and using the formalism which was introduced in Chapter 1 for the formation and interaction of point defects. First this theoretical treatment will be extended to include defect migration and to yield an expression for the ionic conductance. The experimental technique will also be described as will the analysis of data and the relationship between conductance and self-diffusion data. Finally AgCl displays an unexpectedly large increase in conductivity at temperatures approaching its melting point and which is now attributed to the temperatures dependence of the Frenkel energy. The use of theoretical values for this parameter in the data analysis will be shown to substantially improve our understanding of the defect nature of this material.

Theoretical

The specific conductance, σ, of a substance is given by the expression

$$\sigma = \sum_{r=1}^{i} n_r |q_r| \mu_r \qquad (1)$$

where n_r is the number per unit volume of the conducting species r, q_r and μ_r its charge and mobility, respectively, and the summation is taken over all the species which contribute. In the high temperature (intrinsic) region of a pure AgCl crystal the number per unit volume of cation vacancies that results from the Frenkel equilibrium, discussed as equation (7) of Chapter 1 is given by

$$n_{V'_{Ag}} = x_{V'_{Ag}} N = N\sqrt{2} \exp\left[{-g_F}/{2kT} \right] \qquad (2)$$

where N is the number of ion pairs per unit volume in AgCl. An expression for the jump frequency, w, of the vacancy may be found using formalism developed by Wert (1950) and Vineyard (1957) and the same result

$$w_{V'_{Ag}} = \nu \exp\left[\frac{\Delta g_{V'_{Ag}}}{kT} \right] \qquad (3)$$

where ν is an attempt frequency and $\Delta g_{V'_{Ag}}$ the change in the free energy on passing from the initial configuration to that at the saddle point, is given by transition-state theory (Zener; 1952) and by the dynamical theory of Rice (1958). The imposition of an electric field, E, causes the jumps to be biased and the frequency at which they occur in the direction of the field is

$$\vec{w}_{V'_{Ag}} = w_{V'_{Ag}} (|q_{V'_{Ag}}| Ea/kT) S_{V'_{Ag}} \qquad (4)$$

where a is the distance transversed in the field direction and $S_{V'_{Ag}}$ is the symmetry number: that is, the number of equivalent sites into which the vacancy can jump (Corish and Jacobs; 1973). This result, which requires that $|q_{V'_{Ag}}|$, $|E_a| \ll kT$, gives the mobility of the vacancy as

$$\mu_{V'_{Ag}} = S_{V'_{Ag}} |q_{V'_{Ag}}| a^2 w_{V'_{Ag}} / kT \tag{5}$$

and the full expression for the contribution to the conductivity from the cation vacancy migration is therefore

$$\sigma_{V'_{Ag}} = \nu \sqrt{2} \, N S_{V'_{Ag}} \, q^2_{V'_{Ag}} \, a^2 \, \exp\left[\frac{-g_F}{2kT}\right] \exp\left[\frac{-\Delta g_{V'_{Ag}}}{kT}\right] \tag{6}$$

where both the free energy changes can be expressed as their corresponding enthalpy and entropy terms. For accurate work the defect concentrations should be replaced by their activities using the Debye-Hückel-Lidiard approximation, as discussed in Chapter 1, and the mobilities should be adjusted for drag using the Onsager-Pitts correction (Pitts; 1953).

Expressions analogous to equation (6) can be derived for the contributions to the transfer of charge which arise from the migration of the interstitial silver ions in AgCl. When these contributions are summed, as in equation (1), they provide an expression for the specific conductance of the crystal in terms of the enthalpy and entropy of defect formation and of the enthalpies and entropies for migration of the mobile defects. At the lower temperatures in AgCl:Cd^{2+} the principal charge-carrying species are the charge-compensating cation vacancies, but their effectiveness will be gradually reduced as the temperature decreases since they form neutral impurity-vacancy pairs, $(Cd^{\bullet}_{Ag}V'_{Ag})$. The equilibrium for this association of defects is similar in form to that of equation (10) in Chapter 1: it occurs simultaneously with the Frenkel equilibrium and they must be treated and solved in this way for a crystal containing the appropriate level of dopant to yield the concentrations of defect species at each temperature. Thus equation (6) forms the basis for the non-linear least squares fitting of experimental conductance data to determine the thermodynamic parameters that govern formation, migration and interaction.

Experimental

Conductance measurements are made on well-annealed single crystals which have been cut and microtomed, and measured to determine their geometric factor: the ideal dimensions are approximately 10 mm square and 5 mm in thickness. Conducting coatings are then applied on the square faces which will be in contact with the measuring electrodes. A number of paints containing fine dispersions of metals or of graphite have been used for this purpose but a coating of graphite applied by rubbing the crystal into a finely ground powder of the pure material is preferable. This technique avoids the difficulties which may arise from spurious constituents in the paints and from the possible ingress of metal ions into the crystal at high temperatures.

The specimen is then placed between very thin, usually platinum, electrodes each of which incorporates a platinum/platinum rhodium thermocouple. In this way the temperatures at both faces of the specimen can be determined before and after the conductance measurement to check that the crystal is being maintained at a uniform and constant temperature. The electrodes and specimen are placed in a high-purity quartz conductance cell within a furnace: typically it is supported between long quartz rods since, at the lowest temperatures, very large resistances must be measured. Care is necessary in the design of the support system so that the specimens are not subjected to any strain when the temperature is raised. The specimens are maintained under an inert atmosphere, or in vacuo if sublimation does not occur. For this the conductance cell is coupled with a vacuum system capable of attaining pressures of the order of 10^{-6} torr and to a gas supply system to provide a dried, purified and inert atmosphere.

The bridge used for the measurements should include a phase-shifter so that the capacitance and the conductance of the specimen with its electrode assembly and leads can be measured: it should operate at variable frequencies to avoid polarization resistance. It is usually necessary to anneal the specimen at $\sim \frac{7}{8}$ths of its Kelvin melting temperature and to cycle it slowly between this anneal temperature and close to its melting point to obtain a consistent plot of log σT against the inverse temperature. When this line has been established the complete curve is measured at equal intervals of $1/T$ (K^{-1}) from close to the melting temperature to as low a temperature as is possible: typically some two hundred points would be measured in this range with sufficient time being allowed at each temperature to establish the defect equilibria. Finally the geometric factor is remeasured after the experiment. Conductance curves of this type are determined for a range of pure and doped crystals: Fig. 1 shows a set of

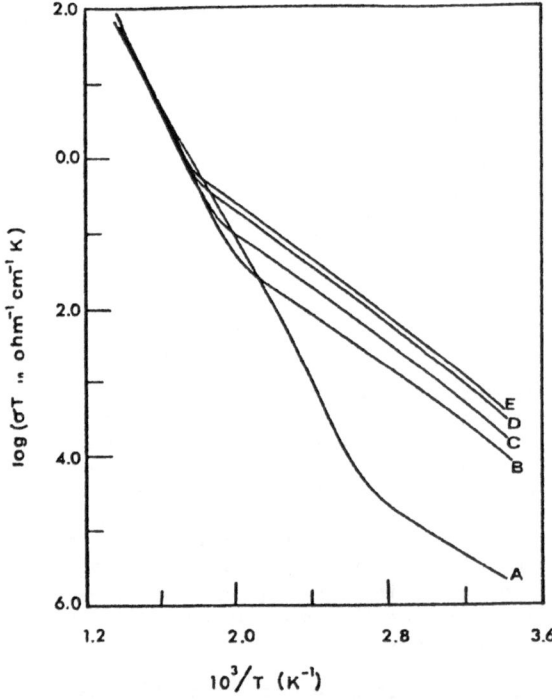

Figure 1. Experimental conductivity data from five single crystals measured by Corish and Mulcahy (1980). Curve A refers to a nominally pure sample and the analytical concentrations of Cd^{2+} in the remaining crystals were: Crystal B, 60×10^{-6}; Crystal C, 150×10^{-6}; Crystal D, 370×10^{-6}; and Crystal E, 480×10^{-6}.

data of this kind for the AgCl:Cd^{2+} system (Corish and Mulcahy; 1980). These curves have been corrected for the temperature dependence of the geometric factors and for the contribution of the lead resistance which were measured separately as a function of temperature.

Analysis of data

In the high temperature regions of the curves in Fig. 1 the defect structure of the crystals is dominated by intrinsic Frenkel defects. Transport of charge is effected by the movement of cation vacancies and by two different interstitialcy migration mechanisms by which the interstitial silver ions can move. On the basis of some early, and by modern standards rather crude, calculations (Hove; 1956) and measured correlation factors (Weber and Friauf; 1969) these two latter mechanisms have traditionally been assigned to interstitialcy collinear and interstitialcy non-collinear motions. As mentioned previously, conduction in the extrinsic region occurs via the charge-compensating vacancies: downward curvature is evident at the lowest temperatures and indicates the removal of these carriers as impurity-vacancy, $(Cd_{Ag}^{\bullet}V_{Ag}')$, dipoles. In this intrinsic region the conductivities of the samples increase in line with their doping levels: the analytical concentrations of the Cd^{2+} impurity are given in the figure caption.

These data are analyzed by fitting an expression for log σ T based on equations (1) and (6) and incorporating defect formation, interaction and migration energies, to the experimental values measured over the full temperature range. The value of

$$\phi = \Sigma\{\log(\sigma T_{Calculated}) - \log(\sigma T_{Experimental})\}^2 \qquad (7)$$

is then minimized in a non-linear fitting routine in which the adjustable parameters are the enthalpies and entropies which govern the defect equilibria and migration. For AgCl:Cd^{2+} there are eleven parameters and the values obtained from free fitting of the curves from the doped crystals in Fig. 1 are listed in Table 1. h_F and s_F represent the enthalpy and entropy of Frenkel defect formation and h_a and s_a are the analogous terms for impurity-vacancy association; Δh_+ and Δs_+ are the vacancy migration parameters, and Δh_{ic} and Δs_{ic}, and Δh_{inc} and Δs_{inc} those for the interstitialcy collinear and interstitialcy non-collinear, motions, respectively; finally, the total-concentration of the dopant, C, is also included as an adjustable parameter in the minimization.

There are a number of alternatives to the standard model and a full discussion of these is given in the original publication (Corish and Mulcahy; 1980). Earlier work (Corish and Jacobs; 1975, Aboagye and Friauf; 1975) had shown the most serious inadequacy in the standard model to be the use of a Frenkel enthalpy, h_F, which was independent of temperature. Conductance data and also the measured diffusion coefficient of sodium ions in AgCl (Batra and Slifkin; 1975) showed the concentration of intrinsic vacancies to increase substantially more rapidly than would be expected on the basis of the usual temperature dependence of g_F even when defect interactions are included. Experimentally this is evident as an upward curvature in the Arrhenius plots: in the analyses reported in Table 1 it is signified by the role assigned to the conduction by the interstitialcy non-collinear mechanism. The rather high value of Δh_{inc} means that this minority mechanism makes a contribution to the overall ionic conduction which increases rapidly with temperature as the melting temperature is approached.

The additional, sometimes termed anomalous, increase in intrinsic

Table 1   Parameters derived for the doped crystals using the standard model for $AgCl:Cd^{2+}$

Parameter	Crystal			
	B	C	D	E
h_F(eV)	1.468	1.486	1.480	1.489
s_F/k	9.56	10.11	9.75	10.03
Δh_+(eV)	0.279	0.273	0.276	0.279
$-\Delta s_+/k$	0.533	0.468	0.499	0.470
Δh_{ic}(eV)	0.035	0.043	0.048	0.044
$-\Delta s_{ic}/k$	2.71	2.64	2.37	2.45
Δh_{inc}(eV)	0.572	0.548	0.543	0.553
$\Delta s_{inc}/k$	5.01	4.09	4.19	4.10
$10^6 C$	65.4	134.2	348.1	481.4
$-h_a$(eV)	0.289	0.317	0.313	0.313
$-s_a/k$	0.835	1.576	1.552	1.715

defect concentrations were modelled by Aboagye and Friauf (1975) by introducing an empirical correction term, $-\Delta g$, to g_F. This brought calculated and experimental conductances curves into agreement and its origin was attributed to a general softening of the lattice which would facilitate defect formation. Attempts to derive temperature-dependent defect formation and migration parameters directly from conductance data (Corish and Jacobs; 1975) proved to be inconclusive. These difficulties have now been partly resolved by the use of a theoretically calculated temperature-dependent energy of defect formation in the fitting of the experimental conductance data. The atomistic simulation methods yield a value for the energy at constant volume, u_v, and this is calculated as a function of temperature in the quasi-harmonic approximation. When this energy for AgCl was combined with a temperature-independent entropy to yield $f_v(T)$, which is shown in equation (15) of Chapter 1 is equal to $g_F(T)$, then it was found to reflect the experimentally observed temperature variation in the intrinsic defect population extremely well (Catlow et al; 1979). Table 2 shows the results of a least squares fitting of the same data for $AgCl:Cd^{2+}$ as was used in Table 1 but now the theoretical value for the Frenkel energy has been used: the ten remaining parameters were allowed to vary as before. In spite of the imposition of a theoretical value for one parameter the quality of the fitting is essentially retained. The only major differences from the earlier result are the values determined for Δh_{inc} and for s_F. The much lower value for the activation energy for the second interstitialcy motion means that the mechanism now makes a more reasonable contribution to the conductance at all temperatures: it is also substantially the same value as was determined by Aboagye and Friauf (1975) from their treatment of pure crystal data. The value of s_F now refers to the constant volume condition and is not therefore directly comparable to that given in Table 1: when corrected to

Table 2 Parameters derived for the doped crystals when the theoretically calculated values for the Frenkel defect formation energy (Catlow et al; 1979) was incorporated into the standard model for $AgCl:Cd^{2+}$

Parameter	Crystal B	Crystal C	Crystal D	Crystal E
$^{*}s_F/k$	5.743	5.362	5.345	5.336
Δh_+ (eV)	0.280	0.273	0.277	0.284
$-\Delta s_+/k$	0.532	0.429	0.476	0.425
Δh_{ic} (eV)	0.0041	0.0024	0.0023	0.0024
$-\Delta s_{ic}/k$	4.18	3.75	3.96	3.64
Δh_{inc} (eV)	0.105	0.120	0.125	0.103
$\Delta s_{inc}/k$	4.53	6.32	3.86	6.14
10^6	67.5	132.3	357.4	521.9
$-h_a$ (eV)	0.287	0.314	0.307	0.307
$-s_a/k$	0.732	1.422	1.391	1.631

*Refers to constant volume conditions (see text)

yield the corresponding constant pressure term the values are in good agreement with those for other similar crystals (Corish and Jacobs; 1973).

This introduction of a temperature-dependent defect formation energy using complementary atomistic simulation techniques has increased our understanding of the $AgCl:Cd^{2+}$ system and our ability to interpret experimental data from both diffusion (see also below) and conductance data. The effect is particularly evident in AgCl and AgBr and it is clear that the theoretical methods which have recently become available (Harding; 1985) to calculate defect entropies, and hence free energies, would be expected to improve the type of analysis described above and to allow it to be extended to other materials.

For the classical ionic conductors such as the alkali and silver halides, and some of the fluorites for which the meticulous experimental measurements and detailed data analysis described here have been carried out, the thermodynamic parameters which govern their defect structures are now rather well established. Both the nature of the defects, their interactions and the mechanisms by which they move are understood. This is particularly evident, for example, when the parameters are used to estimate through the Nernst-Einstein equation the self-diffusion coefficients for the host lattice ions. For impurity ions which move by a vacancy mechanism the numbers of such vacancies, as given by the relevant defect parameters, can be used as a basis for the calculation of their diffusion coefficients. For AgCl, Corish and Jacobs (1972) have calculated the temperature dependence of the diffusion coefficient of the silver ion, cf. Figure 2, with remarkable accuracy considering that contributions from all three mechan-

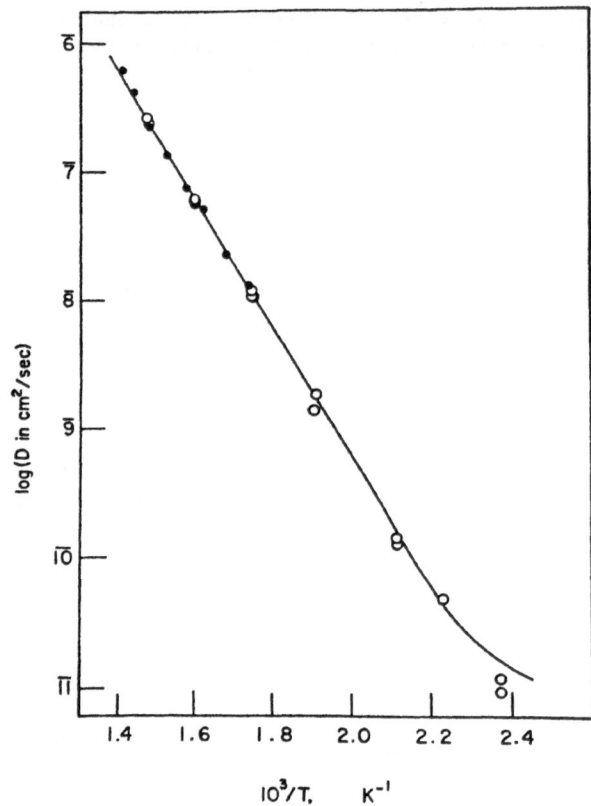

Figure 2. The continuous line represents the diffusion coefficient of the
Ag$^+$ ion in AgCl as calculated by Corish and Jacobs (1972):●
are experimental points from Weber and Friauf (1969) and ○
experimental points from Reade and Martin (1960).

isms, each with its own correlation factor and temperature-dependent transport number, have been included.   Jacobs et al (1980) have shown the AgCl
defect parameters to be consistent with anion diffusion by single vacancy
and vacancy pair mechanisms and Corish et al (1983) have also modelled the
diffusion of a number of univalent impurity ions in AgCl and AgBr.

COMPLEX IMPEDANCE MEASUREMENTS

In this technique the impedance, Z, or the admittance, Y (Y = 1/Z),
of the cell is measured over a wide range of frequencies and the results
plotted in the complex impedance, or admittance, plane.   The principle of
the technique is that if some interfacial process impedes the charge flow
an effective capacitance will be present, in parallel to the interface,
and this will cause a displacement current.   At sufficiently high
frequencies the interface will be by-passed.   In a system containing a
number of types of interface the complex impedance plane diagram will
contain contributions from each type and in favourable cases these can be
resolved.   An early application of "complex impedance spectroscopy" was

in the study of polarization processes in aqueous cells (Sluyters; 1960) and it was first applied to solids by Bauerle (1969). The use of the technique has grown with the increasing interest in ceramics and fast-ion conductors, materials which are often only readily available as sintered or compacted pellets. Several reviews have been published on complex impedance spectroscopy (Heyne; 1983, Rickert; 1982; Archer and Armstrong; 1980). Here we will outline the basic theory and the practical aspects of this technique.

Theory

Application of small, sinusoidally oscillating potential, $\Delta E \sin \omega t$ (where $\Delta E < 0.005V$), across a cell will give rise to a current, $\Delta i \sin (\omega t + \theta)$, which is phase-shifted by an angle θ. The impedance, Z, of the cell will have a magnitude given by:

$$|z| = \Delta E/\Delta i \qquad (8)$$

and a direction given by the phase angle θ. It is convenient to represent Z at a given ω as a point on the complex impedance plane, where the y-axis is the imaginary component, Z", and the x-axis is the real component, Z'. Thus $|z|$ is the length of the line from the origin to the point and θ is the angle between this line and the x-axis. The locus of Z as ω is varied will depend on the elements in the equivalent electrical circuit. The curves for simple elements are shown in Figure 3. The advantage of the complex impedance representation is that a series combination of elements can be described by the vectorial addition of the curves for the individual elements. The situation is more complicated for a parallel combination of elements, where a simple resistance, R, and capacitance combination leads to a semi-circle in the complex impedance plot (as shown in Figure 3). For a parallel combination of elements it is often advantageous to plot the admittance on the complex plane ie., the imaginary part, B (the susceptance), versus the real part, G (the conductance).

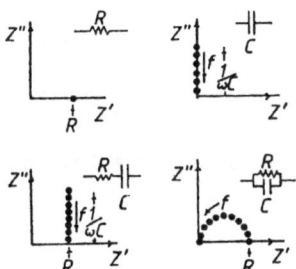

Figure 3. Complex impedance plots for simple circuits. (The arrow indicates the direction of increasing frequency, f)

The analysis of complex plane spectra can be considered as a two step procedure. The first step involves assigning an equivalent circuit, based on lumped RC components, to represent the experimental spectrum. The second step is to relate the components to physical features or chemical processes of the cell. Assignment of the equivalent circuit involves a comparison with theoretical models and examples can be found in the review by Archer and Armstrong (1980). A starting point for the analysis is that each RC component will give rise to a semi-circle in the spectrum. The example taken here will be that of an ionically conducting ceramic pellet with partially blocking electrodes.

The ideal complex impedance spectrum of a ceramic pellet would consist of three semicircles and this is shown in Figure 4 along with the equivalent circuit. The electrical behaviour of the bulk material, i.e., within the grains, is represented by a resistance R_b in parallel with geometric capacitance C_b. The contribution from grain boundaries is represented by a resistance R_{gb} in parallel with a capacitance C_{gb}. For partially blocking electrodes there will be a resistance R_e, the charge transfer resistance, in parallel with a double-layer capacitance C_e. The three RC elements are taken as being connected in series. As indicated on Figure 4 the resistances can be evaluated from the intercepts on the Z' axis. The capacitances can be evaluated from the frequency at the top of the semi-circle (Archer and Armstrong; 1980). Thus in principle it is possible to separate the intra-grain from the inter-grain components and use the former as a probe of defect structures and transport mechanisms (Kilner and Steele; 1981).

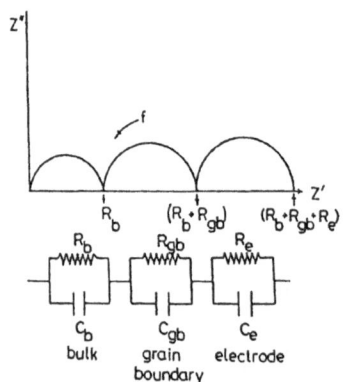

Figure 4. The idealised circuit and complex impedance plot for a ceramic oxide sample.

In practice the spectra are rarely as simple as that shown in Figure 3; the semi-circles can be distorted, they shift with temperature, diffusion may be a limiting step giving rise to Warburg impedance, etc. The various possibilities are treated in the reviews already cited and some of the experimental tests used to assign equivalent circuits will be outlined below.

Practical aspects

Impedance measurements can be made with an a.c. bridge and signal generator. However, it is very laborious and time-consuming as the resistance and capacitance have to be balanced manually at each frequency. Some improvement can be achieved by incorporating phase-sensitive detection to measure directly the in-phase and quadrature components of ΔE and Δi. Complex impedance spectroscopy became a much more widely used technique with the advent of experimental systems based on frequency-response analysers, such as one of the Solartron 1170 series. An experimental set-up has been described by Armstrong et al. (1977). The advantages that are gained are a large frequency range (0.1 mHz to 1MHz), rapid data collection and the possibility of automating the system by interfacing to a computer. Recent developments have included the commercial availability of instruments which give a direct read-out of the required parameters and are very simple to interface to microcomputers (for example, the Hewlett-Packard LF impedance analyser 4192A).

The choice of equivalent circuit to represent the measured complex impedance spectrum can present problems. There are some strategies that can be employed to assist this choice and aid identification of the origin of the semi-circles. Variation of the measuring temperature can help to separate overlapping semi-circles and often the shifts in position can be associated with the temperature dependence of a particular resistive component. Another useful variable is the sample geometry, ie., length, L, and cross-sectional area, A. Electrode effects will be independent of L and their associated semi-circle can be assigned. Other parameters that can usefully be varied are the electrode preparation, impurity concentration and, for oxide ion conductors, the oxygen pressure. Good examples of the analysis of complex impedance spectra can be found in the original work of Bauerle (1969) and the applications of the technique are described in conference proceeding on fast-ion conductors (see, for example Vashishta et al; 1979, Kleitz et al; 1983).

It should be noted that the precision that is possible on the bulk resistance, R_b, for a polycrystalline sample does not approach that usually obtainable for measurements on a single crystal. Thus the level of sophistication that can be used in modelling the defect processes is restricted.

SUMMARY

The current good understanding of defects and transport in ionic solids is founded on ionic conductivity measurements. For single crystals of simple materials these can be made with a very high degree of precision; to the extent that least-squares computer fitting of the data with models containing several defect parameters is now routine and reliable. In the case of more complicated materials, such as ceramics, where the samples are usually polycrystalline the introduction of complex impedance measurements has allowed evaluation of the bulk conductivity. Although the degree of precision is lower it does provide the basis for models of the defect processes in technologically important materials like the fluorite-structured oxides (Kilner and Steele; 1981).

REFERENCES

Aboagye, J.K. and Friauf, R.J., 1975, Phys. Rev., B11, 1654.
Archer, W.I. and Armstrong, R.D., 1980, in "Electrochemistry" , Vol.7, ed. H.R.Thirsk, Specialist Periodical Report of the Chemical Society, (London), p.157.
Armstrong, R.D., Bell, M.F. and Metcalfe, A.A., 1977, J. Electroanal. Chem. Interfacial Electrochem., 77, 287.
Batra, A.P. and Slifkin, L.M., 1975, Phys. Rev., B12, 3473.
Bauerle, J.E., 1969, J. Phys. Chem. Solids, 30, 2657.
Catlow, C.R.A., Corish, J. and Jacobs, P.W.M., 1979, J. Phys. C., 12, 3433.
Corish, J. and Jacobs, P.W.M., 1972, J.Phys. Chem. Solids, 33, 1799.
Corish, J. and Jacobs, P.W.M., 1973, in "Surface and Defect Properties of Solids", Vol.2, eds. M.W. Roberts and J.M. Thomas, Specialist Periodical Report of the Chemical Society, (London), p.184.
Corish, J. and Jacobs, P.W.M., 1975, Phys. Stat. Solidi b 67, 263.
Corish, J. and Mulcahy, D.C.A., 1980, J. Phys. C., 13, 6459.
Frenkel, J., 1926, Z. Physik, 35, 652.
Harding, J.H., 1985, Phys. Rev. (in press); also AERE Harwell Report TP. 1113
Heyne, L., 1983, "Mass Transport in Solids", eds. F. Beniere and C.R.A. Catlow, (Plenum Press, New York), ch. 17.
Hove, J.E., 1956, Phys. Rev., 102, 915.
Jacobs, P.W.M., Corish, J. and Catlow, C.R.A., 1980, J. Phys. C., 13, 1977.
Kilner, J.A. and Steele, B.C.H., 1981, in "Non-Stoichiometric Oxides", ed. O.T. Sorenson, (Academic Press, New York), p. 233.
Kleitz, M., Sapoval, B. and Ravaine, D., 1983, eds. "Solid State Ionics-83", (North-Holland, Amsterdam).
Pitts, E., 1953, Proc. Roy. Soc., A217, 43.
Reade, R.F. and Martin, D.S., Jr., 1960, J. Appl. Phys., 31, 1965.
Rice, S.A. 1958, Phys. Rev., 112, 804.
Rickert, H., 1982, "Electrochemistry of Solids", (Springer-Verlag, Berlin), ch. 6.
Schottky, W., 1935, Z. Phys.Chem.,(Leipzig), B29, 335.
Sluyters, J.H., 1960, Rec. Trav. Chim. Pays-Bas, 79, 1092.
Vashista, P., Mundy, J.N. and Shenoy, G.K., 1979, eds. "Fast-Ion Transport in Solids", (North-Holland, Amsterdam).
Vineyard, G.H., 1957, J. Phys. Chem. Solids, 3, 121.
Weber, M.D. and Friauf, R.J., 1969, J. Phys. Chem. Solids, 30, 407.
Wert, C.A., 1950, Phys. Rev., 79, 601.
Zener, C., 1952 in 'Imperfections in Nearly Perfect Crystals', eds. W. Shockley, J.H. Hallomon, R. Maurer and F. Seitz, John Wiley (New York), p. 289.

THERMALLY STIMULATED DEPOLARIZATION STUDIES OF IONIC SOLIDS

Rosanna Capelletti

Dipartimento di Fisica - CISM-GNSM
University of Parma
Via M. D'Azeglio, 85 - 43100 Parma - Italy

I. INTRODUCTION

Relevant physical properties of crystalline solids (for instance electric resistivity, color, mechanical strength, etc.) and applications (integrated electronics, laser etc.) are determined by the presence of lattice defects (vacancies, interstitials, impurities and complexes built by them). According to the specific nature of defects, suitable techniques have been developed. In ionic solids, defects often bear an electric charge (cation and anion vacancies, interstitial ions, aliovalent impurities). As a consequence of the Coulomb interaction between defects of opposite charge, complexes may be formed, which exhibit an electric dipole moment. Other defects, even in non-ionic solids, may be endowed by their own dipole moment. Hence electrical methods such as dielectric losses ($tg\delta$), isothermal depolarisation currents and more recently ionic thermocurrents (ITC) are suitable to monitor such defects. This last method, introduced by Bucci and Fieschi in 1964, deals with the detection and analysis of thermostimulated depolarization currents arising from ion redisplacements in solids whose electronic conductivity is neglegible (1,2). In a sense ITC enters in the wide class of methods based on the thermal release of stored energy, as for instance thermoluminescence (TL), thermally stimulated currents (TSC) and thermally stimulated depolarization currents (TSDC). ITC deals with TSDC as well, but is restricted to polarization mechanisms, in which only ion redisplacements are involved ruling out a wide class of phenomena which occur in electrets such as carrier injection, electron and/or hole trapping. In this way ITC is more descriptive and specific term than the more general TSDC, which is used by some authors. Due to this restriction and its successful application to simple model systems, such as ionic crystals, it has opened the way for a quantitative study of dipolar processes also in more complex systems of technological interest.

In these lectures the method and the experimental apparatus are described and discussed. Moreover its application to a variety of systems and processes involving ionic defects is reviewed.

2. ITC METHOD

Let us consider the plot vs reciprocal temperature T of the relaxation time $\tau = \tau_o \exp \varepsilon_j/kT$ for an ionic dipole: ε_j is the activation energy for

dipole reorientation, i.e. the energy barrier which divides two equivalent
dipole orientations, A and B in fig. 1a. The actual example is related to
the impurity-vacancy dipole (see fig.5 and § 3.1) in LiF:Mg, for which
ε_j = .654 eV and τ_o=3.6x10^{-14} s.

1) At temperature T_p (polarisation temperature), at which $\tau(T_p)$ is rather
short (in fig.1a: τ=3s at $T_p \simeq 233$ K) a static electric field E_p (polarization
field) is applied to the sample for a time interval t_p (polarization time)
much longer than $\tau(T_p)$, for instance $t_p \sim 180$s, see fig.1b. In this way dipoles
are oriented at saturation, in fact the polarization P(t) is (3)

$$P(t_p)=P_o(T_p,E_p) \cdot \{1-exp-t_p/\tau(T_p)\} \tag{1}$$

In the present case $P(t_p)$ is $P_o\{1-exp-180/3\} \sim P_o$.

2) The sample, with the field always on, is cooled at temperature T_f (see
fig.1b) at which τ is very long (in fig. 1a at T_f=145 K, $\tau(T_f) \sim 1$ century).

3) The field is turned off, see fig. 1b. The dipoles, previously oriented re
main in their preferred orientations, since $\tau(T_f)$ is extremely long: the polarisation is frozen in.

4) The sample is connected to the electrometer and heated at a constant rate: i.e. dT=bdt, see fig.1b. The dipoles experience relaxation times, which
become shorter and shorter, see fig.1a, they gain mobility and hence can attain random orientations. This means that the sample polarisation is changing
vs time: i.e. according to Maxwell equations, a displacement depolarisation
current j(T) is detected, see fig.1b. It increases by increasing the temperature, until the frozen in polarisation is exhausted, hence j(T) drops to
zero, see fig.1b. The j vs T plot and the peak are defined as ITC plot and
ITC peak.

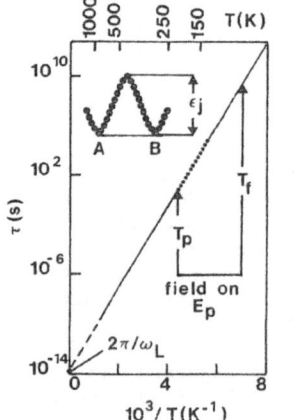
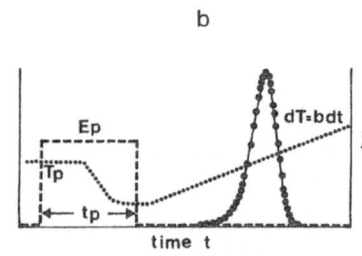

Fig.1. ITC method applied to (IV)$_c$ dipoles in LiF:Mg. a) τ vs 10^3/T plot,
the dots are the experimental data according to eq.12; b) ITC proce-
dure- dashed line: electric field; dotted line: temperature; full li
ne: displacement current density given by (6; dots: experimental
data.

2.1. Basic ITC equations for non interacting dipoles with a single τ. Let us consider a set of non-interacting dipoles, 1) characterised, as above, by a single relaxation time, see fig.1a,

$$\tau(T) = \tau_0 \exp \varepsilon_j/kT \tag{2}$$

and 2) which disorient following a first order monomolecular kinetics, i.e.

$$\frac{dn}{dt} = - \frac{n}{\tau(T)} \tag{3}$$

where n is the concentration of dipoles, which are still oriented. By introducing the dipole moment p in (3 and considering the heating rate b=dT/dt one obtains:

$$\frac{dP}{dt} = - \frac{P}{\tau(T)} = - \frac{bdP}{dT} = -j(T,t) \tag{4}$$

$$\frac{dP}{P} = - \frac{1}{b} \frac{dT}{\tau(T)} \tag{5}$$

On integrating (5 and taking into account (4 one gets:

$$j(T)=P_0(\tau_0 \exp \varepsilon_j/kT)^{-1} \exp- \int_0^T (b\tau_0)^{-1} \exp(-\varepsilon_j/kT')dT' \tag{6}$$

where P_0 is the frozen in polarisation (3), i.e.

$$P_0 = P_0(T_p,E_p) = \alpha N_d p^2 E_p \cdot (kT_p)^{-1} \tag{7}$$

N_d is the dipole concentration, α is a geometrical factor which takes into account the possible dipole orientations (α=1/3 for free dipoles and for IV dipoles in alkali halides, see § 3.1). Eq. (6 gives the typical asymmetric peak (full line) in fig. 1b, which fits quite well the experimental data (dots) for IV dipoles in LiF:Mg.

The peak maximum $j(T_M)$ and the temperature T_M at which it occurs are:

$$j_M=P_0 b\varepsilon_j(kT_M^2)^{-1} \exp- \int_0^{T_M} (b\tau_0)^{-1} \exp(-\varepsilon_j/kT')dT' \tag{8}$$

$$T_M = \{b\varepsilon_j\tau(T_M)/k\}^{1/2} \tag{9}$$

Hence T_M depends on heating rate b, but not on T_p, E_p and t_p; j_M is a linear function of E_p (see eqs.7 and 8) and of N_d, see fig.13. If the condition $t_p \gg \tau(T_p)$ is not fulfilled, P_0 in eqs.6 and 8 has to be substituted with $P(t_p)$ given by eq.1. These observations can be exploited for ITC peak diagnostics (see § 4.1).

The area A delimited by the ITC peak gives P_0, or $P(t_p)$ if $t_p \ll \tau(T_p)$, in fact:

$$A = \int_0^\infty j(t)dt = \int_0^\infty j(T)b^{-1} dT=P_0=N_d\alpha p^2 E_p/kT_p \tag{10}$$

Hence from the peak area A one can evaluate N_d, see sect.5 and § 6.4.

2.2. Determination of ε_j and τ_0: area method. Eq. 4 can be rewritten as:

$$\tau(T)=P(T)/j(T) = \int_T^\infty b^{-1}j(T')dT'/j(T) \tag{11}$$

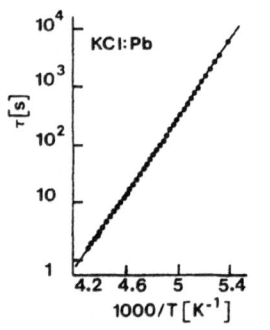 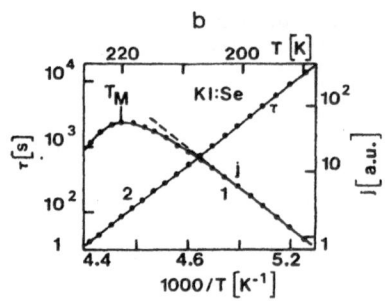

Fig.2. Experimental ε_j and τ_o determination for $(IV)_c$ dipoles in KCl;Pb (a, and for $(IV)_a$ in KI:Se (b, ref.8. τ vs $10^3/T$ plots according to eq.12 in (a and in (b, curve 2; j vs $10^3/T$ in (b, curve 1: deviation from straight line occurs for $T\sim T_M$.

since P(T) is the polarisation still frozen in at temperature T and can be evaluated from the area under ITC curve from T to the end of the curve. From (2 and (11 one obtains

$$\ell n\tau_o + \varepsilon_j/kT = \ell n \{\int_T^\infty b^{-1}j(T')dT' / j(T)\} \tag{12}$$

Hence a semilog plot of last member of (12 vs T^{-1} gives a straight line, from whose slope and intercept at the origin one obtains ε_j and τ_o respectively with high accuracy since information is extracted from the whole curve, see figs.2a, 2b in detail, and more generally fig.1a, 3a and 4 where ITC results were used to build the τ plots vs T^{-1}.

2.3. Determination of ε_j : initial rise method. If one considers only the initial current rise, i.e. for $T\ll T_M$, the integral in (6 is neglegible, hence:

$$j(T) \sim P_o \tau_o^{-1} exp-\varepsilon_j/kT \tag{13}$$

$$\ell n\ j(T) \sim \ell n\{P_o/\tau_o\} - \varepsilon_j/kT \tag{14}$$

Again a semilog plot of j(T) vs T^{-1} gives a straight line, from whose slope ε_j is determined. This straight line (for $T\ll T_M$) is the mirror image of that obtained with the area method (see fig.2b). From (14, τ_o can be evaluated as well, but not in the straightforward way as above, see also § 6.4.

2.4. Cleaning technique. ITC technique allows to resolve overlapping peaks by following the procedure here illustrated for KBr:Sr where two peaks occur at 183 K (peak I) and at 217 (peak II), as shown in fig.3. The related relaxation times τ_I and τ_{II} are plotted vs T^{-1} in fig.3a. The ITC plot of fig.3b is obtained in the usual way (see § 2) by polarizing the sample at T_1=235 K where τ_I=2.7x10^{-2}s and τ_{II}=5s and turned off at T_2=167 K where $\tau_I\sim10^3$s and $\tau_{II}\sim3.5\times10^6$s; hence both types of dipoles are polarized at saturation. In order to get an ITC plot where only a "clean" peak II is displayed, the field is turned on again at T_1=235 K: if, as a rule t_p is $\sim$180s, both types of dipoles are polarized at saturation. The field is then turned off at T_3=210 K where $\tau_I\sim$1s and $\tau_{II}\sim$270: if the subsequent cooling is not too fast, the type I dipoles are still rather mobile and are able to randomise, while type II dipoles remain frozen in their preferred orientations, since their relaxation time τ_{II} is rather long. Hence the ITC plot displayed during the subsequent heating shows only peak II, see fig. 3c.

410

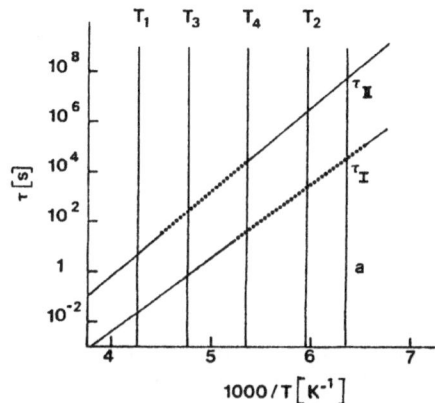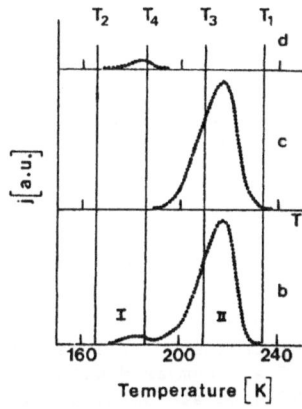

Fig.3. Cleaning technoque for overlappping peaks: (IV)$_c$ dipoles in KB;Sr.
a) τ vs $10^3/T$ plots for the two relaxations as obtained according to
eq.12 (dots); b),c) and d) ITC spectra as obtained by proper choice
of polarisation conditions, see text.

In order to get an ITC plot where only a "clean" peak I is displayed, the
field is turned on at T_4=189 K where $\tau_I \sim 43$s and $\tau_{II} \sim 2.5 \times 10^4$ s. If t_p=180s,
<u>only</u> type I dipoles are polarised (98.5% according to eq.1). The field is
turned off at T_2=167 K. The ITC plot displayed during the subsequent heating
shows only peak I (see fig.3d). The application of the procedure of § 2.2 to
the clean peaks, allowed to obtain $\tau_{o,I}$=1.35$\times 10^{-14}$s, $\varepsilon_{j,I}$=.573 eV; $\tau_{o,II}$=
=2$\times 10^{-14}$s and $\varepsilon_{j,II}$=670 eV and as a consequence τ_I and τ_{II} vs T^{-1} plot of
fig.3a.

2.5. <u>Advantages of ITC technique</u>. ITC technique provides a straightforward
way for a very accurate determination of reorientation parameters ε_j and τ_o
(see § 2.2 and 2.3 and fig.2). Moreover only one measurement is required to
obtain the τ vs.T plot (see fig.2). On the contrary, by using tgδ or isother
mal depolarisation currents techniques, many isothermal measurements at dif
ferent T are required vs. frequency or vs.time, in order to get a reliable
τ vs. T plot.

Small dipole concentrations can be detected by means of ITC: for instan
ce, in LiF:Mg IV dipole concentration as low as 5×10^{-2} ppm are detectable.
The ITC sensitivity is $\sim 10^2$ times better than that offered by tgδ.

The ITC run is, as a rule, performed at lower temperatures than tgδ,
(see fig.4). Hence the annealing processes, related to long range diffusion
of defects, are inhibited during the measurement itself. This feature is
particularly appreciable for the careful analysis of aggregation process of
IV dipoles in alkali halides, which in some systems is very fast even at
room temperature (see fig.14).

By exploiting "cleaning" technique (see § 2.4) it is possible to isola
te overlapping peaks (the difference between the related ε_j being as low as
.01 eV) and then proceed successfully to the evaluation of their reorientation parameters (see § 2.2 and 2.3).

Information is obtained on the low frequency response of ionic dipoles.
For instance in CaF$_2$:Gd, RE-FI dipoles (see § 3.8) display an ITC peak at
128 K for a b=5$\cdot 10^{-2}$ K/s and with ε_j=.420 eV (4$^+$:89): by using eq.9 the jump
frequency $\nu_{j,M}$ is $\sim 1.5 \times 10^{-2}$ Hz at 128 K. Such low frequency response is shown
in fig.4, where the ESR and tgδ results are compared with the ITC ones.

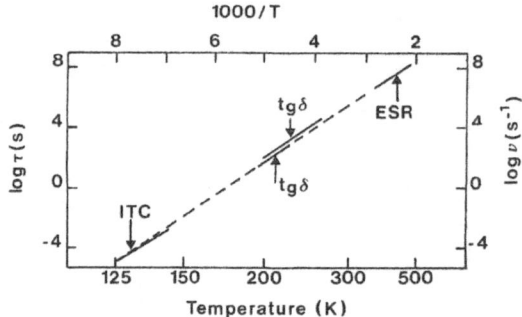

Fig.4. Low frequency vs T response of RE-FI dipoles in CaF$_2$:Gd obtained by ITC and compared with that obtained by means tgδ and ESR, ref.4[+]:89. ITC data are obtained according eq.12.

3. ANALYSIS OF DIPOLAR POINT DEFECTS

Since ITC provides a straightforward way to evaluate the reorientation parameters ε_j and τ_0 with high accuracy (see § 2.2), it has been exploited to study many kinds of dipolar point defects, chiefly in ionic crystals. The most of work has dealt with model systems, such as alkali halides, where many examples of dipolar defects induced by doping are reported. Let us recall 1) I.V. dipoles, which are formed either in the cation sublattice, (IV)$_c$, or in the anion one, (IV)$_a$, whenever a substitutional divalent impurity cation (or anion) is charge compensated by a cation (or anion) vacancy (see fig.5, § 3.1 and 3.2); 2) the off-center monovalent cations (see fig.5 and § 3.3); 3) the substitutional molecular anions, endowed with their dipole moment (see fig.5 and § 3.4). In such crystals and, at less extent, in silver halides ε_j and τ_0 were analysed carefully as a function of the impurity size and of the lattice constant a_0. The aim was to obtain information about the elementary jumps, through which the dipole reorientation takes place, the structure and symmetry of the defect and the mismatch induced into the lattice by the impurity ion.

Similar purposes have been pursued in the case of fluorite-structure crystals, where dipolar defects arise as a consequence of doping with 1) trivalent rare-earth cations, which are charge compensated by an interstitial fluorine (see fig.8 and § 3.8), 2) 0^{2-} or alkali ions which are charge compensated by a fluorine vacancy (see fig.8 and § 3.9).

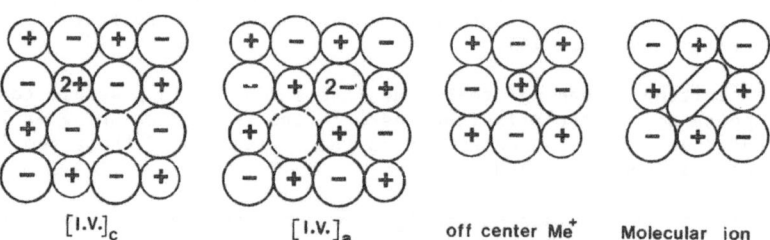

Fig.5. Models of some dipolar defects in alkali halides: the dashed circles indicate vacancies.

412

3.1. (I.V)$_c$ dipoles in alkali and silver halides. The reorientation parameters ε_j and τ_o of (IV)$_c$ dipoles were measured in a wide class of alkali halides such as LiF, NaF, NaCl, KCl, RbCl, KBr, RbBr, CsBr, NaI, KI and CsI, doped with a variety of Me^{2+} such as alkali earths, transition elements (Mn, Fe,Co,Ni,Cr), rare earths (Sm,Eu,Yb) and others (Zn,Cd,Pb,Sn) (4$^+$:17;5$^+$:29;6). As a rule only one ITC peak is detected as due to the reorientation of (IV)$_c$, see fig.6b and is well described by eq.6, see fig.1b. The ε_j values range between .5 and .7 eV, the smaller figures being related to Be (4$^+$:33) (see fig.7a) and τ_o range between 10^{-14} and 10^{-12}s. Among silver halides, mainly

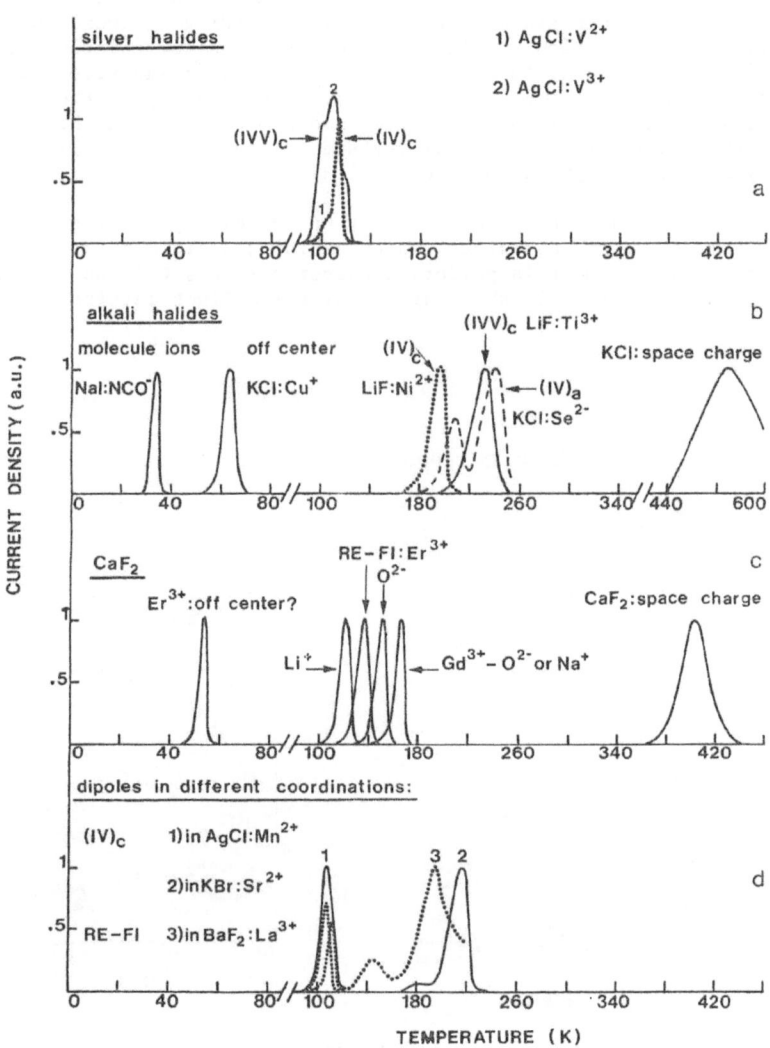

Fig.6. ITC plots for simple dipolar defects and space charge in alkali and silver halides and in fluorite-like crystals.

AgCl doped with Ca,Mn,Fe,Co,Ni,Cr,V,Zn,Cd and Pb (5^+:27;4^+:105) was studied (see fig.6a and d): the motion parameters are smaller than those of alkali halides, i.e. ε_j ranges between .27 and .35 eV and τ_o between 10^{-13} and 10^{-14}s. In AgCl two overlapping peaks are found (7), see fig.6d and § 3.10 and 3.11.

In alkali halides ε_j is function of the Me^{2+} radius r^{2+}, see fig.7a:it increases with r^{2+}, reaches a maximum for $R\sim1$ (being $R\equiv r^{2+}/r^+$ and r^+ the host cation radius), then decreases (5,6), anyway remaining smaller than the activation energy for the free cation vacancy motion ε_{fv}, see fig.7a. Such definite trend cannot be guessed from the available results in AgCl, even if the ε_j values remain close to $\varepsilon_{fv}\sim.28$ eV, as for alkali halides. These results bring to the following conclusions: 1) the $(IV)_c$ reorientation takes place through the cation vacancy jumps, rather than Me^{2+} jumps; 2) the lowest values of ε_j occur, whenever the impurity is either much smaller or much larger than the host cation, since the lattice mismatch induced by it makes easier the vacancy jump around the impurity. According to Cussò (6), the $(IV)_c$ dipole reorientation in alkali halides takes place chiefly through the vacancy jumps from n.n.n. to n.n. position (see fig.9) and viceversa for R<1, and from n.n. to n.n. for R>1 (see fig.7a). This is along to the Dansas'speculation, who argued that the dominating n.n.n. coordination changes to n.n. on going through $R\sim1$ for alkali halides and $\sim.7$ for AgCl (5^+:27) and is supported by binding energy calculations (5^+:31).

The same Tharmalingham model used by Cussò (6) explains, at least qualitatively, the τ_o vs r^{2+} behaviour, which exhibits a minimum for $R\sim1$. As a rule if the ITC measurement is performed properly (see § 6.1 and 6.3) the values of $\omega_o\equiv2\pi(\tau_o)^{-1}$ fall in the range of alkali halides lattice frequencies,

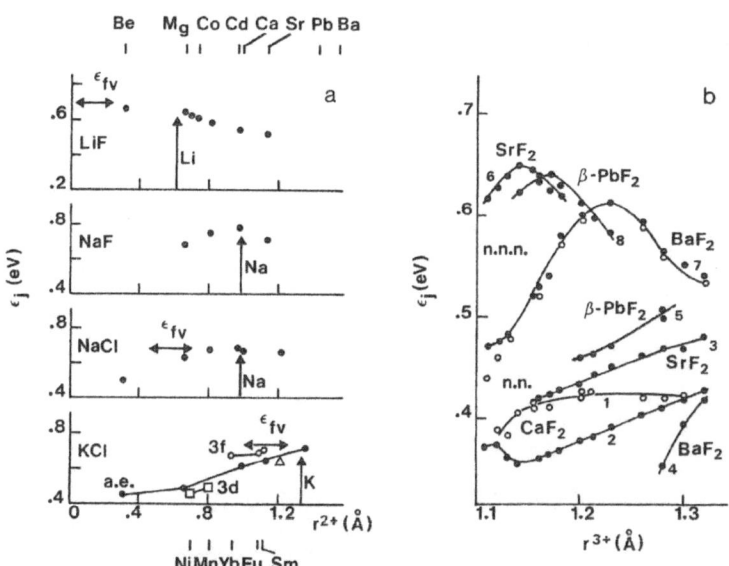

Fig.7. Activation energy ε_j vs impurity ionic radius: a) for $(IV)_c$ dipoles in alkali halides, b) for RE-FI dipoles in fluorite-like crystals: n.n. complexes, curves 1-5; n.n.n. complexes, curves 6-8. For data of curve 1 see ref.24; for other curves see ref.10^+:10.

414

being always smaller than the longitudinal optical frequency ω_L, see for instance fig.1a.

3.2. (IV)$_a$ in alkali halides. Fischer and coworkers (8) reported ITC peaks attributed to (IV)$_a$ dipole reorientation in potassium halides doped with O, S,Se and Te, see figs.5 and 6b. The ITC peaks are well described by eq.6, ε_j range between .48 and .8 eV,see fig.2b, while τ_o is of the order of 10^{-13}s. The results show that ε_j increases with anion impurity radius r^{2-} (except for KI:Te): hence Fischer attributed the (IV)$_a$ reorientation to an exchange between the impurity and the anion vacancy, at variance with (IV)$_c$, see § 3.1. However this interpretation is questionable, because it should be supported by results related to a wider class of impurities and matrices, as for (IV)$_c$. Moreover all the ε_j values reported for (IV)$_a$ are smaller than the migration energies either of free anion vacancy (.85 eV) and of divalent anions (5) as for (IV)$_c$ (see § 3.1).

3.3. Off center ion dipoles in alkali halides. Monovalent cation impurities (M$^+$) of smaller size than that of the host cation, when entering the lattice, can occupy off center positions along <111> directions, originating a permanent electric dipole moment, which can be oriented by an external electric field, see fig.5. ITC peaks due to off center M$^+$-dipoles in potassium and rubidium halides doped with Cu$^+$ (5$^+$:44) and in RbI:Ag$^+$ have been observed (see fig.6b) in the temperature range 20-100 K, suggesting that the reorientation is ruled by a thermally activated relaxation time (eq.2) and does not occur by a tunneling. The peaks are well described by eq.6; ε_j values are rather small ranging from 5×10^{-2} to .31 eV, being the smaller figure related to the larger ion, Ag$^+$, as expected (9).

3.4. Dipolar molecular ions in alkali halides. Foreign molecules (molecular ions) endowed with their own dipole moments can assume only discrete orientations, if they are embedded in a lattice with a defined symmetry, see fig.5. If the jump between the discrete orientations is thermally activated, ITC peak is expected. Such peaks were observed in Na-, K- and Rb-halides doped with AsO_2^-, NO_3^-, NO_2^-, PO_2^-, (10$^+$:7,8) NCS$^-$ and NCO$^-$ (5$^+$:47) (see fig.6b) in the range 15-70 K. They are well described by eq.6. ε_j values are very small, ranging from 70 to 172 meV and the most of τ_o values falls in the range 10^{-11}-10^{-13}s.

3.5. (IVV)$_c$ dipoles in alkali and silver halides. If trivalent cations enters the halide lattice, a dipolar defect (IVV)$_c$ can be formed if two cation vacancies compensate the impurity charge excess. As for the simpler (IV)$_c$, see § 3.1, the (IVV)$_c$ reorientation may occur through the thermally activated jump of the vacancy (or vacancies). Different reorientation paths, more than for (IV)$_c$, are available and hence complex broad peak is foreseen (see § 3.10). This has been reported in AgCl doped with Ti,V,Cr (4$^+$:106), see fig.6a and very recently in LiF:Ti (11), see fig.6b.

3.6. Dependence of ε_j on a$_o$. The dependence of ε_j of simple dipolar defects in alkali halides has been studied vs the lattice constant of the host matrix a$_o$ in three ways: 1) by changing the host matrix, 2) by applying to a given system hydrostatic pressure (5$^+$:45), and 3) by changing the relative concentration of the two halides in mixed alkali halides (see § 3.7). The results obtained are summarized as follows (5). a) for (IV)$_c$, (IV)$_a$ and molecular anions ε_j decreases by increasing a$_o$ and b) for off center cations ε_j increases with a$_o$. The former result can be accounted for by the larger room available for dipole reorientation whenever a$_o$ increases; the latter can be explained since equivalent off center positions become more far apart, hence the energy wall, which separates them, increases.

3.7. (IV)$_c$ dipoles in mixed alkali halides. By mixing salts with the same cation and different anions, two main effects are found: 1) the original peak shifts and ε_j decreases by increasing a_0, (in agreement with § 3.6); 2) chiefly in the case of foreign anion larger than the host one, additional peaks are present which suggest the degeneracy lifting among equivalent cation vacancy-impurity coordinations due to the presence of the nearby perturbing foreign anion (5+:41).

3.8. RE-FI dipoles in fluorite-like crystals. The most investigated matrices were CaF_2, SrF_2, BaF_2 and at less extent PbF_2, CdF_2 and $SrCl_2$ doped with rare earths, Y,Sc,U,Al and In (4+:86-93,95-103;10+:10,12,24). Trivalent rare earth (R.E.) substitutes the divalent host cation. Due to the excess charge compensation, operated by an interstitial fluorine (F.I.) in n.n. or n.n.n. coordination, dipolar defects are formed, i.e. RE-FI dipoles, with tetragonal (type I complex) or trigonal (type II complex) symmetry, see fig.8. ITC peaks due to reorientation of both kinds of dipoles have been observed: at rather low T (∿125-150 K) for peak I (n.n. dipoles) at higher T (∿165-200 K) for peak II (n.n.n. dipoles), see fig.6c and d and § 3.10 and 3.11. The dominating defects are type I complexes in CaF_2 and type II in BaF_2, while in SrF_2 both are observed. The ratio between peak I and peak II amplitudes depends on the impurity radius r^{3+} and increases on decreasing r^{3+}. ε_j and τ_0 have been evaluated for both complexes: ε_j ranges between .36 and .65 eV for complex I and between .44 and .65 eV for complex II, see fig.7b; τ_0 ranges between 10^{-5} and 10^{-14}s. For a given host matrix ε_j increases, as a rule, as a function of r^{3+} for complex I, while ε_j exhibits a maximum for complex II, see fig.7b. Spread of ε_j values is reported as well: this is due to different experimental conditions, see § 6.1 and 6.3, and concentrations, see § 3.13.

3.9. Other dipolar defects in fluorite-like crystals. On doping fluorite-like crystals with monovalent cations Me^+, dipoles are built by the substitutional M^+ and the compensating fluorine vacancy F_v, see fig.8. Related ITC peaks have been observed in CaF_2,SrF_2,BaF_2,CdF_2 and PbI_2 (12+:14,80,113; 4+:95) doped with Li,Na,K and Ag in the 85-175 K range, see fig.6c. ε_j values range between .25 and .55 eV. On oxygen-doping CaF_2, dipoles are formed in the anion sublattice by O^{2-}, substituting F^-, and the compensating vacancy F_v, see fig.8; related ITC peak has been found (12+:5) at ∿150 K with ε_j .47 eV, see fig.6c. ITC peaks attributed to off-center Er^{3+} and Gd^{3+} occur at low T (50-60 K) in CaF_2 with ε_j∿.15 eV (4+:86), see fig.6b. A very complex dipolar defect, responsible for an ITC peak at 166 K, see fig.6c with ε_j∿.49 eV is reported in CaF_2 doped by either RE(Gd,Y) and O^{2-}: the reorientation takes place through the fluorine vacancy jumps in a complex made by RE, three F_v and four O^{2-} (4+:96) see fig.8. In oxides such as CeO_2 with fluorite structure, doped with divalent or trivalent cations, ITC peak occurs due to the reorientation of a dipolar complex formed by the foreign cation and an O^{2-} vacancy: in CeO_2:Y^{3+} ITC peak appears at ∿210 K with ε_j∿.64 eV (10+:13), see fig.8.

3.10. Phenomenological and microscopic reorientation model for simple dipoles in ionic crystals. Let us discuss more in detail the sequence of elementary ionic jumps, which leads to the dipole reorientation. For IV dipoles in alkali and silver halides, Dreyfus (5+:2) has shown that for reorientation paths which involve n.n. and n.n.n. dipole configurations, see fig.9, two modes of relaxation are operating under the effect of the applied electric field, with normal frequencies $\lambda_{1,2}$ given by

$$\lambda_{1,2} = 2\omega_3 + \omega_0 + \omega_4 \pm \{(\omega_0 + \omega_4 - 2\omega_3)^2 + 4\omega_3\omega_4\}^{1/2} \qquad (15$$

where $\omega_0 \equiv \omega_1 + \omega_2$ and ω_1 is the jump frequency of the vacancy from a n.n. to n.n.n. site, ω_2 is the exchange frequency between impurity and vacancy, ω_3 and ω_4 give the jump frequency of the vacancy from n.n.n. to n.n. site and viceversa. The ratio between the polarisations P_1 and P_2 related to λ_1 and λ_2 is given by:

416

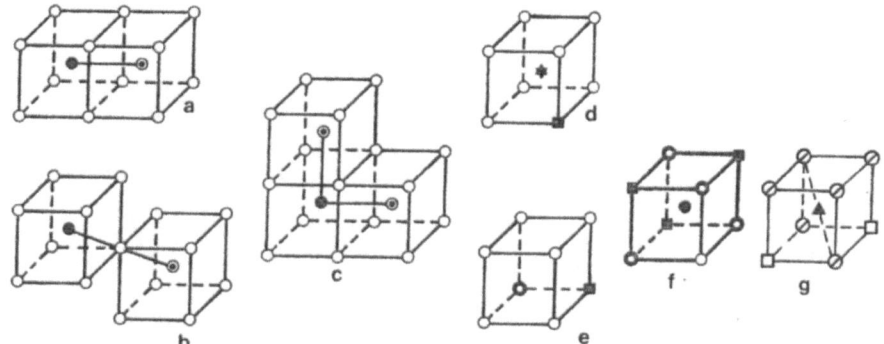

Fig.8. Models of some dipolar defects in fluorite-like crystals: o F^-; ⊚ interstitial F^-; ● Rare Earth; ■ fluorine vacancy F_v; ✱ alkali ion; ❍ O^{2-}; □ O^{2-} vacancy in CeO_2, O_v; ▲ dopant cation (Me^{2+} or Me^{3+}); a) RE-FI complex I; b) RE-FI complex II; c) cluster; d) M^+-F_v dipole; e) O^{2-}-F_v dipole; f) $RE^{3+}·O_4^{2-}(F_v)_3$ complex; g) Me^{2+} (or Me^{3+})-$(O_v)_2$ in CeO_2, where ⊘ is the host oxygen.

$$P_1/P_2 = -\{2(\omega_o+\omega_4)-\lambda_2(1+\omega_4/\omega_3)\}·\{2(\omega_o+\omega_4)-\lambda_1(1+\omega_4/\omega_3)\}^{-1} \quad (16$$

Simplifications can be obtained as follows:

a) for $2\omega_3>>\omega_o,\omega_4$

$$\lambda_1 \backsim 4\omega_3; \quad \lambda_2 \backsim 2\omega_o+\omega_4; \quad P_1/P_2 \backsim \omega_4/4\omega_3$$

$$(17$$

b) for $2\omega_3<<\omega_o,\omega_4$

$$\lambda_1 \backsim 2(\omega_o+\omega_4); \quad \lambda_2 \backsim 2\omega_3\{2-(1+\omega_o/\omega_4)^{-1}\}; \quad P_1/P_2 \backsim \omega_o^2\omega_3\{\omega_4(\omega_o+\omega_4)^2\}^{-1}$$

The same model has been applied to RE-FI dipole reorientation in fluorite-type crystals (12^+:104). Again two operating frequencies are found, λ_I and λ_{II} given by

$$\lambda_{I,II} = 2\omega_{11}+2\omega_{12}+1.5\omega_{21} \pm \{(2\omega_{11}+2\omega_{12}-1.5\omega_{21})^2+4\omega_{12}\omega_{21}\}^{1/2} \quad (18$$

where ω_{11} is the jump frequency from n.n. → n.n. coordination of the interstitial fluorine (with respect to RE^{3+}), ω_{12} and ω_{21} are the jump frequency from n.n. → n.n.n. coordination and viceversa respectively, see fig.9. Simplification can be obtained as follows:

for $\omega_{11}>>\omega_{12},\omega_{21}; \quad \lambda_I \backsim 4\omega_{11}, \quad \lambda_{II} \backsim 3\omega_{21}$ $\qquad$ (19

It turns out that the general solutions $\lambda_{1,2}$ and $\lambda_{I,II}$ are given in both cases by combination of more than one elementary ion jump. This complexity is not eliminated, even if simplifications are made in the case of I.V. dipoles at variance with the case of RE-FI, see (17 and (19. It is to outline that the above microscopic approach and the simpler phenomenological one given by eq.2 lead to the same ITC peak, as shown for $(IV)_c$ dipoles in NaCl:Mn (5^+:17).

Since two frequencies are operating, one expects two corresponding ITC peaks. Two well separated peaks have been observed for RE-FI dipoles, chiefly in SrF_2, see § 3.8 and fig. 6d, two strongly overlapping peaks have been detected for $(IV)_c$ dipoles in AgCl, see § 3.1 and fig.6d. On the contrary, in the case of $(IV)_c$ dipoles in alkali halides, notwithstanding the large

417

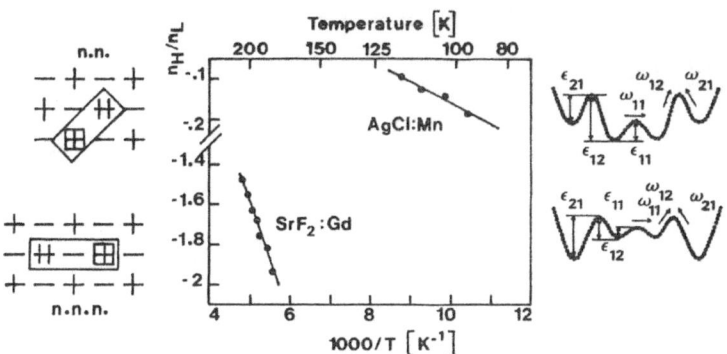

Fig.9. Thermodynamical equilibrium of n.n. and n.n.n. dipoles vs $10^3/T$:in the center semilog plots of the ratio between high T and low T peak amplitudes in AgCl:Mn (curve 1,ref. 7) for IV_c dipoles (n.n. and n.n.n. models on the left) and in SrF_2:Gd (curve 2, ref.12[+]:104) for RE-FIdipoles (energy diagrams on the right for n.n. complex, top, and for n.n.n. complex, bottom; the elementary jumps ω_{ij} are indicated).

number of available data, only one peak is detected as a rule, see fig.6b and § 3.1. In some cases two ITC peaks have been attributed to $(IV)_c$ in n.n. and n.n.n. coordination (5[+]:29). However often one of them has to be ascribed more likely to a dipolar cluster, since it disappears as a consequence of sample quenching. Only in the case of KBr:Sr the two peaks, whose amplitude ratio does not change upon thermal treatments (12[+]:73), are attributable to different relaxation paths of the $(IV)_c$ dipole (see fig.6d).

3.11. Thermodynamical equilibrium between n.n. and n.n.n. coordination dipoles in ionic crystals. In SrF_2 and AgCl, where two ITC peaks are reported as due to different dipole coordinations, see § 3.10 and fig. 6d, it is possible to study the temperature dependence of equilibrium between the two types of dipoles and to determine the difference $\Delta\varepsilon_b$ between the dipole binding energies in the two different coordinations. The ratio between the two peak amplitudes is reported vs T_p^{-1}, where T_p is the temperature at which the sample is annealed in polarising field, usually in the T range spanning the high temperature ITC-peak (type II for SrF_2). As shown in fig.9 for RE-FI dipoles in SrF_2:Gd (12[+]:104) and for $(IV)_c$ in AgCl:Mn (7), this ratio is thermally activated. From the slope of the straight line, $\Delta\varepsilon_b$ is evaluated: it is very small i.e..046 eV and .005 eV respectively.

3.12. Dipolar clusters in ionic crystals. If the concentration of simple dipolar defects exceeds the thermodynamical equilibrium concentration,aggregates are formed, see § 5.1. Some of them exhibit dipole moment and can be oriented by an electric field due to the jump of, at least, one of their more mobile components, such as vacancies or interstitials. Many ITC peaks due to clusters were detected in alkali halides doped with Me^{2+} and in fluorite-type crystals doped with RE: see examples in fig.10. The double doping of ionic crystals allows the formation of mixed dimers, i.e. containing two different impurities, which were detected by the related ITC peaks, for instance in NaCl:Mg,Mn; in CaF_2:Er,Sm and in CaF_2:Tb,Nd (12). An ITC peak can be attributed to clusters if 1) it disappears upon sample quenching; 2) it grows as a consequence of simple defects aggregation (possibly monitored by the related ITC peaks) and 3) it increases with a supralinear law vs the impurity concentration, see § 5.1 and fig.15.

418

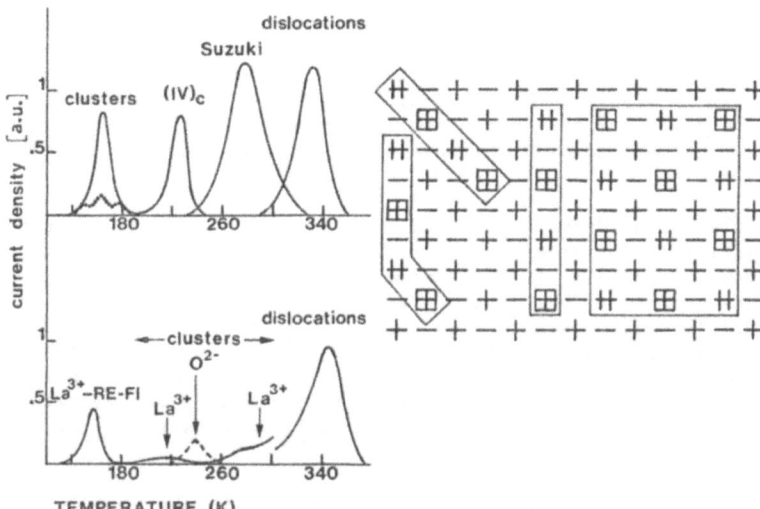

Fig.10. ITC plots to monitor ionic defects of increasing complexity in ionic crystals. Top: KCl:Pb (dotted line is related to quenched sample); bottom: SrF_2:La (full line), CaF_2:O^{2-} (dashed line). On the
right:models of possible clusters and of Suzuki phase in alkali
halides.

3.13. Dipole-dipole interaction and dependence of ε_j on the defect concentration. When impurity concentration overcomes (10^{18}-10^{19} cm^{-3}, electric and
elastic interaction is expected to occur among impurity related dipoles.
These effects are quoted as responsible of RE-FI peak broadening in SrF_2:Ce
(5^+:48). The ITC peak is no longer fitted by the simple expression (6 but
by:

$$j^*(T) = j(T) \cdot F(\varepsilon_j, \tau_o, w, T) \tag{20}$$

where F is a function which assumes a Gaussian distribution of ε_j, whose
halfwidth is w. F has been evaluated for different ε_j values, between .3
and .7 eV (13) and hence for different temperature ranges. It affects the
simple ITC formula (6 for non interacting dipoles, only in the case of low
ε_j values and low temperature ranges. Hence the correction introduced by
(20 is significant only for those dipoles whose ITC peak occurs at low temperatures, for example RE-FI in fluorite-type crystals, off-center ions and
molecular ions in alkali halides, but isn't for IV dipoles in alkali halides. As matter of fact, in the case of SrF_2:Ce^{3+}, w was found to increase
linearly with the dipole concentration,   a typical value being w=7.5x10^{-3}
eV, ε_j=.48 eV, N_d=1.06x10^{19} cm^{-3}. On the contrary in the case of $(IV)_c$ in
alkali halides, for instance in LiF:Ni, no meaningful broadening has been
detected up to concentrations as high as 6.7x10^{19} cm^{-3} (13). Moreover, in
some cases of peak broadening, as in alkali halides doped with NCO^- (5^+:46)
and in SrF_2:La (5^+:49) it has been attributed rather to internal stresses
induced by the large impurity concentration: in fact the broadening scales
rather with the total impurity concentration than with the dipole one.

The dipole-dipole interaction, according to den Hartog, is responsible
also for the apparent decrease of ε_j vs dipole concentration N_d, when (5^+:48)

ε_j is evaluated by the method illustrated in § 2.2. and 2.3 for non-interacting dipoles. Such a decrease is appreciable for N_d higher than $\sim 10^{18}$ cm^{-3} in SrF$_2$:Ce (RE-FI dipoles), but starts only for N_d higher than $\sim 3 \times 10^{19}$ cm^{-3} in LiF:Ni ((IV)$_c$ dipoles) (13). This different behaviour between RE-FI dipoles and (IV)$_c$ dipoles has been again attributed to the different weight of $F(\varepsilon_j, \tau_o, w, T)$ in the temperature range where ITC peak occurs in the two systems (13).

3.14. **Dipolar point defects in other crystals**. ITC has been extended to study reorientation of simple dipolar defects in crystals with structure different from CaF$_2$ and NaCl. ITC measurements in CaX^{6+}O$_4$ (where X^{6+}=W,Mo) suggested that two X^{5+} ions (originating from a reduction process) and an O^{2-} vacancy build dipoles, characterized by ε_j=.44 and .33 eV respectively and unusually low values of τ_o ($\sim 10^{-9}$s) (12$^+$:14,15). In the tetragonal-rutile-type ZnF$_2$ doped with LiF an ITC peak (14) has been observed characterized by τ_o=1.7x10^{-13}s and ε_j=.32 eV and attributed to the dipolar relaxation induced, by the motions of interstitials Li$^+$ (i.e. Li$_i^+$) in a pair (Li$^+$-Li$_i^+$) along the c axis of the crystal.

3.15. **Dipolar relaxations induced by water molecules**. Water molecules, endowed with dipole moment, are present either in the bulk or adsorbed on the surfaces of many solids. Their reorientation and the related peaks have been monitored by means of ITC in variety of systems, see fig.11. Examples for bulk H$_2$O are given in fig.11 and are ice (15), inorganic hydrated compounds as La(SO$_4$)$_3$ (16), where peak shifts as a consequence of deuteration (curves b and c),biopolymers such as melanin (17), lysozyme (18), and keratin (19). The ITC application to biopolymers allowed to identify different water sites and organisation and to analyse the dehydration kinetics (17,18). Often broad peaks, characterised by τ distribution, see § 4.1, are meaningful for slightly different environments which water experience.

ITC peaks due to reorientation of H$_2$O molecules adsorbed on solid surfaces are displayed in fig.11, bottom for SiO$_2$ (20) and NaI, which is quite hygroscopic. This result is interesting since suggests a tool for studying adsorbed dipolar molecules, but is a restriction whenever such peaks overlap to bulk relaxations under investigation, see § 6.1 and compare b curve in fig.11, bottom, with fig.6b in the range 180-260 K.

4. ITC TO MONITOR COMPLEX RELAXATION PHENOMENA

ITC can be exploited also to monitor and analyse complex relaxation phenomena, such as processes characterized by a distribution of relaxation times or originated from polarization mechanisms, different from dipole reorientation, such as interfacial or Maxwell-Wagner polarisation and space charge. The unique requirement for ITC application is that the ionic relaxation process is thermally activated.

4.1. **Distribution of relaxation times**. In complex materials simple dipoles can experience slightly different environments or in simple materials dipolar aggregates of defects can exhibit a variety of configurations slightly different one from each other. In both cases a relaxation time distribution is expected and, as a consequence, a broad ITC peak. The change of the peak position T_M on the polarization parameters (T_p and t_p, see sect.2) is meaningful for a τ distribution, as shown in fig.12 in two ways: 1) by polarizing at a fixed T_p for different t_p's, and 2) by polarizing for a fixed short t_p at different T_p's. In both cases T_p is chosen on the low T side of the peak in order to avoid the polarisation at saturation of all dipoles, but rather to select different sets of dipoles (different τ's). As a result, the peak position T_M and amplitude j_M change, see fig.a and c, at variance with the behaviour of IV dipole peak (characterised by a single τ), whose

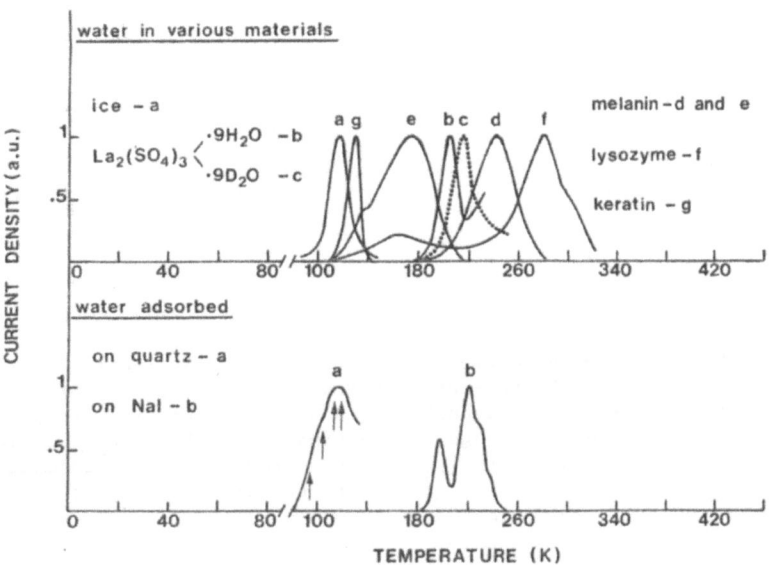

Fig.11. ITC plots induced by H_2O dipoles in various materials. Top: bulk
water; bottom: adsorbed water.

amplitude changes on changing t_p and T_p, as expected from eqs.1 and 6, but
whose T_M remains unchanged.

4.2. Maxwell-Wagner interfacial polarisation. Inhomogeneous dielectric may
give rise in the tgδ vs frequency plot to a Debye peak, characterized by a
pseudo-relaxation time τ_p, which is related to the geometrical shape, electrical conductivity and dielectric constant either of the embedded inhomogeneities and of the host medium. For ellypsoidic occlusions Sillar (5[+]:97)
showed that

$$\tau_p = \{\varepsilon_1(\lambda-1)-\varepsilon_2\} /\sigma_2 \tag{21}$$

where ε_1 and ε_2 are the dielectric constants of the host medium and of the
embedded occlusions respectively, λ is the shape factor and σ_2 is the conductivity of the occlusions, which is assumed to be much higher than that
of the host medium. If τ_p is thermally activated, an ITC peak is expected
to occur (12[+]:63,64). The polarisation is no longer due to the reorientation of simple non interacting dipoles but rather to the accumulation of
mobile carriers, for instance vacancies, at the boundary between the "conducting" occlusions and the "insulating" host matrix.

Such ITC relaxations have been found in the following systems: a) alkali halides (KCl and NaCl) doped with Pb (12[+]:63,64,10[+]:16) and Cd (5[+]:53),
where occlusions of Suzuki phase, i.e. an ordered arrangement of $(IV)_c$ dipoles in n.n.n. coordination, see fig.10, grow as a consequence of a proper thermal treatment; b) alkali halides doped with Me^{2+}, where the Cottrell atmosphere (C.A.), namely the ensemble of defects and impurities
which surrounds the dislocations can be considered as occlusions whose
electrical behaviour is different from that of the host medium (5[+]:64) see
fig.10; c) sintered materials, which are inhomogeneous, since they show

421

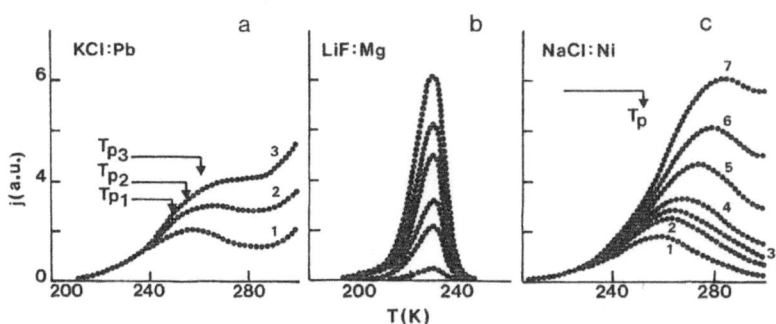

Fig.12. ITC plots to evidentiate relaxation time distribution a) KCl:Pb annealed at 150°C, role of different T_p's indicated by arrows and t_p=const=180s on Suzuki-peak, see § 4.2; b) $(IV)_c$ in LiF:Mg: the peak position doesn't change on changing either T_p or t_p (not indicated for the sake of clarity), c) aggregates in NaCl:Ni annealed at 140°C, role of different t_p's (T_p=const, see arrow): t_{p1}=1'; t_{p2}=2'; t_{p3}=3'; t_{p4}=5'; t_{p5}=10'; t_{p6}=20'; t_{p7}=30'.

grains and micropores, for instance CeO_2 (10+:18), ThO_2 (10+:19) and Si_3N_4 (10+:20). As a general feature, ITC peak related to Maxwell-Wagner (M-W) relaxation is huge and broad. In fact the peak amplitude scales with N_s, given by:

$$N_s = q\lambda^2\varepsilon_1 \cdot \{\varepsilon_1(\lambda-1)+\varepsilon_2\}^{-1} \tag{22}$$

where q is the volume fraction occupied by the occlusions. The amplitude depends not only on q, but also on the shape factor λ, which can be very large, chiefly in the case of elongated occlusions (as in the case of dislocations).

The Suzuki peak is broad and shifts by changing T_p and t_p as in the case where a τ distribution is contributing to the relaxation (see § 4.1 and fig.12a): this is consistent with the fact that the occlusions may exhibit different shapes and/or orientations with respect to the electric field, hence different λ's are expected in (21. For Suzuki phase occlusions in Pb doped KCl and NaCl, the activation energy, as obtained from the analysis of the related and properly "cleaned" (see § 2.4) ITC peak, is very close to the activation energy for the motion of "free" cation vacancy ε_{fv}, supporting the hypothesis that the carriers, responsible for the interfacial polarisation inside the occlusions, are the cation vacancies, as expected from the model (see fig.10).

All alkali halides doped with Me^{2+} investigated up to now, exhibit a huge ITC peak at temperatures higher than RT, see fig.10, which has been attributed to a M-W interfacial polarisation due to C.A. surrounding dislocations (12+:64;21). This band has peculiar features, since its position T_M depends on a variety of parameters, such as thermal history, strain and thickness of the sample, concentration of impurities and type of contacts. The most of these results are consistent with a M-W relaxation. The ITC peak is attributed to dislocations since 1) it cannot be suppressed, even by a severe sample quenching from temperatures close to the melting point; 2) it is heavily reduced by sample deformation and 3) it slowly recovers after deformation (21). The first result rules out that it is due to aggregates, the last two are accounted for as follows. The deformation sweeps the dislocations away, leaving behind their C.A. (hence impurities or vacancies cannot exploit any longer the preferential path along dislocation for dif-

422

fusing when the field is applied). After the deformation the defects migrate slowly to the bare dislocation to build around it C.A.again: so the ITC peak is recovered. An ITC peak occurring in SrF_2:La at ~ 350 K, see fig.10, was attributed (22) to dislocations, however its features have not been analysed in detail enough to confirm the attribution.

4.3. Space charge. The motion of free ionic charge carriers and their accumulation near the electrodes give rise to space charge and have been recognized as responsible for the huge and broad ITC peak occurring at high T in KCl (see fig.6b) since the early work on ITC (2). Additional examples are reported in AgCl (4^+:105,106;10^+:21), CdF_2 (4^+:95) and CaF_2 (4^+:85). In the last case a symmetric peak is found at 404 K (see fig.6c), whose amplitude exhibits only a limited range of linear dependence on the polarizing field E_p, at variance with the behaviour of ITC peaks related to non interacting dipoles (see eq.7 and fig.13a where the two behaviours are compared). The band shape is well described by a bilinear decay (4^+:85), namely the depolarisation rate is proportional to the square of the still frozen in polarisation, i.e.

$$\frac{dP}{dt} = -P^2/\tau_c(T) \tag{23}$$

at variance again with eq.3 for dipolar reorientation. The rate constant τ_c is thermally activated and characterized by a single activation energy ε_c, hence one has

$$j(T)=P_o^2\tau_{o,c}^{-1} \exp(-\varepsilon_c/kT)\{1+P_o \int_o^T \tau_o^{-1} b \exp(-\varepsilon_c/kT')dT'\} \tag{24}$$

The initial rise portion of the ITC band has the same temperature dependence of the d.c. ionic conductivity supporting the hypothesis that ITC peak is related to the long range motion of ionic carriers. The activation energy for both processes is ε_c=1.3 eV.

The thermally activated redisplacements of ionic carriers through the volume of the sample insulated from the electrodes by thin insulating layers, see § 6.2, and previously polarised, were assumed to be responsible for ITC peaks (10^+:22). In this case, at variance with what reported above, the ITC peak is again described by a first order kinetics, i.e.

$$- \frac{dP}{dt} = (f/\varepsilon_o\varepsilon_s)\sigma \cdot P \tag{25}$$

where ε_o is the permittivity of vacuum, f is a factor depending on the dielectric constants ε_s and ε_i and on the thicknesses d_s and d_i of the sample and of the insulating layers respectively; σ is the ionic conductivity. The expressions of f and σ are:

$$f = 2\varepsilon_s d_i (\varepsilon_i d_s + 2\varepsilon_s d_i)^{-1} \tag{26}$$

$$\sigma(T) = (\sigma_o/T)\exp{-\varepsilon_c/kT} \tag{27}$$

From (25 and (27 one obtains:

$$j(T)=P_o f \frac{\sigma(T)}{\varepsilon_o\varepsilon_s} \exp{-\int_o^T b^{-1} \frac{f \sigma(T')}{\varepsilon_o\varepsilon_s} dT'} \tag{28}$$

which is formally identical to the expression found for reorientation of non interacting dipoles (6, if one substitutes τ^{-1} to $f\sigma(T)/\varepsilon_o\varepsilon_s$: here the thermally activated parameter which rules the depolarisation process is the ionic conductivity rather than the relaxation time. Since $P_o=\varepsilon_o\varepsilon_1,V_p/2d_i$, the peak amplitude is again proportional to the polarisation voltage

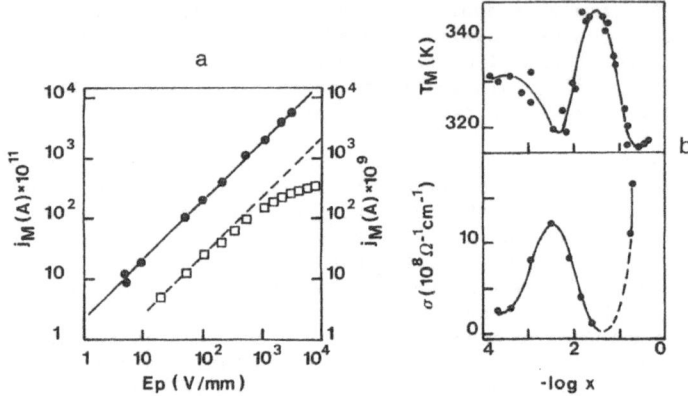

Fig.13. Space charge in fluorite-like crystals. a) log-log plot of the amplitude of space charge peak at 404 K in pure CaF_2 (open squares, ref.4[+]:85) and of dipolar peak at 187 K in CdF_2:Eu (full circles, ref.4[+]:100) vs. polarising field E_p. b) Position of the high T peak T_M (top) and conductivity σ (bottom) vs Yb concentration x in $Sr_{1-x}Yb_xF_{2+x}$, ref.10[+]:23.

V_p as for dipolar reorientation, see eqs.6 and 7, but at variance with space charge. The temperature T_M at which the maximum of the ITC peak occurs is given by:

$$T_M = \varepsilon_c/k \ \ln(fk\sigma_o/b\varepsilon_s) \qquad (29$$

i.e. it turns out that the position of the peak is function of the bulk conductivity of the sample σ_o and of the thicknesses of the sample and of the insulating layers (through f, see eq.26). This is very peculiar feature. Such kind of relaxations have been detected in fluorite type crystals heavily doped with RE, for instance in $Sr_{1-x}RE_x^{3+}F_{2+x}$, where RE was Yb,Nd,Gd (10[+]: 23,24). In $Sr_{1-x}Yb_xF_{2+x}$ the changes of T_M and σ (at constant T) versus x were found "specular", since T_M is increasing whenever σ is decreasing (as expected from eq.29), while in $Sr_{1-x}Gd_xF_{2+x}$ with $x=3.34 \times 10^{-3}$ a dependence of T_M on sample thickness was found as well. Usually the peaks which are related to this kind of relaxation occur at temperatures much higher than the dipole ones do. It should be remarked that such peaks can be detected even if bare sample is put between electrodes: air gaps which, are always present between sample and electrodes act as insulating layers, see § 6.2.

5. STUDIES OF DEFECT DYNAMICS

Since ITC provides a very sensitive method 1) to detect even low dipole concentrations, see § 2.5, 2) to put in evidence relaxations induced by small, see § 3.12, or large aggregates, see § 4.2, it gave a substantial contribution to the understanding of clustering, aggregation, nucleation and dissolution of impurities in ionic crystals, see § 5.1. Moreover it has been exploited to analyze other phenomena in which a dipolar defect concentration is expected to change as a consequence of sample irradiation with ionizing radiation or light, see § 5.2 and 5.3.

5.1. <u>Aggregation and dissolution phenomena in ionic crystals</u>. The solubili<u>ty</u> of aliovalent impurities in ionic crystals is restricted and increases with increasing temperature, see fig.14b. In doped crystals however the actual concentration of dispersed impurities (in the form of simple defects such as for instance IV and RE-FI dipoles) is not the equilibrium concentration, mainly at rather low temperatures at which the migration rate is very low. The actual concentration of simple defects depends on the total amount of doping and on the thermal history of the sample: for instance in heavily doped and quenched samples the IV (or RE-FI) dipole concentration is higher than the equilibrium concentration and few aggregates are present. Since the area delimited by ITC band is proportional to the number N_d of dipoles dispersed in the sample, see eq.10., ITC technique, due to its sensitivity, see § 2.5, is suitable for monitoring: 1) the concentration of dipoles in equilibrium (solubility limit) at different temperatures and 2) the kinetics of the isothermal processes by which the equilibrium is reached, starting from non-equilibrium configurations. By suitable thermal treatments a careful description was obtained for the solubility limit as a function of temperature (12[+]:73), the kinetics of aggregate dissolution (12[+]:72), of the clustering of dipoles, of the nucleation of the separate phase occlusions. The most of the ITC work in this field has dealt with $(IV)_c$ dipoles in alkali halides for which careful analysis of the clustering kinetics, see fig.14a, could be carried out in many systems (12). This brought to the identification of the clusters, such as dimers and trimers, see fig.10, which are formed when IV dipoles start to leave the solid solution. In some cases the originated clusters are detected directly by means of ITC technique since they are endo<u>wed</u> with dipolar moment (see § 3.12 and fig.10). The solubility limit was found to be thermally activated and independent on the total impurity concen<u>tration</u>: from the Arrhenius plot of the equilibrium concentration vs T^{-1} the dissolution energy can be easily found, see fig.14b. Depending on the total impurity concentration and temperature in some systems, for instance KCl:Pb, nucleation of a new phase (Suzuki phase) occlusions occurs at the expenses of dispersed impurities. In this case ITC is a particularly suitable technique since it can monitor from the early beginning and along the whole process (12[+]:63) either the loss of simple defects,$(IV)_c$ dipoles,and the M-W relaxation related to the growing occlusions (see § 4.2 and fig.10 and 12). From the analysis of the kinetics it has been concluded that the nucleation is a diffusion controlled process according to the Zener model and the oc-

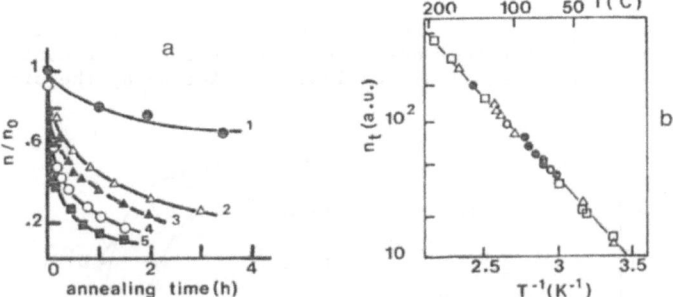

Fig.14. Clustering and dissolution of $(IV)_c$ in NaCl:Cd a) clustering kinetics: ratio of the actual concentration n and the initial one n_o (49 ppm) vs annealing time at different temperatures T_a: 1) 275 K; 2) 295 K; 3) 304 K; 4) 314 K and 5) 325 K, ref.12; b) semilog plot of thermodynamical equilibrium concentration n_t vs $10^3/T$, ref.12.

clusion shape is cylinder-like(12[+]:63).In fluorite-like crystals no detailed study of the aggregation kinetics has been performed until now. However the restricted solubility of RE-FI dipoles and hence the occurrence of aggregation processes have been put in evidence by means of ITC since 1) the ratio ρ_s between the RE-FI dipole concentration and the total RE concentration n_c decreases by increasing n_c, see fig.15 , and 2) the ratio ρ_p between the dipolar cluster concentration (as monitored by the related ITC peak, see for example fig.10) and that of RE-FI dipoles increases by increasing n_c, see fig. 15 . From fig.15 it turns out that the aggregation process is remarkable in CaF_2:Gd, since it is already operating at rather low concentrations. More detailed discussion of the above problems can be found in review paper (5,12).

5.2. <u>Reorientation of dipoles in electronic excited state</u>. The reorientation of impurity induced ionic dipoles has been extensively studied in the electronic ground state (e.g.s.) of the impurity, see sect.3. Whenever the impurity exhibits localized electronic levels in the energy gap, it is possible to study the reorientation of dipole in its electronic excited state (e.e.s.), by photon excitation. This has been previously done for F_A, M and V_k color centers by means of optical measurements with polarized light (4[+]:61). If the activation energy for the dipole reorientation in e.e.s. ε_j' is lower than that in e.g.s. ε_j, as for the above color centers, see fig.17a, the relaxation time in e.e.s. $\tau'(T)$ is shorter that that in e.g.s. $\tau(T)$, see fig.16a. When the sample is kept at rather low T_p and polarised, the orientation of dipoles in e.g.s. does not occur, but if proper light irradiation brings dipoles in e.e.s., the orientation takes place, since $\tau'(T_p)<<\tau(T_p)$ (see fig.16a). In the former case practically no ITC peak is detected while in the latter appreciable peak is monitored. In this way it was possible to observe reorientation of $(IV)_c$ in e.e.s. at low T, at which the reorientation in e.g.s. is hindered. For instance in NaCl:Pb the orientation yield of $(IV)_c$ in e.e.s. depends on the excitation wavelength (rectangules in fig.16b) and follows the Pb absorption spectrum of NaCl (dotted curve). Moreover the reorientation in e.e.s. is thermally activated. The results obtained until now and reviewed in ref.10[+]:25 point out that the reorientation parameter ε_j is deeply affected by the electronic configuration. This is furtherly supported by the observation that in KCl:Eu no preferential reorientation occurs in e.e.s. as for KCl or NaCl:Pb, since the light excitation of Eu^{2+} involves rather inner electrons (4f) in comparison with those involved in Pb^{2+}(6s). The role of electronic state on the ionic motion has been shown as well for the M-W relaxation related to the Suzuki phase in KCl:Pb (5[+]:9,10): in this case the light irradiation lowers the activation energy for the cation vacancy migration.

5.3. <u>Radiation damage</u>. Ionising radiation strongly interacts with matter producing lattice defects, for instance color centers such as F centers. The early stage of coloration in alkali halides is affected by the presence of

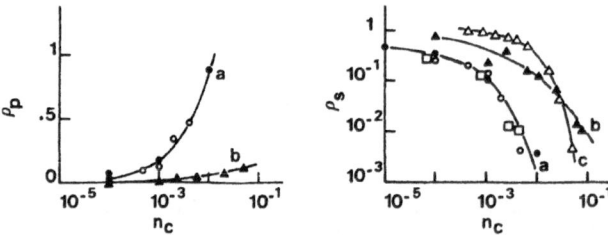

Fig.15. Aggregation phenomena in fluorite-like dihalides doped with RE. Left: plot of ρ_p vs nominal RE concentration n_c: curve a) CaF_2:Gd, ref.12[+]:82; b) SrF_2:La, ref.12[+]:86. Right: plot of ρ_s vs n_c:curve a) CaF_2:Gd; b) SrF_2:La (ρ_s given in a.u.) and c) SrF_2:Eu,ref.10[+]: 24.

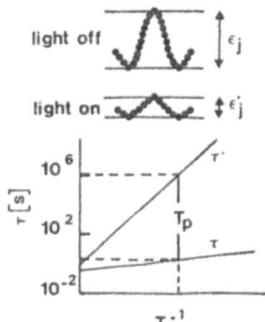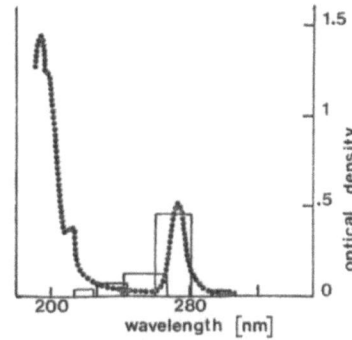

Fig.16. Orientation of $(IV)_c$ dipoles in e.e.s.. Left, top:energy barriers for dipole reorientation in e.g.s. and e.e.s.; bottom: related semilog plot of τ' and τ vs T^{-1}. Right: dotted line: optical absorption spectrum of NaCl:Pb; rectangules: orientation yield of $(IV)_c$ dipoles in e.e.s. as function of the exciting wavelength (ref.10[+:25]).

impurities. ITC measurements performed on KCl:Sr systems stressed that $(IV)_c$ concentration decreases upon irradiation (4[+:49]): the decrease was found proportional to the F center production (23) (in the range 195-293 K), see fig. 17. This suggests that $(IV)_c$ dipoles act as efficient traps for halogen interstitials, hence favoring the F coloration. ITC gave also a contribution to understand the effect in pure crystals, since dipolar defects were detected in KCl irradiated with 1.5 MeV electron (4[+:56]).

6. EXPERIMENTAL DETAILS

6.1. Experimental set-up. The experimental set-up consists of 1) the cryostat, a metal container which allows the thermal cycling of the sample, 2) the electrometer, which detects currents as weak as 10^{-15} A or lower and 3) the recorder which displays both the displacement current and the temperature signal. The sample is sandwiched between two plane parallel massive metal electrodes. The electrical insulation of the electrodes is provided by a highly insulating material flange, which should be kept far away from the electrode-sample sandwich, in order to avoid its thermal cycling, which could cause unwanted thermally induced currents which overlap the sample signal. The best insulating material is sapphire, but is rather expensive; teflon, which is cheaper and easily machineable, is currently used with good results. Sample cooling and heating is obtained in different ways 1) by connecting the sample-electrode sandwich to a cold (or hot) finger and keeping them in dynamic vacuum; 2) by exchanging heat with an external source (heater or dewar vessel filled with cryogenic liquid surrounding the cryostat) through an exchange gas (Ar, N_2). The former arrangement provides a quicker way to perform the measurements, since one has not to change the gas pressure (for the temperature control) and not to deal with unwanted impurities in the gas. The latter solution however minimizes the thermal gradients across the sample, hence allowing the most precise evaluation of the reorientation parameters ε_j and τ_o, see § 6.3. Thermal gradients can be made neglegible. Any contamination of the sample chamber by even the slightest traces of water or other dipolar molecules must be prevented in order to avoid spurious signals. Hence if exchange gas is used , it has to be purged and allowed to flow in a pipe, cooled in liquid nitrogen, in order to remove water traces. This caution is necessary, chiefly if one deals with hygroscopic materials

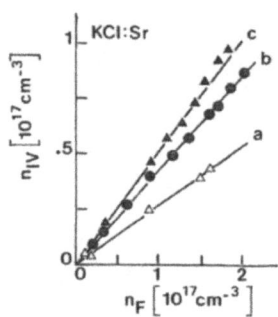

Fig.17. ITC to monitor radiation effects in KCl:Sr. Concentration of $(IV)_c$ dipoles destroyed as function of F-center produced by X rays (ref. 23): 1) "as cleaved" sample and X-irradiated at 293 K; 2) quenched sample and X-irradiated at 273 K; 3) quenched sample and X-irradiated at 293 K.

(or slightly hygroscopic as the most of alkali halides) since water molecules are adsorbed on the sample edges and give additional unwanted contributions to the ITC spectrum (see for instance § 3.15 and fig.11). Mechanical vibrations should be avoided and very good electrical shielding and grounding are required in order to lower the noise level and hence to allow the detection of very weak currents.

6.2. Samples and contacts. The sample is, as a rule, cut (or cleaved) as a thin slice, typical sizes being 1 cm^2 area and 1 mm thickness. The contacts with the electrodes can be improved either by painting colloidal graphite (or silver paint) on the wider surfaces, or by evaporating thin gold films on them. In the last case, semitransparent conducting contacts can be obtained to allow optical measurements on the same sample, see § 5.3. The reproducibility of the measurements is, in this way, very good, better than 2%. Insulating or blocking contacts are used for ITC peak diagnostics. Two thin layers of insulating material (teflon, mylar) of thickness d_i and dielectric constant ε_i are introduced between electrodes and sample, whose thickness and dielectric constant are d_s and ε_s. The j signal is reduced by a factor ρ (26):

$$\rho = (1+2\varepsilon_s d_i/\varepsilon_i d_s)^{-2} \qquad (30$$

Such a reduction is expected also whenever a bare sample is introduced between electrodes: since the crystalline sample surfaces are never ideally flat, between sample and electrodes air gaps occur, which act as insulating layers. From eq.30 it turns out that a severe signal reduction takes place whenever thin samples of high ε_s are considered. If as a consequence of the application of insulating contacts, the signal is reduced much more than expected from eq.30, one has to suspect that carrier injection is responsible for the ITC peak, as for instance in the case of CdF_2:Gd $(4^+$:99).

6.3. Influence of heating rate. As it turns out from (8 and (9 the heating rate b affects both the ITC peak amplitude j_M and position T_M: both increase by increasing b. This is shown in fig.18 either for $(IV)_c$ dipole peak, see § 3.1, and for M-W relaxation peak related to dislocations, see § 4.2. The values of j_M and T_M normalized j_M^* and T_M^* (the values for b^*=.1 K/s) are

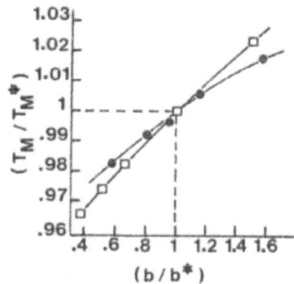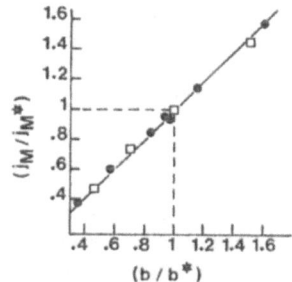

Fig.18. Influence of heating rate on T_M (left) and on j_M (right) vs normalized heating rate b/b^*, for $b^*=.1$ K/s: □ related to $(IV)_c$ peak in AgCl:Ni (ref.4[+]:105); ● related to M-W dislocation peak in KCl:Eu.

plotted vs the ratio between the actual b and b^+. The effect on T_M are moderate (less than 1% for 20% change in b), more appreciable in the case of j_M. In both cases the T_M and j_M changes are directly proportional to the b changes in a rather wide range ($\pm 20\%$). This result is useful for correcting the experimental data, whenever unwanted b changes occur among measurements which have to be compared. Anyway it is better to use rather low values of b in order to avoid additional thermal gradients across the sample (see also § 6.2) since they affect badly the evaluation of ε_j and τ_o. If gradients occur, τ_o may assume unusually low values, which are not consistent with the lattice dynamics. For the sake of example let us quote and compare results for RE-FI dipoles in CaF$_2$:Er obtained by different authors:

Ref.	T_M(K)	ε_j(eV)	τ_o(s)	b(K/s)	exchange gas in the cryostat
24	126	.406	4.1×10^{-15}	.05	yes
4[+]:89	147	.55	3.5×10^{-18}(!)	high[(+)]	no

The unusual figures reported above for Er, extend to a wide series of impurities (24;4[+]:89).

6.4. Dipole concentration determination. Eq.10 provides a way for evaluating N_d from the area A under the ITC peak. However the absolute determination of N_d depends also on the accuracy by which α, p, T_p and E_p are known. α and p are easily evaluated if the dipole model is known and if the unperturbed point ion (UPI) approximation is accepted, namely if the polarisation and relaxation of the surrounding lattice are neglected. In few cases p has been determined experimentally by investigating the electric field effects on ESR, for instance for RE-FI dipoles in CaF$_2$ and SrF$_2$ doped with Gd (5[+]:14; 12[+]:104). In CaF$_2$ the observed p is .77 of the UPI value, suggesting a strong inward relaxation around Gd^{3+} and a possible shift of the charge distribution around type I complex. Very recent calculations (25) point out a reduction of the UPI value up to 30% for the dipole moment of $(IV)_c$. In some cases the local field E_L, as calculated from the Onsager formula has been substituted for E_p in (10). Also the role of T_p has been considered, by using the effective polarisation temperature T_{eff} (5[+]:15): if T_p is close to T_M and the sample is then cooled quickly, T_{eff} coincides with T_p. The correct evaluation of A depends on the contacts, see § 6.3: gold evaporated electrodes give satisfactory results. As for many other techniques, ITC suffers limitations in the absolute concentration determination, unless calibration is performed on

((+) not quoted, but probably very high as can be guedded from the higher T_M value.

systems with a well known dipole concentration. However when relative changes of N_d are required as in aggregation, solubility and radiation damage studies, see sect.5, the ITC is very fruitful due to its sensitivity and reliability, see § 2.5.

7. ITC AS A TOOL FOR DEFECT SPECTROSCOPY

It has been shown that ITC can be applied successfully to a variety of materials and defects. In fact it allows 1) to identify simple point defects endowed with dipole moment (see sect.3 and figs.5,6,8,10); 2) to obtain with high accuracy their reorientation parameters ε_j and τ_0, see fig.7 and sect.3, and 3) to get a deeper insight of the microscopic mechanisms ruling their reorientation, see sect.3 and § 5.2. Moreover it allows the detection of larger defects such as dipolar clusters, see § 3.12, extended aggregates as in the case of occlusions of separate phase, for instance Suzuki phase, see § 4.2, the Cottrell atmosphere surrounding dislocations, see § 4.2. In this frame fig.10 shows how ITC can be regarded as ionic defect spectroscopy, since it displays in different systems peaks due to defects of increasing complexity. For such more complex defects the operating mechanism is no longer the dipolar reorientation but for instance the interfacial polarisation release, see § 4.2, which can be monitored by means of ITC, being thermally activated. A further contribution brought by ITC to the understanding of lattice defect nature is represented by the dynamical studies of defects in which the dipole concentration is modified as a consequence of thermal treatments radiation damage, light irradiation, see sect.5.

+ For space saving, reference quoted in the present paper, for an example, as (4+:5) means: see ref.5 quoted in ref.4 of the following list.

8. REFERENCES

1. C.Bucci and R.Fieschi - Phys.Rev.Lett. 12, 16 (1964).
2. C.Bucci, R.Fieschi and G.Guidi - Phys.Rev. 148, 816 (1966).
3. A.J.Dekker - Solid State Physics - Prentice Hall, Inc. 1958.
4. R.Capelletti and R.Fieschi - Intern.Symp.on Electrets and Dielectrics. Ed.Academia Brasileira de Ciencias - Rio de Janeiro 1977, p.131.
5. R.Capelletti, R.Fieschi, A.Gainotti, C.Mora, L.Romanò and E.Zecchi - Defects in Insulating Crystals. Ed.V.M.Tuckevich and K.K.Shvarts - Springer Verlag - Berlin 1981 (ISBN 3-540-10782-7) p.675.
6. F.Cussò and F.Jaque - J.Phys.C:Solid St.Phys. 15, 2875 (1982).
7. D.A.Golopentia and L.M.Slifkin - phys.stat.sol.(a) 72, 123 (1982).
8. Ch.Kokott and F.Fischer - phys.stat.sol.(b) 106, 141 (1981).
9. M.SiuLi, M.De Souza and S.E.Kapphan - phys.stat.sol.(b) 112, 685 (1982).
10. R.Capelletti, R.Fieschi, G.Lenzi, M.Manfredi, C.Mora and R.Reverberi - Proc.5th Intern.Simp.Electrets, Heidelberg 1985 (Avail.from IEEE,NY), p.463.
11. R.Capelletti, M.G.Bridelli, M.Friggeri, G.Ruani, I.Foldvari, L.Kovacs and A.Watterich - ibid. p.294.
12. R.Capelletti - Radiation Effects 74, 119 (1983).
13. P.Aceituno and F.Cussò - Phys.Rev. 13, 7577 (1982).
14. T.Roth - J.Appl.Phys. 44, 1056 (1973).
15. P.Dansas, S.Mounier, P.Sixou - C.R.Acad.Sc.Paris 267B, 1223 (1968).
16. F.Rull, L.F.Sanz and J.A.de Saja - J.Electrostatics 8, 221 (1980).
17. M.Bridelli, R.Capelletti and P.R.Crippa - Bioelectrochem.and Bioenerg. 8, 555 (1981).
18. M.G.Bridelli, R.Capelletti, G.Ruani, A.Vecli - Proc.5th Intern.Symp.Electrets, Heidelberg 1985 (Avail.from IEEE,NY), p.831.
19. J.L.Leveque, J.C.Garson and G.Boudouris - Biopol. 16, 1725 (1977).

20. F.Ehrburger and J.B.Donnet - J.Appl.Phys. 50, 1478 (1979).
21. R.Capelletti, A.Gainotti and M.Suszynska - Proc.5th Intern.Symp.Electrets, Heidelberg 1985 (Avail.from IEEE,NY), p.151.
22. N.Suarez, E.Laredo, D.Figueroa and M.Puma - Radiation Effects 75, 105 (1983).
23. J.N.Marat-Mendes and J.D.Comins - J.Phys.C. 37, C7-132 (1976).
24. J.H.Crawford,Jr. and E.L.Kitts,Jr. - in Recent Advances in Science and Technology of Materials - ed.A.Bishay - vol.I, p.15.
25. P.B.Fitzsimons and J.Corish - private communication.
26. P.Müller - phys.stat.sol.(a) 67, 11 (1981).

APPENDIX

COHERENT DIFFUSE NEUTRON SCATTERING STUDY OF VACANCY CORRELATIONS IN

YTTRIA-STABILISED ZIRCONIA

M.A.Hackett*, N.H. Andersen+, K.Clausen+, W. Hayes*
M.T. Hutchings$, J.E. Macdonald* and R. Osborn*

*Clarendon Laboratory, Parks Road, Oxford,OX1 3PU, UK
+Dept.of Physics,Riso Nat.Lab,DK-4000,Roskilde,Denmark
$Materials Physics & Metallurgy Division, AERE,Harwell
 OX11 ORA, UK

Following our investigation of the fast ion conducting phase of pure crystals with the fluorite structure[1], we have begun a study of the vacancy doped system: yttria-stabilised zirconia. Pure zirconia (ZrO_2) is monoclinic at room temperature but doping with trivalent oxides (e.g. Y_2O_3) stabilised a cubic fluorite structure, with vacancies at some anion sites. There is widespread interest in the nature of vacancy correlations and their effect on high temperature transport properties. In particular, efforts have been made to understand why there is a maximum in the ionic conductivity with increasing vacancy concentration.

We have measured the room temperature elastic coherent diffuse neutron scattering in single crystals of 9.4, 12,15, and 18 mol % Y_2O_3 - stabilised ZrO_2 and the high temperature diffuse quasielastic scattering for the 9.4 mol % sample[2,3]. The measurements suggest three principal contributions to the scattering:-
  (i)   In the 9.4 and 12 mol % Y_2O_3 samples: vacancy free tetrahedrally distorted regions.
 (ii)   Large, relatively stable vacancy aggregates.
(iii)   Small dynamic clusters or single vacancies.
The analysis provides a simple explanation for the maximum in ionic conductivity with increasing vacancy concentration.

References

1. M.T. Hutchings, K. Clausen, M.H. Dickens, W. Hayes, J.K. Kjems, P.G. Schnabel, C. Smith, J. Phys. C. 17, 3903-40, (1984).
2. R. Osborn, N.H. Andersen, K. Clausen, M.A. Hackett, W. Hayes, M.T. Hutchings and J.E. Macdonald, Mat.Sci.Forum 5 (1985) (to be published).
3. N.H. Andersen, K. Clausen, M.A. Hackett, W. Hayes, M.T. Hutchings, J.E. Macdonald and R. Osborn, Fast Ion and Mixed Conductors, Proc. 6th Intl. Symp. on Mater. Science, Riso (1985) p 279.

435

STRUCTURAL DEFECTS IN THE NH_4CdCl_3-TYPE TERNARY HALIDES

D. Visser

Inorganic Chemistry Laboratory
University of Oxford
South Parks Road, Oxford, OX1 3QR, UK

Recently it has been shown that the NH_4CdCl_3-structure type (1) occupies a central position in the group of ABX_3 ternary halides, A = alkali metal, In, Tl or NH_4^+, B = first-row transition metal, Mg, Cd or Sn and X = Cl, Br or I (2). Polymorphism between the NH_4CdCl_3-type and the other two structure types of this family, the hexagonal and cubic perovskite type, has been found in several cases and is related to the boundaries of the structure-type areas of their structure-field maps.

For some compounds, like $RbCdCl_3$ (3), the existence of the NH_4CdCl_3-structure has been related to the presence of H_2O in the system which means that the NH_4CdCl_3-type compounds are slightly different in composition, non-stoichiometric or have extended structural defects.

Till now three types of defect have been identified:

1. The intergrowth of a hydrate in NH_4CdCl_3 based on a close structural relation of the two crystal structures as observed in $KCdCl_3.H_2O$ and $KCdCl_3$ (4).

2. The replacement of the X-anions on the outside of the double-chain by H_2O or OH^- ions. In this case the possible accommodation is based on the structural relation between the $AX_2.H_2O$ hydrates (5,2) and NH_4CdCl_3. Examples can be found in $TlMnCl_3$ and $KMnCl_3$ (6,2).

3. The third type of defect has been observed in samples of Mn^{2+} doped $RbCdCl_3$ and $RbCdBr_3$ grown from solution. Here Mn^{2+} does not substitute for Cd^{2+} but is incorporated on a perfect cubic-site: the EPR spectra have 6 rotation independent resonance lines. Electron diffraction shows that the defects are located along the double-chain direction and not in the expected ac plane like in (4).

These observations indicate that in certain cases the NH_4CdCl_3 structure may only be incorporation of traces of water. The existence of stoichiometric NH_4CdCl_3-phase is still debatable.

References

1. M.M. Robies & C.J. de Ranter, Acta Cryst A34, 3057 (1978).
2. D. Visser, in preparation.
3. P. Bohac, A. Gäumann & H. Arend, Mat.Res.Bull.8, 1299 (1973).
4. M. Ledesert & J.C. Monier, Acta Cryst B40, 73 (1984).
5. H. Leligny & J.C. Monier, Acta Cryst B30, 305 (1974).
6. A. Horowitz, M. Amit, J. Makovsky, L. Ben Dor & Z.H. Kalman, J. Solid State Chem. 43, 107 (1982).

THE STRUCTURE OF 2-AMINO 3-METHYL PYRIDINE AMMINE CADMIUM

TETRACYANONICKELATE

T. Hökelek, D. Ülkü

Department of Physics Engineering
Hacettepe University
Beytepe-Ankara, Turkey

The crystal structure of $Cd(NH_3)(C_6N_2H_8)Ni(CN)_4$ was determined from three dimensional intensity data measured with MoK_α radiation on a Nonius CAD-4 diffractometer. The crystals are orthorhombic, space group $Pna2_1$ with a = 13.535 (1)Å, b = 13.607 (1)Å, c = 7.645 (1)Å and z = 4. The parameters obtained by direct methods using SHELXS-84 (Sheldrick, 1984) were full-matrix least-squares refined to an R value of 0.034 for 1632 independent reflections. 81 reflections were considered to be unobserved. The structure consists of corrugated polymeric two-dimensional networks made up of tetracyanonickelate ions coordinated to cadmium. The bending in the networks occurs because of a departure of the Ni-C-N-Cd sequence of atoms from linearity at the carbon and nitrogen positions. Similar corrugated polymeric layers are also observed in some other related compounds (Rayner and Powell, 1952; Ülkü, 1975; Büyükgüngör and Ülkü, 1983). The coordination around cadmium is a slightly distorted octahedron involving four cyanide nitrogens and the nitrogens of the ammonia group as well as the pyridine derivative. The average distance from cadmium atom to these six neighbours is 2.36Å. The 2-amino 3-methyl pyridine and ammonia molecules bound to cadmium in trans positions project from both sides of the network. The projections of one network fit into the spaces of the other, so that the layers are arranged as closely as possible. We thank Prof. C. Krüger and his colleagues of Max-Planck-Institut für Kohlenforschung for their intensive help during the International Summer School on Crystallographic Computing 1984.

References

Büyükgüngör, O. and Ülkü, D. (1983). In prep.
Rayner, J.H. and Powell, H.M. (1952). J. Chem. Soc. 42, 319-328.
Sheldrick, G. (1984). Private communication.
Ülkü, D. (1975). Z. für Kristallogr. 142, 271-280.

PLASTIC DEFORMATION OF ENERGETIC MATERIALS

J.C. Miller, H.G. Gallagher and J.N. Sherwood

University of Strathclyde
Dept. of Pure and Applied Chemistry
Glasgow G1 1XL, UK.

The mechanical and chemical properties of single crystals are influenced by the defects present in the bulk of the crystal. This is particularly relevant to energetic materials. The basic mechanism for the impact initiation of explosion involves the formation of very high temperatures within a specific region ('hot spot') in the crystal. One suggestion is that localised heating occurs due to the collapse of a dislocation pile up in an avalanche mode. However, little is known of the properties and characteristics of mechanically induced dislocations in this type of material, so we have commenced a detailed analysis of the defect properties of two secondary organic explosives; cyclotrimethylene trinitramine (RDX) and pentaerythritol tetranitrate (PETN).

Microhardness indentation and solvent etching indicates that slip is the major mode of deformation for both materials. A correlation of the anisotropy of the surface hardness with a theoretical analysis of the orientation dependence of the Effective Resolved Shear Stress (ERSS), acting on the bulk of the crystal, has enabled us to define the slip systems {110} <111> for PETN and {021} [100], (010) [001] for RDX.

Tensile straining in conjunction with X-ray topography has also been used, initially on PETN, for detailed examination of the dislocation configurations present during deformation.

438

ROLE OF WATER AND IRON ON OPTICAL ABSORPTION OF SODIUM DISILICATE

GLASSES

M. Friggeri[*], A. Montenero[*], R. Capelletti[+], L.D. Pye[$]

[*]Istituto di Strutturistica Chimica
v. M. D'Azeglio 85, 43100 Parma, Italy
[+]Dipartimento di Fisica,v.M.D'Azeglio 85, 43100 Parma
[$]Institute of Glass Sciences & Engineering, N.Y.State
College of Ceramics,Alfred University, Alfred, N.Y. USA

The coordination of iron ions in $Na_2O-2SiO_2$ glasses plays a key role in determining the mechanical and chemical stability of such materials, which are good candidates as nuclear waste glasses. For this purpose we investigated at RT the optical absorption spectra of $Na_2O-2SiO_2$ glasses containing different concentration of Fe^{2+} and Fe^{3+}, in the energy range 0.22 - 3.4 eV.

In the i.r. region (0.22 - 0.5 eV) three main bands are observed peaking at 0.42, 0.35 and 0.25 eV. The first two broad bands are tentatively attributed to water, which was not completely removed during the glass melting, while the sharp rise at E<.22 eV is due to the Si-O vibrations.

In the n.i.r. region (0.5 - 1.7 eV) two bands, again broad, appear peaking at 0.62 eV and 1.2 eV: their intensity exhibit nearly a linear dependence on the Fe^{2+} concentration. The 1.2 eV-band arises from a crystal field transition in Fe^{2+} sitting in an octahedral coordination. However, the peak position shift to lower energy (from 1.3 to 1.1 eV) on increasing the Fe^{2+} concentration and the asymmetric band shape suggest that different environments of octahedral coordinated Fe^{2+} are contributing to the absorption at 1.2 eV. The 0.62 eV-band can be tentatively attributed to crystal field transition of Fe^{2+} in tetrahedral coordination: it cannot be related to any vibrational overtone of water, as reported in literature, since it scales with the Fe^{2+} concentration.

In the visible region (1.7 - 3.45 eV), the presence of tetrahedral coordinated Fe^{3+} is monitored by shoulders overlapping the steep rise of the charge transfer absorption of the glass. On these bases we cannot however exclude the presence of Fe^{3+} in six-fold coordination.

ENDOR INVESTIGATION OF V^{3+} in GaAs

J. Hage, J.R. Niklas and J.-M Spaeth

University of Paderborn
Fachbereich Physik, Warburger Str. 100 A
4790 Paderborn, FRG

Vanadium can be used instead of chromium to grow semi-insulating GaAs. The observed ESR due to V in GaAs was assigned to an isolated V^{3+} ($3d^2$) centre with S = 1 [1]. In an ENDOR investigation we could resolve the superhyperfine and quadrupole interactions with the central ^{51}V nucleus (I = 7/2) and with several shells of neighbour nuclei. ENDOR lines were found in the frequency range between 1 MHz to 180 MHz at T = 3.5 K. The lines due to the ^{51}V nucleus show no angular dependence in agreement with a cubic symmetry of the centre. The isotropic ^{51}V hyperfine constant (a/h (^{51}V) = 165.3 MHz)) derived from the ENDOR lines is consistent with the ESR line shape only for S = 1, proving that the electronic structure is indeed V^{3+} ($3d^2$).

There are many lines due to the nearest As nuclei with a very complicated angular dependence. The reason for this is a large quadrupole interaction, which gives also rise to forbidden $\Delta m_I = 2$ transitions in the ENDOR spectra.

The vanadium resides either in a Ga substitutional or in the interstitial site with T_d symmetry surrounded by 4 equivalent nearest As nuclei. The interpretation of the quadrupole interaction gives no conclusive information about the lattice site of the vanadium centre. The obtained spin density, however, points to a substitutional position of the vanadium.

References

1. U. Kaufmann, H. Ennen, J. Schneider and R. Wörner, Phys. Rev. B25, 5598 (1982).

ODMR AND ODENDOR INVESTIGATIONS OF ANTISITE DEFECTS IN GaAs

D.M. Hofmann, B.K. Meyer and J. -M. Spaeth

University of Paderborn
Experimentalphysik, Warburger Str. 100 A
4790 Paderborn, FRG

The ESR of paramagnetic $AsAs_4$ antisite defects in semi-insulating (s.i.) GaAs can be measured with high sensitivity by optical detection, monitoring the microwave-induced change of the magnetic circular dichroism (MCD) of the optical absorption bands discovered to peak at 1.05 eV and 1.29 ev[1]. By simultaneous application of rf radiation, electron nuclear double resonance (ODENDOR) can be measured in which ligand hyperfine interactions are resolved[2]. In s.i. as-grown doped and undoped GaAs regular and distorted antisite defects with inequivalent four nearest As-ligands are observed. In p-type GaAs:Zn the double donor $AsAs_4$ is not in the ESR/MCD active charge state since the midgap level is empty. In a double beam experiment the empty level is populated by exciting electrons from the valence band with light into it. The rise of the MCD signal is monitored as a function of the exciting photon energy. With this technique the two donor levels of the $AsAs_4$ could be determined to be at $E_v + 0.52$ eV for $D^{+/++}$ and at $E_v + 0.74$ eV for $D^{0/+}$.

References

1. B.K. Meyer, J.-M. Spaeth and M. Scheffler,
    Phys. Rev. Lett. 52, 851 (1984).
2. D.M. Hofmann, B.K. Meyer, F. Lohse and J.-M. Spaeth,
    Phys. Rev. Lett. 53, 1187 (1984).

ENDOR INVESTIGATION OF THERMAL DONORS IN SILICON

J. Michel, J.R. Niklas and J.-M. Spaeth

University of Paderborn
Fachbereich Physik, Warburger Str. 100 A
4790 Paderborn, FRG

In Czochralski grown silicon is a high concentration of oxygen (10^{18} cm^{-3}). Heat treatment at 460°C leads to the formation of oxygen-related donors, which are called thermal donors (TD). It is believed that the TD's consist of oxygen clusters, but so far no structure model for these defects could be established.

Depending on annealing time several ESR lines appear[1]. The ESR line "NL8" was ascribed to be due to a thermal donor. It shows no shf structure and is 0.4 mT wide. This centre was investigated by electron nuclear double resonance (ENDOR). Only the superhyperfine (shf) interactions with Si^{29} nuclei could be resolved because the natural abundance of the magnetic oxygen isotope O^{17} is with 4.10^{-2} % too low.

Between 2.8 and 8 MHz many ENDOR lines were found at T = 25 K. The angular dependence of most of the lines shows that the anisotropic shf interaction, especially of the lines at high frequencies, is very weak (~ 1% of the isotropic interaction). Therefore, the variation of the wavefunction of the TD must be very small across the measured Si atoms.

In earlier ESR experiments[1] a shift of the g-values of NL8 up to very long annealing times (~ 200 h) was observed. The ENDOR experiments show, however, that NL8 contains several centres, which develop during the heat treatment already at much lower annealing times (below 2 hours). We suppose that the TD's have all essentially the same inner cluster structure and that they differ in the number of added oxygen atoms in the outer region.

References

1. S.H. Muller, M. Sprenger, E.G. Sieverts and C.A.J. Ammerlaan, Solid State Comm. 25 (1978), 987, and S.H. Muller, Dissertation, Amsterdam, 1981.

COMBINATIONS OF TECHNIQUES: OPTICAL ABSORPTION, IONIC THERMOCURRENT

RAMAN SPECTROSCOPY IN RADIATION DAMAGE OF ALKALI HALIDES

J.D. Comins

Department of Physics
University of Witwatersrand
Johannesburg, South Africa

Studies of irradiation damage processes in alkali halides will be described in which combinations of techniques have provided useful insights into the processes and defect structures involved. The use of optical absorption, defect growth and annealing kinetics and ITC measurements have led to a new understanding of the early stage F-centre production process[1-4]. In particular, the well-known square root dependence of the early stage F-centres on divalent cation impurity concentration follows directly from a model of F-centre production involving trapping of H-centres at I.V. dipoles. The dynamic equilibrium established between defect creation, annihilation, trapping and de-trapping is demonstrated.

The structures of trapped halogen interstitial defects have been studied using both optical absorption and Raman scattering, during growth and annealing processes[5,6]. The results show a variation of defect structure with irradiation temperature. The structures of various halogen interstitial clusters have been identified. Defect annealing experiments show correlations between the Raman spectra associated with the trapped halogen interstitials and the defect absorption spectra[7].

References

1. J.N. Marat-Mendes and J.D. Comins, Cryst. Lattice Defects 6, 141 (1975).
2. J.N. Marat-Mendes and J.D. Comins, J.Phys. Chem. Solids 38, 1003 (1977).
3. J.N. Marat-Mendes and J.D. Comins, J. Phys. C. 10, 4425 (1977).
4. J.D. Comins and B.O. Carragher, J. Phys.(Paris) 41, C6-166 (1980); Phys. Rev. B24, 283 (1981).
5. S. Lefrant and E. Rzepka, J. Phys. C. 12 L573 (1979).
6. J.D. Comins, A.M.T. Allen, P.J. Ford and D.A. Matthews, Radiation Effects 72, 107 (1983).
7. A.M.T. Allen, J.D. Comins and P.J. Ford, J. Phys. C. 18, 5783 (1985).

INFLUENCE OF RADIATION INTENSITY AND LEAD CONCENTRATION ON THE ROOM

TEMPERATURE COLOURING OF KBr and NaCl

Carolina Medrano P.

Institute of Physics, U.N.A.M.
P.O. Box 20-364
01000 Mexico, D.F.

First stage of F-center colouring in quenched samples of lead doped KBr and NaCl has been investigated at room temperature as a function of impurity concentration and dose rate using optical absorption, photoluminescence and ionic thermocurrent techniques. The colouring increases monotonically with concentration, with exposure of the crystals to a radiation intensity of 10 Roeng/min, and it was found that the amount of first stage colouration is proportional to the square root of impurity concentration in agreement with the theoretical model recently proposed[1]. If a higher radiation intensity is used, (50 Roeng/min), the colouring increases only at the very beginning of the first stage and more couloring is produced during stage II of the F-centre growth curve for the lower concentration. For NaCl the situation is completely reversed. A possible explanation is given in terms of the solubility of the impurity into the matrix, room temperature annealing of the samples and impurity precipitation induced by irradiation.

Reference

1. S. Ramos B., J. Hernandez A., H. Murrieta S.,
   J. Rubio O., F. Jaque. Phys. Rev. B 31, 8164 (1985).

STOICHIOMETRY AND OH⁻ DEFECTS IN $LiNbO_3$ CRYSTALS STUDIED BY

INFRA RED SPECTROSCOPY

L. Kovacs

Research Laboratory for Crystal Physics of the Hungarian
Academy of Sciences, 1502 Budapest, POB 132, Hungary

Due to its excellent piezoelectric and optical properties lithium niobate has found a wide variety of practical applications in acoustics and optics. The properties used in the various applications can be influenced by stoichiometry and OH⁻ defects.

High quality single crystals of $LiNbO_3$ are usually grown from a congruent melt composition (48.6 mole % Li_2O, 51.4 mole % Nb_2O_5). Different models were presented to explain the defect structure responsible for the deviation from stoichiometry in the crystal. Our density measurements as a function of Li/Nb mole ratio verify the model where excess niobiums are charge compensated by lithium vacancies.

Protons in the form of OH⁻ ions can be incorporated into $LiNbO_3$ crystals during the growth process. The infra red absorption band near 2.87 μm due to the stretching vibration of the OH⁻ defects has been studied as a function of temperature, polarization and stoichiometry. The weak temperature dependence implies that the OH⁻ vibrations are not strongly coupled to phonons. According to optical anisotropy measurements one can conclude that the O-H bonds are situated in the oxygen planes perpendicularly to the crystallographic c axis. The structure of the band found in nearly stoichiometric crystals indicates three slightly different proton sites which can be explained on the basis of the $LiNbO_3$ crystal structure.

The OH⁻ absorption band has been studied in crystals doped with MgO as well. Above a given concentration of magnesium depending on the Li/Nb ratio the 2.87 μm band disappears and a narrower band peaking at 2.83 μm occurs indicating an energetically more favourable site for the proton.

UNUSUAL EPR SPECTRA OF $Mn^{2+}Zn(NH_4)H\ C_2O_4 \cdot 1/2\ H_2O$ SINGLE CRYSTALS

B. Aktas, M. Korkmaz

Hacettepe University
Engineering Faculty Dept. of Phys.
Ankara, Turkey

Ammonium hydrogen oxalate single crystals doped with Mn^{2+} ions have been studied by the EPR technique. The crystals have been grown by slow evaporation of ammonium hydrogen oxalate solution prepared by mixing equivalent amounts of ammonium oxalate and oxalic acid solutions.

EPR spectra have been taken with a Varian E-9 in. spectrometer over the temperature range 77-450°K. Mn^{2+} ions enter the lattice substitutionally and give two equivalent magnetic sites. The spectra are very complex at ordinary orientation of the magnetic field. However, they are similar and resoluble at the special directions corresponding to principal Z-axes of the two magnetically equivalent sites of Mn^{2+} ions. At these directions, $M_s = -5/2$, $-3/2$ electronic transitions cannot be observed since the fine structure parameter D is relatively large. In the Y-principal directions, which are the same for the two sites, spectra are very clear and $M_s = -3/2$, $-1/2$ transitions cannot take place owing to the large fine structure parameter D.

The spectra have been evaluated by a computer program written in FORTRAN IV language to determine the spin Hamiltonian parameters which are given below.

$D = 1350$ (G), $E = 25$ (G), $g = 2.0006$, $g_y = 2.0012$, $g_z = 1.9947$, $A_x = 92.4$ (G), $A_y = 94.1$ (G), $A_z = 89.9$ (G).

The fine structure parameters increase while temperature decreases and no phase transition has been observed during the temperature change. It has experimentally been found that crystal electric fields of the Mn^{2+} sites are of tetragonal structure which are slightly distorted towards rhombic symmetry.

CATION VALENCE DISTRIBUTION IN THE TERNARY-OXIDE SPINEL, $NiMn_2O_4$

M.S. Islam[*] and C.R.A. Catlow[+]

* Chemistry Department, University College London
  20 Gordon Street, London WC1H OAJ, UK.
+ Chemistry Department, University of Keele
  Staffordshire, ST5 5BG, UK

We present a theoretical study of the cation distribution in the spinel $NiMn_2O_4$, a material of technological importance owing to its NTCR (negative temperature coefficient of resistance) property and use in thermistor devices. The study is based on computer simulation techniques, which have previously been used with success for binary oxides and now extended to the treatment of ternary oxides. An investigation of the relative energetics of the normal and inverse structures of $NiMn_2O_4$ is carried out (and in parallel a similar survey of MnX_2O_4 and NiX_2O_4, X = Al,Cr,Fe). Both electrostatic and short-range energies are considered as well as crystal-field terms. A crucial feature of our approach is the inclusion of the dependence of short-range potentials on coordination number. By including this latter effect there is good agreement between theory and the experimentally observed cation distributions. We proceeded with an investigation of the cation valencies in $NiMn_2O_4$. Our calculations show that only trivalent manganese is present, and the formation of Mn^{2+} and Ni^{3+} can be thermally activated. The results suggest that the observed semiconducting behaviour can be explained by the "hopping" of electrons from Mn^{2+} to Mn^{3+} at octahedral sites.

RECENT SIMULATION STUDIES OF ZEOLITES AND CLAYS

R.A. Jackson

Department of Crystallography
Birkbeck College, University of London
Malet Street, London WC1E 7HX, UK

Two contrasting aluminosilicate systems have been studied: zeolites, which are framework structures, and clays, which are layer structures.

It has been found that the experimentally determined structure of a given zeolite may be very successfully reproduced using a potential model which includes a bond bending term to take some account of three-body interactions. For structural studies, ionic polarisability does not seem to be important, and rigid ion models may be used.

For clays, however, the situation is different. Calculations similar to those that have been so successful for zeolites fail, apparently because they do not take proper account of interlayer interactions. Preliminary calculations, using a shell-model description of oxygen ions, have been more successful, indicating that ionic polarisability is important in this case.

Calculations are also in progress, using Hartfree-Fock SCF methods, to calculate potentials to describe ion-water interactions. These will enable hydration to be simulated in zeolite and clay minerals.

A COMPARISON OF DEFECT ENERGIES OBTAINED FROM FULLY AND PARTIALLY IONIC POTENTIAL MODELS OF OLIVINE (Mg_2SiO_4)

M. Doherty

Geology Department
University College London
20 Gordon Street, London WC1H 0AJ, UK

We have used and compared five different potentials, in a 2-body rigid-ion model of Olivine ($MgSiO_4$). This has shown that all models can obtain 'reasonable' values for the various physical properties; however, the more recent the model, the more accurate it tends to be.

Of these five models, three were fully ionic, and two were partially ionic, and the model of each type which fit the best to these properties do so equally well. The study was therefore continued, to defect properties, i.e. vacancy, interstitial, Schottky and Frenkel Pair energies. We find in each case that the fully ionic model gives a better fit to the experimental data than does the partially ionic model. This has been assumed to be due in part to the inability of this type of model to take accounbt of the non-central force component of covalence, as covalence is assumed to play a large part in the bonding of these minerals.

Finally, the fully ionic potential above was used to determine Arrhenius activation energies for an interstitialcy and vacancy mechanisms. The interstitialcy mechanism has been identified with the intrinsic regime, the vacancy mechanism with the extrinsic regime. We will compare these values with experimental values, and will show the following deviations from experiment, the calculated activation energy for:-

        the interstitialcy mechanism is 20% higher,
        the vacancy mechanism is   35% lower,
        the Frenkel Pair energy is   20% lower.

In conclusion, we feel that this study has shown that better results are obtained via this approach if a fully ionic rather than a partially ionic model is used. The next step is to perform the same study with a 3 body shell model.

A MONTE CARLO SIMULATION OF PRECIPITATION PHENOMENA IN Si:P

G. Ruani[*], A. Desalvo[+]

* Scuola di Perfezionamento in Fisica dello Stato
  Solido, Dipartimento di Fisica, Parma, Italy
+ Istituto Lamel-CNR-Bologna

A Monte Carlo computer code for the precipitation phenomena simulation in binary alloys with nearest-neighbour (n.n.) interactions has been set up for square, cubic and diamond lattices[1]. We have done it by using the kinetic Ising model, in which the Ising composition variables are formally analogous to the Ising spin variables, and the time evolution is simulated through the direct exchange of two n.n. atoms[2]. We assume that (i) the unique energy involved in the system evolution is the interaction energy between n.n. atoms, and (ii) the transition probability W from i- to j-configuration is:

$$W_{ij} = \exp(-\Delta E/kT) / [1 + \exp(-\Delta E/kT)]$$

where ΔE is the energy difference between the two configurations (the difference being due to the exchange of two n.n. atoms). Starting from a random distribution of A and B atoms in the lattice (completely disordered alloy), one goes along the following steps: (a) A pair of n.n. atoms is chosen by selecting two random numbers. (b) If the two atoms are equal, one is sent back to (a); in the other case the transition probability W_{ij} (i means the actual state of the system and j the one in which the two selected atoms are exchanged) is determined. (c) A third random number R is then selected; the interchange occurs if $R < W_{ij}$. The procedure is repeated from (a) to (c).

The model was applied to silicon heavily doped with phosphorus; the value of the energy interaction was taken from the experimental temperature dependence of the solubility of P in Si, in the range of 450° to 850°C [3,4] assuming for it an exponential dependence. For comparison with the experiments, the precipitated impurity fraction and the cluster size distribution were examined. The time evolution of the precipitated fraction did not depend apparently on the lattice structure. On the other hand, the evolution of the precipitate size turned out to be strongly dependent on the lattice structure. In particular, for diamond-lattice a single giant cluster of 1000 atoms was obtained after 12500 attempted interchanges per site at 850°C, this size already being in an experimentally observable range. The results are compared with the experimental ones and discussed critically.

Reference

1. A. Desalvo and G. Ruani, Phil. Mag. B, 50, 505, 1984.
2. Monte Carlo Methods in Statistical Physics, ed. K. Binder (Berlin: Springer-Verlag), 1979.
3. M. Masetti, P. Negrini, S. Solmi and D. Nobili, J. Electrochem. Soc., 128, 1313, 1981.
4. G.Masetti, D.Nobili, S.Solmi, Semiconductor Silicon, ed.H.R.Huff and E.Sirtl (Princeton, NJ: The Electrochem.Soc.), p.648, 1977.

AN INVESTIGATION INTO THE STRUCTURAL AND TRANSPORT PROPERTIES OF LaF_3

USING COMPUTER SIMULATION

W.M. Jordan and C.R.A. Catlow*

Department of Chemistry, University College London, London WC1
England *Department of Chemistry, University of Keele
Keele, Staffordshire, England

LaF_3 attracts attention for two reasons: firstly, it is a fast ion conductor (conductivity of 0.01 $\Omega^{-1}cm^{-1}$ at 1000K), and secondly it is one of the first crystals identified as showing multi-site behaviour. It is also a subject of controversy with conflicting experimental evidence for both its structure (space group $P6_3cm$ v $P\bar{3}c1$) and transport mechanism over the different F^- sublattices. Also the question arises as to what causes the predominant disorder: Schottky or Frenkel defects? To help resolve these problems, a static computer simulation study has been carried out, using shell model potentials to calculate point defect formation and migration energies with the computer code CASCADE.

Calculations show little difference between space groups and $P6_3cm$, a four site model (2:4:6:6), has been chosen. Models used to describe LaF_3 are based on inequivalent F^- sites which according to the Goldman and Shen model (1966) can be said to reside on two sublattices, A and B. For transport, A is identified with the (6+6) α sites, whilst B is identified with $2\beta + 4\gamma$ sites situated in channels along the c axis. Interpretation of ^{19}F nmr results lead to two opposing views: one suggests the faster motion occurs between α sites over the A sublattice predominantly in the (a-b) plane, the other favours motion over B along the c axis. Both agree about the conductivity anisotropy: $\sigma_{//c} = 2\sigma_{\perp c}$

Experimental estimates suggest formation energies of 0.5 ev per defect. Calculated values give Schottky energies 3-4 times too large and Frenkel pair energies twice as large as expected, the latter disorder is favoured. Energies obtained depend critically on the choice of potential.

Migration calculations based on a simple 'direct' mechanism where the migrating F^- ion is equidistant from two vacant F^- lattice sites, show that jumps involving $\beta - \beta$ or $\gamma - \gamma$ sites along the c direction are unfavourable. Jumps predominantly in the a-b plane, $\alpha - \alpha$, $\alpha - \beta$ and $\alpha - \gamma$ jumps yielded more favourable activation energies, several < 0.3 ev agreeing with the conductivity value of 0.28 ev. $\beta - \alpha$ jumps were lowest in energy.

The following mechanism is proposed: a net motion occurs in the c direction, but not by a direct path, it involves a sideways step via $\beta - \alpha$ sites. Once on an α site continued motion could involve α, β or γ sites. An $\alpha - \alpha$ jump, the lowest energy pathway, supports the view of motion over A. The next favoured $\alpha - \gamma$, sets up an indirect motion in the c direction, thus supporting motion over B, as would $\alpha - \beta$. A net motion is thus set up in the c direction which also agrees with the observed conductivity anisotropy.

The results are at present preliminary and further work is in progress.

Reference

Golman, M. and Shen, L., 1966, *Phys. Rev.* 144, 321.

THEORETICAL CALCULATIONS OF DIPOLE MOMENTS ASSOCIATED WITH IMPURITY-

VACANCY COMPLEXES IN THE ALKALI AND SILVER HALIDES

P.B. Fitzsimmons, J. Corish and P.W.M. Jacobs*

Dept. of Chemistry, Trinity College, Dublin, 2, Ireland
*Dept. of Chemistry, University of Western Ontario, London
Ontario, Canada, N6A 5B7

Atomistic simulation techniques have been used to calculate the effective dipole moments of impurity-vacancy (IV) complexes in the nearest neighbour (nn), next nearest neighbour (nnn) and third nearest neighbour (3nn) configurations in the following alkali and silver halide systems:

Mg^{2+}, Mn^{2+}, Cd^{2+}, Ca^{2+}, Sr^{2+} and Pb^{2+} in NaCl, KCl and KBr and Zn^{2+}, Mn^{2+}, Cd^{2+}, Ca^{2+} and Pb^{2+} in AgCl and AgBr.

A modified version of the HADES program (Norgett, 1974) was used for all the calculations. The short-range interactions were represented by Buckingham type potentials and polarization effects were simulated by the shell model of Dick and Overhauser (1958). The calculations assumed that the atomistic simulation gives a realistic description of the dielectric properties of a relaxed lattice and that the calculated dipole moment is given by the sum of the small induced moments in the relaxed lattice and that of the perfect lattice point ion dipole moment.

The results show that the effective dipole moment of an IV complex is considerably less than would be given by an ideal perfect lattice point ion calculation. This is due to the lattice relaxation and the dipoles induced on the ions surrounding the defect. The decrease in the dipole moment in going from a perfect lattice to a relaxed lattice is approximately 30 per cent for the alkali halides and approximately 40 per cent for the silver halides. The nn configurations in KCl and KBr are exceptions and show a decrease nearer to 15 per cent.

References

Dick, B.G. and Overhauser, A.W., 1958, Phys. Rev. 112, 90.
Norgett, M.J., 1974, AERE Harwell Report R7650.

INTERPRETATION OF THE PTCR EFFECT IN $BaTiO_3$ CERAMICS USING BULK AND SURFACE DEFECT ENERGY CALCULATIONS

G.V. Lewis and C.R.A. Catlow

University College London, Dept. of Chemistry, London WC1H 0AJ
*University of Keele, Dept. of Chemistry Staffordshire

Polycrystalline, donor-doped samples of $BaTiO_3$ undergo a transition from semiconducting n-type to insulating type electrical behaviour around the ferroelectric Curie temperature, T_c. Both theory and experiment indicate that this effect, the positive temperature coefficient of resistance (PTCR) effect varies with dopant concentration, dopant type and the conditions under which the sample is prepared, e.g. oxygen partial pressure (pO_2), cooling rate, etc.

We have developed a model of the defect chemistry responsible for the PTCR effect which is based upon the results of surface and bulk defect calculations. The calculations suggest that in normally 'undoped' $BaTiO_3$ intrinsic disorder is masked by oxygen vacancies which compensate acceptor impurities. At high pO_2 holes are trapped by acceptors whereas at lower pO_2 values, doubly ionised oxygen vacancies lead to insulating layers above T_c. However, in donor-doped $BaTiO_3$ the bulk of the grain will be n-type conducting. Our calculations suggest that both thermodynamic and kinetic factors favour segregation of acceptors to the grain surfaces, which in predominantly n-doped $BaTiO_3$ will lead to the development of n-i-n junctions at intergranular contacts. These barriers are lowered below T_c, leading to PTCR behaviour.

453

RAMAN SPECTROSCOPY FOR NON-DESTRUCTIVE DEPTH PROFILE STUDIES OF

ION IMPLANTATION IN SILICON

A.C. deWilton, M. Simard-Normandin and P.T.T. Wong

Nothern Telecon Electronics Ltd. Ottawa, Canada, K1Y 4H7
*Chemistry Division Nat. Research Council, Ottawa, Canada

Raman spectroscopy is now recognized as a valuable technique for characterization of semiconductor materials, particularly for studies of ion implantation and annealing processes[1]. The Raman phonon spectrum of a solid is sensitive to lattice disorder caused by damage, defects or incorporation of dopants or impurities. For example:

(a) Crystalline, microcrystalline and amorphous phases of silicon can be readily distinguished by their lattice phonons.
(b) Incorporation of dopants gives rise to additional vibrational local modes which yield information on dopant concentration, chemical binding and electrical activation.
(c) Free carrier concentrations can be determined from the effects of electron-phonon interactions on bandshapes and intensities.

Examples of Raman spectra of B^+, BF_2^+ and Si^+ implanted silicon are presented[2]. In addition to providing information on the bulk sample, it is shown that by varying the angle of incident, θ_i, of the laser on the sample to obtain depth resolution, it is possible to measure a depth profile of ion implantation damage. Although the range of penetration depths as a function of θ_i is limited by the high refractive index of silicon, depths in the range 50 nm to 500 nm are accessible with different wavelengths from an Ar^+ ion laser. The main advantage of obtaining depth resolution optically is that it avoids the etching or sputtering damage common to depth profile measurements with SIMS, XPS or AES. In addition spectra can be measured rapidly, under ambient conditions, with no special sample preparation. This technique is therefore of particular interest for non-destructive characterization of sub-micron pn junctions for VLSI applications.

Reference

1. F.H. Pollak, Proc. S.P.I.E., Vol. 352 Spectroscopic Characterization Techniques in Semiconductor Technology, eds. F.H. Pollak and R.S. Bauer, pp. 26-43 (S.P.I.E. 1983).
2. A.C. deWilton, M. Simard-Normandin, P.T.T. Wong, submitted for publication in J. Electrochem. Soc.

HYDROGEN DEFECTS IN Y_2O_3 AT HIGH TEMPERATURES

Truls Norby

Department of Chemistry
University of Oslo
Blindern, 0315, Oslo 3, Norway

Hydrogen defects have been found to dominate the defect structure of yttria to high temperatures (<1500°C in "wet" atmospheres and <1100°C in "dry"). They are present mainly as interstitial protons bonded to oxygen ions and are denoted by $H_i^{\cdot}$ or $OH_O^{\cdot}$. In pure yttria the positively charged protons are probably counterbalanced by oxygen interstitials. Protons can also be counterbalanced by lower-valent foreign cations, and thus the proton content can be increased and fixed by dopants.

Yttria is a p-conductor at near-atmospheric oxygen pressures. The role of protons in the defect structure has been evaluated from the characteristic dependences of the p-conductivity on the water vapour pressure.

At low oxygen activities yttria is a mixed electronic/ionic conductor. Emf-measurements have been employed in order to separate the conductivity into individual contributions from electronic defects, native ionic defects (metal and oxygen species), and hydrogen defects. From other studies it is known that hydrogen is transported in oxides at elevated temperatures mainly as "free" protons, H^+, moving from one oxygen ion to the next. For yttria, with its channels in the anion sublattice, we have also considered the possibility of easy diffusion of interstitial hydroxide groups, OH^-, H_2O molecules, and hydronium ions, H_3O^+. These are all smaller than interstitial oxygen ions which are presently considered to be the most mobile native defects in yttria. The emf over an oxide specimen comprising the above-mentioned species can be expressed by

$$E_{II-I} = \frac{RT}{4F}\, \bar{t}_{ion}\, \ln \frac{p_{O_2}^{II}}{p_{O_2}^{I}} - \frac{RT}{2F}\, \bar{t}_{(H)}\, \ln \frac{p_{H_2O}^{II}}{p_{H_2O}^{I}}$$

where $t_{ion} = t_M + t_O + t_{H^+} + t_{OH^-} + t_{H_3O^+}$ and $t_{(H)} = t_{H^+} + 3t_{H_3O^+} - t_{OH^-}$. From measurements with a gradient in P_{O_2} only, t_{ion} can be found, and with a gradient in P_{H_2O} only, $t_{(H)}$ can be found. Based on the sign of $t_{(H)}$ and its magnitude compared with t_{ion} it is concluded that free proton migration is the major transport mechanism of charged hydrogen species in yttria above 500°C. The proton conductivity constitutes a large part of the total ionic conductivity in yttria. The diffusion coefficient for protons evaluated from the emf/conductivity data agree reasonably well with tritium tracer diffusion coefficients. In polycrystalline yttria enhanced diffusion of protons along grain boundaries is observed up to around 1100°C.

THE ELECTRICAL CONDUCTIVITY OF ZnO CRYSTALS IN CO GAS

J.M. Steele, A.V. Chadwick and J.D. Wright

University Chemical Laboratory
University of Kent
Canterbury, Kent CT2 7NJ
U.K.

The chemisorption of a gaseous species onto the surface of a semiconductor is an electronic process which produces changes in the electrical properties of the solid (Garner, 1955). This process involves the transfer of charge between the solid and adsorbed species (Wolkenstein, 1960) resulting in charges in the surface conductivity. These changes have been used in studies of adsorption processes, active species on surfaces and the mechanisms of heterogeneous catalysis (Goodwin et al, 1972). This effect is now widely employed as the basis of gas sensors using semiconducting oxides (Bott et al, 1974), Firth et al, 1974, Taguchi, 1972).

The electrical conductivity of ZnO pellets at elevated temperatures is enhanced by the presence of CO gas and this provides a means of monitoring this atmospheric pollutant. Recent work (Jones et al, 1982, Bott et al, 1983 has shown that improved selectivity and sensitivity to CO is achieved by the use of a single crystal of ZnO. We have made a study of the effect of growth conditions of ZnO crystals on their response to CO. Crystals grown from a flux of $KOH-H_2O$ (Kashyap, 1973) were in the form of small hexagonal cross-sectioned needles. Their typical resistance was 10-50 ohms and they had a good response to CO, confirming earlier studies. Introducing LiOH to the flux produced hexagonal platelets similar to those reported from $NaOH-LiOH-H_2O$ fluxes (Hashimoto, 1982). These had a resistance of 10^6 - 10^7 ohms and did not show any response to CO.

Reference

Bott, B., Firth, J.G., Jones, A. and Jones, T.A., 1974, British Patent, 1,374,575.

Bott, B., Jones, T.A. and Mann, B., 1983, Sensors and Actuators, 5, 65.

Firth, J.G., Jones, A., and Jones, T.A., 1974, Proc. I.E.R.E. Conf. on Environmental Sensors, p. 57.

Garner, W.E., 1955 ed, "Chemistry of the Solid State" (Butterworth, London).

Goodwin, T.A. and Mark, P., 1972, in "Progress in Surface Science" ed S.G. Davidson (Pergamon, London), Vol. 1.

Jones, T.A., 1982, Sensors Review, 4, 14.

Hashimoto, H., Hayashi, F. and Vematsu, T., 1982, Mat. Sci. Letters, 1, 4.

Kashyap, S.C., 1973, J. App. Phys. 44, 4381.

Wolkenstein, Th., 1960, Adv. Catalysis, 12, 189.

Taguchi, N., 1972, British Patent, 1, 280, 809.

CONNECTION OF THE STATIC DIELECTRIC CONSTANT AND THE DEFECT

VOLUMES IN ALKALINE EARTH FLUORIDES

S. Patapis, P. Varotsos and K. Eftaxias

University of Athens, Department of Physics
Solid State Section, 104 Solonos Street
106 80 Athens, Greece

Conductivity measurements under hydrostatic pressure lead to the formation volume v^f of an anion Frenkel defect in CaF_2, SrF_2 and BaF_2 [1]. By following the classical Jost model, as slightly modified by Flynn[2], we show that v^f-values can be successfully obtained for CaF_2 and SrF_2 (and less accurately for BaF_2) when the pressure derivative of the static dielectric constant ε is used. Considering that the main source of the experimental value of $(d\varepsilon/dP)_T$ comes from anharmonic effects[3] we conclude that anharmonicity plays a major role in v^f-calculations in agreement to earlier suggestions by Varotsos et al[4].

Conductivity or dielectric loss studies at various pressures lead also to the volume v^m for the following migration processes: free anion vacancy, free anion interstitial, bound vacancy and bound interstitial motion. A fact very recently realised is that the quantity $-h^m(\delta\ln\varepsilon/\delta P)_T$ - where h^m denotes the corresponding migration enthalpy - gives in many cases a reliable estimate of the v^m-value. In view of the fact that the Jost model does not have any justification for migration processes we conclude that the above connection might be understood in the frame of the thermodynamical $cB\Omega$-model developed by Varotsos and Alexopoulos[5].

References

1. G.A. Samara, Solid State Physics 38 (1984) 1-80.
2. C.P. Flynn in Point Defects and Diffusion (Clarendon Press, Oxford, 1972).
3. P. Varotsos, Phys. Stat. Sol.(b) 90 (1978) 339-343.
4. P. Varatsos, W. Ludwig and K. Alexopoulos, Phys. Rev. B18 (1978) 2683-2691; P. Varotsos, J. Physique Lett. 38 (1977) L455-459.
5. P. Varotsos and K. Alexopoules, in Thermodynamics of Point Defects and their Connection to Bulk Properties (North Holland 1985) chapters 13 and 14.

NMR OBSERVATIONS IN THE FAST IONIC CONDUCTOR PbSnF$_4$

D. van der Putten, J.H.Strange and A.V.Chadwick

University of Kent
Canterbury
Kent, U.K.

In the temperature range 250K-500K the temperature dependence of both the spin-lattice and the spin-spin relaxation times T_1 and T_2 in α and β PbSnF$_4$ exhibit two minima. The absolute T_1 and T_2 values are low compared to those of pure PbF$_2$. Relaxation is due to a defect-induced self-diffusion mechanism. The two minima suggest fluoride ion motion on at least two non-equivalent sites (1). The existence of a large number of interstitials may be inferred from the low absolute values of the T_1 minima, originating from a relatively strong dipole-dipole coupling.

On approaching the β-γ phase transition (630K), T_2 increases dramatically by almost two orders of magnitude. Upon cooling, T_2 exhibits a single activated behaviour down to room temperature and the recovery of the z-magnetization becomes non-exponential. Upon heating further into the γ-phase, the relaxation behaviour is found to be identical to that of the pure PbF$_2$. This experiment was performed in an open glass tube under a helium atmosphere. Consequently, SnF$_2$ was observed to separate from PbF$_2$. This did not happen when vacuum sealed tubes were used. Reproducible relaxation times were then obtained.

We conclude:

(1) fast multiple F$^-$ site diffusion exists in PbSnF$_4$,
(2) a large number of interstitial sites are occupied,
(3) non-reproducible relaxation times at high temperature in previous work are explained, (2)
(4) the non-exponentiality of the recovery of the z-magnetization serves as a tool to examine the correct homogeneous sintering of samples.

More detailed experiments are in progress.

References

1. G.A.Jaroszkiewicz and J.H.Strange, J.Phys.C. 18, 2331 (1985).
2. G.Villeneuve, P.Echegut, C.Lucat, J.M.Reau and P.Hagenmuller, Phys.Stat.Sol. 97b, 295 (1980).

TRANSPORT OF NICKEL THROUGH A MIXED LAYER OF NiO and Ni_3S_2

Arnfinn Andersen and Per Kofstad

Department of Chemistry
University of Oslo
Norway

In high temperature corrosion of nickel and nickel alloys in sulphur-containing gases the corrosion product often is a mixed layer of NiO and Ni_3S_2. The defect structure of NiO is well known; diffusion in this oxide is slow. The high temperature modification of Ni_3S_2 has a large homogenity range and little is known about its defect structure; diffusion in this phase is rapid. When this phase is distributed in the form of continuous channels through the scales on nickel and nickel alloys, corrosion rates are rapid. Such channels have not as yet been directly observed, but calculations of diffusion rates through mixed layers have suggested that the thickness is in the range of 30 to 50 Angstroms along the grain boundaries of NiO in the scale. In this work electrical conductivity measurements at room temperature have been used to determine the effective cross section of the sulphide phase in several mixed layers. These determinations are based on the fact that the sulphide is a metallic conductor and the oxide is an insulator. The measurements show that the effective cross sectional area of the sulphide in the layer is smaller than previously suggested, and estimates of diffusion rates in the sulphide phase indicate that diffusion becomes the rate limiting process after extended exposure to SO_2.

The crystal and defect structure in the high temperature Ni_3S_2 phase is not well understood. More accurate estimates of diffusion in mixed layers will require more thorough studies of this phase.

THE IONIC SPACE CHARGE POTENTIAL IN SILVER CHLORIDE

R.A. Hudson, L. Slifkin

University of North Carolina
Physics Department
Phillips Hall, Chapel Hill, NC 27514, USA

Using radiotracer and dissolution techniques, depth profiles of the ionic space charge region in oriented, bulk, single crystals of pure AgCl have been obtained. These potential profiles over the temperature range of 25°C to 230°C give the Gibbs free energies of formation of the silver vacancy and the silver interstitial.

DISLOCATION ENHANCED DIFFUSION IN SODIUM CHLORIDE CRYSTALS

J.A. Archer and A.V. Chadwick

University Chemical Laboratory
University of Kent
Canterbury, Kent CT2 7NH
U.K.

A recent study of ^{22}Na diffusion in NaCl crystals by Ho and Pratt (1983) provided detailed information on dislocation diffusion. This work was interpreted with the theoretical model of dislocation diffusion by LeClaire and Rabinowitch (1981). A feature of the experimental data was a very fast diffusion along dislocations and a very large effective diameter of the dislocations.

We have repeated some of these experiments using the same radiotracer serial sectioning technique with Czochralski grown single crystals. Short anneal times were employed to accentuate the region of the diffusion profiles due to enhanced diffusion. These experiments revealed the same qualitative features of Ho and Pratt. Similar experiments using ^{36}Cl suggest that the Cl$^-$ ion also diffuses rapidly along dislocations.

References

Ho, Y.K. and Pratt, P.L., 1983, Radiation Effects, 75, 183.
LeClaire, A.D. and Rabinowitch, A., 1981, J. Phys. C., 3863.

THERMAL DEPOLARIZATION CURRENTS IN DOPED T1Cl

A.N. O'Reilly and J. Corish

Department of Chemistry
Trinity College
Dublin 2, Ireland

Recent conductivity studies by Corish et al. (1984) on TlCl doped with divalent cation and anion impurity ions have shown that the effects of the divalent charges of the impurity ions are quite evident at the lower temperatures. Extensive least squares fitting of this conductivity data indicated almost all of the impurity ions to be associated with a charge compensating vacancy. The present study investigates this association in TlCl using the thermal depolarization current (TDC) technique on freshly grown and annealed samples. Data were collected using a transient recorder and were analyzed using least-square fitting techniques. Two cation-doped samples $TlCl:Sr^{2+}$ and $TlCl:Pb^{2+}$ and two anion-doped samples $TlCl:SO_4^{2-}$ and $TlCl:CO_4^{2-}$ were studied.

The results show all of these systems to have characteristic TDC spectra. The peak temperatures for both cation-doped samples were at least 60 K higher than for the anion-doped specimens which is to be expected on the basis of the greater mobility of the anion vacancy in TlCl. The analysis of the data from $TlCl:Sr^{2+}$ showed a reorientation energy of 0.49 eV and a relaxation time of 0.5×10^{-10} which were unchanged after annealing at 575 K for three weeks. The freshly grown sample of $TlCl:Pb^{2+}$ produced a single TDC peak at 206 K with a reorientation energy of 0.46 eV and a relaxation time of 0.30×10^{-10} s, while an aged sample showed a major peak at 216 K and a second rather small peak at 224 K. The behaviour of these peaks on annealing and aging has also been studied. Data from aged samples of both of the anion-doped crystals showed peaks in the region of 145 K and the spectra were best analyzed as two overlapping peaks.

These systems have also been studied using atomistic simulation computer techniques.

References

Corish, J., Parker, B.M.C., Quigley, J.M., Allnatt, A.R.A., and Mulcahy, D.C.A., 1984, J. Phys. C. 17, 2689.

MATERIALS INDEX

Ag, 158, 353
Ag-Cu alloys, 358
Ag-Zn alloys, 364, 370
AgBr, 308
AgCl, 3, 396, 461
AgCl:Cd^{2+}. 396-402, 452
AgCl:Cr, 415
AgCl:Mn^{2+}, 418, 452
AgCl:Ti, 415
AgCl:V, 415
AgI, 22, 291
Ag_2S, 22, 291
Al, 153, 158, 192, 289, 353
Alkali halides, 13, 413-415, 443, 452
Alkali halide:Tl^+, 208
Alkaline earth fluorides, 19, 37-50, 414, 458
Alkaline earth oxides, 13, 14
Al_2O_3, 76, 133, 377
$AlPO_4$, 195
β-Alumina, 22, 291-295
β"-Alumina, 22, 291-295
Au, 158, 353
Au-Ni alloys, 366

BaF_2, 247, 260
BaF_2:K^+, 42, 261
BaF_2:La^{3+}, 42, 261
BaF_2:O^{2-}, 265
BaFCl, 214, 223
BaO, 288
$BaTiO_3$, 287, 453
Bipolymers, 420

CaF_2, 3, 277-282, 288, 308
CaF_2:Er^{3+}, 4
CaF_2:La^{3+}, 4
CaF_2:O^{2-}, 419
CaF_2:Rare $earth^{3+}$, 20, 43, 244, 288, 411, 416
$CaMoO_4$, 420
CaO, 330
CaO:Mg^{2+}, 230
$CaPuTi_2O_7$, 88

$CaWO_4$, 420
$Ca_{14}Y_5F_{43}$, 44
$Cd(NH_3)(C_6N_2H_8)Ni(CN)_4$, 437
CeO_{2-x}, 18
CeO_2:Y^{3+}, 23, 56, 296-299, 416
Clays, 448
CoO, 310, 383
Cu, 158, 353
Cu:Kr, 163
Cu-Zn alloys, 366
Cu_2O, 382
Cyclooctane, 167
Cyclotrimethylene trinitramine (RDX), 438

Fe, 354
Fe:C, 11, 164
Fe:Si, 196
Fe-Cr-Al alloys, 11, 129
Fe-Ni alloys, 368
$Fe_{1-x}O$, 18, 19, 86, 377, 379
Fe_9S_{10}, 84

GaAs, 123, 191, 199, 235-441
GaAs:Cr, 238
GaAs:V, 440
GaP:Ni^{3+}, 218, 222
GaP:Zn, 231
$Gd_2(MoO_4)_3$, 187
Ge, 322

Hexamethylethane, 252

Ice, 168
InSb, 202

K, 289
KBr:Pb, 444
KBr:Sr, 410
$KCaF_3$, 258-260
KCl, 307
KCl:Pb, 410, 452
KCl:F^-, 220, 222
KCl:H, 213-218, 221
KCl:Li^+, 5
KCl:Tl, 211, 233, 236

465

SUBJECT INDEX

467

470

472